Statistical Analysis

Statistical Analysis
for Induction and Decision

CHAIHO KIM
University of Santa Clara

THE DRYDEN PRESS INC.
Hinsdale, Illinois

Copyright © 1973 by The Dryden Press Inc.
All rights reserved
Library of Congress Catalog Number: 72-78875
ISBN: 0-03-089102-7
Printed in the United States of America
9 8 7 6 5 4 3 2 1 038 5 4 3 2

To my wife, Taeock, and sons, James and Michael

Preface

I have written this book as a text for quarter or one-semester courses in beginning statistics for students who are majoring in Administrative or Management Science, Business, Economics, Industrial Engineering, Operations Research, or other related disciplines.

A number of reasons have motivated the style of presentation in this textbook. My experience in using several standard textbooks written for these students has revealed that the students often become confused between an idea which is merely an arbitrary definition and an idea which requires a logical justification. I have, therefore, decided to present the material by means of a series of definitions and theorems. Moreover, the level of the textbooks written for these students requires that, whereas some statistical concepts can be proved, most of the remaining ones have to be accepted, more or less, on faith. If that is the case, then I believe that the students should be told clearly that they are being asked to accept something on faith. Therefore, I have made an attempt to point out for each theorem I present that either I am proving it or merely illustrating its plausibility with some numerical examples.

Many of the theorems in the textbook are not really proved. However, I have tried to provide several examples for almost all theorems so that the students can relate them to their own experiences. In most instances, the theorems are illustrated first with frivolous (and perhaps trivial) examples, such as tossing coins or picking marbles from an urn. I realize that such illustrations may eventually become repetitive, boring, and even annoying, but as

Richard Bellman points out, "Repetition, however, no matter how dismaying as a social or literary attribute, is no mathematical sin."* My experience also reveals that the students are at their best in grasping a very difficult statistical idea when it is illustrated with an example. At the same time, however, I have made a serious attempt not to be side-tracked from the main goal of teaching the subject matter to students: to enable them to relate the ideas that they learn from this textbook to some of the real phenomena or problems they encounter. I have, therefore, supplemented, whenever appropriate, these frivolous examples with somewhat more realistic ones. In many of the chapters, I have added an extra section specifically to describe some of the applications of the statistical concepts discussed earlier in that chapter.

My treatment of the subject matter in many of the chapters is straightforward. I would like to believe, however, that I have taken a somewhat more unified approach to the entire question of *classical* versus *Bayesian* controversy than the one that can be found in many other elementary statistics texts. In Chapter 12, I have established a basic framework for any decision making under the conditions of uncertainty. In Chapter 13, I have discussed how the classical school of thought fits into the framework established in Chapter 12. In Chapter 14, I have discussed how the Bayesian school of thought fits into the framework established, again, in Chapter 12. I hope that the students who read these chapters will acquire a birds-eye view of the status of modern decision theory as well as the precise nature of the controversy surrounding the classical and Bayesian statisticians.

As a pedagogical experiment, I have added a computer program for a simple regression analysis at the end of Chapter 16. I realize that computer programs for regression analysis are likely to be in quite abundant supply in most university computer centers. However, I have found that the beginning students usually have quite a difficult time using these programs either because they cannot figure out how to prepare input for the programs or because the terminology used in the output of these programs usually does not correspond exactly to the terminology that they have learned in a textbook. I have tried to overcome these difficulties first, by using the same terminology in the computer program as in the textbook and, second, by writing the program in a conversational mode so that the students can interact with the computer during the input-output phase, as well as during the computational phase.

Santa Clara, Calif. Chaiho Kim
April 1972

---

*Bellman, Richard *Dynamic Programming*, Princeton University Press, 1957, p. 12.

Acknowledgements

Many persons have helped me in the process of completing this book. I would like to express my special thanks to Dean Charles J. Dirksen of the Graduate School of Business and Professor Zeb Vancura, chairman of my department, both at the University of Santa Clara, for providing me with secretarial and graduate student assistants. Without their generous assistance the completion of this text would not have been possible.

I would like to thank my colleague John Heineke, my former colleague Stan Fromovitz, now at the University of Maryland, and a number of anonymous reviewers for Dryden Press for reading various portions of the manuscript and offering valuable criticisms and suggestions. Many of their suggestions have been incorporated in the text, although I am entirely responsible for any errors and shortcomings that may remain.

I would like to thank Mikes Sisois, my graduate student assistant, for writing the interactive computer program which appears in the appendix section of Chapter 16.

My thanks are due also to Judith Lyding and Michelle Merckel for editing as well as typing the manuscript.

To my wife Taeock, I express my appreciation for being a constant source of encouragement during the years the manuscript has been in preparation.

Contents

Preface   *vii*

1 Introduction   1
    1.1   Statistical Description   1
    1.2   Statistical Induction   2
    1.3   Statistical Decision   2
    1.4   Plan of the Textbook   4

2 Sample Space and Probability   5
    2.1   Sample Space of a Random Phenomenon   5
    2.2   Events Defined in a Sample Space   9
    2.3   Occurrence of an Event   10
    2.4   Types of Events   12
    2.5   Probability of an Event   15
    2.6   Converting a Likelihood-Assessment into a Probability   15
    2.7   Controversy with Regard to Likelihood-Assessment   19

3 Probability Calculations   22
    3.1   Probability of a Compound Event   22
    3.2   Probability of Two Compound Events   23
    3.3   Joint Probability   25
    3.4   Conditional Probability   28

xii   Contents

    3.5 Independent Events   33
    3.6 Bayes' Theorem   38

4 Random Variables   42
    4.1 Function   42
    4.2 Random Variable   45
    4.3 Discrete versus Continuous Random Variable   48
    4.4 Probability Function of a Discrete Random Variable   50
    4.5 Probability Function of a Continuous Random Variable   56
    4.6 Expected Value of a Random Variable   62
    4.7 Variance and Standard Deviation   64
    4.8 Application of Expectation and Variance   67
    4.9 Tchebycheff Inequality   69

5 Joint Probability Function   72
    5.1 Joint Probability Function of Two Discrete Random Variables   72
    5.2 Conditional Probability Function   78
    5.3 Statistical Independence   80
    5.4 Covariance   83
    5.5 Statistical Independence and Covariance   87
    5.6 Correlation   89
    5.7 Applications of Covariance and Correlation   92

6 Functions of Random Variables   96
    6.1 Introduction   96
    6.2 Scalar Product of Random Variables   97
    6.3 Sum of Two Random Variables   100
    6.4 Sum of Many Independent Random Variables   105
    6.5 Distinguishing the Sum of Random Variables and a Scalar Product of One Random Variable   109
    6.6 Adding a Constant to a Random Variable   111
    6.7 Practical Applications of Some Statistical Concepts   113

7 Special Probability Functions   119
    7.1 Bernoulli-Distributed Random Variable   119
    7.2 Binomially Distributed Random Variable   122
    7.3 Poisson-Distributed Random Variable   128
    7.4 Uniformly Distributed Random Variable   135
    7.5 Exponentially Distributed Random Variable   137
    7.6 Normally Distributed Random Variable   142

8 Central Limit Theorem 150
   8.1 Central Limit Theorem: Bernoulli Case   150
   8.2 Central Limit Theorem: General Case   156
   8.3 Applications   162

9 Sampling Techniques 165
   9.1 Simple Random Samples   165
   9.2 Techniques for Selecting a Simple Random Sample   169
   9.3 Use of Random Numbers   178

10 Sampling Distribution 181
   10.1 Probability Function of a Sample Proportion   181
   10.2 Probability Function of a Sample Mean   188
   10.3 Sampling Errors   194
   10.4 Expectation and Variance of Sampling Errors   197
   10.5 Law of Large Numbers   200
   10.6 Central Limit Theorem for Sampling Errors   202
   10.7 Sample Size and Sampling Errors   207

11 Statistical Estimation 212
   11.1 Criteria of a Good Estimator   213
   11.2 Point Estimate   222
   11.3 Maximum Likelihood Estimate   226
   11.4 Interval Estimates: Mean   232
   11.5 Interval Estimates: Proportions   237

12 Testing a Hypothesis 242
   12.1 Introduction   242
   12.2 A Simple Bernoulli Hypothesis   244
   12.3 Randomization of a Decision Function   249
   12.4 Admissible Decision Function   254
   12.5 Loss Function   256
   12.6 Minimax Decision Function   260
   12.7 Bayes Decision Function   264
   12.8 Illustrative Applications   271

13 Classical Decision Functions 279
   13.1 Likelihood-Ratio Test   279
   13.2 Simple Hypothesis: Continuous Case   285
   13.3 The Neyman-Pearson Lemma   290
   13.4 Composite Hypothesis: Bernoulli Case   294

13.5 Composite Hypothesis: Continuous Case   305
13.6 Composite Hypothesis: Two-sided   313

14 Bayes Decision Functions   325
14.1 Prior Probability Function   325
14.2 Posterior Probability Function   327
14.3 Finding a Bayes Decision Function: Alternate Method   331
14.4 Equivalence Relationship between Two Bayesian Procedures   338
14.5 Value of Information Obtained from a Sample   339
14.6 Preposterior Analysis   348
14.7 Composite Hypothesis   351
14.8 Bayes Decision Function for Normal Probability Function without Sampling   362
14.9 Bayes Decision Function for Normal Probability Function with Sampling   368
14.10 Preposterior Analysis of a Normal Probability Function   371

15 Chi-Square, Student-t, and F Distributions   376
15.1 Chi-Square Distribution   376
15.2 Chi-Square and Statistical Inference Pertaining to Variance   382
15.3 Student-t Distribution   388
15.4 t Distribution and Statistical Inference for a Small Sample   390
15.5 F Distribution   395
15.6 F Distribution and Statistical Inference Pertaining to Two Variances   399
15.7 Chi-Square and Theory of Large Sampling   402
15.8 Testing "Goodness-of-Fit"   404

16 Regression Analysis   413
16.1 Multivariate Sampling   413
16.2 Simple Regression   414
16.3 Underlying Statistical Assumptions   415
16.4 Estimation of Parameters   418
16.5 Alternate Method of Estimating Parameters   425
16.6 Mean and Variance of an Estimator   429
16.7 Some Properties of Estimators   433
16.8 Inferences for Parameters   434
16.9 Predictions with Regression Equations   438
16.10 Re-Examination of Regression Assumptions   445
16.11 An Application of Regression Analysis   446

A.16 Computer Program for Simple Regression Analysis   456
    A.16.1   Description of the Program   456
    A.16.2   Data-Input Phase   456
    A.16.3   Calculation Phase   461
    A.16.4   Interval-Prediction Phase   462

Bibliography   *465*

Appendix A   Sets and Their Operations   467
    A.1   Definitions and Notations   467
    A.2   Relationships between Sets   469
    A.3   Operations with Sets   471

Appendix B   List of Tables   475
    Table B.1   Area under the Normal Curve   477
    Table B.2   Chi-Square   478
    Table B.3   Student-t Distribution   479
    Table B.4 F Distribution   480
    Table B.5   Exponential Functions   484
    Table B.6   Binomial Probabilities   491
    Table B.7   Poisson Probabilities   496
    Table B.8   Unit Normal Loss Function   501
    Table B.9   Random Numbers   503
    Table B.10   Natural Logarithms of Numbers (Base e)   505
    Table B.11   Common Logarithms of Numbers (Base 10)   509
    Table B.12   Powers and Roots   511

Index   521

Chapter 1
Introduction

1.1 Statistical Description

What is the meaning of the term "statistics"? This question perhaps may be answered in several different ways. First, we might propose the following definition.

Definition 1.1.1
Statistics are classified facts with regard to a particular class or interest in terms of numbers.

Facts pertaining to a class or interest may be described in numerous ways. However, according to the preceding definition, a statistical description requires that the facts be stated in terms of numbers. For example, to state that Americans are rich is not a statistical description; but if we state that the annual per capita income in America is $3000, it is a statistical description. To state that Willy Mays is a superior batter is not a statistical description; but if we state that he hits 60 home runs a year and has a batting average of .400, it is a statistical description.

The reader is probably familiar with the preceding definition. It is probably the most widely accepted meaning of the term in our daily use of language. Since people must have been using some type of numbering scheme to describe things of interest for a very long time, the application of statistics as we have defined the term probably is as old as human history.

1.2 Statistical Induction

Today the term "statistics" means much more than merely a description of things of interest in terms of numbers, charts, or graphs. Within the last hundred years or so, the term has acquired a new meaning.

Definition 1.2.1

Statistics is the science of collection and classification of facts in terms of numbers as a ground for induction.

Induction is an act of making a generalization pertaining to a class of objects after observing only a subclass of that class. The class about which such a generalization is being made is called the *population*, and the subclass which is actually being observed is called the *sample*. Thus, we may alternatively propose that statistics is the science of making numerical valued inductions on the basis of sample observations.

Some useful examples of such numerical valued inductions are not difficult to conceive. A political pollster might survey over 1000 voters across the country and then predict who will be the next President of the United States. A manufacturer of television picture tubes might test a certain number of them from his production line and then conclude what is the mean life of all tubes produced by this production line. An educator may take a sample from his student body to determine the average number of hours that the students spend in studying outside the classroom.

It must be kept in mind that any numerical induction pertaining to a population, based on a sample, whether it was made one hundred years ago or now, may be subject to error.

What is new, however, in the science of making such numerical valued inductions since the turn of this century is that we are now able to assess with a great degree of accuracy the probable margin of error which may be committed in making such inductions. When a numerical valued induction includes the probable margin of error associated with it, we call it a *statistical induction*.

1.3 Statistical Decision

Within the past 25 years or so, the meaning of the term "statistics" has acquired a major new dimension. Now we can claim the following:

Definition 1.3.1

Statistics is the science of making decisions under the condition of uncertainty.

1.3 Statistical Decision

By making decisions under the condition of uncertainty, we mean loosely that the decision maker has to choose a course of action among various alternatives available to him, at the same time that he keeps in mind that the desirable or undesirable consequences of choosing a particular course of action will depend on the turn of events which are not predictable by him.

We will list some familiar examples of decision making under the condition of uncertainty. A man at a racetrack must decide on which horse to place his bet. The desirable or undesirable consequences of his decision will depend on the outcome of the race, which he cannot predict. A quality-control manager of a manufacturing process must decide whether to stop the process for a corrective adjustment or let the machine run. The desirable or undesirable consequences of his decision will depend on whether the process is actually in control or out of control, something that usually he does not know for certain. A space headquarters monitoring a flight to the moon must make the final decision as to whether the flight crew should land on the moon or not. Again, the desirable or undesirable consequences of that decision will depend on whether the landing craft in fact will work successfully if it were to land on the moon, something that the space headquarters cannot predict with absolute certainty.

Even though the scenarios for these examples differ, the decision maker, confronted with each of these problems, is likely to go through some of the following analytical steps to come to his decision. First, he is likely to evaluate, for each possible turn of events, the desirable or undesirable consequences of choosing a particular course of action. Having done this, if he knows which way the events will turn, then he should not have any difficulty in identifying the optimal course of action. However, we have pointed out that the problem arose in the first place precisely because the decision maker cannot predict the turn of these events. What should he do then? One thing he can do is to choose a course of action so that, regardless how the events turn out, he can live with the consequences of having chosen that particular course of action. Another thing that he can do is to guess the probable outcome of events and then choose a course of action which will yield the most desirable consequences if his guess turns out to be correct.

We have so far broadly outlined some of the steps which one must go through to reach a rational decision. However, outlining these steps raises another set of questions: for example, how should the decision maker specify the desirable and undesirable consequences; how should he determine the consequences that are livable; or how should he go about guessing the probable outcome of the events?

In recent years, statisticians have devoted a great deal of effort to answering some of these questions. The fruit of their efforts has brought about a formation of a new branch of science, called *statistical decision theory*, which specifically deals with decision making under the conditions of uncertainty.

1.4 Plan of the Textbook

The main objective of this textbook is to elaborate on the philosophy and techniques of modern statistical concepts as they relate to numerical valued induction or decision making under the conditions of uncertainty. The basic idea of statistical induction is discussed in Chapters 9, 10, and 11; decision making under the conditions of uncertainty is discussed in Chapters 12, 13, and 14; Chapters 15 and 16 contain extensions of the basic ideas presented in the preceding six chapters.

Even though the main objective of this textbook is to explore the concept of statistical induction and of statistical decision making, we also provide in this textbook a self-contained treatment of the probability theory, for the following reasons: first, both statistical induction and statistical decision making derive their logical justification from the probability theory; second, an understanding of the probability theory is valuable for its own sake. Every branch of physical and social science utilizes it extensively today. Some basic concepts of the probability theory are discussed in Chapters 2 through 8.

Chapter 2
Sample Space
and Probability

In this chapter we will begin the discussion of probability theory.

2.1 Sample Space of a Random Phenomenon

Any phenomenon whose outcomes are determined by chance is called a *random phenomenon*. Assume that we are going to toss a coin. The possible outcomes of this experiment are clearly known. On the other hand, we do not know whether we will obtain "heads" or "tails," because the actual outcome of the toss depends on chance. The experiment, therefore, can be considered to be a random phenomenon. Further, let us assume that a given manufacturing process is known to produce k percent defectives. Suppose we were to take a sample of ten items from the process. Again, the possible sample outcomes are zero to ten defectives. On the other hand, we cannot say exactly how many defectives will be in our sample, since that depends on chance, even though the manufacturing process is producing exactly k percent defectives on the average. Thus, the sampling of ten items from the manufacturing process may be considered a random phenomenon.

The notion of *sample space* of a random phenomenon plays an important role in the development of probability theory. The following definition is offered.

Definition 2.1.1

Sample space is a set whose elements represent all possible outcomes of a random phenomenon and in which each possible outcome corresponds to exactly one element of that set.

In the preceding definition two properties of a sample space should be emphasized. First, if a set is to be a sample space of a random phenomenon, it must contain all possible outcomes. Second, the set should not contain an identical outcome more than once. We offer the following examples to illustrate the notion of sample space.

EXAMPLE 2.1.1

Let the random phenomenon consist of tossing a coin. Suppose we describe the following three sets associated with the phenomenon, as given in Figure 2.1.1.

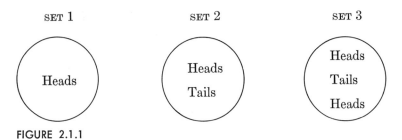

FIGURE 2.1.1

Only Set 2 is the sample space for the given random phenomenon. Set 1 cannot be a sample space because it does not contain all possible outcomes for the phenomenon. Set 3, however, cannot be a sample space even though it contains all possible outcomes for the phenomenon, because it contains the same outcome more than once, for example, "heads."

EXAMPLE 2.1.2

Assume that a hat contains the following three items:

cigarette, lighter, match

and that we are to draw one item from the hat. Suppose we describe the following three sets which list some or all possible outcomes of this experiment, as shown in Figure 2.1.2.

Only Set 2 is the sample space for the given experiment. Set 1 cannot be a sample space, since it does not contain all possible outcomes of the experiment. Set 3, however, cannot be a sample space even though it contains all possible outcomes of the experiment, because it contains one outcome more than once.

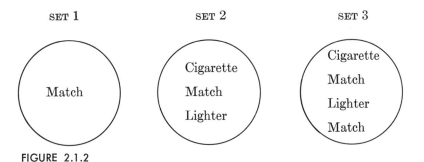

FIGURE 2.1.2

EXAMPLE 2.1.3

Assume that we are to toss two coins. Then, we can describe the sample space for this experiment by the following set:

$$\{HH, HT, TH, TT\}.$$

On the other hand, suppose that what we are interested in is the number of "heads" obtained. Then, the sample space may alternatively be described by the following set of numbers:

$$\{0, 1, 2\}.$$

The relationship between the two different descriptions of the sample space is shown in Figure 2.1.3. This figure also shows that the same physical phenomenon yields two different sample spaces. Thus, the same physical phenomenon can yield a number of different sample spaces depending on the nature of one's interest in the phenomenon.

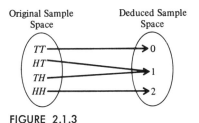

FIGURE 2.1.3

EXAMPLE 2.1.4

Consider the system shown in Figure 2.1.4. It consists of two components, A and B, which operate independently of each other. At any randomly selected time, a component may be in one of two possible states: working or not working. Assume that whether a component is in one state or the other is a result of chance.

8 Sample Space and Probability

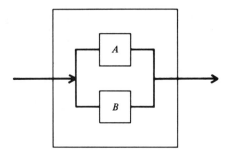

FIGURE 2.1.4

Assume now that we wish to describe various possible states that the system can be in. Suppose that the state of the system is characterized in terms of the states for its two separate components. Then, all possible states for the system are given by the set in Figure 2.1.5.

> A (working) and B (working)
>
> A (working) and B (not working)
>
> A (not working) and B (working)
>
> A (not working) and B (not working)

FIGURE 2.1.5   Possible states of a system.

As we can see, there are four possible states for the system. It must be clear also that chance should be the governing factor determining which state the system is in at any given time. This set then may be considered to be the sample space depicting the states for the system.

EXERCISES

*1. Let a random phenomenon consist of tossing a coin, where "heads" is denoted by H and "tails" by T. Is the set $S = \{H,T,H\}$ a sample space for this phenomenon? Explain.

---

* All exercises preceded by an asterisk throughout this textbook are recommended exercises.

2.2 Events Defined in a Sample Space

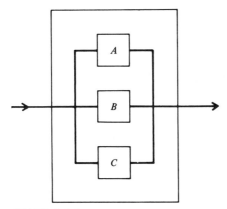

FIGURE 2.1.6

*2. Let a random phenomenon consist of tossing three coins. Describe the sample space for this phenomenon.

*3. Let a random phenomenon consist of tossing a die. Is the set $S = \{1,3,4,6\}$ the sample space for this phenomenon? Explain.

4. A petroleum company decides to drill a ground for oil. Describe the sample space depicting the company's outcome.

5. An electrical system consists of three components, A, B, and C, as shown in Figure 2.1.6. At any randomly selected time a component may be in one of two possible states: working or not working. The state that the system is in is characterized in terms of the states for its three separate systems. Describe the sample space for the electrical system.

2.2 Events Defined in a Sample Space

When a sample space contains many possible outcomes it may be too cumbersome to consider each possible outcome individually. Or, even if such a consideration were not too cumbersome, we might be interested only in the occurrences of certain types of outcomes. Thus, some type of classification of the elements in the sample space may be useful. The following definition is given for this purpose.

Definition 2.2.1
Any subset of a sample space may be called an *event*.

EXAMPLE 2.2.1
Consider the three sets in Figure 2.2.1.

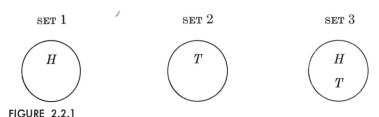

FIGURE 2.2.1

Each set is a subset of the sample space given in Example 2.1.1; therefore, it is an event for the given sample space.

EXAMPLE 2.2.2
Consider the following sets:

$$\{C\}, \{L\}, \{M\}, \{C,L\}, \{C,M\}, \{M,L\}, \{C,L,M\},$$

where C denotes cigarette, L denotes lighter, and M denotes match. Each of the above sets is a subset of the sample space for Example 2.1.2. Thus, each also constitutes an event definable in the sample space.

EXERCISES

*6. Describe the sample space for tossing two coins. Then, describe all events which are definable in the sample space.
7. For the sample space given in Exercise 4, describe all events which are definable for the sample space.
8. You are to select two items from a manufacturing process and count the number of defectives for your sample of two items. Describe the sample space. Then, describe all events which are definable for the sample space.

2.3 Occurrence of an Event

An event is merely a description of a certain type of outcome for a random phenomenon. There is, however, a fundamental difference between the notion of an *event* and the *occurrence of that event*. Whereas the former describes one or more different ways in which an event can occur, the latter describes the exact manner in which it actually occurs; that is,

Definition 2.3.1
Let E denote an event defined in a sample space. Then, the event E is said to have *occurred* if the actual outcome of the random phenomenon corresponds to an element of the set describing the event E.

EXAMPLE 2.3.1

Assume that we are going to toss a dime and a quarter and will win $1 if we obtain exactly one "heads." Then, the sample space for the game is described by

$$\begin{array}{|c|} \hline H_d H_q \\ H_d T_q \\ T_d H_q \\ T_d T_q \\ \hline \end{array}$$

where $H_d T_q$, for example, denotes that the dime lands on "heads" and the quarter on "tails." Let W depict the event corresponding to our win. Then, W may be described by

$$W \rightarrow \begin{pmatrix} H_d T_q \\ T_d H_q \end{pmatrix}$$

Now suppose that we toss the two coins and obtain two "heads." Then, the outcome of our toss may be described by $H_d H_q$. However, we find that $H_d H_q$ is not an element of W. Thus, the event W did not occur. In nonstatistical terms, this means that we did not win $1 since we did not obtain exactly one "heads." On the other hand, instead of two "heads," suppose we obtain "heads" for the dime and "tails" for the quarter. Then, the outcome of our toss is described by $H_d T_q$. We find that it is an element of the set describing W. Thus, the event W has occurred. To state that the event W has occurred is really another way of saying that we have won $1. The fact that the event W has occurred by attaining $H_d T_q$ is equivalent to saying that we have won $1 by obtaining "heads" for the dime and "tails" for the quarter. Of course, we could have won $1 by obtaining "tails" for the dime and "heads" for the quarter: that is, the event W would occur if the outcome of the random phenomenon corresponds to $T_d H_q$.

EXERCISES

*9. Describe all events for the sample space for tossing two coins. Label them E_1, E_2, and so on. Assume that the toss of the two coins has resulted in exactly one "heads." Indicate the events which have occurred and those which have not occurred.

10. A hat contains three numbers: 1, 2, and 3. You are to draw one number from the hat. Describe the sample space for the experiment, and then describe all events

for the sample space. Now assume that you have drawn the number 2. Indicate the events which have occurred.

2.4 Types of Events

Since the set consisting of a sample space may be divided into different subsets, a number of different events may be defined for a sample space. There are a number of special types of events which are of interest to us.

Definition 2.4.1

Let E be an event defined in a sample space. The event E is said to be a *simple event* if the set describing the event contains only one possible outcome of the sample space. If E contains more than one possible outcome, it is said to be a *compound event*.

EXAMPLE 2.4.1

Assume we are going to draw one card from a deck of 52 cards, jokers excluded. Suppose we define the event F as that of drawing an ace of diamonds and the event G as that of drawing any ace. The two events are illustrated by the following diagrams.

F	G
Ace (Diamond)	Ace (Diamond)
	Ace (Heart)
	Ace (Club)
	Ace (Spade)

The event F is a simple event. On the other hand, the event G is a compound event. Clearly, then, if an event is a compound event, two or more simple events may be defined in it.

Definition 2.4.2

Let A and B be two events defined in a sample space. The events A and B are said to be *mutually exclusive* if their intersection is empty.

The definition just given merely states that if A and B are mutually exclusive events, then A and B cannot occur simultaneously. This means that the occurrence of one will preclude the occurrence of the other.

2.4 Types of Events   13

EXAMPLE 2.4.2

Assume again that we are going to draw a card from a deck of 52 cards. Let us now define the following events for the experiment:

$$E_1 = \text{Drawing a diamond}$$
$$E_2 = \text{Drawing a club}$$
$$E_3 = \text{Drawing an ace.}$$

Then, the events E_1 and E_2 are mutually exclusive because drawing a diamond precludes the possibility that the same card may be a club. On the other hand, E_1 and E_3 are not mutually exclusive, since drawing a diamond does not preclude the possibility that the same card may also be an ace.

EXAMPLE 2.4.3

Consider the sample space for Example 2.3.1. Let

$$E_1 = \text{Obtaining either two ``heads'' or two ``tails''}$$
$$E_2 = \text{Obtaining exactly one ``heads''}$$
$$E_3 = \text{Obtaining at least one ``heads.''}$$

Then, E_1 and E_2 are, for example, mutually exclusive, because if we obtain exactly one "heads" in a single toss of two coins, then the possibility of obtaining either two "heads" or two "tails" must be ruled out. In other words, the intersection of E_1 and E_2 is empty, as illustrated in Figure 2.4.1.

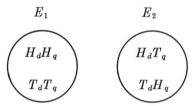

FIGURE 2.4.1 Empty intersection of two events.

On the other hand, E_1 and E_3 are, for example, not mutually exclusive, since the fact that we have obtained at least one "heads" in a single toss of two coins cannot rule out the possibility that we could have obtained two "heads." The intersection of the two events is not empty, as illustrated in Figure 2.4.2.

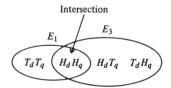

FIGURE 2.4.2 Nonempty intersection of two events.

Sample Space and Probability

Definition 2.4.3

Let A be an event defined in a sample space S. The *complement* of A, denoted $\bar{A}$, is the set consisting of all outcomes of S which are not in A.

EXAMPLE 2.4.4

Let the sample spaces S and A be given by

$$S = \begin{pmatrix} HH \\ TH \\ HT \\ TT \end{pmatrix} \text{ and } A = (HH).$$

Then,

$$\bar{A} = \begin{pmatrix} TH \\ HT \\ TT \end{pmatrix}.$$

EXAMPLE 2.4.5

Let $S = \{1,2,3,4,5,6\}$ and $A = \{2,3\}$. Then,

$$\bar{A} = \{1,4,5,6\}.$$

It is apparent from the preceding discussion that

$$A \cup \bar{A} = S \text{ and } A \cap \bar{A} = \emptyset.$$

EXERCISES

*11. Using the sample space for tossing one coin, describe an example of a simple event.

12. If you toss two coins, one possible outcome is that you may obtain exactly one "heads." Does this outcome correspond to a simple event? Explain.

*13. In the sample space for tossing one coin, identify two events which are mutually exclusive.

14. Toss two coins and let A be the event which occurs if exactly one "heads" is obtained, and B be the event which occurs if at least one "heads" is obtained. Are A and B mutually exclusive events? Explain.

*15. Toss two coins and let A be the event which will occur if exactly one "heads" is obtained. What is the complement of this event in the sample space?

16. A manufacturing process is assumed to be operating satisfactorily if a sample of five items from the process contains no defective item. Let A be the event which occurs when this happens. What is the complement of A in the sample space?

2.5 Probability of an Event

Since an event is a description of a certain kind of possible outcome, we are interested in the likelihood of its occurrence. In this respect, we might, for example, infer that event A is "most likely to occur," whereas event B is "very likely to occur." This statement presumably implies that event A is more likely to occur than event B, because of our usage of these two words. If this is true, we cannot tell to what extent A is more likely to occur than event B. In order to avoid this kind of ambiguity, we prefer to express the likelihood of occurrence of an event in numerical terms. The concept of probability is useful at this point. We might state:

Definition 2.5.1
The *probability* of an event is a number expressing the likelihood of that event.

The definition given above is quite crude. Some additional information is needed to make the definition operationally meaningful.

In defining the probability of an event, we have said nothing about how a number is to be assigned for the probability of that event. There are two problems which arise in the process of assigning a number for the probability of any given event. First, there is the problem of assessing the likelihood that the event actually will occur. Second, once this assessment has been made, there is the problem of converting the assessment into a number. There is a significant amount of controversy with regard to how the first of these two problems should be resolved, however, no major controversy exists with regard to the second problem.

The second problem will be discussed in the next section, and following this discussion, we will take up the controversy concerning the first problem.

2.6 Converting a Likelihood-Assessment into a Probability

Let us assume for the moment that the likelihood-assessment for an event has been made. In converting that likelihood-assessment into a number corresponding to the probability of that event, the following rules will be observed. Let A be an event for which we are to assign a probability. Then, if our likelihood-assessment indicates that

1. it would be impossible to occur, assign 0 as its probability
2. it would be certain to occur, assign 1 as its probability
3. it would be neither impossible nor certain to occur, assign a number between 0 and 1.

An implicit idea underlying these rules is that, given two different events, the one which is most likely to occur will have a larger numerical value as its probability.

Although our primary interest in a sample space may be confined to a specific event, for example A, there may be other events related to A which may be of interest. For example, since we are interested in A, we should also be interested in its complement $\bar{A}$. In turn, we should be interested in $A \cup \bar{A}$, which is the sample space, and in $A \cap \bar{A}$ which is a null set. Thus, the set containing $\{A\}$, $\{\bar{A}\}$, $\{S\}$, and $\{\varnothing\}$ may be said to contain all events related to A which conceivably may be of some interest to us. Such a set is called a *Borel field of events* and satisfies the following definition.

Definition 2.6.1

A *Borel field of events* is a collection of events such that if any event is in the collection, then its complement is in the collection and, furthermore, the union or intersection of any finite number of events is also in the collection.

EXAMPLE 2.6.1

Assume that we are interested in the event $\{A\}$ in the sample space given in Figure 2.6.1. The Borel field may be ascertained in the following manner. First, the collection must contain $\{\bar{A}\}$, which is $\{B,C\}$. In turn, the collection must contain $\{A\} \cup \{B,C\}$, which is the sample space $\{S\}$, and it must contain $\{A\} \cap \{B,C\}$, which is the null set $\{\varnothing\}$.

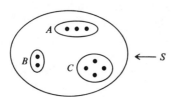

FIGURE 2.6.1 Sample space.

Listing the events which we have enumerated thus far, we have

$$\{\varnothing\},\ \{A\},\ \{B,C\},\ \{S\}.$$

By using the following guidelines, we will observe that the collection of events in Figure 2.6.1 satisfies the definition of a Borel field.

1. Select any event from the collection. We find that the complement of the selected event is also in the collection.
2. Form a union of any number of events in the collection. We find that the union is also in the collection.
3. Form an intersection of any number of events in the collection. We find that the intersection is also in the collection.

EXAMPLE 2.6.2

Suppose for the sample space given in Example 2.6.1 we are interested in the events A and B. Then, a Borel field which contains A and B is given as

$$\{\varnothing\}, \{A\}, \{B\}, \{C\}, \{A,B\}, \{A,C\}, \{B,C\}, \{S\}.$$

The reader should attempt to justify why this collection of events is a Borel field.

We observe also that the collection just given is a Borel field that contains $\{A\}$. On the other hand, if our interest is confined only to $\{A\}$, then the collection of events given in Example 2.6.1—$\{\varnothing\}, \{A\}, \{B,C\}, \{S\}$—may serve as the Borel field of interest. The latter Borel field may be thought of as the smallest Borel field which contains the event A.

Since a Borel field contains all conceivable events of interest, we might assign or make a provision for assigning a probability to each event in the Borel field. The probabilities assigned to each event in the Borel field, however, must be consistent with each of the others, thus reflecting their likelihoods. This means that we must observe a set of rules in assigning the probabilities. (The three rules provided previously for one single event are not adequate for this purpose.)

Probabilists have found that the following three rules, or *axioms*, are sufficient. Other propositions pertaining to probabilities may be derived from these axioms by using the rules of logic. The probability axioms are:

1. Let A be any event in S. Then, $0 \leq P(A) \leq 1$.
2. Let A and B be any two mutually exclusive events in S. Then, $P(A \cup B) = P(A) + P(B)$.
3. Let S depict a sample space. Then, $P(S) = 1$.

Necessity for the first axiom already has been considered. The second axiom states that the probability of either one of two mutually exclusive events occurring is the sum of their separate probabilities. The third axiom points out the fact that the event S will certainly occur, since it contains all possible outcomes of the random phenomenon.

EXAMPLE 2.6.3

Let us assume that we wish to make a probability assignment for a coin which is supposedly biased. We are told that "heads" is three times more likely to be obtained than "tails." Consider the three assignments of probabilities for "heads" and "tails" of this coin in Table 2.6.1. All of the three assignments satisfy the first axiom given. Furthermore, the relative numerical values given in each of the three assignments indicates that "heads" are three times more likely to occur than "tails." However, only the second assignment satisfies the remaining two axioms.

TABLE 2.6.1 Three Likelihood-Assignments for a Coin

Outcome	1st Assignment	2d Assignment	3d Assignment
Heads	.3	.75	.9
Tails	.1	.25	.3
	.4	1.00	1.2

To show why this is the case, we enumerate the Borel field of events for the sample space as $\{\varnothing\}$, $\{H\}$, $\{T\}$, and $\{H,T\}$. Then, if we resort to the first assignment we would have

$$P\{\varnothing\} = .0$$
$$P\{H\} = .3$$
$$P\{T\} = .1,$$

and if we apply the second axiom we would have

$$P\{H,T\} = P(H) + P(T) = .4.$$

On the other hand, $\{H,T\}$ is the sample space S and, therefore, $P\{H,T\}$ should be 1 according to the third axiom. Thus, the assignment cannot be used as a probability assignment.

If we resort to the third assignment, we would obtain

$$P\{\varnothing\} = .0$$
$$P\{H\} = .9$$
$$P\{T\} = .3$$
$$P\{H,T\} = 1.2.$$

Again, $P\{H,T\}$ is not a unity when we resort to the third assignment. Thus, it cannot be used as a probability assignment.

Let us now resort to the second assignment. Then,

$$P\{\varnothing\} = .00$$
$$P\{H\} = .75$$
$$P\{T\} = .25$$
$$P\{H,T\} = 1.00.$$

Thus, the assignment satisfies all three axioms and, therefore, may be considered to be a probability assignment.

EXERCISES

*17. Toss a coin and let H be the event which will occur if "heads" is obtained. Describe the Borel field of events which contains H. Then, assign a probability to each event in the Borel field. Verify the fact that your probability assignments are consistent with the probability axioms.

18. Toss a die and let A be the event which will occur if the number 1 or 2 is obtained. Describe the smallest Borel field which contains A. Then, assign a probability to each event in the Borel field. Verify the fact that your probability assignments are consistent with the probability axioms.

2.7 Controversy with Regard to Likelihood-Assessment

We now come to the question of how the likelihood-assessment actually should be made for a given sample space. To illustrate the nature of this problem, if we were asked what the probability is of obtaining "heads" when a coin is tossed, we would immediately say that it is .5. On the other hand, if we were asked why the probability is precisely .5, we would probably hesitate.

The probability assignment that we have made for tossing the coin may be said to be based on the *principle of insufficient reason*. According to this principle, if a sample space is divided into m mutually exclusive events and if we have no reason to believe that one of the events is more likely to occur than the others, then, we may simply assume that all of m events are *equally likely* to occur and we can assign $1/m$ as the probability of each event in the sample space. The mutually exclusive events for the sample space for tossing a coin are "heads" and "tails," and if we have no reason to believe that one of them is more likely to occur than the other, we would obviously assign .5 for the probability of obtaining "heads" and .5 for the probability of obtaining "tails."

The approach that we have suggested for assigning probabilities for events is called the *equally-likely approach*. We can use this approach to assign probabilities for events for random phenomena, such as in tossing a die or in drawing a card from a deck for which we can assume a symmetry among the events.

FIGURE 2.7.1 Possible positions of a thumbtack.

There are, however, random phenomena for which we cannot make the assumption of symmetry among the events in the sample space. For example, suppose we toss a thumbtack. Assume that it will land in one of two different ways, as illustrated in Figure 2.7.1. Let the sample space for this random phenomenon be given by

20   Sample Space and Probability

Sample Space

where U denotes "up" and D denotes "down."

Suppose now that we wish to make likelihood-assessments for U and D. One way to make the assessments naturally is to toss the thumbtack many times and to count, for example, the number of times it lands "up." Let this number be denoted by N and let the total number of times that the thumbtack has been tossed be denoted by T. Then, we might let

$$P = \frac{N}{T}$$

represent the estimated value of the probability that the thumbtack will land "up" in any given toss. Obviously,

$$1 - \frac{N}{T}$$

would represent the probability that it will land "down."

There is an important school of thought among the probabilists which believes that the procedure just described is the only way that the likelihood-assessments should be made for the events in a sample space. The approach suggested by this school is known as the *relative-frequency approach*, since the ratio N/T is, in essence, a relative frequency between N and T. The school also is known as the *objective school*, because it believes that, if the likelihood-assessments are in fact made by repeated experiments in order to obtain the relative frequencies, the resulting estimates will be free of any subjective beliefs of the person who is attempting to assess the likelihoods for the events in question.

Another school of thought among the probabilists, known as the *subjective school*, claims that the relative-frequency approach cannot enable us to solve a large number of problems for which it is either impossible or too costly to conduct experiments. Consider, for example, that we have been given an option to play the following game: we toss a thumbtack and win $1000 if it lands "up" or lose $500 if it lands "down." We can play this game only once. Furthermore, we are not allowed any trial tosses before deciding whether or not to play the game.

Let us assume that we will definitely play the game with the amount of money involved, if we know that the odds are in our favor. This means that we must make the likelihood-assessments for the given thumbtack. However, the approach suggested by the objective school is not useful in making the assessments, since we are not allowed to make any trial tosses. Instead of

2.7 Controversy with Regard to Likelihood-Assessment

concluding that the likelihood-assessments for the two outcomes are impossible to make, we might decide to examine the physical shape of the thumbtack very carefully and, then, to try to make the likelihood-assessments by utilizing whatever knowledge we have accumulated through our past experiences which might have some bearing upon the problem. For example, if the size of the head is quite large, but the length of the nail is very short, as illustrated in (a) of Figure 2.7.2, we might conclude that the behavior of the thumbtack may be similar to that of a coin and estimate $P(U)$ to be approximately 50 percent. On the other hand, if the shape of the thumbtack looks like that of (b) in Figure 2.7.2, we might estimate $P(U)$ to be very close to 0.

(a)   (b)

FIGURE 2.7.2 Possible shapes of a thumbtack.

The subjective school of thought advocates that the procedure such as the one just described be used to make the likelihood-assessments for the events in the sample space. This approach is called the *personalistic approach*.

The controversy may appear to be of little significance, since we have illustrated it with a frivolous example. However, now consider the problem of an oil company which is in the process of deciding whether or not to drill a piece of ground for oil. Before making a decision, suppose the company wishes to assess the likelihood of hitting oil. The objective school would claim that such a probability of hitting oil would be indeterminable, since the drilling of that particular ground is a unique experiment which is not repeatable elsewhere. Thus, this school would place the oil company's problem outside of its domain of analysis. On the other hand, the subjective school would claim that such a probability could be ascertained by carefully evaluating seismographic records and other geological findings.

There are other ramifications of the controversy. We will explore them in detail in Chapters 12 and 14.

Chapter 3
Probability
Calculations

Even though there exists a controversy concerning how likelihood-assessments should be made, probabilists are in agreement concerning probability calculations. By probability calculations, we mean how the probabilities of the events are to be added or multiplied.

3.1 Probability of a Compound Event

We have distinguished between simple and compound events in Chapter 2. We will now show how the probability of a compound event is related to that of its simple events.

Theorem 3.1.1
Let A be a compound event defined in a sample space where $a_1, a_2, \ldots, a_k$ are the simple events in A. Then,

$$P(A) = P(a_1) + P(a_2) + \cdots + P(a_k).$$

The validity of this theorem is easy to justify. From the second probability axiom it follows that the probability of the union of mutually exclusive events is equal to the sum of the probabilities of the individual events in the union. A compound event A is a union of the simple events $a_1, a_2, \ldots, a_k$, which, in turn, are mutually exclusive. Thus,

$$P(A) = P(a_1) + P(a_2) + \cdots + P(a_k).$$

We will illustrate Theorem 3.1.1 with the following examples.

EXAMPLE 3.1.1

Consider the experiment wherein two coins are tossed. Suppose we want to obtain the probability that exactly one "heads" will be obtained. The event would occur if TH or HT of the sample space is obtained. Thus, we might denote the event by Figure 3.1.1. Let $P(TH) = \frac{1}{4}$ and $P(HT) = \frac{1}{4}$. The probability that A would occur is then

$$P(A) = P(TH) + P(HT) = \frac{1}{4} + \frac{1}{4} = \frac{1}{2}.$$

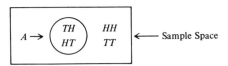

FIGURE 3.1.1

EXAMPLE 3.1.2

Assume that a random phenomenon consists of tossing a die. Let A be the event which occurs if an even number is obtained, for example, 2, 4, or 6. The event A and the sample space are described in Figure 3.1.2. Let $P(2) = \frac{1}{6}$, $P(4) = \frac{1}{6}$, and $P(6) = \frac{1}{6}$. The probability that A will occur is then obtained as

$$P(A) = P(2) + P(4) + P(6) = \frac{1}{6} + \frac{1}{6} + \frac{1}{6} = \frac{1}{2}.$$

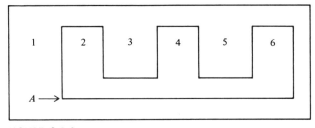

FIGURE 3.1.2

3.2 Probability of Two Compound Events

Let A and B denote two compound events, and let C be the event which occurs if either A or B occurs. We then state that $C = A \cup B$. We will describe how the probability for C may be calculated.

Theorem 3.2.1

Let A and B be two compound events which are mutually exclusive. Then,

$$P(A \cup B) = P(A) + P(B).$$

The reader will observe that Theorem 3.2.1 is a restatement of the second probability axiom.

EXAMPLE 3.2.1

In the experiment consisting of tossing two coins, let $A = \{HH\}$ and $B = \{HT, TH\}$. Suppose we define C as the event which occurs if at least one "heads" is obtained. Then, C would occur if A occurs; C also would occur if B occurs. Thus, $C = A \cup B$. On the other hand, A and B are mutually exclusive events. Consequently,

$$P(C) = P(A) + P(B) = \frac{1}{4} + \frac{2}{4} = \frac{3}{4}.$$

EXAMPLE 3.2.2

Let a random phenomenon consist of drawing one card from a deck of bridge cards. Let A be the event which occurs if a heart is drawn and B be the event which occurs if a diamond is drawn. Let C be the event which occurs if the card drawn is a red card. Obviously, then, C would occur if a card is a heart. On the other hand, C also would occur if the card is a diamond. Thus, $C = A \cup B$. But A and B are mutually exclusive events; if a card drawn is a heart, it cannot at the same time be a diamond. (The reader should keep in mind that we are to draw just one card.) If we assume that the probability of drawing a heart is $\frac{1}{4}$ and that of drawing a diamond is $\frac{1}{4}$, then,

$$P(\text{red card}) = P(\text{heart}) + P(\text{diamond})$$
$$= \frac{1}{4} + \frac{1}{4} = \frac{1}{2}.$$

EXERCISES

*1. Describe the sample space for tossing three coins. Assign a probability to each simple event in the sample space. Then, find the probability of obtaining exactly one "heads."

*2. Refer to the experiment given in Exercise 1 and find the probability of at least one "heads."

3. Describe the sample space for tossing two dice. Assign a probability to each simple event in the sample space. Then, find the probability that the sum of the two numbers obtained from tossing the two dice will be 4.

4. For the experiment given in Exercise 3, find the probability that the sum of the two numbers will be 6.

3.3 Joint Probability

Let A and B be two events defined in a sample space, and let D be the event which occurs if both A and B occur. Then, D is said to be a *joint event* of A and B. A formal definition of a *joint event* may be advanced as follows.

Definition 3.3.1
Let A and B be any two events defined in a sample space. If D is the event given by the intersection of A and B, then, D is said to be the *joint event* of A and B and is denoted by $D = A \cap B$.

EXAMPLE 3.3.1
Assume that we are going to draw one card from a deck of bridge cards. Let A be the event which occurs if the card drawn is a diamond, and let B be the event which occurs if the card is an ace. Let $D = A \cap B$. Then, D would occur if, and only if, the card drawn happens to be an ace of diamonds. Figure 3.3.1 illustrates the joint event.

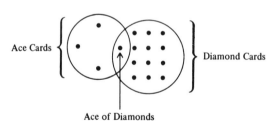

FIGURE 3.3.1 A joint event.

If the joint event of A and B does not exist, A and B are said to be *disjoint events;* in other words, two mutually exclusive events are *disjoint events*.

Theorem 3.2.1 has been given for $P(A \cup B)$, assuming that A and B are mutually exclusive. Now we will describe how $P(A \cup B)$ may be obtained regardless of whether or not A and B are mutually exclusive.

Theorem 3.3.1
Let A and B be any two events defined in a sample space. Then,

$$P(A \cup B) = P(A) + P(B) - P(A \cap B).$$

The preceding theorem states that the probability that $C = A \cup B$ is equal to the sum of the probabilities for the separate events less the probability for their joint event. We might also point out that Theorem 3.3.1 is a generalization of Theorem 3.2.1, since we get Theorem 3.2.1 if $A \cap B = \emptyset$ for Theorem 3.3.1.

EXAMPLE 3.3.2

Assume that we are going to draw one card from a bridge deck and that we will win \$1 if the card drawn is either a diamond or an ace. Let A be the event which occurs if the card drawn is a diamond, B be the event which occurs if the card drawn is an ace, and C be the event which occurs if we win \$1. Then, $C = A \cup B$. We can determine immediately that the probability of winning \$1 is $16/52$, since we will win if any of the 13 diamond cards or any of the 3 nondiamond aces are drawn. Theorem 3.3.1 confirms what we know already in this case, since according to the theorem

$$P(A \cup B) = P(A) + P(B) - P(A \cap B)$$
$$= \frac{13}{52} + \frac{4}{52} - \frac{1}{52} = \frac{16}{52}.$$

$P(A \cap B)$ in this case represents the probability of drawing the ace of diamonds.

EXAMPLE 3.3.3

Assume that a box contains 14 red glass marbles, 26 brown glass marbles, 36 red plastic marbles, and 24 brown plastic marbles. The contents of the box also can be shown by Table 3.3.1.

Let G or P denote the event which occurs if a marble drawn is glass or plastic, respectively, and let R or B denote the event which occurs if the marble drawn is red or brown. Then,

$$P(R) = .5 \qquad P(G \cap R) = .14$$
$$P(B) = .5 \qquad P(G \cap B) = .26$$
$$P(G) = .4 \qquad P(P \cap R) = .36$$
$$P(P) = .6 \qquad P(P \cap B) = .24.$$

Now let C be the event which occurs if the marble drawn is either glass or red. Then, we may denote $C = G \cup R$ and calculate its probability as

$$P(G \cup R) = P(G) + P(R) - P(G \cap R)$$
$$= .40 + .50 - .14 = .76.$$

TABLE 3.3.1

	Red	Brown
Glass	14	26
Plastic	36	24

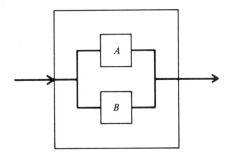

FIGURE 3.3.2 An electrical system.

EXAMPLE 3.3.4

Consider the electrical system in Figure 3.3.2, which works if any one of its two components works. Suppose component A works 80 percent of the time, B works 90 percent of the time, and both A and B work simultaneously 75 percent of the time. In notation, then

$$P(A) = .80,$$
$$P(B) = .90,$$

and
$$P(A \cap B) = .75.$$

Suppose now that we wish to ascertain the probability that the system will work at any given time. In notation, the probability of the system working is given by $P(A \cup B)$. Then,

$$P(A \cup B) = P(A) + P(B) - P(A \cap B)$$
$$= .80 + .90 - .75 = .95$$

is the probability that the system will work at any given time.

EXERCISES

*5. A box contains 6 red glass marbles, 14 brown glass marbles, 24 red plastic marbles, and 16 brown plastic marbles. Find the following probabilities: (a) $P(R)$, (b) $P(B)$, (c) $P(G)$, and (d) $P(P)$, where $P(R)$, for example, denotes the probability that a marble drawn is red.

*6. For the problem given in Exercise 5, find: (a) $P(G \cap R)$, (b) $P(G \cap B)$, (c) $P(P \cap R)$, and (d) $P(P \cap B)$. When does the event corresponding to (a), for example, occur?

*7. For the problem given in Exercise 5, find $P(G \cap P)$. When would the event in question occur?

8. For the problem given in Exercise 5, find: (a) $P(G \cup R)$, (b) $P(G \cup B)$, (c) $P(R \cup R)$, and (d) $P(P \cup B)$. When does the event corresponding to (a), for example, occur?

9. At a university 40 percent of the student body are girls and 10 percent are freshmen girls. The freshmen class composes 30 percent of the student body. If you pick one student at random from this university, what is the probability that the student will be either a freshman or a girl?

*10. An aerospace company needs to develop a special component used in a spaceship. It assigns the task of developing the component to its own research team and, at the same time, to an outside consulting firm. The company believes that there is a .6 probability that the company's own research staff will develop the component and a .8 probability that the consulting firm will develop the component. It also believes that there is a .5 probability that both of them will develop the component. What is, then, the probability that the needed component will indeed be developed?

3.4 Conditional Probability

Let A and B be two events defined in the sample space. At times we might want to know the probability that A will occur, assuming that B already has occurred or is bound to occur. Such a probability is called the *conditional probability* of A. A formal definition is now given.

Definition 3.4.1

Let A and B be two events defined in the sample space. The *conditional probability* of A, given B, and denoted $P(A|B)$ is defined as

$$P(A|B) = \frac{P(A \cap B)}{P(B)},$$

and the *conditional probability* of B, given A, and denoted $P(B|A)$ is defined as

$$P(B|A) = \frac{P(A \cap B)}{P(A)}.$$

EXAMPLE 3.4.1

In a drawing of one card from a bridge deck, let D represent the event which occurs if the card is a diamond, and let R be the event which occurs if the card is red. If we are asked what the probability is that a card drawn would be a diamond, our answer would be $\frac{1}{4}$. On the other hand, suppose we are asked to give the probability that the card drawn would be a diamond, assuming that it would be also in one of the red suits, then our answer would be $\frac{1}{2}$, since it must either be a diamond or a heart. We can calculate this probability, according to the previous definition, in the following way. The probability that the card is $R \cap D$ is $\frac{1}{4}$ and that it is red is $\frac{1}{2}$. Thus,

$$P(D|R) = \frac{P(R \cap D)}{P(R)} = \frac{\frac{1}{4}}{\frac{1}{2}} = \frac{1}{2}.$$

EXAMPLE 3.4.2

Let us return to the box of marbles described in Example 3.3.3. Suppose we are told that the marble drawn is red but we are asked the probability of its being a glass marble. In notation we would express this question as $P(G|R)$. Definition 3.4.1 indicates then that

$$P(G|R) = \frac{P(G \cap R)}{P(R)} = \frac{.14}{.50} = .28.$$

A physical interpretation of the preceding conditional probability may be considered in the following context. Suppose that, instead of drawing a marble from the entire box (see Example 3.3.3), we separate the marbles into two different urns according to color. Then, suppose we draw a marble from the urn containing only the red marbles. We now ask what the probability is that the marble drawn is glass. Since the urn contains 50 marbles, of which 14 are glass, the probability of drawing a glass marble would be .28 which is $14/50$.

EXAMPLE 3.4.3

Let us return to Example 3.3.4. If we are told that component A works, what can we then say about the probability that component B also works? In notation this probability is given as $P(B|A)$. Since $P(A \cap B) = .75$ and $P(A) = .8$, we have

$$P(B|A) = \frac{P(A \cap B)}{P(A)}$$
$$= \frac{.75}{.8}$$
$$= .9375.$$

On the other hand, suppose we are told that B works. Then, the probability that A also works is ascertained as

$$P(A|B) = \frac{P(A \cap B)}{P(B)}$$
$$= \frac{.75}{.90}$$
$$\cong .8333.$$

From Definition 3.4.1, we can derive

Theorem 3.4.1

Let $A \cap B$ be the joint event of A and B. Then,

$$P(A \cap B) = P(A)P(B|A)$$
$$= P(B)P(A|B).$$

30 Probability Calculations

The validity of this theorem may be illustrated as follows. Consider, for example, the conditional probability

$$P(B|A) = \frac{P(A \cap B)}{P(A)}.$$

By multiplying both sides of this equation by $P(A)$, we obtain

$$P(A)P(B|A) = P(A \cap B),$$

which is the first equation of Theorem 3.4.1. The second equation of the theorem may be obtained in a similar manner.

EXAMPLE 3.4.4

Let us return to the box of marbles given in Example 3.3.3. For this box we may calculate the following conditional probabilities:

$$\begin{array}{ll} P(G|R) = .28 & P(G|B) = .52 \\ P(P|R) = \underline{.72} & P(P|B) = \underline{.48} \\ 1.00 & 1.00. \end{array}$$

Now we wish to find $P(G \cap R)$:

$$P(G \cap R) = P(R)P(G|R) = (.5)(.28) = .14.$$

Again we find that $P(G \cap R)$ is the same as that given in Example 3.3.3.

EXAMPLE 3.4.5

Assume that the price of a certain stock has decreased substantially during recent months because of a recession in the general economy. We wish to assess the probability that the price of this stock will increase again within a year.

A well-known economist claims that there is a 60 percent probability that the recession will be over at the end of a year. An investment counselor advises that if the recession is over, there is an 80 percent probability that the price of the stock will increase, and if the recession is not over, there is only a 10 percent probability that the stock will increase. What, then, is the probability that the price of the stock will increase within a year, if the probability assessments of the economist and the investment advisor are correct? We can ascertain the probability in question in the following manner. Let us denote the following:

R = Recession is over within a year
$\tilde{R}$ = Recession is not over within a year

B = Price of stock increases
$\tilde{B}$ = Price of stock does not increase.

Then,
$$P(R) = .6,$$
$$P(\tilde{R}) = .4,$$
and
$$P(B|R) = .8 \quad P(\tilde{B}|R) = .2,$$
$$P(B|\tilde{R}) = .1 \quad P(\tilde{B}|\tilde{R}) = .9.$$

The price of the stock may increase in two different ways:
$$R \cap B$$
and
$$\tilde{R} \cap B.$$

The probabilities are calculated as
$$P(R \cap B) = P(R)P(B|R) = (.6)(.8) = .48$$
$$P(\tilde{R} \cap B) = P(\tilde{R})P(B|\tilde{R}) = (.4)(.1) = .04.$$

The probability of the stock increasing again is then given as
$$P\{(R \cap B) \cup (\tilde{R} \cap B)\} = .48 + .04 = .52.$$

EXERCISES

*11. A box contains 6 red glass marbles, 14 brown glass marbles, 24 red plastic marbles, and 16 brown plastic marbles. Find: (a) $P(R|G)$, (b) $P(R|P)$, (c) $P(B|G)$, and (d) $P(B|P)$.

12. For the problem given in Exercise 11, find: (a) $P(G|R)$, (b) $P(P|R)$, (c) $P(G|B)$, and (d) $P(P|B)$.

13. Refer to Exercise 9. Suppose you pick a girl's name at random from the student roster and call her for a date. What is the probability that she will be a freshman?

*14. Refer to Exercise 10 and find the probability that the company's own research staff will develop the component, given that the outside consulting firm will develop the component. Further, what is the probability that the outside consulting firm will develop the component, given that the company's own staff will develop the component?

Consider the system shown in Figure 3.4.1. The probability that A works is .9; that B works is .7; and that both A and B work is .65. What is the prob-

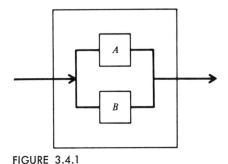

FIGURE 3.4.1

ability that A will work, given that B works. What is the probability that B will work, given that A works?

15. A box contains 30 glass marbles and 70 plastic marbles. If you pick one glass marble, the probability that it will be red is $1/3$. How many red glass marbles are there in the box?

16. For the problem given in Exercise 15, if you pick 1 plastic marble, the probability that it will be brown is $2/7$. How many brown plastic marbles are there in the box?

17. A box contains 60 red marbles and 40 brown marbles. If you pick 1 red marble, the probability that it will be plastic is $5/6$. How many red plastic marbles are there in the box?

18. Use the box given in Exercise 17. If you pick 1 brown marble, the probability that it will be glass is $1/2$. How many brown glass marbles are there in the box?

*19. In the system shown in Figure 3.4.2, the probability that A will work is .9 and that B will work, given that A works, is .8. The system will work only if both A and B work. What is the probability that the system will work?

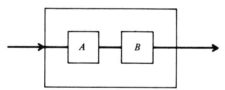

FIGURE 3.4.2

*20. In the system shown in Figure 3.4.3, the probability that A will work is .9. Given that A works, the probability that B will work is .8. On the other hand, given that A does not work, the probability that B will work is .6. What, then, is the probability that B will work, regardless of whether or not A works?

*21. Referring to the system given in Exercise 20, suppose the system works if either A or B works. What is the probability that the system will work?

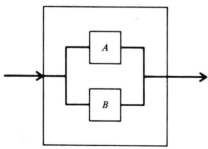

FIGURE 3.4.3

22. Let us assume you believe that the probability that your university's football team will win tomorrow is .6 if the weather is warm and .4 if the weather is not warm. The weatherman forecasts that the probability is .7 that the weather will be warm tomorrow. What is the probability that your university's football team will win tomorrow?

3.5 Independent Events

Given two events A and B, if $P(A|B)$ is different from $P(A)$, then the likelihood-assessment for the occurrence of A is affected by the knowledge that B has occurred. The event A is then said to be *statistically dependent* on B. On the other hand, if the likelihood-assessment of A is not affected by the knowledge that B has occurred, then, $P(A|B)$ must be equal to $P(A)$. The event A is said to be *statistically independent* of B. We offer a formal definition to distinguish the notion of *statistical independence* and *dependence* of two events.

Definition 3.5.1

Let A and B be two events defined in a sample space. A is said to be *statistically independent* of B if

$$P(A) = P(A|B),$$

and B is said to be *statistically independent* of A if

$$P(B) = P(B|A).$$

EXAMPLE 3.5.1

Let D be the event which occurs if a card drawn from a bridge deck is a diamond, and let A be the event which occurs if the card drawn is an ace. Then, $P(D) = 1/4$ and $P(A) = 1/13$. On the other hand, upon reflection we will reach the conclusion that $P(D|A) = 1/4$ and $P(A|D) = 1/13$. Thus, D is independent of A and, conversely, A is independent of D.

EXAMPLE 3.5.2

In contrast to the box of marbles given in Example 3.3.3, consider another box of marbles, the contents of which are shown in Table 3.5.1.

TABLE 3.5.1 Contents of Box of Marbles

	Red	Brown	Total
Glass	20	20	40
Plastic	30	30	60
Total	50	50	100

Let G or P denote the event which occurs if a marble drawn is glass or plastic, respectively, and let R or B denote the event which occurs if the marble drawn is red or brown, respectively. Then, for example, we find that $P(G) = .4$. We also find that $P(G|R) = .4$. Consequently, we may state that the event underlying G is independent of R. A similar computation will show that R also is independent of G.

Definition 3.5.2

Let A and B be two events defined in a sample space. A is said to be *statistically dependent* on B if

$$P(A) \neq P(A|B),$$

and B is said to be *statistically dependent* on A if

$$P(B) \neq P(B|A).$$

If we rephrase this definition, we can state that A is statistically dependent on B if it is not statistically independent of B and, conversely, that B is statistically dependent on A if it is not statistically independent of A.

EXAMPLE 3.5.3

Referring to Example 3.3.3, we find that

$$P(G) = .4 \quad \text{and } P(G|R) = .28.$$

Since they are not equal, event G is not independent of event R.

We also find that

$$P(R) = .5 \quad \text{and } P(R|G) = .35,$$

which indicates that event R is not independent of event G.

EXAMPLE 3.5.4

Let us return to the electrical system given in Example 3.3.4. We have noted that

$$P(A) = .80,$$
$$P(B) = .90,$$

and $$P(A \cap B) = .75.$$

Subsequently, we calculated in Example 3.4.3 that

$$P(B|A) = .9375$$

and $$P(A|B) \cong .8333.$$

The preceding figures indicate that the working of A is statistically dependent on B, since

$$P(A) \neq P(A|B),$$

and, in turn, that the working of B is statistically dependent on A, since

$$P(B) \neq P(B|A).$$

EXAMPLE 3.5.5

Suppose now for the system shown in Figure 3.5.1 we assume that

$$P(A) = .80,$$
$$P(B) = .90,$$
and
$$P(A \cap B) = .72.$$

Then,

$$P(A|B) = \frac{P(A \cap B)}{P(B)} = \frac{.72}{.90} = .80$$

and

$$P(B|A) = \frac{P(A \cap B)}{P(A)} = \frac{.72}{.80} = .90.$$

Thus, $P(A|B) = P(A) = .80$, and $P(B|A) = P(B) = .90$. Therefore, the working of A is independent of B, and, in turn, the working of B is independent of A.

If we know that A is statistically independent of B, does it necessarily follow that B is independent of A? The answer is "yes" if both A and B have positive probabilities. To show the validity of this proposition, assume that A is independent of B. Then,

$$P(A \cap B) = P(B)P(A|B)$$
$$= P(B)P(A),$$

but

$$P(A \cap B) = P(B|A)P(A),$$

which implies that $P(B|A) = P(B)$. Thus, B must be independent of A.

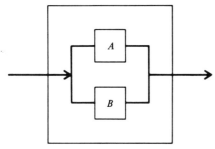

FIGURE 3.5.1

36 Probability Calculations

On the basis of the preceding computations, we have

Theorem 3.5.1

Let A and B be two statistically independent events. Then,
$$P(A \cap B) = P(A)P(B).$$

EXAMPLE 3.5.6

In Example 3.5.1, we have shown that D and A are statistically independent events. Thus, according to Theorem 3.5.1,
$$P(D \cap A) = P(D)P(A) = \left(\frac{1}{4}\right)\left(\frac{1}{13}\right) = \frac{1}{52}.$$

Of course, we already know that the probability of drawing an ace of diamonds is $\frac{1}{52}$.

EXAMPLE 3.5.7

Returning to Example 3.5.2, suppose we wish to find $P(G \cap R)$. Then, according to Theorem 3.5.1,
$$P(G \cap R) = P(G)P(R) = (.40)(.50) = .2,$$

since G and R are shown to be statistically independent events.

EXAMPLE 3.5.8

Consider the system in Figure 3.5.2. The system works if any of its three components work. The probability that each component works is given as
$$P(A) = P(B) = P(C) = .9,$$

assuming each component works independently of the other two.

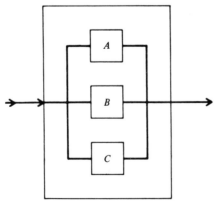

FIGURE 3.5.2

Let $P(\tilde{A})$, $P(\tilde{B})$, and $P(\tilde{C})$ depict the probabilities that A, B, and C will fail. Then,

$$P(\tilde{A}) = P(\tilde{B}) = P(\tilde{C}) = .1.$$

According to our assumption pertaining to the system, the system fails if all of its three components fail. Since we have assumed that A, B, and C operate independently of each other, we have

$$\begin{aligned}P(\tilde{A} \cap \tilde{B} \cap \tilde{C}) &= P(\tilde{A})P(\tilde{B})P(\tilde{C}) \\ &= (.1)(.1)(.1) \\ &= .001,\end{aligned}$$

which depicts the probability that all three components will fail at the same time. Then, the equation

$$1 - P(\tilde{A} \cap \tilde{B} \cap \tilde{C}) = 1 - .001 = .999$$

depicts the probability that the system will work at any given time.

EXERCISES

23. You are going to play the following game: You will toss a coin and a die and will win if "heads" and a 3 appear. If "heads" and a 3 do not appear, you will lose. What is the probability that you will win?

24. Three coins are tossed. What is the probability that all three coins will land on "heads"?

25. A box contains 2 brown marbles and 3 red marbles. You are going to draw two marbles in succession, the first of which you will replace in the box before you draw the second one. What is the probability that you will draw: (a) a red marble twice, (b) a red marble once and a brown marble once, not necessarily in that order, and (c) a brown marble twice?

*26. Three persons designated A, B, and C will shoot at a given target. The probability that A will hit the target is .9, that B will hit the target is .8, and that C will hit the target is .7. Suppose all three of them shoot at the target. What is the probability that two of them will hit the target?

*27. Suppose that the target discussed in Exercise 26 is destroyed by one hit. What is the probability that the target will be destroyed?

*28. You and your golf partner make a bet and agree that the winner will get a gourmet dinner at a good restaurant. You will determine the winner by playing either one 18-hole game or three 18-hole games. If the latter option is chosen, whoever wins two out of three games will win the bet. Suppose you believe that your chance of winning any single game is 60 percent and that the outcome of each game is independent of the others. Then, should you favor the one-game option or the three-game option?

29. For the problem given in Exercise 28, suppose you and your partner have decided

to play three 18-hole games. What is the probability that you will win the bet without having to play the third game?

30. A consumer product company plans to introduce three new products this year. In the past only 30 percent of new products introduced by the company were favorably received by the public. Assume that the success or failure of any one new product introduced by the company is not influenced by that of the other new products. Then, what is the probability that all three new products introduced will be favorably received by the public? What is the probability that at least one of the three will be favorably received by the public?

3.6 Bayes' Theorem

Suppose we divide a sample space into two mutually exclusive events A and $\tilde{A}$, as illustrated in Figure 3.6.1. Let B be any arbitrary event defined in the sample space, as also shown in Figure 3.6.1. Assume that we are given $P(A)$, $P(\tilde{A})$, $P(B|A)$, and $P(B|\tilde{A})$. There are situations in which we may wish to ascertain $P(A|B)$ or $P(\tilde{A}|B)$. A procedure for ascertaining these probabilities is given by the following theorem, known as *Bayes' theorem*.

Theorem 3.6.1

Let A and $\tilde{A}$ represent the decomposition of a sample space into two mutually exclusive events. Let B be any arbitrary event in the sample space. Then,

$$P(A|B) = \frac{P(A)P(B|A)}{P(A)P(B|A) + P(\tilde{A})P(B|\tilde{A})}$$

and

$$P(\tilde{A}|B) = \frac{P(\tilde{A})P(B|\tilde{A})}{P(A)P(B|A) + P(\tilde{A})P(B|\tilde{A})}.$$

The validity of the theorem may be explained as follows. First, according to Definition 3.4.1, we have

$$P(A|B) = \frac{P(A \cap B)}{P(B)}.$$

Event B occurs, however, when either one of the two joint events $A \cap B$ or $\tilde{A} \cap B$ occurs. Therefore, we have

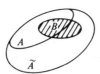

FIGURE 3.6.1 Sample space broken up into three events.

$$P(B) = P\{(A \cap B) \cup (\tilde{A} \cap B)\}$$
$$= P(A \cap B) + P(\tilde{A} \cap B).$$

Thus,
$$P(A|B) = \frac{P(A \cap B)}{P(A \cap B) + P(\tilde{A} \cap B)}.$$

On the other hand, according to Theorem 3.4.1, $P(A \cap B) = P(A)P(B|A)$, and $P(\tilde{A} \cap B) = P(\tilde{A})P(B|\tilde{A})$. In turn,

$$P(A|B) = \frac{P(A)P(B|A)}{P(A)P(B|A) + P(\tilde{A})P(B|\tilde{A})},$$

which is the first equation of Theorem 3.6.1. The second equation of the theorem may be justified in a similar manner.

We will now explore a number of applications of the theorem.

EXAMPLE 3.6.1

Half the marbles in a box are glass marbles and half are plastic marbles. It is also known that 20 percent of the glass marbles are red and 60 percent of the plastic marbles are red.

Suppose we select a marble at random from the red marbles. What is the probability that it is a glass marble? In notation this question may be expressed as

$$P(\text{glass}|\text{red}).$$

According to Bayes' theorem,

$$P(\text{glass}|\text{red}) = \frac{P(\text{glass})P(\text{red}|\text{glass})}{P(\text{glass})P(\text{red}|\text{glass}) + P(\text{plastic})P(\text{red}|\text{plastic})},$$

but we have already given that

$$P(\text{glass}) = .5 \qquad P(\text{plastic}) = .5$$
$$P(\text{red}|\text{glass}) = .2 \qquad P(\text{red}|\text{plastic}) = .6.$$

Thus,
$$P(\text{glass}|\text{red}) = \frac{(.5)(.2)}{(.5)(.2) + (.5)(.6)} = \frac{.1}{.4} = .25.$$

The validity of the preceding calculations may be illustrated in another way. Assume, for example, that the box contains 100 marbles. Then, there must be 50 glass marbles and 50 plastic marbles in the box. Since 20 percent of the glass marbles are red, there must be 10 red glass marbles and 40 nonred glass marbles. Similarly, there must be 30 red plastic marbles and 20 nonred

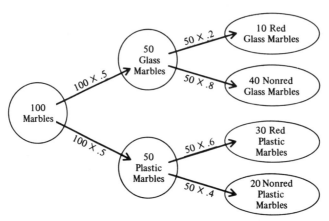

FIGURE 3.6.2

plastic marbles. The calculations leading to these propositions are shown in Figure 3.6.2. The numbers provided in Figure 3.6.2 reveal that there are a total of 40 red marbles in the box,and, furthermore, that 10 of these 40 are glass marbles. Therefore, if we select one marble from the red ones, the probability is .25 that it will be a glass marble. Thus, $P(\text{glass}|\text{red}) = .25$.

EXAMPLE 3.6.2

An oil company is considering whether or not to drill a given area for oil. Before making the decision, the company wants to assess the probability of actually hitting oil under the ground. In order to assess this probability, the company plans to obtain seismographic recordings for the area to be drilled. Based on past experience, the company believes that, if the ground under the area contains oil, the probability of obtaining positive seismographic recordings is .9. On the other hand, it also believes that there is a .3 probability of obtaining positive seismographic recordings even if the ground does not contain any oil.

Suppose the company believes that the probability of hitting oil is .2 even before it obtains any seismographic recordings. Assume now that the company subsequently obtains positive seismographic recordings for the area. What should the assessment be of the probability that the ground under the area contains oil?

We can ascertain the probability in question by applying Bayes' theorem. First, let us denote the following:

O = Ground contains oil
$\tilde{O}$ = Ground does not contain oil

S = Positive seismographic recordings
$\tilde{S}$ = Nonpositive seismographic recordings.

Then, we note that we have been given that

$$P(O) = .2 \quad P(\tilde{O}) = .8$$
$$P(S|O) = .9 \quad P(S|\tilde{O}) = .3.$$

Now, applying Bayes' theorem, we have

$$P(O|S) = \frac{P(O)P(S|O)}{P(O)P(S|O) + P(\tilde{O})P(S|\tilde{O})}$$
$$= \frac{(.2)(.9)}{(.2)(.9) + (.8)(.3)} = \frac{.18}{.42} = .428,$$

which depicts the probability that the ground contains oil, assuming that the seismographic recordings are positive.

EXERCISES

*31. In a box, 60 percent of the marbles are red and the remaining 40 percent are green. Among the red marbles, 70 percent are plastic and the remaining 30 percent are glass. Among the green marbles, 20 percent are plastic and the remaining 80 percent are glass. Suppose we put all the plastic marbles into another box and select one marble from it. What is the probability that the marble selected will be red?

32. Examining his loan files, a banker classifies 80 percent of personal loans made by him as good and the remaining 20 percent as bad. Among the good loans, 70 percent were made to finance necessary household items, such as automobiles and appliances, and the remaining 30 percent were made to finance luxuries, such as a vacation trip. Among the bad loans, 40 percent were made to finance necessary household items and 60 percent, luxuries.

    Suppose now, in processing a new loan application, the banker finds that the purpose of the loan is to finance a vacation trip. What should be his assessment of the probability that the loan, if granted, will be a good loan?

*33. It is known that 90 percent of the time a urine test will provide a positive sign of a certain disease if a person actually has that disease. However, 20 percent of the time the test will indicate a positive sign of the disease, even though the person does not really have the disease. It is known that 1 percent of the population in an area have this disease.

    Suppose we pick one person at random from the population and find that the urine test for that person gives a positive sign of the disease. What is the probability that the person actually has the disease?

Chapter 4
Random Variables

The term *random variable* is probably the most frequently used term in probability and statistics; yet, its meaning is often misunderstood. Such a misunderstanding stems from the fact that the term is a misnomer. A *random variable* is not the type of variable to which we are accustomed in algebra; rather, it is a *function*.

4.1 Function

Before we discuss the characteristics and properties of a *random variable*, we should try to understand what we mean by a *function*.

Definition 4.1.1

Let A and B be two nonempty sets. A *function* is defined for A if there is a rule which associates a unique element of B with each element of A. A is then said to be the *domain* of the function, and B is said to be its *range* or *image set*.

EXAMPLE 4.1.1

Consider the equation $y = x^2$. Suppose we let x assume the values 0, .5, 1, 1.5, and 2. Then, $A = \{0,.5,1,1.5,2\}$ is the domain of the function, and $B = \{0,.25,1,2.25,4\}$ is the image set. The equation $y = x^2$ is the rule which

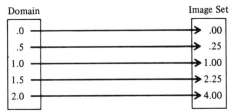

FIGURE 4.1.1 Function.

assigns a unique number in B for each element in A. The sets A and B and the equation $y = x^2$ together constitute a function. Another way of expressing the given function is shown in Figure 4.1.1.

EXAMPLE 4.1.2

For $y = x^2$, let x assume the values $-2, -1.5, -1, -.5, 0, .5, 1, 1.5,$ and 2. Then, the domain of the function is given as

$$A = \{-2, -1.5, -1, -.5, 0, .5, 1, 1.5, 2\}.$$

The image set of the function would, however, remain the same. The function is then illustrated in Figure 4.1.2.

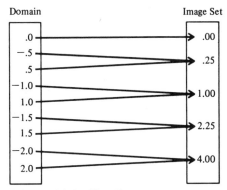

FIGURE 4.1.2 Function.

In Figures 4.1.1 and 4.1.2, both the domain and the image set contain numbers. On the other hand, neither the elements of the domain nor those of the image set have to be numbers, as illustrated by the next example.

EXAMPLE 4.1.3

The coach of a football team tells his players, "If we win this game Saturday, we will celebrate on Sunday; otherwise, we will scrimmage on Sunday."

The statement in the quotation is a function, if we consider the set {celebrate, scrimmage} as the image set and the set {win, tie, lose} as the domain of the function. This function is illustrated further in Figure 4.1.3.

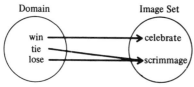

FIGURE 4.1.3

EXERCISES

*1. Suppose a father tells his son, "I will take you fishing tomorrow if it is sunny. However, if it is either cloudy or raining, we will stay home." Show by a diagram that these statements are a function.

2. Let the domain of the equation $y = x^2 + 10$ be the set of integers $\{0,1,2,3\}$. Describe in a diagram the image set of the function, and indicate by directed arrows how the elements of the domain are associated with the elements of the image set.

*3. Let the domain of the equation $y = x^2 + 10$ be the set of integers $\{-3,-2,-1, 0,1,2,3\}$. Describe in a diagram the image set of the function, and indicate by directed arrows how the elements of the domain are associated with the elements of the image set.

4. Figure 4.1.4 shows the relationship between two real numbers x and y.
   a. Can the set of real numbers $\{-4 \leq x \leq 4\}$ serve as the domain of the function whose image set contains the set of real numbers $\{0 \leq y \leq 4\}$?
   b. Can the set of real numbers $\{0 \leq y \leq 4\}$ serve as the domain of the function whose image set contains the set of real numbers $\{-4 \leq x \leq 4\}$?

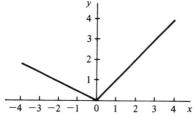

FIGURE 4.1.4 Relationship between two real numbers (x and y).

5. Indicate whether or not each of the following diagrams in Figure 4.1.5 defines a function from the set A to the set B.

6. Figure 4.1.6 shows that $x \in A$ is related to $y \in B$ by $y = \sqrt{x}$. Does the diagram describe a function?

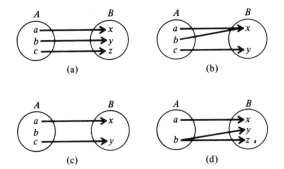

(a) (b) (c) (d)

FIGURE 4.1.5

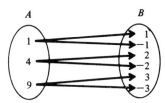

FIGURE 4.1.6

4.2 Random Variable

We will now propose a definition of a *random variable*.

Definition 4.2.1

A *random variable* is a function whose domain is a sample space and whose image set is a set of real numbers.

EXAMPLE 4.2.1

Consider the following statement: "Toss a coin, and win \$1 if 'heads' appears and lose \$1 if 'tails' appears." If we evaluate this statement, we find that it fits the definition of a function. The domain of the function is the set of outcomes generated by tossing a coin {heads, tails}, and the image set is the amount obtained from winning or losing $\{1, -1\}$. The statement also specifies the rule according to which the elements of the domain are mapped to the elements of the image set.

Since the domain of the function is the sample space for tossing one coin, and the image set of the function is the set of real numbers, the statement in the quotation is a random variable according to Definition 4.2.1. The random variable in question is depicted in Figure 4.2.1. The reader should note that the entire figure represents the random variable, not a portion of the figure.

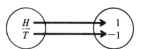

FIGURE 4.2.1 Random variable.

EXAMPLE 4.2.2

An oil company is considering whether or not to drill a certain offshore location for oil. The cost of drilling is estimated to be $1 million. If oil is not discovered, none of the drilling costs can be recovered; however, if oil is discovered, the company expects to make a profit of $10 million after recovering the drilling costs. We might describe the situation faced by the oil company by the function shown in Figure 4.2.2.

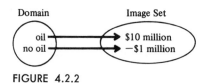

FIGURE 4.2.2

Suppose now the company believes that whether or not it discovers oil at that particular location, it is, in a sense, due to chance. Then, the functions described previously may be considered to be a random variable.

EXAMPLE 4.2.3

Suppose we spin the wheel given in Figure 4.2.3 and win a gold bar whose weight is twice the amount indicated by the arrow when the wheel stops. Even though only the numbers 0, 2.5, 5, and 7.5 are shown on Figure 4.2.3, it is assumed that the arrow can point to any real number between 0 and 10. For example, if it points exactly midway between 0 and 2.5, then we assume that the arrow points to 1.25.

The situation may be depicted by the function in Figure 4.2.4. We may assume that the domain of the function is a sample space, since the number which is indicated by the arrow in any given spin is due to chance. Thus, the function may be considered to be a random variable.

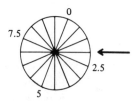

FIGURE 4.2.3

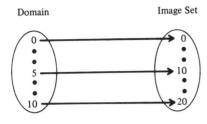

FIGURE 4.2.4

EXAMPLE 4.2.4

A data-processing center of a company has just repaired its computer. The director of the center wonders how long the computer will operate before it breaks down again. The length of time that the computer will operate without any trouble can be expressed as a function of the exact time at which it will break down again, which is illustrated in Figure 4.2.5.

Suppose we assume that the exact time at which the computer will break down again is due to chance. Then, the domain of the function just discussed may be considered to be a sample space. Thus, the function itself is a random variable.

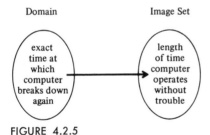

FIGURE 4.2.5

EXERCISES

*7. An instructor tells his class, "I will toss a coin. If 'heads' appears, I will give you a mid-term examination. If 'tails' appears, I will not give you a mid-term examination." Show by a diagram that the statements are a function. Then, explain whether the function is also a random variable.

8. A father tells his son, "Toss a die. If 1 appears, I will take you fishing; otherwise, you will have to mow the lawn." Show that the statement is a function. Then, indicate whether the function is a random variable.

9. John tells James, "I will toss a coin and, if 'heads' appears, I will pay you $1. If 'heads' does not appear, you will pay me $1." Show that the statements are a function. Then, determine whether or not the function is a random variable.

48 Random Variables

*10. A box contains one white and nine black balls. You are to draw one ball from the box. If the ball drawn is white, you will become a member of a fraternal society. If the ball drawn is black, you cannot become a member. Show that the procedure used to admit you into the society is a function. Then, indicate whether or not the function is a random variable.

*11. Refer to Exercise 10, and assume that the initiation fee is $1000. Of course, you will not pay any money to the society if you have been blackballed. In a diagram show the relationship between the ball selected from the box and the amount of money that you have to pay to the society. Then, determine whether or not the diagram describes a random variable.

4.3 Discrete versus Continuous Random Variable

We have provided in the preceding section four different examples of random variables. The first two examples depict a so-called *discrete random variable*, and the last two examples depict a so-called *continuous random variable*.

A formal distinction between a *discreet* and a *continuous random variable* will require fairly abstract mathematical concepts. We will, therefore, provide a somewhat crude distinction which can readily be understood. To do so, we propose first:

Definition 4.3.1

Let T be a set of numbers where s and l are the smallest and largest numbers in T and $s < l$. T is said to be a *continuous space* if it contains all real numbers between s and l.

EXAMPLE 4.3.1

Consider three sets of numbers, T_1, T_2, and T_3, depicted in Figure 4.3.1. Then, T_1 is a continuous space, since it contains all real numbers between s and l. T_2 is not continuous between s and l, since it does not contain the real numbers between a and b. However, the line segments $\overline{sa}$ and $\overline{bl}$ are continuous. It must be fairly obvious, then, that T_3 is not continuous, since it does not contain any continuous segment at all.

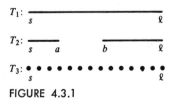

FIGURE 4.3.1

4.3 Discrete versus Continuous Random Variable   49

Definition 4.3.2

Let R be the set of values that the random variable X can assume. The random variable is said to be *discrete* if R does not contain any continuous space.

Definition 4.3.3

Let R be the set of values that the random variable X can assume. The random variable is said to be *continuous* if R consists of a set of continuous spaces.

EXAMPLE 4.3.2

Let us return to Example 4.2.1. R is depicted by Figure 4.3.2. We observe that R does not contain any continuous line segment. Therefore, the random variable is discrete.

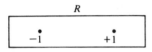

FIGURE 4.3.2

EXAMPLE 4.3.3

Let us return to Example 4.2.3. R for that example is depicted by Figure 4.3.3. We observe that R is a continuous line segment. Therefore, the random variable is continuous.

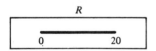

FIGURE 4.3.3 A continuous line segment.

EXAMPLE 4.3.4

Referring again to Example 4.2.3, let us suppose we assume that the weight of the gold bar can be measured only in ounces. Then, the random variable can be depicted as shown in Figure 4.3.4. In turn, we may illustrate R as

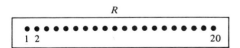

We observe now that R does not contain any continuous segment; therefore, the random variable is discrete.

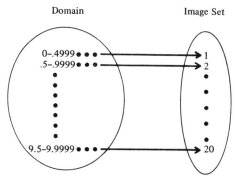

FIGURE 4.3.4

EXERCISES

*12. Suppose you toss 100 coins and win $1 for each "heads" you obtain. Let X be the random variable associated with the dollar value of total wins. Is X discrete or continuous?

13. Suppose you toss a coin and, if "heads" is obtained, the game ends; if "tails" is obtained, you can toss the coin again. The game ends whenever "heads" appears for the first time. Let X be the random variable associated with the number of times that the coin is tossed before the game ends. Is X discrete or continuous?

*14. Assume that you are an automobile salesman. Let X be the random variable associated with the number of automobiles sold in any given day. Is X discrete or continuous?

*15. Assume that you are to throw a javelin. Let X be the random variable associated with the distance of your throw in any given trial. Is X discrete or continuous?

16. Suppose you leave your house at 7 A.M. every morning and walk 1 mile to a subway station. The amount of time you spend waiting for the train varies from day to day. Let X be the random variable associated with the amount of time you wait on any randomly selected day. Is X discrete or continuous?

*17. You have just replaced a burned-out bulb. You wonder how long the new bulb will last. Let X be the random variable associated with the life span of the new bulb. Is X discrete or continuous?

4.4 Probability Function of a Discrete Random Variable

Consider now the image set R of a discrete random variable. Suppose it contains k discrete numerical values. Then, we can partition the sample space into k mutually exclusive events so that the occurrence of one of these events will correspond to the attainment of one of the k numerical values in the image set.

4.4 Probability Function of a Discrete Random Variable

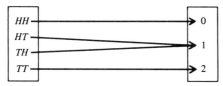

FIGURE 4.4.1 A random variable.

EXAMPLE 4.4.1

Let a random variable be described by the diagram given in Figure 4.4.1. The sample space can be partitioned into the following mutually exclusive events:

$$E_1 = \{HH\},$$
$$E_2 = \{HT, TH\},$$
and
$$E_3 = \{TT\}.$$

The occurrence of one of the mutually exclusive events E_1, E_2, or E_3 will correspond to the attainment of one of the numbers 0, 1, or 2.

Assume that x_i is the ith numerical value in the image set. Let

$$P(X = x_i) = p_i$$

denote the fact that the probability that the random variable X will assume the numerical value x_i is p_i. (Note that we denote a random variable by an italic, capital letter and a value that the random variable will assume by an italic, lower case letter. Unless otherwise indicated in our subsequent discussion, we will adhere to these notational conventions.) Then, the value of p_i may be found in the following manner. Suppose the sample space has been partitioned in such a manner that the occurrence of E_i will correspond to the attainment of x_i. Then, the probability of attaining x_i must be the same as the probability that E_i will occur; that is,

$$p_i = P(E_i).$$

Thus, the value of p_i may be found indirectly by calculating the value of $P(E_i)$.

For the simple events in the sample space just given, let us assign the following probabilities.

Sample Space	Probability
HH	¼
HT	¼
TH	¼
TT	¼

Then,
$$P(E_1) = \frac{1}{4},$$
$$P(E_2) = \frac{2}{4},$$
and
$$P(E_3) = \frac{1}{4}.$$

Since the occurrence of E_1 corresponds to the attainment of 0, the occurrence of E_2 corresponds to the attainment of 1, and the occurrence of E_3 corresponds to the attainment of 2, we have
$$P(X = 0) = \frac{1}{4},$$
$$P(X = 1) = \frac{2}{4},$$
and
$$P(X = 2) = \frac{1}{4}.$$

We will discuss next the notion of a *probability function* of a discrete random variable.

Definition 4.4.1

A *probability function* of a discrete random variable is a function whose domain contains the set of numerical values that the random variable can assume and whose image set contains the probabilities for the elements in the domain.

The preceding definition may be restated in another way: Let x_1, ..., x_n, the elements of R, be the values that the random variables can assume with the associated probabilities $P(X = x_1)$, ..., $P(X = x_n)$, the elements of P. The elements of R may be assigned to the elements of P by the ordered pairs
$$[x_1, P(X = x_1)], \ldots, [x_n, P(X = x_n)].$$
R becomes the domain of a function, and P becomes its image set. The function in question is the probability function of the discrete random variable.

For the sake of simplicity, we will denote hereafter that
$$P(X = x_i) = f(x_i)$$
and, in turn, that the probability function is
$$[x_1, f(x_1)], \ldots, [x_n, f(x_n)].$$

EXAMPLE 4.4.2

Consider the random variable described in Example 4.4.1. Figure 4.4.2 shows the probability function of the random variable.

4.4 Probability Function of a Discrete Random Variable

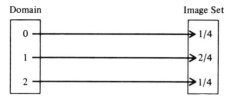

FIGURE 4.4.2 Probability function of a random variable.

We might find it convenient to describe the given probability function in tabular format, as in Table 4.4.1, graphically as in Figure 4.4.3, or by a set of ordered pairs:

$$S = \left\{ \left(0, \frac{1}{4}\right), \left(1, \frac{2}{4}\right), \left(2, \frac{1}{4}\right) \right\},$$

where the first number in an ordered pair represents a value that the random variable can assume, and the second number represents the associated probability.

TABLE 4.4.1 Probability Function of a Random Variable

Values That X Can Assume x_i	Associated Probabilities $f(x_i)$
0	1/4
1	2/4
2	1/4

The preceding example illustrates that the probability function of a discrete random variable is merely a list of all possible values that the random variable can assume with associated probabilities.

We stated in Definition 4.2.1 that a random variable is a function. Since the probability function of a discrete random variable is also a function, we should illustrate the distinction between the two functions.

Suppose we let R be the set containing all possible values that a discrete random variable can assume. Then R is the image set of the function describing

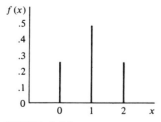

FIGURE 4.4.3

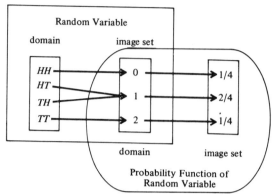

FIGURE 4.4.4 Distinction between a random variable and its probability function.

the random variable. On the other hand, R becomes the domain of the probability function of the random variable. This distinction is illustrated in Figure 4.4.4 for the random variable associated with Example 4.4.2. The left-hand square in Figure 4.4.4 describes the random variable, and the right-hand eclipse describes the probability function of the given random variable.

Definition 4.4.2

Let the domain of a function contain the range of values that the random variable X can assume. Let the image set of the function be such that if x in the domain is mapped to $F(x)$ in the image set, then

$$F(x) = P(X \leq x).$$

The function is said to be the *cumulative probability function* of X.

EXAMPLE 4.4.3

Let us return to Example 4.4.2. From the probability function described in the example we can derive the following:

$$P(X \leq 0) = \frac{1}{4},$$

$$P(X \leq 1) = \frac{3}{4},$$

and

$$P(X \leq 2) = \frac{4}{4}.$$

The cumulative probability function of X is then depicted by

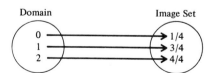

EXAMPLE 4.4.4

Let us suppose we are an automobile salesman. An examination of our sales record for the past one hundred days reveals the following:

Number of Days	Number of Automobiles Sold
60	0
30	1
8	2
2	3
100	

Suppose now we wish to derive a probability function for the random variable associated with a daily sales. If we assume that our past sales record reflects the probability function of our future daily sales, then

x	$f(x)$
0	.60
1	.30
2	.08
3	.02
	1.00

where X denotes the random variable associated with a daily sales. The cumulative probability function, then, is given as

x	$F(x)$
0	.60
1	.90
2	.98
3	1.00

Using one of the preceding probability functions, we derive, for example, that the probability of selling at most two automobiles in any single day is .98 and of selling at least two automobiles in any single day is $1 - .90 = .10$.

EXERCISES

*18. Suppose you are to toss three coins and win $1 for each "heads" obtained; that is, you will win $2 if two "heads" appear. Describe the probability function for the associated random variable.

*19. For the game described in Exercise 18, what is the probability that you will win: (a) at least $1, (b) at least $2, (c) exactly $2, and (d) at most $2?

20. Suppose you are to toss three coins. The equation $W = 10 + 2H^2$ symbolizes the fact that you have won, where W denotes the amount you have won and

56 Random Variables

H denotes the number of "heads" you have obtained. Describe the probability function for the associated random variable.

21. For the game described in Exercise 20, what is the probability that you will win: (a) at least $10, (b) at least $12, (c) exactly $20, and (d) at least $12, but not more than $20?

*22. In the system shown in Figure 4.4.5, the components A, B, and C work independently of each other. The probability that A will work is .9; that B will work is .8; and that C will work is .7. Let X be the number of components working at any given time. Describe the probability function of X.

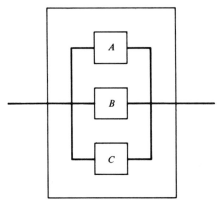

FIGURE 4.4.5

*23. Assume that the following figures depict the daily changes in the price of a certain stock for 20 randomly selected days in the past:

$$\frac{1}{2}, 0, -\frac{1}{4}, -\frac{1}{2}, \frac{1}{4}, \frac{1}{4}, 0, \frac{1}{2},$$

$$-\frac{1}{2}, 0, -\frac{1}{4}, \frac{1}{4}, 0, \frac{1}{4}, \frac{1}{2}, 0$$

$$-\frac{1}{4}, \frac{1}{4}, \frac{1}{2}, -\frac{1}{2}.$$

Let X be the random variable associated with the change in the stock price for a randomly selected day in the future. Construct the probability function of X, utilizing the data provided.

4.5 Probability Function of a Continuous Random Variable

Let us return to the spinning wheel in Example 4.2.3. The domain and image set of the associated random variable is shown again with a slight

4.5 Probability Function of a Continuous Random Variable

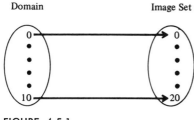

FIGURE 4.5.1

variation in Figure 4.5.1. Assume that X can be any value between 0 and 20. This implies that X is a continuous random variable. Let the line in the following diagram depict the range of values that X can assume.

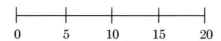

Then, any point on the line depicts a value that X can assume. On the other hand, a point on a line segment should occupy an infinitesimal space. Therefore, the probability that X will assume a numerical value corresponding to any single point on the line segment must be also infinitesimal. Consequently, it is not meaningful to obtain, for example,

$$P(X = 10).$$

since this probability is 0, for all practical purposes. Thus, the only type of meaningful question that should be asked pertains to the probability that the random variable X will assume a value between, say, k_1 and k_2. For example, it is not difficult to see that

$$P(0 \leq X \leq 5) = \frac{1}{4},$$

$$P(0 \leq X \leq 10) = \frac{2}{4},$$

$$P(0 \leq X \leq 15) = \frac{3}{4},$$

and
$$P(0 \leq X \leq 20) = \frac{4}{4}.$$

Clearly, then, Definition 4.4.2, which we have provided for the cumulative probability function of a discrete random variable, is equally applicable for that of a continuous random variable. Figure 4.5.2 shows the cumulative probability function of the continuous random variable X.

Let us now examine the relationship between the elements of the domain and those of the image set in Figure 4.5.2. We observe that if x is an element of the domain, then,

58 Random Variables

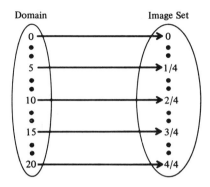

FIGURE 4.5.2 Cumulative probability function of a continuous random variable.

$$F(x) = P(X \leq x) = \frac{x}{20}.$$

For example,

$$F(10) = P(X \leq 10) = \frac{10}{20} = \frac{1}{2}.$$

The relationship between the elements of the domain and those of the image

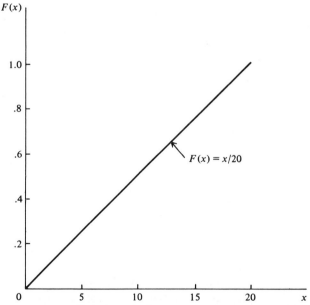

FIGURE 4.5.3 Relationship between elements of the domain and the image set.

4.5 Probability Function of a Continuous Random Variable

set may be illustrated further by the graph in Figure 4.5.3. We now can propose the following definition:

Definition 4.5.1
Let the domain of the function contain the values that a continuous random variable X can assume. Let the image set of the function be such that, if x in the domain is mapped to $f(x)$ in the image set, then $f(x)$ is the *slope* of the cumulative probability function of X at x. Then, the function is said to be the *probability function* of X.

The *slope* of the cumulative probability function of X at x sometimes is referred to as the *density* of the probability function at $X = x$, and, in turn, the probability function is called the *probability density function* of X.

EXAMPLE 4.5.1
For the random variable X discussed in Figure 4.5.1 we now observe that the slope of the cumulative probability function is $1/20$, regardless of the value of X. Thus, the probability density function of X is that given in Figure 4.5.4.

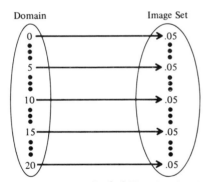

FIGURE 4.5.4 Probability density function of X.

If we were to depict this probability density function of X graphically, we would have

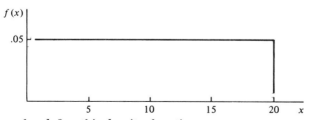

We can also define this density function as

$$f(x) = .05 \quad \text{for } 0 \leq x \leq 20$$
$$f(x) = 0 \quad \text{all other } x.$$

The probability density function is useful because it enables us to calculate the probability that the random variable will assume a value between any two different numbers, for example, k_1 and k_2.

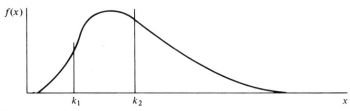

FIGURE 4.5.5

Suppose the curve in Figure 4.5.5 depicts the probability density function of a certain continuous random variable. Then, we can show that the area under the curve is unity. Therefore,

Theorem 4.5.1

Let $f(x)$ depict the density of a continuous random variable X at x. Then,

$$P(k_1 \leq X \leq k_2) = \int_{k_1}^{k_2} f(x)\,dx.$$

To the reader who is familiar with calculus, the preceding theorem proposes that $P(k_1 \leq X \leq k_2)$ is found by integrating the density function between k_1 and k_2. To the reader who is not familiar with calculus, the theorem proposes that $P(k_1 \leq X \leq k_2)$ is equal to the area under the curve depicting the density between k_1 and k_2.

EXAMPLE 4.5.2

Let us return to Example 4.5.1. Suppose we wish to calculate the probability that X assumes a value between 5 and 15. We can answer this question without any calculation if we re-examine the wheel in Example 4.2.3. Since X assumes a value between 5 and 15 if the arrow stops between 2.5 and 7.5, the probability in question must be .5.

We will now show how to calculate this probability by applying Theorem 4.5.1. Since the density function is

$$f(x) = .05 \quad 0 \leq x \leq 20$$
$$f(x) = 0 \quad \text{all other } x,$$

we calculate

$$P(5 \leq X \leq 15) = \int_5^{15} .05\,dx = .5.$$

4.5 Probability Function of a Continuous Random Variable

We will now show that .5 is the area under the density function between 5 and 15. The height of the density function is .05 between 5 and 15. Therefore, the area under the density function between 5 and 15 is

$$.05(15 - 5) = .5,$$

which is the same as that of $P(5 \leq X \leq 15)$.

Although we can always calculate $P(k_1 \leq X \leq k_2)$ of a continuous random variable X by integrating its density function between k_1 and k_2, such a procedure can be quite cumbersome for many random variables. There is, however, another way to calculate the probability, which usually is easier.

Theorem 4.5.2

Let X be a continuous random variable. Assume $k_1 < k_2$. Then,

$$P(k_1 \leq X \leq k_2) = F(k_2) - F(k_1),$$

where $F(k_2) = P(X \leq k_2)$ and $F(k_1) = P(X \leq k_1)$.

EXAMPLE 4.5.3

Let us return to Example 4.5.2. We observe from the cumulative probability function for the random variable that

$$F(5) = \frac{x}{20} = \frac{5}{20} = .25$$

and

$$F(15) = \frac{x}{20} = \frac{15}{20} = .75.$$

Thus,

$$P(5 \leq X \leq 15) = F(15) - F(5)$$
$$= .75 - .25 = .5,$$

which yields the same result as that obtained in Example 4.5.2.

EXERCISES

*24. Suppose you are to spin the wheel shown in Figure 4.5.6, and you are to measure the angle θ counterclockwise between $\overline{OA}$ and $\overline{OB}$ when the wheel stops. Let

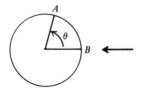

FIGURE 4.5.6 A spinning wheel.

X be the random variable associated with the angle θ. Describe the cumulative probability function of X. Then, ascertain the probability density function of X.

*25. For the random variable given in Exercise 24, using its probability density function, find: (a) $P(90 \leq X \leq 180)$, and (b) $P(100 \leq X \leq 200)$.

*26. For the random variable given in Exercise 24, using its cumulative probability function, find: (a) $P(100 \leq X \leq 250)$, and (b) $P(50 \leq X \leq 150)$.

*27. A continuous random variable has a cumulative probability function

$$F(x) = .1x \qquad \text{for } 0 \leq x \leq 10.$$

Ascertain its density function.

*28. For the random variable given in Exercise 27, find: (a) $P(4 \leq X \leq 8)$, and (b) $P(5 \leq X \leq 6)$.

29. A random variable X has the probability density function

$$\begin{aligned} f(x) &= .01 & 0 \leq x \leq 100 \\ f(x) &= 0 & \text{all other } x. \end{aligned}$$

Find the cumulative probability function of X.

30. For the random variable given in Exercise 29, find: (a) $P(15 \leq X \leq 45)$, and (b) $P(25 \leq X \leq 75)$.

4.6 Expected Value of a Random Variable

Given a probability function of a random variable, we might wish to ascertain a measure of its center. One such measure is called the *expected value of a random variable*. The *expected value of a random variable* is, in a sense, a weighed average of its probability function; that is,

Definition 4.6.1

Let X be a discrete random variable. Then, the expected value of X, denoted $E(X)$, is

$$E(X) = x_1 f(x_1) + \cdots + x_n f(x_n).$$

We will also call $E(X)$ the *mean* of X and denote it by a Greek letter μ.

EXAMPLE 4.6.1

Let the probability function of X be given as

x_i	$f(x_i)$
0	.5
1	.5
	1.0

Then, according to Definition 4.6.1,

$$E(X) = (0)(.5) + (1)(.5) = .5.$$

EXAMPLE 4.6.2

Consider the random variable given in Example 4.2.2. Assume now that the following depicts the probability function of the random variable.

x_i	$f(x_i)$
−$1 million	.8
$10 million	.2
	1.0

Then, the expected value of the random variable is

$$E(X) = (-1)(.8) + (10)(.2) = 1.2,$$

or $1.2 million.

Definition 4.6.2

Let X be a continuous random variable. Then,

$$E(X) = \int_{-\infty}^{\infty} xf(x)\, dx.$$

EXAMPLE 4.6.3

Let us return to the random variable in Example 4.5.1, whose density function was given as

$$f(x) = .05 \quad 0 \leq x \leq 20$$
$$f(x) = 0 \quad \text{all other } x.$$

Then, according to Definition 4.6.2,

$$E(X) = \int_0^{20} .05x\, dx = 10.$$

EXERCISES

*31. Suppose you are to play the following game: toss a coin and win $1 if "heads" appears and lose $2 if "tails" appears. What is the expected value of this game?

32. Suppose you are to toss one die. Your winning in dollars will correspond to the number which appears on the die. What is the expected amount of your winning?

33. A roulette wheel has 38 equally spaced parts numbered 1 through 38. You may bet $10 on any number. If the ball drops into the roulette wheel while it is spinning and stops at your number, you will win $350 and can also keep your $10; otherwise, you will lose your $10. What is the expected value of this game?

34. Suppose you are asked to play the following game: toss a coin and win a quarter if "heads" appears and win nothing if "tails" appears. However, you must pay

64 Random Variables

a dime each game. You will be allowed to play this game as long as you like. Are you willing to play this game? Why?

35. You are asked to play the following game: toss a coin and win $250,000 if "heads" appears and win nothing if "tails" appears. However, you must pay $100,000 to play this game. You will not be allowed to play this game more than once. Are you willing to play this game? Explain why.

36. You are asked to play the following game. Toss a coin and win 2 cents if "heads" appears and nothing if "tails" appears. If you win on the first trial, you can play a second time. If "heads" appears on the second trial, you will win 4 cents and can toss the coin again. If "heads" appears on the third trial, you will win 8 cents; if "heads" appears the fourth time, you will win 16 cents; 32 cents for the fifth "heads," and so on. The game will stop whenever "tails" appears. How much should you be willing to pay for the game? Explain.

*37. An oil company expects to make a profit of $10 million from a drilling operation after recovering its drilling costs if it hits oil. If the oil company does not hit oil, none of the drilling costs of $1 million will be recovered. The company believes that there is a .2 probability of hitting oil if it drills. What is the expected gain (or loss) from the drilling operation?

38. If you are given that a continuous random variable X has the following density function,

$$f(x) = .1 \quad 0 \leq x \leq 10$$
$$f(x) = 0 \quad \text{all other } x,$$

calculate $E(X)$.

39. Assume that you have spun the wheel shown in Figure 4.6.1 and have won the amount indicated by the arrow when the wheel stopped. What is the expected amount of your win?

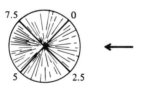

FIGURE 4.6.1 A spinning wheel.

4.7 Variance and Standard Deviation

Given a probability function of a random variable, we might also wish to ascertain a measure of its dispersion. One such measure, which is very important, is called the *variance of the random variable*.

Definition 4.7.1

Let X be a discrete random variable. Then, the *variance* of X, denoted $V(X)$, is

4.7 Variance and Standard Deviation

$$V(X) = [x_1 - E(X)]^2 f(x_1) + \cdots + [x_n - E(X)]^2 f(x_n),$$

and the *standard deviation* of X, denoted σ_x, is

$$\sigma_x = \sqrt{V(X)}.$$

EXAMPLE 4.7.1

Consider the random variable given in Example 4.6.1. The probability function of the random variable is given again in the first two columns in Table 4.7.1. The last two columns in the table illustrate how the relevant terms for the variance may be calculated. The variance is then obtained by adding the values obtained in the last column; that is,

$$V(X) = (0 - .5)^2(.5) + (1 - .5)^2(.5) = .25.$$

The standard deviation of X, therefore, is

$$\sigma_x = \sqrt{.25} = .5.$$

TABLE 4.7.1

x_i	$f(x_i)$	$[x_i - E(X)]^2$	$[x_i - E(X)]^2 f(x_i)$
0	.5	$(0 - .5)^2$	$(0 - .5)^2(.5) = .125$
1	.5	$(1 - .5)^2$	$(1 - .5)^2(.5) = .125$
			.250

EXAMPLE 4.7.2

Consider the random variable given in Example 4.6.2. The probability function is given in the first two columns in Table 4.7.2. Thus, $V(X) = 19.360$ and, in turn,

$$\sigma_x = \sqrt{19.360} = 4.4.$$

TABLE 4.7.2

x_i	$f(x_i)$	$[x_i - E(X)]^2$	$[x_i - E(X)]^2 f(x_i)$
-1	.8	$(-1 - 1.2)^2 = 4.84$	$(4.84)(.8) = 3.872$
10	.2	$(10 - 1.2)^2 = 77.44$	$(77.44)(.2) = 15.488$
	1.0		19.360

Definition 4.7.2

Let X be a continuous random variable. Then,

$$V(X) = \int_{-\infty}^{\infty} [x - E(X)]^2 f(x)\, dx$$

and

$$\sigma_x = \sqrt{V(X)}.$$

EXAMPLE 4.7.3

Let us return to Example 4.6.3. We have already calculated that $E(X) = 10$. Thus,

$$V(X) = \int_0^{20} (x-10)^2(.05)\, dx = \int_0^{20} (x^2 - x + 5)\, dx = \frac{100}{3}$$

and, in turn,

$$\sigma_x = \sqrt{\frac{100}{3}}.$$

EXERCISES

*40. Suppose you toss two coins and win $1 for each "heads" obtained. What is the expected value, the variance, and the standard deviation of the associated random variable?

*41. Suppose you toss a coin and win $2 if "heads" appears and nothing if "tails" appears. What is the expected value, the variance, and the standard deviation of the associated random variable?

*42. Compare the games proposed in Exercises 40 and 41. If you were to select one game between the two, which one would you choose? Explain why.

TABLE 4.7.3 Probability Function of X

x_i	$f(x_i)$
0	.05
1	.10
2	.20
3	.30
4	.20
5	.10
6	.05
	1.00

43. Let X be the random variable associated with the number of accidents in a day for a factory. The probability function of X is given in Table 4.7.3. Calculate the expected value, the variance, and the standard deviation of X.

44. If you are given the following probability density function,

$$f(x) = .1 \quad 0 \leq x \leq 10$$
$$f(x) = 0 \quad \text{all other } x,$$

calculate the variance and the standard deviation of X.

4.8 Application of Expectation and Variance

In this section we will illustrate the usefulness of expectation and variance of a random variable in a simple statistical analysis.

Consider the following game: Toss a coin and win $3 if "heads" appears and lose $1 if "tails" appears. The probability function of the random variable associated with this game is given in Table 4.8.1. The calculations shown in the extreme right column indicate that the expected value of the random variable is $1.

TABLE 4.8.1

x_i	$f(x_i)$	$x_i f(x_i)$
3	.5	1.5
−1	.5	−.5
	1.0	1.0

What is the meaning of this $1? If we state that $1 is what we expect to win by playing the game, we have not really answered the question, for we would still have to explain what we mean by the phrase "expect to win." Although the answer to the question appears to be simple, it is difficult to answer. One approach to answering the question may be as follows. Assume that we are to play the game repetitively. Let W depict the total amount of winnings and T denote the total number of trials. Then, the $1 is the limiting value of the ratio W/T as T becomes larger and larger. This means that if we were to play the game many times, we would win approximately $1 on the average.

If the game in question can be played many times, the limiting concept of interpreting the value of expectation for a random variable is not only justifiable on a logical ground but also verifiable empirically. On the other hand, if the game can be played only once, then such an interpretation is no longer defensible.

Yet, the expected value of the random variable may still provide critical information, which we can evaluate before deciding whether or not to play a game such as the one just discussed. Therefore, for the given game, we might consider the expected value of the associated random variable as a measure by which we can evaluate the odds for the game. For example, the fact that the expected value is positive may be construed to mean that the odds are in our favor.

We do not imply, however, that the expectation should necessarily be the guide line in deciding whether or not to play such a game. Nor do we imply that a rational person will always play such a game whenever the expected value of the associated random variable indicates that the odds are in his favor.

To point out some underlying reasons for the preceding proposition, we will propose the following two games. Let game A be that which has already been proposed, but let game B be played as follows: toss a coin and win $300,000 if "heads" appears and lose $100,000 if "tails" appears. The expected value of the random variable associated with A is $1 and associated with B is $100,000. Thus, the odds are in our favor for both of these games. Although we may be eager to play game A, we may be unwilling to play game B. Does this mean that our decision-making processes are inconsistent? The answer is "no," because the set of factors to be considered in making such decisions may contain not only the odds but also the amount of money at stake which we feel we can afford.

The usefulness of the variance of a random variable in statistical analysis will not be apparent until we become acquainted with such concepts as the *Tchebycheff inequality* and the *central limit theorem*. However, we might present the following examples to illustrate one interesting property of the variance of the random variable. Consider the following three games:

1. Toss a coin and win $1 if "heads" appears and lose $1 if "tails" appears.
2. Toss a coin and win $10 if "heads" appears and lose $10 if "tails" appears.
3. Toss a coin and win $100 if "heads" appears and lose $100 if "tails" appears.

Let X, Y, and W be the random variables used for these three games, in the order given. We find that the expected values of the random variables are all 0, assuming that the probability of obtaining "heads" or "tails" is .5. Does this mean that we will be indifferent to the three games when we are given an option to choose one? The answer is "no." It must be intuitively apparent that a conservative person will favor the first game, whereas an adventurous person might favor the third one. How can we explain this intuition that these are not the same kind of game? If we were to make the calculations, we would find that $V(X) = 1$, $V(Y) = 100$, and $V(W) = 10,000$. These illustrations indicate that the smallest variance is associated with the game that the conservative peron would choose, whereas the largest variance is associated with the game that the adventurous person would choose. Can we then propose that, given the three games discussed, the smaller the vari-

4.9 Tchebycheff Inequality

If we know the probability function of a random variable X, we can make a probabilistic statement that X will assume a number which lies between any two arbitrary numerical values a and b. On the other hand, if we know only the expected value and the variance of the random variable, then, such a statement generally cannot be made. However, a somewhat crude probabilistic statement can be made concerning the random variable X according to the following theorem.

Theorem 4.9.1
Let X be a random variable having a finite mean and variance. Then, for any non-negative number k, we have

$$P\{|X - E(X)| \geq k\sigma_x\} \leq \frac{1}{k^2}.$$

Theorem 4.9.1 is known as the *Tchebycheff inequality*. Restating the theorem in words, we have: The probability that the random variable X will assume a value which is k standard deviations (or farther) away from its expected value is, at the most, $1/k^2$.

A proposition which is equivalent to Theorem 4.9.1 is that for the given random variable X we have

$$P\{|X - E(X)| < k\sigma_x\} > 1 - \frac{1}{k^2}.$$

Restated in words, this proposition says that the probability that the random variable X will assume a value which is less than k standard deviations away from its expected value must be greater than $1 - 1/k^2$.

EXAMPLE 4.9.1
Let X be the random variable associated with the number of "heads" obtained by tossing nine coins. The probability function of X is given in Table 4.9.1. (We will show later how we may calculate such a probability function.) Our calculations also reveal that $E(X) = 4.5$ and $V(X) = 2.25$, which means that $\sigma_x = 1.5$.

Consider now the set of integers $\{0,1,2,7,8,9\}$. Each element in the set is at least 1.7 standard deviations away from the expected value of X. The integers 2 and 7, which are closest to the expected value of X, are about 1.67

TABLE 4.9.1 Probability Function of X

x_i	$f(x_i)$
0	1/512
1	9/512
2	36/512
3	84/512
4	126/512
5	126/512
6	84/512
7	36/512
8	9/512
9	1/512
	512/512

standard deviations away from the expected value. The Tchebycheff inequality implies, then, that the probability that X will assume either one of the following values 0, 1, 2, 7, 8, or 9 is, at the most, $1/(1.67)^2 = .36$. It also implies that the probability that X will assume one of the following values 3, 4, 5, or 6 is greater than $1 - [1/(1.67)^2] = .64$.

If we examine the probability function of X just given, we find that the actual probability that X will assume one of the following values 0, 1, 2, 7, 8, or 9 is $92/512 = .18$. This probability, as implied by the Tchebycheff inequality, is less than $1/(1.67)^2 = .36$. Furthermore, the actual probability that X will assume one of the following values 3, 4, 5, or 6 is $420/512 = .82$, which is greater than the lower limit of $1 - [1/(1.67)^2] = .64$ established by the Tchebycheff inequality.

Consider the integers 1 and 8. Each of them are 2.33 standard deviations away from the expected value of X. Then, each element in the set of integers {0,1,8,9} must be at least 2.33 standard deviations away from the expected value of X. The Tchebycheff inequality now implies that the probability that X will assume one of the following values 0, 1, 8, or 9 is at most $1/(2.33)^2 = .184$; and, in turn, the probability that X will assume a value between 2 and 7 is greater than $1 - [1/(2.33)^2] = .816$. The probability function of X shows that the actual probability that X will assume one of the following values 0, 1, 8, or 9, is $20/512 = .039$. The actual probability that X will assume a value between 2 and 7 is $492/512 = .961$. Thus, again, the actual probabilities are within the bounds established by the Tchebycheff inequality.

Of course, if we already know what the probability function of a random variable is, there is really no reason to call upon the Tchebycheff inequality to establish the upper or lower limits for such a probability, since we can find the exact probability in question. On the other hand, as we have stated before, the Tchebycheff inequality may turn out to be quite useful if we only know the expected value and the variance of a random variable. This is illustrated in Example 4.9.2.

EXAMPLE 4.9.2

Let X denote the random variable associated with the daily sales of a service station. Assume that $E(X) = 800$ gallons and $V(X) = 40{,}000$. This means that $\sigma_x = 200$.

What is the probability that, for example, the sales for a given day will be between 400 and 1200? Since 400 and 1200 are $2\sigma_x$ away from the expected value of X, the Tchebycheff inequality implies that the probability in question must be greater than $[1 - 1/(2)^2] = .75$.

What is the probability that the sales will exceed 1600 gallons in any given day? Since 0 and 1600 are $4\sigma_x$ from the expected value, the probability that the sales will be less than 1600 must be greater than $[1 - 1/(4)^2] = .9375$. This means that the probability that the sales may be 1600 gallons or more must be at most .0625.

EXERCISES

*45. Let X be a random variable and let $E(X) = 50$ and $V(X) = 100$. Using the Tchebycheff inequality, establish either the upper or lower limit for the probability measure for: (a) $30 \geq X \geq 70$, (b) $20 \geq X \geq 80$, (c) $30 < X < 70$, and (d) $20 < X < 80$.

*46. A service station sells an average of 400 gallons of regular gasoline per day. The standard deviation of the daily sales is 100 gallons.
   a. What can we say about the probability that the actual sales in any given day will be between 200 and 600 gallons?
   b. What can we say about the probability that the sales for any given day may exceed 800 gallons?

Chapter 5
Joint
Probability
Function

So far we have been concerned with the behavior of one random variable. However, in many areas of probability and statistics there arises the need to study the joint behavior of two or more random variables. In the following sections we will study the joint behavior of two discrete random variables.

We can extend our study of the joint behavior of two discrete random variables to that of two continuous random variables. However, we will not do so for the following reasons: first, the study of the joint behavior of two continuous random variables will require some difficult calculus manipulations; second, most of the theorems which we derive in this chapter pertaining to the joint behavior of two discrete random variables are also valid for that of two continuous random variables.

5.1 Joint Probability Function
of Two Discrete Random Variables

To commence our study of the joint behavior of two discrete random variables, we offer the following definition.

Definition 5.1.1
Let X and Y be the discrete random variables where $x_1, \ldots, x_m$ and $y_1, \ldots, y_n$ are the possible values that X and Y can assume respec-

5.1 Joint Probability Function

TABLE 5.1.1 Outcomes of Coins

Dime	Quarter
T	T
H	T
T	H
H	H

tively. Let $f(x_i, y_j)$ denote the probability that X will assume the value x_i and Y will assume the value y_j. The function whose domain is the set containing every ordered pair (x_i, y_j) and whose image set contains the corresponding $f(x_i, y_j)$ is called the *joint probability function* of X and Y.

EXAMPLE 5.1.1

Consider the following situation. We toss a dime and a quarter and can keep the coins which land on "heads." Four outcomes are possible, as indicated in Table 5.1.1.

Let X represent the amount that we win from tossing the dime, and let Y represent the amount that we win from tossing the quarter. The outcomes for the coins may be assigned to the ordered pairs as shown in Figure 5.1.1. By examining the sample space, we can ascertain the probability of occurrences for each ordered pair as we have indicated in Figure 5.1.1. The function described in the right-hand square in Figure 5.1.1 is, then, a joint probability function of the two random variables X and Y.

This joint probability function may also be presented in tabular format (see Table 5.1.2).

EXAMPLE 5.1.2

Suppose an urn contains a $10 bill and a $20 bill. We are to draw a bill from the urn twice. We will replace the first bill drawn in the urn before we draw the second bill.

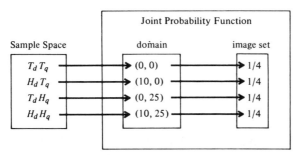

FIGURE 5.1.1 Outcomes for coins assigned to ordered pairs.

74 Joint Probability Function

TABLE 5.1.2

y x	0	25	$f(x)$
0	¼	¼	½
10	¼	¼	½
$f(y)$	½	½	1.0

Let X be the random variable associated with the outcome of the first drawing, and let Y be the random variable associated with the outcome of the second drawing. Then, the joint probability function may be given as shown in Table 5.1.3.

TABLE 5.1.3

y x	10	20	$f(x)$
10	¼	¼	½
20	¼	¼	½
$f(y)$	½	½	1.0

EXAMPLE 5.1.3

In Example 5.1.2 we are to draw a bill twice. However, this time the first bill drawn will not be replaced in the urn. Since the only possible outcomes of this experiment are that we would draw $10 first and $20 second, or $20 first and $10 second, the joint probability function is given diagrammatically in Figure 5.1.2, and tabularly in Table 5.1.4.

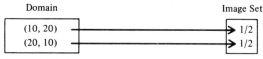

FIGURE 5.1.2

EXAMPLE 5.1.4

After carefully studying the daily price movements of two stocks x and y, we have arrived at the following conclusions, the probabilities of which are:

1. .2 that both x and y will come down a point
2. .2 that both x and y will go up a point
3. .3 that x will go up a point and y will come down a point
4. .3 that x will come down a point and y will go up a point.

TABLE 5.1.4

x \ y	10	20	$f(x)$
10	0	½	½
20	½	0	½
$f(y)$	½	½	1.0

Let X and Y now depict the random variables associated with the price movements of x and y. Then, the conclusion that we have reached concerning their joint behavior may be expressed by the joint probability function given in Table 5.1.5.

TABLE 5.1.5

x \ y	−1	+1	$f(x)$
−1	.2	.3	.5
+1	.3	.2	.5
$f(y)$	.5	.5	1.0

EXAMPLE 5.1.5

Suppose we are an insurance salesman. The number of policies we sell in a day is related to the number of customers who contact us that day (for example, the former cannot exceed the latter). After examining past records, we have arrived at the following conclusions. During any day the probabilities are:

1. .3 that no customer contacts us
2. .2 that one customer contacts us but we make no sale
3. .2 that one customer contacts us and we make the sale
4. .1 that two customers contact us but we make no sale
5. .1 that two customers contact us and we make one sale
6. .1 that two customers contact us and we make two sales.

Let X depict the random variable associated with the number of customers who contact us, and let Y depict the random variable associated with the number of policies we sell. Then, the joint probability function of X and Y is given in Table 5.1.6.

Definition 5.1.2

Let X and Y be two random variables which have a joint probability function. Then, the probability functions of X and Y are called the

Joint Probability Function

TABLE 5.1.6

x \ y	0	1	2	f(x)
0	.3	0	0	.3
1	.2	.2	0	.4
2	.1	.1	.1	.3
f(y)	.6	.3	.2	1.0

marginal probability functions corresponding to the given joint probability function.

EXAMPLE 5.1.6

Consider the joint probability function given in Table 5.1.2. The two marginal probability functions for the given joint probability function are

x	$f(x)$
0	½
10	½
	1.0

and

y	$f(y)$
0	½
25	½
	1.0

These marginal probability functions are also shown in the margins of Table 5.1.2, which illustrates the joint probability function of Example 5.1.1.

EXERCISES

*1. Let random phenomena consist of tossing a dime and a quarter. If the dime lands on "heads," you will win $1 and if it lands on "tails," you win nothing. If the quarter lands on "heads," you win $2, and if it lands on "tails," you win nothing. Describe the joint probability function of the two random variables.

2. Suppose you toss a coin and a die. If the coin lands on "heads," you win $1, and if it lands on "tails," you win nothing. The winnings from tossing the die is equal in dollars to the number which appears. Describe the joint probability function of the two random variables.

*3. A hat contains two $1 bills and one $10 bill. You are to draw two bills in succession. The first one drawn will not be replaced in the hat before you draw the second one. Let each of the two drawings be represented by a random variable. Describe the joint probability function of the two random variables.

4. A hat contains two numbers, 1 and 2. You are to draw one number from the hat. If the number is 1, you will win $10, and if the number is 2, you will win $40. Let X be the random variable associated with the number drawn from the hat. Let Y be the random variable associated with the amount of your winnings. Describe the joint probability function of X and Y.

5. Suppose you toss a coin and win $20 if "heads" is obtained, and the game ends. If "tails" is obtained you win nothing but you can toss that coin again. If "heads" is then obtained, you win another $20, but if "tails" is obtained, you win nothing. The game ends after the second trial, regardless of the outcome. Describe the joint probability function for the two associated random variables.

6. An urn contains 2 red glass marbles, 2 brown glass marbles, 3 red plastic marbles, and 3 brown plastic marbles. You are to draw one marble from the urn. Your winning will be determined by the following rule: you win $10 if you draw a red marble and $20 if you draw a brown marble. In addition, you win $30 if the marble is glass and $50 if it is plastic. Let the amount of your winnings based on drawing a red or brown marble be represented by one random variable, and let the amount of your winnings based on drawing a glass or plastic marble be represented by another random variable. Describe the joint probability function of the two random variables.

7. Refer to Exercise 6. Assume now that the urn contains 20 red glass marbles, 40 brown glass marbles, 10 red plastic marbles, and 30 brown plastic marbles. Describe the joint probability function of the two random variables.

8. A hat contains two $1 bills and one $10 bill. You are to draw two bills in succession. You are not to replace the first one drawn in the hat before drawing the second one. Rather, you get to keep the bills drawn. In addition, you will be given the amount equal to the product of the two bills. Let the two types of wins be represented by two different random variables. Describe the joint probability function of the two random variables.

*9. A hat contains three numbers: 0, 1, and 2. You draw two numbers and win in dollars the sum of the two numbers, in addition to the product of the numbers. Describe the joint probability function of the two types of wins.

10. Suppose you toss three coins and win in dollars the number of "heads" obtained and the number of changes in sequence, where the changes in sequence are illustrated as

$$\underbrace{H \ H}_{\text{change}} \ T$$

$$\underbrace{T \ H \ T}_{\text{changes}}$$

Describe the joint probability function of the two types of wins.

5.2 Conditional Probability Function

In the preceding chapter we considered the notion of conditional probability of an event. We will now extend that idea to the joint probability function of two random variables.

Definition 5.2.1
Let X and Y be two discrete random variables. The function whose domain is the set containing $(Y = y_j | X = x_i)$ and whose image set contains the corresponding probability $P(Y = y_j | X = x_i)$ is called the *conditional probability function* of Y, given that X assumes the value x_i.

In the preceding definition $(Y = y_j | X = x_i)$ implies that Y may assume any one of the values $y_1, \ldots, y_n$, whereas X is to assume a specifically given value, x_i.

We have defined the properties that constitute a conditional probability function. The following definition tells us how to find the elements of a conditional probability function.

Definition 5.2.2
Let $P(Y = y_j | X = x_i)$ depict the probability that Y will assume the value y_j, given that X has assumed the value x_i. Then,

$$P(Y = y_j | X = x_i) = \frac{f(x_i, y_j)}{f(x_i)}.$$

EXAMPLE 5.2.1
Consider the joint probability function given in Example 5.1.2 and in Table 5.2.1.

TABLE 5.2.1

x \ y	10	20	$f(x)$
10	¼	¼	½
20	¼	¼	½
$f(y)$	½	½	1.0

The conditional probability function of Y, given that $X = 10$, is illustrated in Figure 5.2.1.

5.2 Conditional Probability Function

FIGURE 5.2.1 Conditional probability function of Y, given $X = 10$.

The conditional probability function of Y, given that $X = 20$, is shown in Figure 5.2.2.

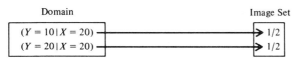

FIGURE 5.2.2 Conditional probability function of Y, given $X = 20$.

EXAMPLE 5.2.2

Consider the joint probability function given in Example 5.1.3 and in Table 5.2.2.

TABLE 5.2.2

x \ y	10	20	$f(x)$
10	0	½	½
20	½	0	½
$f(y)$	½	½	1.0

The conditional probability function of Y, given that $X = 10$, is shown in Figure 5.2.3.

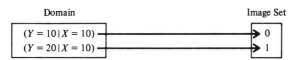

FIGURE 5.2.3 Conditional probability function of Y, given $X = 10$.

The conditional probability function of Y, given that $X = 20$, is shown in Figure 5.2.4.

FIGURE 5.2.4 Conditional probability function of Y, given $X = 20$.

5.3 Statistical Independence

In Chapter 3 we considered the notion of statistical independence for any two events. We will now apply the concept of statistical independence to any two discrete random variables.

Definition 5.3.1
Let X and Y be two discrete random variables. Then, Y is said to be statistically independent of X if every conditional probability function of Y, given that $X = x_i$, is identical to the marginal probability function of Y. Conversely, X is said to be statistically independent of Y if every conditional probability function of X, given that $Y = y_j$, is identical to the marginal probability function of X.

EXAMPLE 5.3.1
Let us return to Example 5.2.1. We find that each of the two conditional probability functions of Y is identical to its marginal probability function. Therefore, Y is statistically independent of X.

On the other hand, if we return to Example 5.2.2, we see that each of the two conditional probability functions is different from its marginal probability function. Therefore, for this example, Y is *not* statistically independent of X.

Let us now re-examine the two examples just given. In Example 5.2.1 we assumed that the first bill drawn would be replaced in the urn. Thus, the outcome of the second drawing would not be affected by that of the first; that is, the conditional probability functions of Y should be identical to each other, since the outcome for X should not have any bearing upon them. If that is the case, the conditional probability functions of Y would be identical not only to each other, but also to the marginal probability function of Y.

In Example 5.2.2, we assumed that the first bill drawn would not be replaced in the urn. Then, the outcome of the second drawing is bound to be influenced by that of the first. For this reason, the conditional probability functions of Y cannot be identical to each other, since they are influenced by the outcome of X. If some of the conditional probability functions of Y are different from each other, this situation would invariably preclude the possibility that all of them may be identical to the marginal probability function of Y.

In general, if Y is said to be statistically independent of X, it means that the outcome of Y is not influenced by that of X. Conversely, if X is said to be statistically independent of Y, it means that the outcome of X is not influenced by that of Y.

EXAMPLE 5.3.2

Let us return to Example 5.1.5. We can intuitively see that Y cannot be independent of X, since the number of new policies sold during a day depends on the number of customers who contact us. Application of Definition 5.3.1 will also show that Y is not independent of X. First, we observe that the marginal probability function of Y is

y_j	$f(y_j)$
0	.6
1	.3
2	.1
	1.0

On the other hand, the conditional probability functions of Y are:

y_j	$P(Y = y_j\|X = 0)$	$P(Y = y_j\|X = 1)$	$P(Y = y_j\|X = 2)$
0	1.0	.5	$\frac{1}{3}$
1	0	.5	$\frac{1}{3}$
2	0	0	$\frac{1}{3}$
	1.0	1.0	1.0

We observe that all three conditional probability functions of Y are different from its marginal probability function. Therefore, Y is not statistically independent of X.

If the random phenomenon for Y takes place after that of X, as was the case in the two examples just given, it is not meaningful to ask whether X is statistically independent of Y. However, there are cases in which the random phenomena underlying X and Y are such that it is meaningful to ask whether X is statistically independent of Y, and to ask whether Y is statistically independent of X. One such example is provided in Example 5.1.1, in which X depicts the random variable associated with the dime and Y depicts the random variable associated with the quarter.

If we were to examine the joint probability function of X and Y without considering the underlying physical phenomena, we would always conclude that, if Y is statistically independent of X, then X is also statistically independent of Y. This proposition may be proved on purely axiomatic grounds. Thus, we propose the following definition.

Definition 5.3.2

Let X and Y be two random variables. The two random variables are said to be statistically independent if one of them is statistically independent of the other.

A logical extension of Definition 5.3.2 is the following.

Definition 5.3.3

Let X and Y be two random variables. Then, X and Y are said to be statistically dependent if they are not statistically independent.

EXERCISES

*11. Refer to Exercise 1 and find the conditional probability functions of the random variable associated with the quarter, given that the random variable associated with the dime has assumed a specific value. Then, indicate whether the two random variables are statistically independent. Can you justify your answer on an intuitive ground?

12. Refer to Exercise 2 and find the conditional probability functions of the random variables associated with the die, given that the random variable associated with the coin has assumed a specific value. Then, indicate whether the two random variables are statistically independent. Can you justify your answer on an intuitive ground?

*13. Refer to Exercise 3 and find the conditional probability function for the random variable associated with the second drawing, given that the random variable associated with the first drawing has assumed a specific value. Then, indicate whether the two random variables are statistically independent. Can you justify your answer on an intuitive ground?

14. Refer to Exercise 4 and find the conditional probability function of the random variable associated with the amount of your winnings, given that the random variable associated with the number drawn from the hat has assumed a specific value. Then, indicate whether the two random variables are statistically independent. Can you justify your answer on an intuitive ground?

15. Refer to Exercise 5 and find the conditional probability function of the random variable associated with the second toss, given that the random variable associated with the first toss has assumed a specific value. Then, indicate whether the two random variables are statistically independent. Can you justify your answer on an intuitive ground?

16. Refer to Exercise 6 and find the conditional probability function of the random variable associated with the color of the marble drawn, given that the random variable associated with the material of the marble drawn has assumed a specific value. Then, indicate whether the two random variables are statistically independent. Can you justify your answer on an intuitive ground?

17. Refer to Exercise 7 and find the conditional probability function of the random variable associated with the color of the marble drawn, given that the random variable associated with the material of the marble drawn has assumed a specific value. Then, indicate whether the two random variables are statistically independent. Can you justify your answer on an intuitive ground?
18. Refer to Exercise 8 and find the conditional probability function for one of the random variables, given that the other random variable has assumed a specific value. Then, indicate whether the two random variables are statistically independent. Can you justify your answer on an intuitive ground?
*19. Refer to Exercise 9 and find the conditional probability function for one of the random variables, given that the other random variable has assumed a specific value. Then, indicate whether the two random variables are statistically independent. Can you justify your answer on an intuitive ground?
20. Refer to Exercise 10 and find the conditional probability function for one of the random variables, given that the other random variable has assumed a specific value. Then, indicate whether the two random variables are statistically independent. Can you justify your answer on an intuitive ground?

5.4 Covariance

Given two random variables X and Y, we may not only want to know whether or not they are statistically dependent, but also the strength of their dependence. One of the measures which indicates the strength of their *linear dependence* is called the *covariance* between the two random variables.

Before we define the term *covariance*, however, we should clarify what we mean by *linear dependence*. Given two mathematical variables x and y, we say that y is linearly dependent on x if we can express

$$y = a + bx,$$

where a is any real number constant and b is a non-zero real number constant. For example, suppose the values of the two variables are those given in Table 5.4.1. Then, we can express that $y = 20 + 2x$. Thus, y is linearly dependent on x. On the other hand, suppose the values of x and y are those given in Table 5.4.2. Then, we can express that $y = 100 - 2x$. Thus, again y is linearly dependent on x.

Assume now that y is linearly dependent on x. We say that y is positively

TABLE 5.4.1

x	10	20	30
y	40	60	80

TABLE 5.4.2

x	10	20	30
y	80	60	40

dependent on x if the coefficient b is positive, and that y is negatively dependent on x if b is negative. Then, for the relationship between x and y given in Table 5.4.1, we can say that y is positively dependent on x, whereas, for the relationship between x and y given in Table 5.4.2, y is negatively dependent on x. Positive or negative linear dependence of y on x may be interpreted in another context. If y is positively linear dependent on x, then, the values of x and y move in the same direction; that is, an increase in the value of x will increase the value of y. However, if y is negatively linear dependent on x, then, the values of x and y move in opposite directions; that is, an increase in the value of x will decrease the value of y.

We can now extend the concept of linear dependence to two random variables X and Y. Assume, for example, that their joint probability function is that given in Table 5.4.3. Then, the probability is .8 that we can express that $y = x$ and the probability is .2 that we can express that $y = 30 - x$. Thus, there is a .8 probability that y is positively linear dependent on x and a .2 probability that y is negatively linear dependent on x. We might argue that, for the given joint probability function, y is predominantly positively linear dependent on x.

TABLE 5.4.3

x \ y	10	20	$f(x)$
10	.4	.1	.5
20	.1	.4	.5
$f(y)$	.5	.5	1.0

The term *covariance between* X *and* Y is defined in such a way that its value is positive if the linear dependence between x and y is predominantly positive and its value is negative if the linear dependence between x and y is predominantly negative.

Definition 5.4.1

Let X and Y be two discrete random variables. The *covariance* of X and Y, denoted Cov(X,Y), is given as

$$\text{Cov}(X,Y) = \sum_{i=1}^{m} \sum_{j=1}^{n} [x_i - E(X)][y_j - E(Y)]f(x_i,y_j).$$

In this definition it is implicitly assumed that X takes the values $x_1, \ldots, x_m$ and that Y takes the values $y_1, \ldots, y_n$.

EXAMPLE 5.4.1

Consider again the joint probability function given in Table 5.4.3. From the marginal probability function of X and Y we can calculate that $E(X) = 15$ and $E(Y) = 15$. The process of calculating the covariance of X and Y is illustrated in Table 5.4.4. The value of the covariance is obtained by adding the elements in the last column. Thus, $\text{Cov}(X,Y) = 15$ and is positive as we should have expected.

TABLE 5.4.4 Calculating Covariance of X and Y

$x_i - E(X)$	$y_j - E(Y)$	$f(x_i,y_j)$	$[x_i - E(X)][y_j - E(Y)]f(x_i,y_j)$
10 − 15	10 − 15	.4	10.0
20 − 15	10 − 15	.1	−2.5
10 − 15	20 − 15	.1	−2.5
20 − 15	20 − 15	.4	10.0
		1.0	15.0

EXAMPLE 5.4.2

Consider the joint probability function given in Example 5.1.2 and repeated here in Table 5.4.5.

TABLE 5.4.5

x \ y	10	20	$f(x)$
10	¼	¼	½
20	¼	¼	½
$f(y)$	½	½	1

We note that the probability is .5 that $y = x$ and .5 that $y = 30 - x$. Thus, there is a .5 probability that y is positively linear dependent on x,

Joint Probability Function

and there is a .5 probability that y is negatively linear dependent on x. Thus, we might expect the covariance between X and Y to be zero.

The process of calculating the covariance of X and Y is illustrated by Table 5.4.6. The value of the covariance is obtained by adding the elements in the last column. We note that $\text{Cov}(X,Y) = 0$ for the two random variables.

TABLE 5.4.6

$x_i - E(X)$	$y_j - E(Y)$	$f(x_i, y_j)$	$[x_i - E(X)][y_j - E(Y)]f(x_i, y_j)$
10 − 15	10 − 15	1/4	25/4
20 − 15	10 − 15	1/4	−25/4
10 − 15	20 − 15	1/4	−25/4
20 − 15	20 − 15	1/4	25/4
		1.0	0

EXAMPLE 5.4.3

Consider now the joint probability function given in Example 5.1.3. Again, the probability function is given in Table 5.4.7. Since the marginal probability functions of X and Y are the same as those for Example 5.1.2, we have $E(X) = 15$ and $E(Y) = 15$.

TABLE 5.4.7

x \ y	10	20	$f(x)$
10	0	1/2	1/2
20	1/2	0	1/2
$f(y)$	1/2	1/2	1.0

The process of calculating the covariance of X and Y is shown in Table 5.4.8. Thus, we have $\text{Cov}(X,Y) = -25$.

TABLE 5.4.8

$x_i - E(X)$	$y_j - E(Y)$	$f(x_i, y_j)$	$[x_i - E(X)][y_j - E(Y)]f(x_i, y_j)$
10 − 15	20 − 15	1/2	−25/2
20 − 15	10 − 15	1/2	−25/2
		1.0	−25

The fact that the covariance between X and Y is negative should not be surprising, however. We note from the joint probability function of X and Y that the probability is 1.0 that $y = 30 - x$. Thus, the probability is 1.0 that y is negatively linear dependent on x. Consequently, the covariance between X and Y should be a negative number.

EXERCISES

*21. Find the covariance between the two random variables given in Exercise 1. Does your answer agree with that found for Exercise 11?

22. Find the covariance between the two random variables given in Exercise 2. Does your answer agree with that found for Exercise 12?

*23. Find the covariance between the two random variables given in Exercise 3. Does your answer agree with that found for Exercise 13?

24. Find the covariance between the two random variables given in Exercise 4. Does your answer agree with that found for Exercise 14?

25. Find the covariance between the two random variables given in Exercise 5. Does your answer agree with that found for Exercise 15?

26. Find the covariance between the two random variables given in Exercise 6. Does your answer agree with that found for Exercise 16?

27. Find the covariance between the two random variables given in Exercise 7. Does your answer agree with that found for Exercise 17?

28. Find the covariance between the two random variables given in Exercise 8. Does your answer agree with that found for Exercise 18?

*29. Find the covariance between the two random variables given in Exercise 9 Does your answer agree with that found for Exercise 19?

30. Find the covariance between the two random variables given in Exercise 10. Does your answer agree with that found for Exercise 20?

5.5 Statistical Independence and Covariance

The notions of statistical independence and covariance have been considered as separate topics thus far. The following theorem defines a certain relationship between these two concepts.

Theorem 5.5.1

Let X and Y be two jointly distributed random variables. If X and Y are statistically independent, then $\text{Cov}(X,Y) = 0$.

The theorem states that if X and Y are statistically independent, then the covariance between the two random variables must necessarily be zero.

However, this does not imply that if the covariance between any two random variables is zero, then the two random variables invariably must be statistically independent. Two random variables can be nonlinearly dependent and yet the covariance between them can be zero, as we will illustrate in Example 5.5.3.

A logical extension of the preceding theorem is:

Theorem 5.5.2

Let X and Y be two jointly distributed random variables. If $\text{Cov}(X,Y)$ is not zero, then X and Y must be statistically dependent.

EXAMPLE 5.5.1

Consider the two random variables first introduced in Example 5.1.2. We have shown subsequently that these two random variables are statistically independent. Further calculations indicate that $\text{Cov}(X,Y) = 0$. This illustrates the validity of Theorem 5.5.1.

EXAMPLE 5.5.2

Consider the two random variables given in Example 5.1.3. For these random variables we have calculated that $\text{Cov}(X,Y) = -25$. This implies, according to Theorem 5.5.2, that the two random variables must be statistically dependent. We have, however, already found that the two random variables are statistically dependent.

EXAMPLE 5.5.3

We have pointed out that the covariance between two statistically dependent random variables does not have to be zero. We will present here an example in which the two random variables are statistically dependent and, yet, have a covariance of zero.

The joint probability function of the two random variables is given in Table 5.5.1.

TABLE 5.5.1

x \ y	1	2	3	$f(x)$
1	0	⅓	0	⅓
2	⅓	0	⅓	⅔
$f(y)$	⅓	⅓	⅓	1.0

The conditional probability function of Y, given that $X = 1$, is

y_j	$P(Y = y_j\|X = 1)$
1	0
2	1
3	0
	1.0

and the conditional probability function of Y, given that $X = 2$, is

y_j	$P(Y = y_j\|X = 2)$
1	½
2	0
3	½
	1.0

Thus, X and Y are statistically dependent random variables.

On the other hand, the calculations in Table 5.5.2 reveal that $\text{Cov}(X,Y) = 0$.

TABLE 5.5.2

$x_i - E(X)$	$y_j - E(Y)$	$f(x_i,y_j)$	$[x_i - E(X)][y_j - E(Y)]f(x_i,y_j)$
$1 - 5/3$	$2 - 2$	⅓	0
$2 - 5/3$	$1 - 2$	⅓	$-1/9$
$2 - 5/3$	$3 - 2$	⅓	$1/9$
			0

5.6 Correlation

In calculating the values of covariance between the two random variables X and Y, we did not take into account the fact that X and Y may be measured by two different scales, which are not commensurate to each other. For example, suppose X is the random variable depicting the height of a person, and Y is the random variable depicting his weight. Then, X is measured in terms of feet, whereas Y is measured in terms of pounds. On the other hand, if the values of X and Y are expressed in terms of their respective standard deviations, then the resulting values will be measured by the same scale, since the unit of measurement for both X and Y is now the standard deviation.

Instead of calculating the covariance between X and Y, we can calculate a similar coefficient after the values of X and Y and the values of $E(X)$ and

90 Joint Probability Function

$E(Y)$ have been divided by their respective standard deviations. The coefficient thus obtained is called the *correlation coefficient*, the formal definition of which is given as:

Definition 5.6.1
Let X and Y be two jointly distributed random variables. The *correlation* between X and Y, denoted ρ, is given as

$$\rho = \sum_{i=1}^{m} \sum_{j=1}^{n} \left[\frac{x_i - E(X)}{\sigma_x}\right]\left[\frac{y_j - E(Y)}{\sigma_y}\right] f(x_i, y_j).$$

A little reflection on Definition 5.6.1 will reveal that

$$\rho = \frac{1}{\sigma_x \sigma_y} \sum_{i=1}^{m} \sum_{j=1}^{n} [x_i - E(X)][y_j - E(Y)] f(x_i, y_j),$$

since σ_x and σ_y are constants. Thus, we derive the following theorem:

Theorem 5.6.1
Let X and Y be two random variables. Then,

$$\rho = \frac{\text{Cov}(X, Y)}{\sigma_x \sigma_y}.$$

Theorem 5.6.1 indicates that the correlation between X and Y can be calculated without transforming the values of X and Y, and, in turn, those of $E(X)$ and $E(Y)$ in terms of their standard deviations if the covariance between the two random variables is known.

EXAMPLE 5.6.1
Consider the two random variables X and Y given in Example 5.1.2. Table 5.6.1 illustrates the process of calculating the value of the correlation

TABLE 5.6.1

$\dfrac{x_i - E(X)}{\sigma_x}$	$\dfrac{y_j - E(Y)}{\sigma_y}$	$f(x_i, y_j)$	$\left[\dfrac{x_i - E(X)}{\sigma_x}\right]\left[\dfrac{y_j - E(Y)}{\sigma_y}\right] f(x_i, y_j)$
−1	−1	¼	¼
1	−1	¼	−¼
−1	1	¼	−¼
1	1	¼	¼
		1.0	0

coefficient of X and Y. The correlation coefficient is found by adding the elements in the extreme right-hand column. Thus, $\rho = 0$.

Let us now calculate the value of ρ according to Theorem 5.6.1. The covariance between X and Y was calculated to be zero in Example 5.4.2. Our calculations also reveal that $V(X) = 25$ and $V(Y) = 25$, which means that $\sigma_x = 5$ and $\sigma_y = 5$. Then,

$$\rho = \frac{\text{Cov}(X,Y)}{\sigma_x \sigma_y} = \frac{0}{(5)(5)} = 0.$$

EXAMPLE 5.6.2

Let us consider the two random variables given in Example 5.1.3. Table 5.6.2 illustrates the process of calculating the value of the correlation coefficient of X and Y. The correlation coefficient is found by adding the elements in the extreme right-hand column. Thus, $\rho = -1$.

TABLE 5.6.2

$\dfrac{x_i - E(X)}{\sigma_x}$	$\dfrac{y_j - E(Y)}{\sigma_y}$	$f(x_i, y_j)$	$\left[\dfrac{x_i - E(X)}{\sigma_x}\right]\left[\dfrac{y_j - E(Y)}{\sigma_y}\right] f(x_i, y_j)$
-1	1	$\frac{1}{2}$	$-\frac{1}{2}$
1	-1	$\frac{1}{2}$	$-\frac{1}{2}$
		1.0	-1

Let us now calculate the value of ρ, according to Theorem 5.6.1. The covariance between X and Y in Example 5.4.3 was calculated to be -25. Our calculations also reveal that $\sigma_x = 5$ and $\sigma_y = 5$. Thus,

$$\rho = \frac{\text{Cov}(X,Y)}{\sigma_x \sigma_y} = \frac{-25}{(5)(5)} = -1.$$

EXAMPLE 5.6.3

Consider the two random variables given in Example 5.4.1. Our calculations revealed that the $\text{Cov}(X,Y) = 15$.

If we calculate the variances of X and Y, we find that $V(X) = 25$ and $V(Y) = 25$, which means that $\sigma_x = 5$ and $\sigma_y = 5$. Thus,

$$\rho = \frac{15}{(5)(5)} = .6.$$

We have illustrated by Example 5.6.1 that the correlation coefficient between the two statistically independent random variables is equal to zero. Theorem 5.5.1 implies that, if X and Y are, for example, statistically independent, then, the correlation coefficient between them must necessarily be zero.

What can we state about the correlation coefficient between two statistically dependent random variables? We can show that if the two random variables are linearly dependent on each other, then, the coefficient will be a nonzero value between 1 and -1. This is illustrated in Examples 5.6.2 and 5.6.3. On the other hand, if the two random variables are nonlinearly dependent on each other, then, it is possible that the correlation coefficient between them may be zero. This is illustrated by the fact that if we were to calculate the correlation coefficient for the two random variables in Example 5.5.3, it would be zero.

EXERCISES

*31. Find the correlation between the two random variables given in Exercise 1.
32. Find the correlation between the two random variables given in Exercise 2.
*33. Find the correlation between the two random variables given in Exercise 3.
34. Find the correlation between the two random variables given in Exercise 4.
35. Find the correlation between the two random variables given in Exercise 5.
36. Find the correlation between the two random variables given in Exercise 6.
37. Find the correlation between the two random variables given in Exercise 7.
38. Find the correlation between the two random variables given in Exercise 8.
*39. Find the correlation between the two random variables given in Exercise 9.
40. Find the correlation between the two random variables given in Exercise 10.

5.7 Applications of Covariance and Correlation

At this point, the reader may question the relevance of the notions of statistical independence, covariance, and correlation for practical problems. We will, therefore, illustrate one simple application of the statistical concepts elaborated in this chapter.

EXAMPLE 5.7.1

Let X, Y, W, and Z depict the random variables associated with the daily price fluctuations of four different stocks. Let us assume the joint probability functions between X and the three remaining random variables, as shown in Tables 5.7.1, 5.7.2, and 5.7.3.

5.7 Applications of Covariance and Correlation

TABLE 5.7.1

x \ y	−2	2
−2	.4	.1
2	.1	.4

TABLE 5.7.2

x \ w	−2	2
−2	.25	.25
2	.25	.25

TABLE 5.7.3

x \ z	−2	2
−2	.1	.4
2	.4	.1

Let us first examine the joint behavior of X and Y in Table 5.7.1. We observe that the prices of both stocks have moved in the same direction 80 percent of the time and in opposite directions 20 percent of the time. It seems, then, that the price movements of the two stocks are dependent on each other.

Ascertaining the conditional probability functions of Y, for example, will show that Y is dependent on X. The conditional probability functions of Y, in addition to its marginal probability function, are shown in Table 5.7.4.

TABLE 5.7.4 Conditional Probability Functions of Y

y	Marginal Probability $f(y)$	Conditional Probability $P(Y=y\mid X=-2)$	$P(Y=y\mid X=2)$
−2	.5	.8	.2
2	.5	.2	.8
	1.0	1.0	1.0

We observe in Table 5.7.4 that the conditional probability functions of Y are not identical to its marginal probability function. Thus, Y is statistically dependent on X. Similarly, it can be shown that X is statistically dependent on Y.

Next let us calculate the covariance between X and Y. The calculations in Table 5.7.5 reveal that $\text{Cov}(X,Y) = 2.4$. The fact that the covariance is not zero again implies that X and Y must be statistically dependent.

TABLE 5.7.5 Covariance between X and Y

$[x_i - E(X)]$	$[y_j - E(Y)]$	$f(x_i, y_j)$	$[x_i - E(X)][y_j - E(Y)]f(x_i,y_j)$
(−2 − 0)	(−2 − 0)	.4	= 1.6
(2 − 0)	(−2 − 0)	.1	= −.4
(−2 − 0)	(2 − 0)	.1	= −.4
(2 − 0)	(2 − 0)	.4	= 1.6
			2.4

Next let us calculate the correlation coefficient between X and Y. To do so, we must first calculate the variance and the standard deviations of X and Y, as

$$V(X) = (-2-0)^2 \left(\frac{1}{2}\right) + (2-0)^2 \left(\frac{1}{2}\right) = 4$$

and

$$V(Y) = (-2-0)^2 \left(\frac{1}{2}\right) + (2-0)^2 \left(\frac{1}{2}\right) = 4,$$

so that $\sigma_x = 2$ and $\sigma_y = 2$. Then,

$$\rho_{xy} = \frac{\text{Cov}(X,Y)}{\sigma_x \sigma_y} = \frac{2.4}{(2)(2)} = .6,$$

where ρ_{xy} denotes the correlation coefficient between X and Y.

Let us now examine the joint behavior of X and W, as illustrated in Table 5.7.2. We observe that the prices of both stocks have moved in the same direction 50 percent of the time and in opposite directions 50 percent of the time. It seems, then, that the prices of the stocks move independently of each other.

By ascertaining the conditional probability functions, we can show that X and W are statistically independent. Further calculations would reveal that

$$\text{Cov}(X,W) = 0$$

and

$$\rho_{xw} = 0.$$

Finally, let us examine the joint behavior of X and Z, illustrated in Table 5.7.3. We observe that the prices of both stocks have moved in the same direction 20 percent of the time and in opposite directions 80 percent of the time. It seems, then, that the price movements of the two stocks are inversely dependent.

By ascertaining the conditional probability functions, we can show that X and Z are statistically dependent. Further calculations would show that

$$\text{Cov}(X,Z) = -2.4$$

and

$$\rho_{xz} = -.6.$$

TABLE 5.7.6

Stocks	Predominant Movement Pattern	Statistical Dependence/ Independence	Covariance	Correlation
X, Y	together	dependent	2.4	.6
X, W	independent	independent	0	0
X, Z	opposite	dependent	-2.4	-.6

5.7 Applications of Covariance and Correlation

Let us now compare the coefficients, summarized in Table 5.7.6. We observe that when the two stock prices move together, their covariance and correlation coefficients are positive, but when the two stock prices move in opposite directions, their covariance and correlation coefficients are negative. Furthermore, when the prices move independently of each other, their covariance and correlation coefficients are zero.

To conclude our analysis, let us assume that

x \ u	-2	2
-2	.5	0
2	0	.5

We observe that the prices of both stocks have moved in the same direction 100 percent of the time. We should, therefore, expect that X and U are statistically dependent and, furthermore, that their covariance and correlation are positive. Our calculations would reveal that all of these propositions are, in fact, true. They would also reveal that the correlation coefficient is unity. The given joint probability function, therefore, provides an example of two random variables which are said to be perfectly correlated to each other.

EXERCISES

*41. Assume that X, Y, and Z depict the daily price movements of three different stocks. Assume the following joint probability functions. Ascertain the co-

x \ y	-1	1
-1	.3	.2
1	.2	.3

x \ z	-1	1
-1	.2	.3
$.1$	.3	.2

variance and the correlation coefficients between X and Y and between X and Z. On the basis of your calculations, what conclusion do you reach with regard to the joint behavior of X and Y, and of X and Z?

42. Select any two stocks listed in the New York Stock Exchange. Construct a joint probability function for the two stocks, using the price quotations for the past 20 days. Calculate the covariance and correlation coefficients between the two price movements. What conclusions do you reach with regard to the joint behavior of the two stocks?

Chapter 6
Functions
of Random
Variables

6.1 Introduction

Suppose we let a_1 and a_2 represent real number constants and x_1 and x_2 represent variables. Then, in algebraic terms, we may let

$$y = a_1 x_1,$$
$$y = x_1 + x_2,$$
$$y = a_1 x_1 + a_2 x_2,$$
$$y = x_1 x_2,$$

and
$$y = \frac{x_1}{x_2}.$$

Each y in the preceding equations is, then, a function of the independent variables x_1 and x_2.

Such functions may also be formed with random variables. For example, if we let X_1 and X_2 be two random variables and a_1 and a_2 be arbitrary real number constants, we can express that

$$Y = a_1 X_1,$$
$$Y = X_1 + X_2,$$
$$Y = a_1 + X_1,$$
$$Y = X_1 X_2,$$

and
$$Y = \frac{X_1}{X_2}.$$

In each of the preceding equations Y may be thought of as a function of random variables. In the following sections we will study the first three of the functions given. The last two types of functions require more complex treatment, which is beyond the scope of this textbook.

6.2 Scalar Product of Random Variables

Suppose we let $Y = aX$, where X is a random variable and a is an arbitrary real number. Then, Y is also a random variable.

To illustrate this proposition, let X be the random variable which assumes the value 0 if "heads" is obtained and assumes the value 1 if "tails" is obtained from tossing one coin. The random variable is described, then, by the rectangle in Figure 6.2.1. Now let $Y = 100X$. Then, Y assumes the value 0 if X assumes the value 0, and Y assumes the value 100 if X assumes the value 1.

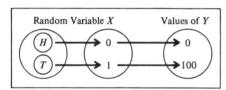

FIGURE 6.2.1

In turn, we can relate the values that Y can assume to the sample space of X as shown in Figure 6.2.2. This figure clearly indicates that Y satisfies the definition of a random variable, since its image set is a set of real numbers and its domain is a sample space.

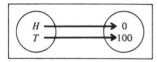

FIGURE 6.2.2

For a random variable such as Y, we may be interested in knowing its probability function. If we were interested, we would have to find the probability function by enumerating all possible values of Y and, in turn, associating each value of Y to the sample space of X. On the other hand, in many instances all we may need to know is the expected value and the variance of Y. In such a case, the following theorem would enable us to calculate the expected value and the variance of Y without finding the probability function of Y, provided we know the values of the expected value and the variance of X.

Theorem 6.2.1

Let X be a random variable, and let a be an arbitrary real number constant. Then, $Y = aX$ is a random variable whose expected value and variance are given as

$$E(Y) = aE(X)$$
and
$$V(Y) = a^2 V(X).$$

The theorem is valid for both discrete and continuous random variables. In fact, all of the theorems to be presented in this chapter are valid for both types of random variables.

EXAMPLE 6.2.1

Let X be a random variable whose probability function is given as

x	$f(x)$
0	½
1	½

Then, $E(X) = \frac{1}{2}$ and $V(X) = \frac{1}{4}$. Now define $Y = 100X$. Then, the probability function of Y is given in the first two columns in Table 6.2.1. The calculation of the expected value is shown in the third column and of the variance, by the last column. Thus, $E(Y) = 50$ and $V(Y) = 2500$.

TABLE 6.2.1

y	$f(y)$	$yf(y)$	$[y - E(Y)]^2 f(y)$
0	½	0	1250
100	½	50	1250
	1.0	50	2500

Theorem 6.2.1 enables us to calculate $E(Y)$ and $V(Y)$ in a simplified manner; that is,

$$E(Y) = aE(X) = (100)\left(\frac{1}{2}\right) = 50,$$
and
$$V(Y) = a^2 V(X) = (100)^2 \left(\frac{1}{4}\right) = 2500,$$

which gives us the same results as those obtained previously.

EXAMPLE 6.2.2

When the probability function of X is known, the expected value and the variance of Y may be obtained by explicitly enumerating its probability function, as we have illustrated in Example 6.2.1. At times, however, the only data that we have pertaining to X may be its expected value and variance. Theorem 6.2.1, nevertheless, enables us to calculate the expected value and the variance of Y, as we will illustrate.

Let X represent the random variables associated with the quantity of a certain item sold in a day. Assume that $E(X) = 200$ and $V(X) = 100$. Suppose now that the price of the item is \$100 per unit. Then, $Y = 100X$ is the random variable associated with the total daily revenue from sales of this item. According to Theorem 6.2.1, we have

$$E(Y) = 100E(X) = 100 \times 200 = 20{,}000,$$
$$V(Y) = 100^2 V(X) = 100^2 \times 100 = 1{,}000{,}000,$$
and
$$\sigma_y = \sqrt{1{,}000{,}000} = 1000.$$

Assume now that the profit margin per unit is \$10. Then, $W = 10X$ is the random variable associated with the total daily profits from sales of this item. The expected value, variance, and standard deviation of W are found by the following calculations:

$$E(W) = 10E(X) = 2{,}000,$$
$$V(W) = 10^2 V(X) = 10{,}000,$$
and
$$\sigma_w = \sqrt{10{,}000} = 100.$$

EXERCISES

*1. Suppose you toss one coin and will win \$2 if "heads" appears and nothing if "tails" appears. Find the expected value, the variance, and the standard deviation of the random variable associated with the amount of winnings by explicitly enumerating its probability function.

*2. Find the expected value, the variance, and the standard deviation of the random variable given in Exercise 1 by expressing it as a function of another random variable whose probability function is given as

x	$f(x)$
0	$\tfrac{1}{2}$
1	$\tfrac{1}{2}$

3. Suppose you toss one coin and will win \$100 if "heads" appears and nothing if "tails" appears. Find the expected value, the variance, and the standard deviation of the random variable associated with the amount of winnings by explicitly enumerating its probability function.

100   Functions of Random Variables

4. Find the expected value, the variance, and the standard deviation of the random variable given in Exercise 3 by expressing the random variable as a function of another random variable whose probability function is given as

x	$f(x)$
0	½
1	½

5. Find the expected value, the variance, and the standard deviation of the random variable given in Exercise 3 by expressing the random variable as a function of another random variable whose probability function is given as

x	$f(x)$
0	½
10	½

*6. Suppose a service station sells an average of 300 gallons of regular gasoline per day. The standard deviation of the daily sales is 100 gallons. Assume that the profit margin on the sales of regular gasoline is 10 cents per gallon. Then, what is the expected value, the variance, and the standard deviation of the random variable associated with the total daily profits from sales of regular gasoline?

6.3 Sum of Two Random Variables

Suppose that we are to play the following game: we toss a dime and a quarter and win $1 for each "heads" obtained. This means that if both coins land on "tails," we will win nothing; if both coins land on "heads," we will win $2; if one coin lands on "heads" and the other on "tails," we will win $1. Thus, the amount that we can win ranges from zero to $2. From our study in preceding chapters, we can derive the probability function associated with the amount of winnings, as shown in the following chart.

Amount of Winnings	Probabilities
0	¼
1	2/4
2	¼

Now let X_1 represent the random variable associated with the amount of winnings from tossing the dime, and let X_2 represent the amount of winnings from tossing the quarter. Then, the probability functions for X_1 and X_2 are those given in Table 6.3.1.

6.3 Sum of Two Random Variables

TABLE 6.3.1

Values of X_1	Probability	Values of X_2	Probability
0	½	0	½
1	½	1	½
	1.0		1.0

Suppose now that we define $Y = X_1 + X_2$. Then, the value of Y depends on those of X_1 and X_2. If both X_1 and X_2 assume the value of 0, then the value of Y must also be 0. On the other hand, if X_1 assumes the value of 0 and X_2 assumes the value of 1, then the value of Y must be 1.

The functional relationship between X_1, X_2, and Y is illustrated in Figure 6.3.1. Y, in this case, represents the random variable associated with the amount won from tossing the two coins. Thus, Y may be stated as the *sum of the two random variables*.

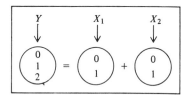

FIGURE 6.3.1

For this simple example, we can calculate the probability function of Y and, in turn, the expected value and the variance of Y. Such calculations are shown in Table 6.3.2. However, the probability function of $Y = X_1 + X_2$ may be difficult to obtain if X_1 and X_2 assume many different values. On the other hand, we may wish to know only the expected value and the variance of Y. The following two theorems will enable us to find the expected value and the variance of Y without explicitly enumerating the probability function of Y, provided that we know the expected values and the variances of X_1 and X_2.

TABLE 6.3.2

y_i	$f(y_i)$	$y_i f(y_i)$	$[y_i - E(Y)]^2 f(y_i)$
0	¼	0	¼
1	2/4	2/4	0
2	¼	2/4	¼
	1.0	1.0 $E(Y)$	½ $V(Y)$

Theorem 6.3.1

Let X_1 and X_2 represent two random variables. Then, $Y = X_1 + X_2$ is a random variable whose expected value and variance are given as

$$E(Y) = E(X_1) + E(X_2)$$
and
$$V(Y) = V(X_1) + V(X_2) + 2\,\text{Cov}(X_1, X_2).$$

In this theorem we have not made any assumptions with regard to the statistical dependence between X_1 and X_2, but if we suppose that X_1 and X_2 are statistically independent, then, the covariance between X_1 and X_2 must be zero. Consequently, Theorem 6.3.1 may be modified as follows:

Theorem 6.3.2

Let X_1 and X_2 be two random variables. If X_1 and X_2 are statistically independent, then, $Y = X_1 + X_2$ is a random variable whose expected value and variance are given as

$$E(Y) = E(X_1) + E(X_2)$$
and
$$V(Y) = V(X_1) + V(X_2).$$

EXAMPLE 6.3.1

Consider the example of tossing two coins discussed previously. $\text{Cov}(X_1, X_2)$ must be zero, since the outcome of one coin neither affects nor is affected by the outcome of the other coin.

Further calculations reveal that $E(X_1) = \frac{1}{2}$, $E(X_2) = \frac{1}{2}$, $V(X_1) = \frac{1}{4}$, and $V(X_2) = \frac{1}{4}$. Then, $E(Y)$ and $V(Y)$ may be calculated according to Theorem 6.3.2:

$$E(Y) = E(X_1) + E(X_2) = \frac{1}{2} + \frac{1}{2} = 1,$$
and
$$V(Y) = V(X_1) + V(X_2) = \frac{1}{4} + \frac{1}{4} = \frac{1}{2}.$$

These results confirm those we have obtained previously with explicit enumeration of the probability function.

EXAMPLE 6.3.2

Suppose an urn contains a $10 bill and a $20 bill. We are to draw two bills from the urn. We will replace the first bill drawn in the urn before we draw the second one. The total amount we win will be equal to the sum of the two drawings.

Let X_1 and X_2 be the random variables associated with the amount of winnings for the first and second drawings, respectively. The joint probability function of X_1 and X_2 is given in Table 6.3.3, from which we find that

TABLE 6.3.3

x_1 \ x_2	$10	$20	$f(x_1)$
$10	¼	¼	½
$20	¼	¼	½
$f(x_2)$	½	½	1.0

$\text{Cov}(X_1, X_2) = 0$. Our calculations also reveal that

$$E(X_1) = 15 \quad E(X_2) = 15$$
and
$$V(X_1) = 25 \quad V(X_2) = 25.$$

Now let Y be the random variable associated with the total amount of winnings. Then, $Y = X_1 + X_2$. According to Theorem 6.3.2, we find that

$$E(Y) = 15 + 15 = 30$$
and
$$V(Y) = 25 + 25 = 50.$$

We can also determine $E(Y)$ and $V(Y)$ by explicitly enumerating the probability function of Y, which is given in Table 6.3.4. The values obtained in the third and fourth columns for $E(Y)$ and $V(Y)$, respectively, are the same values obtained according to Theorem 6.3.2.

TABLE 6.3.4

y_i	$f(y_i)$	$y_i f(y_i)$	$[y_i - E(Y)]^2 f(y_i)$
20	¼	5	25
30	2/4	15	0
40	¼	10	25
	1.0	30	50
		$E(Y)$	$V(Y)$

EXAMPLE 6.3.3

Suppose, for the games given in Example 6.3.2, we do not replace the first bill drawn. Then, the joint probability function may be given as shown in Table 6.3.5. Our calculations reveal that $\text{Cov}(X_1, X_2) = -25$, $E(X_1) = 15$, $E(X_2) = 15$, $V(X_1) = 25$, and $V(X_2) = 25$. If we let $Y = X_1 + X_2$, then, according to Theorem 6.3.1,

$$E(Y) = 15 + 15 = 30$$
and
$$V(Y) = 25 + 2(-25) + 25 = 0.$$

TABLE 6.3.5

x_1 \ x_2	10	20	$f(x_1)$
10	0	½	½
20	½	0	½
$f(x_2)$	½	½	1.0

Let us now explicitly enumerate the probability function of Y. It is obvious that every time we play the game we will win $30. The probability function of Y is given in the first two columns in Table 6.3.6. The expected value and the variance of Y are calculated in the last two columns.

TABLE 6.3.6

y	$f(y)$	$yf(y)$	$[y - E(Y)]^2 f(y)$
30	1	30	0
		$[E(Y)] = 30$	$[V(Y)] = 0$

EXERCISES

7. Suppose you toss two coins and will win $1 for each "heads" obtained. Find the expected value, the variance, and the standard deviation of the random variable associated with the amount of winnings by explicitly enumerating its probability function.

8. Find the expected value, the variance, and the standard deviation of the random variable in Exercise 7 by expressing the random variable as a sum of two random variables.

*9. Assume a hat contains two bills: $10 and $30. You are to draw one bill from the hat twice. The first bill drawn will be replaced in the hat before you draw the second one. Your total winnings will be equal to the sum of the two separate drawings.
   (a) Find the expected value and the variance of your total winnings by explicitly enumerating the probability function.
   (b) Find the expected value and the variance of your total winnings without explicitly enumerating the probability function.

*10. Refer to Exercise 9 and assume that you do not replace the first bill drawn before you draw the second one.
   (a) Find the expected value and the variance of your total winnings by explicitly enumerating the probability function.

(b) Find the expected value and the variance of your total winnings without explicitly enumerating the probability function.

*11. Compare and contrast the two games proposed in Exercises 9 and 10. In what respect are they similar and different?

12. Assume an urn contains 2 red glass marbles, 2 brown glass marbles, 3 red plastic marbles, and 3 brown plastic marbles. You are to draw one marble from the urn. You will win $10 if the color of the marble drawn is red; $30 if the color is brown. In addition, you will win $10 if the material of the marble drawn is glass; $60 if the material is plastic.
    (a) Find the expected value and the variance of your total winnings by explicitly enumerating the probability function.
    (b) Find the expected value and the variance of your total winnings without explicitly enumerating the probability function.

13. Refer to Exercise 12 and assume that the urn contains 1 red glass marble, 3 brown glass marbles, 4 red plastic marbles, and 2 brown plastic marbles. The rules for winning are the same.
    (a) Find the expected value and the variance of your total winnings by explicitly enumerating the probability function.
    (b) Find the expected value and the variance of your total winnings without explicitly enumerating the probability function.

14. Compare and contrast the two games proposed in Exercises 12 and 13. Statistically, in what respect are they similar and different?
    Suppose you were offered an option to choose between the two games. Which one would you choose if you had not made any statistical analysis? Would you still choose the same one in light of your analysis?

6.4 Sum of Many Independent Random Variables

Let us now define $Y = X_1 + \cdots + X_n$. Y must be a random variable if $X_1, \ldots, X_n$ are random variables. Y, thus, is said to be the sum of n random variables.

If $X_1, \ldots, X_n$ are statistically dependent random variables, then $E(Y)$ and $V(Y)$ may be found by applying Theorem 6.3.1 when $n = 2$. Unfortunately, extending Theorem 6.3.1 to any number of n variables, where $n \geq 3$, is complex, particularly with respect to finding $V(Y)$. We will not consider the extension of the theorem for $n \geq 3$.

On the other hand, if $X_1, \ldots, X_n$ are statistically independent, then $E(Y)$ and $V(Y)$ may be found by extending Theorem 6.3.2 for $n \geq 3$. Such an extension is given by the following theorem.

Theorem 6.4.1

Let $X_1, \ldots, X_n$ be statistically independent random variables whose expected values and variances are denoted by $E(X_1), \ldots, E(X_n)$, and $V(X_1), \ldots, V(X_n)$. Let $Y = X_1 + \cdots + X_n$. Then,

and
$$E(Y) = E(X_1) + \cdots + E(X_n),$$
$$V(Y) = V(X_1) + \cdots + V(X_n).$$

EXAMPLE 6.4.1

Assume that we are to toss 100 coins and will win \$1 for each "heads" obtained. Let $X_1, \ldots, X_{100}$ be the random variables associated with the amount of winnings for each "heads" obtained. Then, the random variable associated with the total amount of winnings may be given as

$$Y = X_1 + \cdots + X_{100};$$

that is,

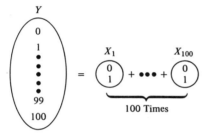

The relationship between the outcome of tossing the individual coins and the aggregate amount of winnings is shown in the preceding diagram. Clearly, the least that we can win is zero and the most that we can win is \$100. Thus, the range of values that Y can assume is 0 to 100.

For each value between 0 and 100 there must be a probability that Y will assume that value. By calculating the probability for each of the 101 integer values that Y can assume, we can obtain the probability function of Y.

Once the probability function of Y has been explicitly enumerated, we can calculate $E(Y)$ and $V(Y)$ in the usual manner, which is relatively cumbersome. However, Theorem 6.4.1 gives us a simplified method of calculating $E(Y)$ and $V(Y)$.

Although each coin tossed is distinct, we assume that they have identical probability functions. We may, then, calculate the following:

$$E(X_1) = \frac{1}{2}, \ldots, E(X_{100}) = \frac{1}{2}$$

and
$$V(X_1) = \frac{1}{4}, \ldots, V(X_{100}) = \frac{1}{4}.$$

According to Theorem 6.4.1, we can determine that

$$E(Y) = E(X_1) + \cdots + E(X_{100}) = 50$$
and
$$V(Y) = V(X_1) + \cdots + V(X_{100}) = 25,$$

and, subsequently, that $\sigma_y = 5$.

EXAMPLE 6.4.2

Assume that we are to toss 100 dice and will win an amount equal to the sum of the numbers obtained. Let $X_1, \ldots, X_{100}$ be the random variables associated with the amount of winnings from tossing each individual die. Then, the random variable associated with the total amount of winnings is given as

$$Y = X_1 + \cdots + X_{100};$$

that is,

$$\begin{pmatrix} Y \\ 100 \\ 101 \\ \vdots \\ 599 \\ 600 \end{pmatrix} = \begin{pmatrix} X_1 \\ 1 \\ 2 \\ 3 \\ 4 \\ 5 \\ 6 \end{pmatrix} + \cdots + \begin{pmatrix} X_{100} \\ 1 \\ 2 \\ 3 \\ 4 \\ 5 \\ 6 \end{pmatrix}$$

$$\underbrace{\hspace{4cm}}_{100 \text{ Times}}$$

The relationship between the outcome of tossing each individual die and that of the total amount of winnings is shown in the preceding diagram. The least that we can win is $100 and the most that we can win is $600.

For each value between 100 and 600 we can ascertain the probability that Y will assume that value. However, to calculate the probabilities for all values between 100 and 600 is quite involved.

Fortunately, the calculations for finding the expected value and the variance of Y are simplified by applying Theorem 6.4.1. If we assume that the probability function of each die is given as

x	$f(x)$
1	$1/6$
2	$1/6$
3	$1/6$
4	$1/6$
5	$1/6$
6	$1/6$

then, $E(X_1) = 3.5, \ldots, E(X_{100}) = 3.5$, and $V(X_1) = 2.91, \ldots, V(X_{100}) = 2.91$. Thus,

$$E(Y) = E(X_1) + \cdots + E(X_{100}) = 350,$$

and

$$V(Y) = V(X_1) + \cdots + V(X_{100}) = 291.$$

EXAMPLE 6.4.3

Let $X_1, \ldots, X_{100}$ be the random variables associated with the daily demand for a certain item for the next 100 days. Then,

$$Y = X_1 + \cdots + X_{100}$$

depicts the random variable associated with the total demand for the next 100 days. If we assume that the demand for any given day is not influenced by that of the other days, then, $X_1, \ldots, X_{100}$ must be statistically independent. Given these conditions, we may calculate that

$$E(Y) = E(X_1) + \cdots + E(X_{100})$$
and
$$V(Y) = V(X_1) + \cdots + V(X_{100}).$$

For example, assume that the expected daily demand is 80 units with a standard deviation of 20 units. Then, the variance of the daily demand is 400. According to Theorem 6.4.1, we have

$$E(Y) = 8000,$$
$$V(Y) = 40,000,$$
and
$$\sigma_y = 200.$$

Thus, the expected total demand for the next 100 days is 8000 units, and the standard deviation associated with the total demand for the given period is 200 units.

EXERCISES

15. Suppose you are to toss four coins and will win $1 for each "heads" obtained. Find the expected value, the variance, and the standard deviation of the random variable associated with the amount of winnings without explicitly enumerating its probability function.

16. Suppose you are to toss one coin and will win $4 if "heads" appears and nothing if "tails" appears. Find the expected value, the variance, and the standard deviation of the random variable associated with the amount of winning without explicitly enumerating its probability function.

*17. Suppose you are to toss 100 coins and will win $1 for each "heads" obtained. Find the expected value, the variance, and the standard deviation of the random variable associated with the amount of winning without explicitly enumerating its probability function.

*18. Suppose you are to toss 100 coins and will win $10 for each "heads" obtained. Find the expected value, the variance, and the standard deviation of the random variable associated with the amount of winning.

*19. Assume a service station sells an average of 300 gallons of regular gasoline per day. The standard deviation of the daily sales is estimated to be 100 gallons. Find the expected value, the variance, and the standard deviation of the random variable associated with the annual sales for the station, assuming that the station will be open for business 360 days in the year and that the daily sales are statistically independent.

6.5 Distinguishing the Sum of Random Variables and a Scalar Product of One Random Variable

Consider the following two algebraic functions:

$$f = x_1 + \cdots + x_k,$$

and

$$g = kx.$$

Assume that $x_1 = x_2 = \cdots = x_k = x$. Then, f and g are equivalent functions, which means that one function may be substituted for the other.

Consider now the following two functions,

$$F = X_1 + \cdots + X_k$$

and

$$G = kX,$$

where $X, X_1, \ldots, X_k$ are random variables. Assume now that $X, X_1, \ldots, X_k$ have identical probability functions; however, these two functions are not equivalent. In other words, a random variable associated with the sum of k identical random variables is not the same as a random variable obtained by multiplying one of the k identical random variables by a constant k.

The distinction between the sum of k identical random variables and a random variable multiplied by a constant k is fundamental. To illustrate the distinction, we present the following two options for analysis.

Option A: Toss 100 coins and win $1 for each "heads" obtained.
Option B: Toss one coin and win $100 if "heads" is obtained and nothing if "tails" is obtained.

Intuitively, we can see that a conservative person would prefer Option A, whereas a speculative person would prefer Option B.

Let us now evaluate the difference between the two options. Let Y be the random variable associated with the amount of winnings for Option A. Then, Y can assume any integer value between 0 and 100. If we let X_i depict the random variable associated with the amount of winnings from tossing the ith coin, then,

$$Y = X_1 + \cdots + X_{100}.$$

Thus, Y is a sum of 100 identical random variables.

On the other hand, let W be the random variable associated with the amount of winnings for Option B. In this case, W can assume only one of two values: either 0 or 100. Suppose we let X be the random variable having the following probability function:

x	$f(x)$
0	½
1	½

Then, W is a random variable obtained by multiplying X and a constant k, which equals 100; that is,

$$W = 100X.$$

The distinction between the two processes by which Y and W have been obtained is illustrated in Figure 6.5.1.

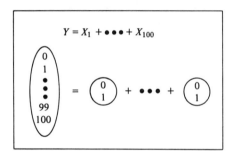

FIGURE 6.5.1

The expected value and the variance of Y are calculated in the following manner:

$$E(Y) = E(X_1) + \cdots + E(X_{100}) = 50$$
and
$$V(Y) = V(X_1) + \cdots + V(X_{100}) = 25;$$

and those of W are calculated as

$$E(W) = 100E(X) = 50$$
and
$$V(W) = 100^2 V(X) = 2500.$$

We see that the variance of W is 100 times that of Y, which indicates that

Option A would be preferred by a conservative person. We may recall that we have already reached this conclusion on the basis of intuition.

EXERCISES

*20. Suppose you are offered the following options: (1) Toss a die and win in dollars the number which appears on top. You are to play this game 100 times. You have to pay $3.50 before each toss; (2) Toss a die and win in dollars the number which appears on top multiplied by 100. You have to pay $350 to play this game, and you can only play this game once.

Compare the expected values, the variances, and the standard deviations of the random variables associated with the amount of winnings for each of these options. Then, indicate which option you would prefer if you were a conservative player.

*21. Suppose you are to toss 100 coins and will win $1 for each "heads" obtained. What is the probability that your winnings will be: (a) between $40 and $60, (b) between $30 and $70, and (c) less than $20 or more than $80?

22. Suppose you are to toss 10 coins and will win $10 for each "heads" obtained. What is the probability that your winnings will be: (a) between $40 and $60, (b) between $30 and $70, and (c) less than $20 or more than $80?

23. Compare the two games proposed in Exercises 21 and 22. Which game do you prefer, assuming that you will have to pay $50 to play either game?

*24. Suppose a service station sells an average of 300 gallons of regular gasoline per day. The standard deviation of the daily sales is estimated to be 100 gallons. What can you say about the probability that the actual sales in any given day will be within 200 gallons of the average sales; that is, between 100 and 500 gallons?

*25. Assume that the daily sales of gasoline in Exercise 24 is statistically independent. What is the probability that the total sales for the next 100 days will be between 10,000 and 50,000 gallons?

*26. Refer to Exercise 25. What is the probability that the total sales for the next 100 days will be between 25,000 and 35,000 gallons?

27. Assume that the nylon cords produced by a firm have a mean breaking strength of 10 pounds and a standard deviation of 2 pounds. Suppose a rope is made with 100 of these cords. If you assume that the breaking strength of the rope is the sum of the strengths of the cords, what can you say about the probability that the rope will support a weight of 1100 pounds or more?

6.6 Adding a Constant to a Random Variable

Suppose we let $Y = a + X$, where X is a random variable and a is a real number constant. Then, Y is a random variable. We propose:

Theorem 6.6.1

Let X be a random variable and a be a real number constant. Then, $Y = a + X$ is a random variable, and

$$E(Y) = a + E(X)$$
and
$$V(Y) = V(X).$$

EXAMPLE 6.6.1

Let X be a random variable whose probability function is given as

x	$f(x)$
0	.5
1	.5
	1.0

Then, $E(X) = .5$ and $V(X) = .25$.

Now define

$$Y = 100 + X.$$

The relationship between X and Y is illustrated in Figure 6.6.1. The prob-

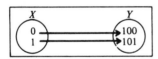

FIGURE 6.6.1

ability function of Y is given as

y	$f(y)$
100	.5
101	.5
	1.0

From this probability function we calculate that

$$E(Y) = (100 \times .5) + (101 \times .5) = 100.5$$
and
$$V(Y) = (100 - 100.5)^2(.5) + (101 - 100.5)^2(.5) = .25.$$

We could have calculated $E(Y)$ and $V(Y)$ without explicitly ascertaining the probability function of Y. Since $Y = 100 + X$, where $E(X) = .5$ and $V(X) = .25$, according to Theorem 6.6.1,

6.7 Practical Applications of Some Statistical Concepts

and
$$E(Y) = 100 + E(X) = 100.5$$
$$V(Y) = V(X) = .25.$$

We observe that the two methods of calculating $E(Y)$ and $V(Y)$ yield the same numerical values.

EXERCISES

*28. Suppose you are to toss two coins and will win $10, in addition to $1 for each number of "heads" obtained. Find the expected value and the variance of the random variable associated with the amount of winnings by explicitly enumerating its probability function.

*29. Refer to Exercise 28 and find the expected value and the variance of the random variable by expressing the random variable as a function of another random variable whose probability function is given as

x	$f(x)$
0	.25
1	.50
2	.25
	1.00

6.7 Practical Applications of Some Statistical Concepts

In this section we will describe some practical applications of the statistical concepts presented in the preceding sections.

EXAMPLE 6.7.1

Let X, Y, W, and Z depict the random variables associated with per-share annual earnings of four different stocks. Let us assume the joint probability functions between X and the three remaining random variables are those shown in Tables 6.7.1, 6.7.2, and 6.7.3.

TABLE 6.7.1

x \ y	4	8
4	.4	.1
8	.1	.4

TABLE 6.7.2

x \ w	4	8
2	.25	.25
4	.25	.25

TABLE 6.7.3

x \ z	4	8
4	.1	.4
8	.4	.1

Suppose we are given the following three options for our investment portfolio:

1. one share of X and one share of Y
2. one share of X and one share of W
3. one share of X and one share of Z.

If we were conservative investors, which portfolio would we choose?
We note that
$$E(X) = E(Y) = E(W) = E(Z) = 6.$$
If we give the following definitions:
$$I_1 = X + Y,$$
$$I_2 = X + W,$$
and
$$I_3 = X + Z,$$
then, according to Theorem 6.3.1,
$$E(I_1) = E(I_2) = E(I_3) = 12.$$
Thus, the expected earnings for the three portfolios are the same.

Does this mean that we should be indifferent to the three portfolios? An evaluation of the three joint probability functions would reveal that this should not be so. We observe that the earnings of

1. X and Y tend to move in the same direction
2. X and W tend to move independently of each other
3. X and Z tend to move in opposite directions.

These observations are also verified by our calculations that
$$\text{Cov}(X,Y) = 2.4,$$
$$\text{Cov}(X,W) = 0,$$
and
$$\text{Cov}(X,Z) = -2.4.$$

It must be apparent, then, that a conservative investor would choose the portfolio whose earnings tend to move in opposite directions, which would be X and Z.

The fact that X and Z constitute the most conservative portfolio may be illustrated in another context. To do so, we first calculate that
$$V(X) = V(Y) = V(W) = V(Z) = 4.$$
Then,
$$V(I_1) = V(X) + V(Y) + 2\text{Cov}(X,Y)$$
$$= 4 + 4 + 2(2.4) = 12.8,$$
$$V(I_2) = V(X) + V(W)$$
$$= 4 + 4 = 8.0,$$
and
$$V(I_3) = V(X) + V(Z) + 2\text{Cov}(X,Z)$$
$$= 4 + 4 + 2(-2.4) = 3.2.$$

6.7 Practical Applications of Some Statistical Concepts

Thus, the smallest variance is associated with the sum of the earnings for X and Z. This means that, if we chose X and Z, it would be more likely that the actual earnings from the portfolio would not deviate too much from the expected earnings than if we chose either of the remaining two portfolios.

The example just given is not very realistic. However, the type of analysis we have presented in the example constitutes a very important segment of modern portfolio management theories.

EXAMPLE 6.7.2

Let us suppose that a service station sells 500 gallons of gasoline per day. The standard deviation of the daily sales is 100 gallons. The station makes a 5-cent profit on every gallon of gasoline sold. The owner of the station would like to project total profits for the next 100 days, assuming that the daily sales are statistically independent.

Let X_i depict the random variable associated with daily sales. Then, according to the problem given,

$$E(X_i) = 500,$$
and
$$V(X_i) = 10,000.$$

Now let

$$Y_i = 5X_i.$$

Then, Y_i depicts the random variable associated with daily profits in cents. According to Theorem 6.2.1,

$$E(Y_i) = 5E(X_i) = 2500$$
and
$$V(Y_i) = (5)^2 V(X_i) = 250,000.$$

Now let

$$W = Y_1 + \cdots + Y_{100}.$$

Then, W depicts the random variable associated with the sum of daily profits for the next 100 days. According to Theorem 6.4.1,

$$E(W) = E(Y_1) + \cdots + E(Y_{100}) = 250,000$$
and
$$V(W) = V(Y_1) + \cdots + V(Y_{100}) = 25,000,000,$$

and, in turn, $\sigma_w = 5000$. The expected value and the standard deviation are expressed in cents. If we convert them into dollars, we have

$$E(W) = \$2500$$
and
$$\sigma_w = \$50.$$

Thus, the owner of the service station should expect to profit $2500 during the next 100 days. Furthermore, by applying the Tchebycheff inequality, he might, for example, infer that there is at least a .99 probability that the actual profits for this period will be between $2000 and $3000. (Note that $2000 and

$3000 are both 10 standard deviations away from the expected profits for this period.)

EXAMPLE 6.7.3

Assume that a publishing company signs a contract to publish a college textbook. The editor of the company would like to predict the date when the textbook will be ready for distribution. He considers that four activities will have to be completed in the sequence shown by the following diagram. The

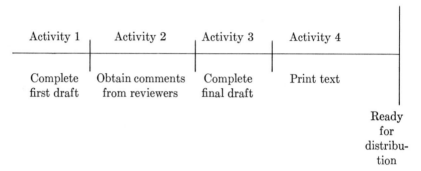

editor knows from experience that the time required for each of these activities is a variable quantity. He has made estimates of time needed for these activities, as shown in Table 6.7.4.

TABLE 6.7.4

Activity	Expected Completion Time	Standard Deviation of Completion Time
1	24 months	4 months
2	2 months	1 month
3	6 months	2 months
4	8 months	2 months

Let X_1, X_2, X_3, and X_4 be the random variables associated with the completion time for each of these activities. Then, if we let

$$Y = X_1 + X_2 + X_3 + X_4,$$

we have a random variable associated with the total length of time required for the textbook to be published. The expected total length, then, may be given as

$$E(Y) = E(X_1) + E(X_2) + E(X_3) + E(X_4)$$
$$= 24 + 2 + 6 + 8 = 40,$$

or 40 months.

Let us next calculate the variance of Y. To do so, we first obtain that

and
$$V(X_1) = 16,$$
$$V(X_2) = 1,$$
$$V(X_3) = 4,$$
$$V(X_4) = 4,$$

and, in turn, that

$$V(Y) = V(X_1) + V(X_2) + V(X_3) + V(X_4)$$
$$= 16 + 1 + 4 + 4 = 25.$$

Also, $\sigma_y = \sqrt{25} = 5$ months.

The editor may, for example, predict that the probability is at least .75 that the textbook will be published between 30 and 50 months. (Note that 30 and 50 months are two standard deviations away from the expected publication date.)

EXERCISES

*30. Suppose that the random variables X, Y, W, and Z represent the annual per-share earnings of two different stocks (in which the price gains or losses are reflected in the earnings), whose joint probability functions are given in the following tables.

x \ w	1	5
1	.25	.25
5	.25	.25

x \ y	1	5
1	.2	.3
5	.3	.2

x \ z	1	5
1	.3	.2
5	.2	.3

You have an option to include in your portfolio:

1. two shares of X
2. one share of X and one share of Y
3. one share of X and one share of W
4. one share of X and one share of Z.

If you were a conservative investor, which portfolio would you choose? Explain your reasons. If you were a speculative investor, which portfolio would you choose?

31. Suppose X, Y, and W are the random variables associated with the earnings of three different companies. Your investigation has revealed that the average earnings, as well as the standard deviation, are the same. On the other hand, the covariance between X and Y is zero, between X and W is positive, and between Y and W is negative. What can you say about the relationship between the earnings of: (a) the companies X and Y, (b) the companies X and W, and (c) the companies Y and W?

118 Functions of Random Variables

32. Assume that the price movements of the stocks for the companies given in Exercise 31 reflect earning movements. You wish to select two stocks for your portfolio.
    (a) If you were a conservative investor, which two stocks would you choose? Why?
    (b) If you were a speculative investor, which two stocks would you choose? Why?

*33. Suppose a service station sells an average of 1000 gallons per day, with a standard deviation of 200 gallons. What is the probability that the total sales for the next 100 days will be between 95,000 and 105,000 gallons, assuming that the daily sales are statistically independent? The service station makes 5 cents per gallon. What is the probability that the profits for the period will be between $4700 and $5300?

*34. Among many activities required in building a house, the three activities given in Table 6.7.5, for example, must be completed in the sequence given. The length of time required to complete each of the activities is a variable quantity. The expected completion time and the standard deviation of completion time for each activity is provided.

What is the probability that the siding will be completed between 22 and 58 days after the foundation has begun?

TABLE 6.7.5

Activity	Expected Completion Time	Standard Deviation of Completion Time
foundation	10 days	4 days
frame	20 days	7 days
siding	10 days	4 days

Chapter 7
Special
Probability
Functions

The probability function of a random variable does not need to have any special mathematical structure; however, if it does possess a special mathematical structure, then our knowledge of that structure can help us with computational problems connected with the evaluation of the underlying random phenomena.

We will, therefore, describe in this chapter several types of random variables whose probability functions possess very useful mathematical structures. The first three of the random variables we will discuss are discrete, and the last three are continuous.

7.1 Bernoulli-Distributed Random Variable

Suppose we need to divide the sample space for a given random phenomenon into two disjointed subsets. Assume that we have defined two events corresponding to these two subsets. These two events are, then, mutually exclusive.

Let the associated random variable assume the numerical value of 1 if one of these two events occurs and 0 if the other event occurs. In this case, 1 and 0 are arbitrarily selected values, but the assignment of these two particular numerical values to the random variable simplifies the evaluation of the underlying random phenomena. The random variable is said to be *Bernoulli dis-*

120   Special Probability Functions

tributed. A formal definition of a *Bernoulli-distributed random variable* is now given.

Definition 7.1.1
Let X be a random variable for which 1 and 0 are the only values that it can assume with the probabilities π and $(1 - \pi)$. The random variable is then said to be *Bernoulli distributed*.

EXAMPLE 7.1.1
Consider the following situation. We are to toss a coin and will win $1 if "heads" appears and nothing if "tails" appears. We can describe the associated random variable by the following probability function:

x	$f(x)$
0	½
1	½
	1.0

Thus, according to Definition 7.1.1, the random variable is Bernoulli distributed.

EXAMPLE 7.1.2
Consider the following situation. We are to toss a die and will win $1 if 6 appears and nothing if any other number appears. The probability function of the random variable may be described as

x	$f(x)$
0	⅚
1	⅙
	1.0

Thus, the random variable is Bernoulli distributed.

EXAMPLE 7.1.3
We know that a manufacturing process produces 10 percent defectives. Assume that we are to sample one item from this process. Then, the probability that it is defective is .1 and that it is nondefective is .9. Instead of specifying the outcome of the sample as defective or nondefective, we can let the outcome be depicted by 1 or 0 according to the following rule of association:

$$\text{Defective} \rightarrow 1$$
$$\text{Nondefective} \rightarrow 0.$$

Then, the probability function for the associated random variable may be described as

x	$f(x)$
0	.9
1	.1
	1.0

Once again we encounter a Bernoulli-distributed random variable.

Theorem 7.1.1
Let X be a Bernoulli-distributed random variable. Then,
$$E(X) = \pi$$
and
$$V(X) = \pi(1 - \pi).$$

EXAMPLE 7.1.4
Consider the Bernoulli-distributed random variable given in Example 7.1.1. Its expected value and variance are calculated as shown in Table 7.1.1.

TABLE 7.1.1

x	$f(x)$	$xf(x)$	$[x - E(X)]^2 f(x)$
0	.5	0	.125
1	.5	.5	.125
	1.0	.5	.250

Thus, $E(X) = .5$ and $V(X) = .25$. Using Theorem 7.1.1, we find that
$$E(X) = \pi = .5$$
and
$$V(X) = \pi(1 - \pi) = .25.$$
Thus, the theorem simplifies the computations needed to find $E(X)$ and $V(X)$.

EXAMPLE 7.1.5
Consider the Bernoulli-distributed random variable given in Example 7.1.3. Its expected value and variance are calculated as shown in Table 7.1.2.

TABLE 7.1.2

x	$f(x)$	$xf(x)$	$[x - E(X)]^2 f(x)$
1	.1	.1	.081
0	.9	0	.009
		.1	.090

Thus, $E(X) = .1$ and $V(X) = .09$. Alternatively, we can calculate the following, according to Theorem 7.1.1:

$$E(X) = \pi = .1$$
and
$$V(X) = \pi(1 - \pi) = (.1)(.9) = .09.$$

EXERCISES

*1. A manufacturing process is known to generate 10 percent defectives. Assume that you are to select one item from the process and decide whether or not it is defective. Describe the outcomes associated with the sampling in terms of a Bernoulli phenomenon. Then, find the probability function for the random variable.

*2. For the random variable given in Exercise 1, find the expected value and the variance.

3. It has been claimed that 80 percent of college students favor a certain political proposition. Assume that you will take a sample of one student and will ask him for his views on the proposition. Describe the outcome of the interview in terms of a Bernoulli phenomenon. Then, find the probability function for the random variable.

4. For the random variable given in Exercise 3, find the expected value and the variance.

5. A sponsor of a television program claims that 20 percent of the viewers watch his program. Assume that you are to select a sample of one viewer and determine whether or not he is actually watching the program. Describe the outcome of the survey in terms of a Bernoulli phenomenon. Then, find the probability function for the random variable.

6. For the random variable given in Exercise 5, find the expected value and the variance.

7.2 Binomially Distributed Random Variable

Frequently, we may repeat an experiment in which each individual trial of that experiment may be described by a Bernoulli-distributed random variable. Each trial generates either 1 or 0 and, in turn, a set of n trials must generate k number of 1's and $n - k$ number of 0's, where $0 \leq k \leq n$. We may then be interested in knowing the probability that a set of n repeated Bernoulli trials will produce exactly k occurrences of the value 1. The following definition and theorem will aid us in solving this problem.

Definition 7.2.1

Let $Y = X_1 + \cdots + X_n$, where $X_1, \ldots, X_n$ are n statistically independent, Bernoulli-distributed random variables which have identical probability functions. Then, Y is said to be a *binomially distributed random variable*.

EXAMPLE 7.2.1

Assume that we are to toss a coin twice. After each toss we will win $1 if "heads" appears and nothing if "tails" appears. We may describe the outcome of each toss by a Bernoulli-distributed random variable, as we have done in Example 7.1.1. Let X_1 and X_2 be the random variables associated with the two tosses. The total amount of winnings may be given, then, by the equation,

$$Y = X_1 + X_2,$$

whose probability function is given as

y	$f(y)$
0	1/4
1	2/4
2	1/4
	1.0

Definition 7.2.1 merely states that Y is a binomially distributed random variable.

Theorem 7.2.1

Let $Y = X_1 + \cdots + X_n$ be a binomially distributed random variable, where π denotes the probability that X_i will assume the value 1. Let $b(k,n,\pi)$ denote the probability that Y will assume the value k. Then,

$$b(k,n,\pi) = C(n,k)\pi^k(1 - \pi)^{n-k},$$

where $C(n,k) = n!/[(k!)(n - k)!]$.

EXAMPLE 7.2.2

We obtained the probability function of Y in Example 7.2.1 by enumerating the sample space for Y. However, Theorem 7.2.1 indicates that this same probability function may be derived without elaborating the sample space for the associated random variable. The alternate method of finding the probability function is illustrated in Table 7.2.1. We find that by using this method the probability function is identical to that given in Example 7.2.1.

EXAMPLE 7.2.3

Let us assume that three coins are to be tossed and we will win $1 for each "heads" obtained. Then, $Y = X_1 + X_2 + X_3$ depicts the random variable associated with the total winnings if X_1, X_2, and X_3 are the random vari-

TABLE 7.2.1

y	Application of Theorem 7.2.1	$f(y)$
0	$\frac{2!}{0!2!}(.5^0)(.5^2)$	$\frac{1}{4}$
1	$\frac{2!}{1!1!}(.5^1)(.5^1)$	$\frac{2}{4}$
2	$\frac{2!}{2!0!}(.5^2)(.5^0)$	$\frac{1}{4}$

TABLE 7.2.2

y	Application of Theorem 7.2.1	$f(y)$
0	$\frac{3!}{0!3!}(.5^0)(.5^3)$	$\frac{1}{8}$
1	$\frac{3!}{1!2!}(.5^1)(.5^2)$	$\frac{3}{8}$
2	$\frac{3!}{2!1!}(.5^2)(.5^1)$	$\frac{3}{8}$
3	$\frac{3!}{3!0!}(.5^3)(.5^0)$	$\frac{1}{8}$

ables associated with the individual coins. According to Theorem 7.2.1, we calculate the probability function of Y as shown in Table 7.2.2.

EXAMPLE 7.2.4

Assume that we are to take a sample of four items from the manufacturing process given in Example 7.1.3. Let Y depict the total number of defectives in the sample. Then, Y can assume any value between 0 and 4. Let $X_1, \ldots, X_4$ be the random variables associated with the outcome of each sampled item. If the process is stable and, at the same time, if there is no pattern in the production of defective and nondefective items, then we may assume that $X_1, \ldots, X_4$ are identically distributed and also statistically independent. This implies that Y is a binomially distributed random variable. The probability function of Y may then be calculated as shown in Table 7.2.3.

The probability function of Y, however, may be obtained from the binomial-probability distribution table given in Appendix B (Table B.6). Examine the block of numbers found at the intersection of the column which

TABLE 7.2.3

y	Application of Theorem 7.2.1	$f(y)$
0	$\frac{4!}{0!4!}(.1^0)(.9^4)$	.6561
1	$\frac{4!}{1!3!}(.1^1)(.9^3)$	.2916
2	$\frac{4!}{2!2!}(.1^2)(.9^2)$	.0486
3	$\frac{4!}{3!1!}(.1^3)(.9^1)$	.0036
4	$\frac{4!}{4!0!}(.1^4)(.9^0)$	.0001+

TABLE 7.2.4

k	$\pi = .1$
0	.6561
1	.2916
2	.0486
3	.0036
4	.0001+

corresponds to $\pi = .10$ and the row which corresponds to $n = 4$. This block of numbers is also reproduced in Table 7.2.4 and represents the probability function of the random variable Y.

At this point, we should look into the principle underlying Theorem 7.2.1. Suppose we ask, "What sequence of 0 and 1 in a sample of 4 will yield the value of Y to be 0?" There is only one sequence which yields 0 as a value of Y: 0, 0, 0, 0. The fact that there is only one such sequence is implied by the equation

$$\frac{4!}{0!4!} = 1.$$

The probability of obtaining this sequence is $.9^4$. Since $.1^0 = 1$, the probability that the value of Y is 0 is calculated as

$$\frac{4!}{0!4!}(.1^0)(.9^4).$$

This calculation has already been shown in Table 7.2.3.

Next, let us ask, "What sequence of 0 and 1 will yield 2 as the value of Y?" There are six such sequences, as illustrated in the following group of numbers.

$$
\begin{array}{cccc}
1, & 1, & 0, & 0 \\
0, & 1, & 1, & 0 \\
0, & 0, & 1, & 1 \\
1, & 0, & 0, & 1 \\
1, & 0, & 1, & 0 \\
0, & 1, & 0, & 1
\end{array}
$$

The fact that there are six such sequences is given by the equation

$$\frac{4!}{2!2!} = 6.$$

The probability of obtaining any one of the six sequences is $(.1)^2(.9)^2$, which means that the probability that the value of Y is 2 may be ascertained by multiplying these two terms; that is,

$$\frac{4!}{2!2!}(.1^2)(.9^2) = .0486.$$

Again, this calculation has already been shown in Table 7.2.3.

Theorem 7.2.2

Let $Y = X_1 + \cdots + X_n$ be a binomially distributed random variable, where π denotes the probability that X_i will assume the value 1. Then,

$$E(Y) = n\pi$$
and
$$V(Y) = n\pi(1 - \pi).$$

The validity of this theorem depends on that of Theorems 6.4.1 and 7.1.1. It might be instructive for the reader at this point to review these two theorems and, then, to decide why Theorem 7.2.2 must be true. We will, however, illustrate the application of Theorem 7.2.2 in the following examples.

EXAMPLE 7.2.5

Consider the random variable associated with Example 7.2.1. The expected value and the variance of the random variable are calculated as shown in Table 7.2.5. Thus, $E(Y) = 1$ and $V(Y) = \frac{1}{2}$. However, according to Theorem 7.2.2, we could have calculated that

$$E(Y) = n\pi = (2)(.5) = 1$$
and
$$V(Y) = n\pi(1 - \pi) = (2)(.5)(.5) = .5,$$

which yields the same results as those just given.

TABLE 7.2.5 Expected Value and Variance of a Random Variable

y	$f(y)$	$yf(y)$	$[y - E(Y)]^2 f(y)$
0	$\frac{1}{4}$	0	$\frac{1}{4}$
1	$\frac{2}{4}$	$\frac{2}{4}$	0
2	$\frac{1}{4}$	$\frac{2}{4}$	$\frac{1}{4}$
	1.0	1	$\frac{1}{2}$

EXAMPLE 7.2.6

Assume that we are to toss 100 coins and will win $1 for each "heads" obtained. Let $Y = X_1 + \cdots + X_{100}$ be the random variable associated with the total winnings. Then, according to Theorem 7.2.2, we have

$$E(Y) = n\pi = (100)(.5) = 50$$

and

$$V(Y) = n\pi(1 - \pi) = (100)(.5)(.5) = 25,$$

which are identical to those we obtained for the same random variable in Example 6.4.1.

EXERCISES

*7. Three coins are tossed. Find the probability function of the random variable associated with the number of "heads" obtained, using Theorem 7.2.1.

8. Four coins are tossed. Find the probability function of the random variable associated with the number of "heads" obtained, using Theorem 7.2.1.

9. Five coins are tossed. Find the probability function of the random variable associated with the number of "heads" obtained, using Theorem 7.2.1.

*10. A manufacturing process is known to generate 10 percent defectives. Suppose you plan to control the process by selecting a random sample of four items. Find the probability function associated with the number of defectives found for your sample.

11. It has been claimed that 20 percent of college students favor a certain political proposition. Assume that you will take a random sample of five students from the college student population. Find the probability function of the random variable associated with the number of students who would favor the proposition in your sample of five, assuming that the claim is true.

12. It is claimed that 40 percent of California voters are Republican. Suppose you are to take a sample of four voters. What is the probability that 50 percent or more of those in your sample will be Republicans, assuming the claim is true?

13. One hundred coins are tossed. Find the expected value, the variance, and the standard deviation of the random variable associated with the total number of "heads" obtained.

128 Special Probability Functions

*14. Refer to the manufacturing process given in Exercise 10 and, assuming that you will take a sample of 100 items, find the expected value, the variance, and the standard deviation of the random variable associated with the total number of defectives in the sample.

*15. Refer to Exercise 14. What can you say about the probability that your sample of 100 items might contain between 4 and 16 defectives?

*16. Refer to Exercise 12. Suppose you are to take a random sample of 2400 voters in California. Find the expected value, the variance, and the standard deviation of the random variable associated with the number of Republicans in the sample.

17. A sponsor of a television program claims that 20 percent of the viewers watch his program. Suppose that 1600 viewers are selected at random from the nation's television audience. Find the expected value, the variance, and the standard deviation for the random variable associated with the number of viewers in the sample who watch his program.

18. Suppose you are a bright young man in charge of the political campaign of a candidate for state-wide office, and your candidate wants to know how the voters generally feel about a certain political proposition. You take a random sample of 2500 voters in the state. Suppose that the voters are equally divided for and against the proposition. Then, what would be the expected value, the variance, and the standard deviation for the random variable associated with the number of voters in the sample who will be for the proposition.

7.3 Poisson-Distributed Random Variable

One of the most useful probability functions is that of the *Poisson-distributed random variable*. We will first describe the structure of the probability function and then explore its usefulness.

Definition 7.3.1
Let X be a random variable, where $k = 0, 1, 2, \ldots$ are the possible values that X can assume with the associated probabilities

$$P(X = k) = \frac{\lambda^k e^{-\lambda}}{k!},$$

where λ is a positive value.* Then, X is said to be a *Poisson-distributed random variable*.

EXAMPLE 7.3.1
Assume that X is a Poisson-distributed random variable with $\lambda = 1$. Then, for example,

---

* In Definition 7.3.1 e is the base for natural logarithm; that is, $e = 2.71828 \ldots$

7.3 Poisson-Distributed Random Variable

$$P(X = 0) = \frac{(1)^0 e^{-1}}{0!} = \frac{1}{e} = \frac{1}{2.71828} = .3679,$$

$$P(X = 1) = \frac{(1)^1 e^{-1}}{1!} = \frac{1}{e} = \frac{1}{2.71828} = .3679,$$

and $\quad P(X = 2) = \frac{(1)^2 e^{-1}}{2!} = \frac{1}{2e} = \frac{1}{5.43656} = .1836.$

The remaining probabilities $P(X = 3), P(X = 4)$, and so on, can be calculated in a similar manner. The complete probability function of X for $\lambda = 1$ is given in Table 7.3.1. Table 7.3.1 indicates that $P(X = 8)$ is 0. On the other hand,

$$P(X = 8) = \frac{(1)^8 e^{-1}}{8!} = \frac{1}{8!e}$$

is not 0. However, since the number is very small, we can, for all practical purposes, assume that it is 0. This is what is implied by the table.

TABLE 7.3.1

x	$f(x)$
0	.3679
1	.3679
2	.1839
3	.0613
4	.0153
5	.0031
6	.0005
7	.0001
8	.0000
	1.0000

If $P(X = 8)$ is so small that we can ignore it, then, we can also ignore $P(X = 9)$, $P(X = 10)$, and so on, since these probabilities are even smaller than $P(X = 8)$. For example, we find that

$$P(X = 8) = \frac{1}{8!e} > \frac{1}{9!e} = P(X = 9).$$

A table of Poisson probability functions for different values of λ are provided in Appendix B. (See Table B.7.) The probability function that we have just calculated may be found in this table under the column headed $\lambda = 1.0$.

One use of the Poisson probability function is reflected by the following theorem.

Theorem 7.3.1

Let X be a random variable associated with the sum of successes from n independently repeated Bernoulli trials. If π is small, the probability function of X approaches that of a Poisson-distributed random variable with $\lambda = n\pi$ as n becomes larger and larger.

Theorem 7.3.1 can be stated in another way. According to Definition 7.2.1, X given in the preceding theorem must be binomially distributed. Therefore, we can propose that, for a given binomially distributed random variable, if π is very small but n is very large, then, the probability function of X may be approximated by that of a Poisson-distributed random variable with $\lambda = n\pi$.

Instead of giving a proof of the theorem, however, we will illustrate its plausibility with the following example.

EXAMPLE 7.3.2

Assume that a process generates an average of 10 percent defectives. Let X be the random variable associated with the number of defectives in a sample of 10 items. We may assume that the drawing of each item from the process constitutes a Bernoulli trial and, furthermore, that the value of X is determined by the sum of 10 independently repeated Bernoulli trials. Then, X is a binomially distributed random variable with $\pi = .1$ and $n = 10$. The probability function of X, obtained from Table B.6 in Appendix B is given in the middle column in Table 7.3.2.

TABLE 7.3.2

x	Binomial Probability ($\pi = .10; n = 10$) $f(x)$	Poisson Probability ($\lambda = 1$) $f(x)$
0	.3487	.3679
1	.3874	.3679
2	.1937	.1839
3	.0574	.0613
4	.0112	.0153
5	.0015	.0031
6	.0001	.0005
7	.0000	.0001
8	.0000	.0000
9	.0000	.0000
10	.0000	.0000
	1.0000	1.0000

7.3 Poisson-Distributed Random Variable

The Poisson approximation for the given binomial probability function may be obtained as follows. Since $\pi = .10$ and $n = 10$, we know that $\lambda = 1$. Next, we obtain the Poisson probability function from the table of Poisson probabilities (Table B.7 in Appendix B). The corresponding Poisson probability function for $\lambda = 1$ is shown in the right-hand column of Table 7.3.2. If the reader compares the two functions, he will find that they are almost identical to each other. This is what Theorem 7.3.1 proposes.

Use of the Poisson probability function is not, however, confined to approximating the binomial probability function. It has been found that occurrences of many empirical phenomena may be considered to be Poisson-distributed random variables. Some well-known examples are:

1. number of telephone disconnections in a given time span
2. number of accidents (such as coal-mine disasters) in a given time span
3. number of spare parts of one type needed for an aircraft in a given time span
4. number of bacteria in a Petri plate
5. number of surface defects of a material
6. number of misprints on a given page of a newspaper.

The reader may note that the first three examples in the preceding list pertain to phenomena which occur in a given time span and the last three examples pertain to phenomena which occur in a given area of space. We now propose:

Theorem 7.3.2

Let X be a random variable associated with a phenomenon which occurs in a given time span or in a given area of space. X is a Poisson-distributed random variable if the following assumptions are satisfied for X.

1. Let h depict the length of the time span or the size of the area of space in which X assumes its value. As h approaches zero, X approaches a Bernoulli-distributed random variable.
2. For a small, positive h there exists μ such that $P(X = 1) = \mu h$.
3. The probability function of X for a given h is independent of the location of the time span or of the area of the space chosen.

We will illustrate the significance of the preceding assumptions in the following examples.

EXAMPLE 7.3.3

Let X depict a random variable associated with the number of disconnections at a telephone exchange in a given time span. We have already pointed out that X is known to be a Poisson-distributed random variable.

We will, therefore, evaluate the implications of the three assumptions listed for the physical process generating the disconnections.

Let h depict the length of time span chosen for X. If h is very large, X may assume any value 0, 1, 2, and so on. However, if h is very small, we can rule out the possibility of more than one disconnection during the time span. Thus, for a very small h, X assumes either 1 or 0 and, in turn, becomes a Bernoulli-distributed random variable. Thus, X satisfies the first assumption of Theorem 7.3.2.

Assume now that, on the average, μ number of disconnections occur during a given unit of time. Then, for a very small value of h it is reasonable to assume that the probability of one disconnection occurring during h length of time is equal to μh. For example, suppose one disconnection occurs every hour, on the average. It would be reasonable to assume that the probability of one disconnection occurring in a minute is $1/60$ and that of one disconnection occurring in any given two minutes is $2/60$. Thus, X satisfies the second assumption of Theorem 7.3.2. Note, however, that the assumption is valid only for a very small h. For example, it does not imply that if h is 2 hours, the probability of one disconnection occurring during the time span is 200 percent.

The implication of the third assumption, using the telephone exchange example, may be stated as follows: The frequency of disconnections is not influenced by the time of the day, by the day of the week, and so on. In other words, if one disconnection occurs, on the average, between 8 A.M. and 9 A.M. every day, then one disconnection occurs, on the average, during any other randomly selected 1-hour time span during the day.

It is unlikely that all of the three assumptions would be strictly satisfied for any telephone exchange. Suppose, however, that they are satisfied reasonably well for an exchange. Then, we can apply the Poisson probability function to the number of disconnections occurring during any given time span.

In Example 7.3.3, we have illustrated the significance of the assumptions given in Theorem 7.3.2 for a Poisson-distributed random variable which occurs in a given time span. We can also illustrate the significance of these assumptions for a Poisson-distributed random variable which occurs in a given area of space.

EXAMPLE 7.3.4

Let X be a random variable associated with the number of surface defects in a given area of a material. We have already pointed out that X is known to be a Poisson-distributed random variable. We will, therefore, examine the implications of the three assumptions listed in Theorem 7.3.2 for this new example.

The first assumption implies that if the area chosen is very small, the likelihood of finding more than one defect in the area should be negligible. The second assumption implies that for a small area the probability of finding a defect is proportional to the size of the area. The third assumption implies that the number of defects found in any area of a given size should not be dependent on the particular location of the area.

It would seem that for most materials what we have just stated are reasonable assumptions. Therefore, we can apply a Poisson probability function to the number of defects found in a given area of such materials.

Although we have described the types of empirical phenomena which may be explained in terms of the Poisson-distributed random variable, we still must illustrate how we can obtain the value of λ for the Poisson probability function. We propose now:

Theorem 7.3.3

Let X be the Poisson-distributed random variable for a given h where it depicts the length of a time span or the size of an area. Then,

$$E(X) = \lambda$$
and
$$V(X) = \lambda.$$

The significance of this theorem is illustrated in the following examples.

EXAMPLE 7.3.5

Assume that two disconnections occur, on the average, per hour at a telephone exchange. Suppose we let X depict the random variable associated with the number of disconnections occurring during any given 1-hour time span at the exchange. Then, by assumption,

$$E(X) = 2$$

and, in turn, $\lambda = 2$. Therefore, the probability function of X is given as

$$P(X = k) = \frac{\lambda^k e^{-\lambda}}{k!} = \frac{2^k e^{-2}}{k!}.$$

For example, the probability of exactly one disconnection occurring during the hour is

$$P(X = 1) = \frac{2^1 e^{-2}}{1!} = .2707.$$

Assume that X depicts the random variable associated with the number of disconnections during h hours and Y depicts the random variable associated with the number of disconnections during t times h hours where t is a positive number. If the average number of disconnections is μh during h hours, then, the average number of disconnections during th hours must be $t\mu h$. Thus, if $\lambda = \mu h$ for X, then $\lambda = t\mu h$ for Y.

EXAMPLE 7.3.6

On the average, two disconnections occur in a telephone exchange during any hour. We wish to find the probability of exactly one disconnection occurring in any given 2-hour period.

Let Y depict the random variable associated with the number of disconnections in any given 2-hour period. Then, since the average number of disconnections per hour is 2, the average number of disconnections per 2 hours must be 4. Thus, $\lambda = 4$ for Y, and, in turn,

$$P(Y = 1) = \frac{4^1 e^{-4}}{1} = .0733.$$

EXERCISES

*19. A production process is known to generate one defective per thousand, on the average. You are to select a random sample of 100 items and count the number of defectives in your sample. Let X be the random variable associated with the number of defectives. Describe the probability function of X, using the Poisson probability function.

*20. The production manager for the process given in Exercise 19 claims that the process is producing not more than one defective per thousand. Suppose you are the quality-control manager. You have selected a sample of 100 items from the process and have found that the sample contains four defectives. What conclusion might you draw with regard to the claim of the production manager? Answer the question by applying the Poisson probability function.

21. A telephone switchboard receives 30 calls, on the average, per minute. Describe the probability function associated with the number of incoming calls for any given second, assuming that the incoming calls are Poisson distributed.

22. Refer to Exercise 21. What is the probability that the switchboard will receive: (a) more than one call in any given second, (b) more than two calls in any given 2 seconds, and (c) more than three calls in any given 3 seconds?

*23. Suppose you are an automobile salesman, who has been selling an average of one car per day. Assume that your sales can be explained by the Poisson probability function. Then, what is the probability that: (a) you will sell more than one car in any given day, and (b) you will sell more than 7 cars during any 7 days?

24. A casualty insurance company finds that an average of one house among 1000 houses insured by the company is damaged by fire in any given year. The company has insured 5000 houses. Describe the probability function associated with the number of claims that it might have to pay due to fire damages in a given year, assuming that it is Poisson distributed.

25. Refer to Exercise 24 and find the probability that the company might have to pay more than ten claims within the year.

*26. An instructor has been flunking two students per course. Suppose the instructor's behavior, with respect to flunking students, can be explained by the Poisson probability function. What is the probability that in a given course: (a) he might not flunk anyone, (b) he might flunk two students, and (c) he might flunk more than five students?

27. Suppose you are the manager of a store, which has been selling an average of one unit of a certain item per day. Suppose you wish to accept not more than a 5-percent risk of not having the item when there is a demand for it. Then, how many units of that item should the store have in stock at the beginning of each day?

28. Refer to Exercise 27 and assume that the item in question may be ordered once a week. Suppose you wish to accept not more than a 5-percent risk of not having the item when there is a demand for it in any given week. Then, how many units of the item should be in the store at the beginning of each week, assuming that there are 7 working days in a week?

29. On the average, two misprints per page are found in a certain newspaper. Assume that the misprints can be defined by the Poisson probability function. What is the probability that a randomly selected page of the newspaper will contain more than five misprints?

*30. The manufacturer of a certain wallpaper finds that its manufacturing process produces one defective spot per 100 feet of wallpaper. Each roll of wallpaper is 20 feet long. Suppose one roll of wallpaper is selected at random. What is the probability that it contains a defective spot, assuming that the number of defective spots can be defined by the Poisson probability function?

7.4 Uniformly Distributed Random Variable

Among the random variables possessing continuous probability functions, the one possessing the simplest structure is the *uniformly distributed random variable*.

Definition 7.4.1

Let X be a continuous random variable which can assume a value between a and b. The random variable is said to be *uniformly distributed* if its density is constant between a and b.

EXAMPLE 7.4.1

Let X be the random variable associated with spinning the wheel shown in Figure 7.4.1. According to our discussion in Chapter 4, we observe that the density function of X is given by

$$f(x) = .1 \quad 0 \leq x \leq 10.$$

Thus, the density is constant between 0 and 10, and the random variable is uniformly distributed.

The reader may now note that we have already considered a similar example in Chapter 4. In fact, all the continuous random variables discussed in Chapter 4 are uniformly distributed random variables. We will not, therefore, repeat what we have already discussed in Chapter 4 pertaining to the probability function. We have yet to point out:

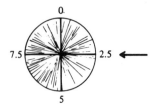

FIGURE 7.4.1   Spinning wheel.

Theorem 7.4.1
Let X be a uniformly distributed, continuous random variable which can assume a value between a and b where $a < b$. Then, its density function is

$$f(x) = \frac{1}{b-a} \quad \text{for } a \leq x \leq b.$$

EXAMPLE 7.4.2
Let us return to Example 7.4.1. We observe that $a = 0$ and $b = 10$. Thus,

$$f(x) = \frac{1}{b-a} = \frac{1}{10-0} = .1 \quad \text{for } 0 \leq x \leq 10.$$

We note that this density function is the same as that given in Example 7.4.1.

Theorem 7.4.2
Let X be a uniformly distributed random variable which can assume a value between a and b. Then,

$$E(X) = \frac{a+b}{2}$$

and

$$V(X) = \frac{(b-a)^2}{12}.$$

EXAMPLE 7.4.3
Let us return to Example 7.4.2. For the random variable we calculate that

$$E(X) = \frac{a+b}{2} = \frac{0+10}{2} = 5$$

and

$$V(X) = \frac{(b-a)^2}{12} = \frac{(10-0)^2}{12} = \frac{25}{3}.$$

Note, however, that we could have calculated $E(X)$ and $V(X)$ by using calculus; that is,

$$E(X) = \int_0^{10} xf(x)\, dx = \int_0^{10} .1x\, dx = 5,$$

and
$$V(X) = \int_0^{10} [x - E(X)]^2 f(x)\, dx$$
$$= \int_0^{10} (x - 5)^2 .1\, dx = \frac{25}{3}.$$

EXERCISES

31. Let X be a random variable associated with spinning the wheel shown in Figure 7.4.2. Find the density function of X.

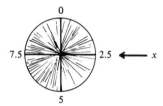

FIGURE 7.4.2 Spinning wheel.

32. Using the density function of the random variable given in Exercise 31, find $E(X)$ and $V(X)$.

33. Suppose X is uniformly distributed between 5 and 25. Find the density function of X and, in turn, ascertain $E(X)$ and $V(X)$.

*34. Assume that the daily demand for a certain item is uniformly distributed between 0 and 100. Find the density function for the demand and, in turn, ascertain the expected value and the variance of the demand.

*35. Suppose you have invited a guest for dinner, and the guest has indicated that he will arrive between 5 P.M. and 6 P.M. Let Y be the random variable associated with the arrival time of the guest, and, furthermore, assume that it is uniformly distributed. Derive the density function of Y and, then, ascertain $E(Y)$ and $V(Y)$.

7.5 Exponentially Distributed Random Variable

Another continuous random variable which has a simple structure and is very useful is the *exponentially distributed random variable*.

Definition 7.5.1

Let X be a continuous random variable whose density function is given as

138   Special Probability Functions

and
$$f(x) = \lambda e^{-\lambda x} \quad \text{for } 0 \leq x < \infty$$
$$f(x) = 0 \quad \text{for all other } x.$$

The random variable is said to be *exponentially distributed*.

The shape of the density function of the exponentially distributed random variable is illustrated in Figure 7.5.1.

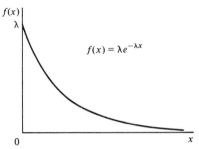

FIGURE 7.5.1   Shape of an exponential density function.

Given an exponential density function, we can ascertain $P(a \leq X \leq b)$, where $0 < a < b$:

$$P(a \leq X \leq b) = \int_a^b \lambda e^{-\lambda x} \, dx,$$

according to Theorem 4.5.1. However, the calculations yielding $P(a \leq x \leq b)$ are simplified considerably once we obtain the cumulative probability function of X.

Theorem 7.5.1
Let $F(k) = P(0 \leq X \leq k)$ for the exponentially distributed random variable. Then,

$$F(k) = 1 - e^{-\lambda k}.$$

The proof for the preceding theorem is straightforward and understandable to the reader who is familiar with calculus. He will find that

$$F(k) = \int_0^k \lambda e^{-\lambda x} \, dx = 1 - e^{-\lambda x}.$$

We will now illustrate some applications of this theorem.

EXAMPLE 7.5.1
Assume that $\lambda = 1$, so that the exponential density function is

$$f(x) = e^{-x} \quad 0 \leq x < \infty$$
$$f(x) = 0 \quad \text{all other } x.$$

Suppose we wish to ascertain $P(0 \leq X \leq 2)$. Then, according to Theorem 7.5.1,

$$F(2) = 1 - e^{-2} = 1 - .1353 = .8647.$$

EXAMPLE 7.5.2

Assume that we are given the following exponential density function:

$$f(x) = 2e^{-2x} \quad 0 \leq x < \infty$$
$$f(x) = 0 \quad \text{all other } x.$$

Assume now that we wish to ascertain $P(1 \leq x \leq 3)$. We have pointed out by Theorem 4.5.2 that, given a continuous random variable,

$$P(k_1 \leq X \leq k_2) = F(k_2) - F(k_1).$$

Therefore, we ascertain first that

$$F(1) = 1 - e^{-2x} = 1 - e^{-2} = .8647$$
and
$$F(3) = 1 - e^{-2x} = 1 - e^{-6} = .9975$$

and, in turn, that

$$P(1 \leq X \leq 3) = F(3) - F(1)$$
$$= .9975 - .8647 = .1328.$$

Theorem 7.5.2

Let X be an exponentially distributed random variable whose density function is

$$f(x) = \lambda e^{-\lambda x} \quad 0 \leq x < \infty$$
$$f(x) = 0 \quad \text{all other } x.$$

Then,

$$E(X) = \frac{1}{\lambda}$$
and
$$V(X) = \frac{1}{\lambda^2}.$$

We will now explore some practical applications of exponential probability functions, but to do so, we first propose:

Theorem 7.5.3

Let X be a random variable associated with the number of occurrences of certain events during a given time span. Then, we can define a related random variable Y, which is associated with the elapsed time between any two successive occurrences of the events. If X is Poisson distributed with a probability function of

140   Special Probability Functions

$$P(X = k) = \frac{\lambda^k e^{-\lambda}}{k!} \qquad k = 0, 1, 2, \ldots,$$

then Y is exponentially distributed with a probability function of

$$f(y) = \lambda e^{-\lambda y} \qquad 0 \le y < \infty$$
$$f(y) = 0 \qquad \text{all other } y.$$

Instead of proving the preceding theorem, we will illustrate its meaning by the following examples.

EXAMPLE 7.5.3

Assume that X is a random variable associated with the number of disconnections at a telephone exchange. We have already pointed out that X is a Poisson-distributed random variable. Assume further that $\lambda = 2$ per hour, which implies that, on the average, 2 disconnections occur in an hour. Now let Y depict the length of elapsed time between any 2 successive disconnections. Since 2 disconnections occur per hour, on the average, it must be apparent that the average elapsed time between the successive disconnections must be 30 minutes. On the other hand, the actual duration of the elapsed time for any 2 successive disconnections is likely to be a variable quantity. Theorem 7.5.3 indicates that Y is an exponentially distributed random variable with a mean of 30 minutes or .5 hours. Thus, the probability function of X is

$$P(X = k) = \frac{\lambda^k e^{-\lambda}}{k!} = \frac{2^k e^{-2}}{k!}$$

and of Y is

$$f(y) = \lambda e^{-\lambda y} = 2e^{-2y} \qquad 0 \le y < \infty$$
$$f(y) = 0 \qquad \text{all other } y.$$

We can now make the following probability analysis. Assume that a disconnection has just occurred at a telephone exchange. What is the probability that another disconnection will occur within the next 2 hours, for example? This question may be answered by calculating

$$P(0 \le Y \le 2) = \int_0^2 2e^{-2y}\, dy$$

and, in turn,

$$F(2) = 1 - e^{-4} = .982.$$

EXAMPLE 7.5.4

Assume that a computer breaks down twice a month, on the average. If we assume that the number of breakdowns in a given time span is Poisson distributed, then we can equally assume that the elapsed time between any two successive breakdowns is exponentially distributed.

7.5 Exponentially Distributed Random Variable

Let Y depict the random variable associated with the elapsed time. Then, the density function of Y is given as

$$f(y) = 2e^{-2y} \quad 0 \leq y < \infty$$
$$f(y) = 0 \quad \text{all other } y.$$

Assume now that the computer has just been repaired. What is the probability that it will break down again, for example, within a month? The answer to this question may be obtained by calculating

$$P(0 \leq Y \leq 1) = \int_0^1 2e^{-2y}\, dy$$

and, in turn,

$$F(1) = 1 - e^{-2} = .865.$$

The probability that the computer will last longer than 1 month without breaking down must be .135.

In the preceding examples, the exponential probability functions illustrated are explicitly related to a Poisson-distributed random variable. It has been found, however, that the exponential probability function may be applied to the time variables which are not explicitly related to a Poisson-distributed random variable. Some of these time variables are: (1) failure-time of electronic components, and (2) time needed to service a customer at a service counter. We will illustrate how to derive a density function for such examples.

EXAMPLE 7.5.5

Assume that the failure-time of a certain type of television picture tube produced by a company is exponentially distributed. The company wishes to find the density function pertaining to the failure-time.

Assume that the company has found from its past experience that the average life of the picture tube is 2 years. Let Y be the random variable associated with the failure-time, whose density function is given as

$$f(y) = \lambda e^{-\lambda y} \quad 0 \leq y < \infty$$
$$f(y) = 0 \quad \text{all other } y.$$

Then, $E(Y) = 2$ where Y is measured in terms of years. According to Theorem 7.5.2, however,

$$E(Y) = \frac{1}{\lambda}.$$

This means that

$$\lambda = \frac{1}{E(Y)} = \frac{1}{2} = .5,$$

and, in turn, that the density function of Y is

$$f(y) = .5e^{-.5y} \quad 0 \le y < \infty$$
$$f(y) = 0 \quad \text{all other } y.$$

Assume now that the manufacturer gives a warranty that any picture tube which fails within a year will be replaced free of charge. What is the probability that a picture tube sold will have to be replaced free of charge? This probability is equal to that of the picture tube failing within a year. Thus,

$$P(0 \le Y \le 1) = \int_0^1 .5e^{-.5y}\, dy$$

or
$$P(1) = 1 - e^{-.5} = .393.$$

The significance of the preceding analysis is that, if the manufacturer does give such a warranty, he will have to replace approximately 40 picture tubes out of every 100 sold.

EXERCISES

36. Assume that the number of airplanes arriving at the landing area of an airport is Poisson distributed with a mean of ten per hour. Let X be the random variable associated with the elapsed time between any two successive arrivals. Derive the probability density function of X.

37. Refer to Exercise 36 and assume that a plane has just arrived. What is the probability that another plane will arrive: (a) within the next 6 minutes, and (b) within the next 30 minutes?

*38. Suppose you are a real-estate salesman, who believes that the number of houses sold during a given time span obeys the Poisson probability law. On the average, you have been selling one house per week. Suppose you have sold one house today. What is the probability that you will sell another house: (a) within the next 7 days, and (b) within the next 2 weeks?

39. Suppose a certain equipment breaks down once per month, on the average. Assume that the number of breakdowns of the equipment during a given time span obeys the Poisson probability law. Suppose that the equipment has just been repaired. What is the probability that the equipment will have to be repaired again within the next 3 months?

*40. Assume that the failure-time of television picture tubes produced by a company is exponentially distributed and that the average life of each picture tube is 30 months. Suppose the company gives a 6-month warranty so that any tube not lasting 6 months is replaced free of charge. What proportion of tubes sold by the company will have to be replaced free of charge?

7.6 Normally Distributed Random Variable

Among the random variables possessing continuous probability functions, the most important one is that which is *normally distributed*. We will consider such a random variable in this section.

Definition 7.6.1

A random variable is said to be *normally distributed* if its density function is given as

$$f(x) = \frac{1}{\sqrt{2\pi}\sigma} e^{-1/2[(x-\mu)/\sigma]^2} \qquad -\infty < x < \infty.$$

In this definition μ and σ denote the expected value and the standard deviation of X, respectively.

We will illustrate in Figure 7.6.1 the density function of a normally distributed random variable for several different values of μ and σ. First, however, two very important properties which hold for every normally distributed random variable should be pointed out. (1) The height of the density function at $-k$ is equal to that at k where k is an arbitrary, non-negative real

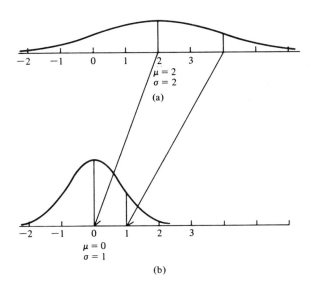

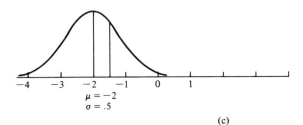

FIGURE 7.6.1 Density function of a normally distributed random variable for different values of μ and σ.

number. The validity of this proposition will be apparent when we examine the density function in Figure 7.6.1. One significance of this proposition is that if we wish to find the area under the curve for a normal density function between μ and $\mu - k$, then, we can obtain it by finding the area between μ and $\mu + k$. (2) Given any two normal density functions, the area under the density functions between μ and $\mu + k\sigma$ are the same for the two functions. This means, for example, that the shaded areas in Figure 7.6.1 must be the same, since each of them consists of the area between μ and μ plus one standard deviation. According to the preceding statement, the areas between μ and $\mu - k\sigma$ must also be the same. The significance of this proposition is that if we wish to find the area under a normal density function between any two arbitrary values, for example, k_1 and k_2, we can do so by finding the corresponding area for the curve, such as that given in (b) of Figure 7.6.1, after making a proper transformation of k_1 and k_2 to the axis for (b).

Examining (b) of this figure, we find that $\mu = 0$ and $\sigma = 1$. The random variable depicted by (b) must play an important role, and we offer the following definition for it.

Definition 7.6.2

Let X be a normally distributed random variable. If $E(X) = 0$ and $V(X) = 1$, the random variable is called a *standard normal random variable*, and its density function is given as

$$f(x) = \frac{1}{\sqrt{2\pi}} e^{-x^2/2}.$$

We will denote the standard normal random variable by Z instead of X.

Suppose now that we wish to find the area under the standard normal density function between two arbitrary points a and b, where $a < b$. We can find the area in question by resorting to calculus and by integrating

$$\int_a^b \frac{1}{\sqrt{2\pi}} e^{-z^2/2} \, dz.$$

However, for all practical purposes, we do not have to resort to this technique. Table B.1 in Appendix B provides the area between 0 and various positive values of Z. For example, we can find from this table that the area under the curve between

$$0 \text{ and } 1.00 = .3413,$$
$$0 \text{ and } 1.50 = .4332,$$
$$0 \text{ and } 1.64 = .4495,$$
$$0 \text{ and } 1.96 = .4750,$$
and
$$0 \text{ and } 2.00 = .4772.$$

The preceding figures, for example, imply that

7.6 Normally Distributed Random Variable

and
$$P(0 \leq Z \leq 1) = .3413$$
$$P(0 \leq Z \leq 2) = .4772.$$

EXAMPLE 7.6.1

Let Z be the standard normal random variable. Assume that we are interested in finding: (a) $P(-1 \leq Z \leq 2)$ and (b) $P(1 \leq Z \leq 2)$. The probabilities sought for (a) and (b) are illustrated in Figure 7.6.2. The two probabilities are calculated as

$$P(-1 \leq Z \leq 2) = P(-1 \leq Z \leq 0) + P(0 \leq Z \leq 2)$$
$$= .3413 + .4772 = .8185,$$

and
$$P(1 \leq Z \leq 2) = P(0 \leq Z \leq 2) - P(0 \leq Z \leq 1)$$
$$= .4772 - .3413 = .1359.$$

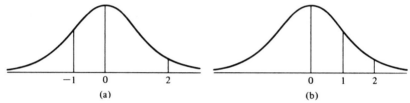

FIGURE 7.6.2 Probabilities for (a) $P(-1 \leq Z \leq 2)$ and (b) $P(1 \leq Z \leq 2)$.

EXAMPLE 7.6.2

Let X be a normally distributed random variable where $\mu = 200$ and $\sigma = 50$. Assume that we wish to find $P(200 \leq X \leq 275)$. By dividing $(275 - 200)/50$, we find that 275 is 1.5 standard deviations away from the expected value of X, which means that 200 and 275 on the X-scale correspond to 0 and 1.5, respectively, on the Z scale. Thus,

$$P(200 \leq X \leq 275) = P(0 \leq Z \leq 1.5).$$

Since $P(0 \leq Z \leq 1.5) = .4332$, we conclude that $P(200 \leq X \leq 275) = .4332$.

EXAMPLE 7.6.3

Let X be a random variable associated with the daily sales of gasoline for a service station. Let us assume that X is normally distributed and that $\mu = 1000$ gallons and $\sigma = 200$ gallons. Suppose, for some reason, we wish to find the probability that the sales will exceed 1500 gallons. In notation, this probability becomes $P(1500 \leq X)$, which, in turn, is equal to $P(2.5 \leq Z)$, since 1500 is 2.5 standard deviations away from the expected value of X. Since

$$P(2.5 \leq Z) = P(0 \leq Z) - P(2.5 \leq Z)$$
$$= .5000 - .4938 = .0062,$$

we conclude that $P(1500 \leq X) = .0062$.

Theorem 7.6.1

Let X be a normally distributed random variable. If we let

$$Y = a + bX,$$

where a and b are real number constants and b is not zero, then Y is also a normally distributed random variable.

The theorem proposes that a linear function of a normally distributed random variable is also a normally distributed random variable.

EXAMPLE 7.6.4

Assume that X is a normally distributed random variable and that $E(X) = 2$, $V(X) = 4$, and, in turn, $\sigma_x = 2$. The shape of the probability function of X is depicted in (a) of Figure 7.6.1. Suppose we let

$$Y = -1 + .5X.$$

Then,

$$E(Y) = -1 + .5E(X) = -1 + (.5)(2) = 0$$

and

$$V(Y) = (.5)^2 V(X) = (.25)(4) = 1,$$

and, in turn, $\sigma_y = 1$. Theorem 7.6.1 proposes that Y is also normally distributed. Thus, the shape of the probability function of Y is depicted in (b) of Figure 7.6.1. Suppose we let

$$W = -2 + .5Y.$$

Then,

$$E(W) = -2 + .5E(Y) = -2 + (.5)(0) = -2$$
$$V(W) = (.5)^2 V(Y) = (.25)(1) = .25,$$

and, in turn, $\sigma_w = .5$. Theorem 7.6.1 proposes that W is also a normally distributed random variable. Thus, the probability function of W is depicted in (c) of Figure 7.6.1.

EXAMPLE 7.6.5

Let X be a random variable associated with the number of units of a certain item sold during a day. Assume that X is normally distributed and that $E(X) = 200$, $V(X) = 2500$, and $\sigma_x = 50$.

Assume that the cost per unit of this item is $10. Suppose we ask the following question: "What is the probability that the total revenue from the sales of this item in 1 day will exceed $1500?" We may proceed to answer this in the following way. The revenue will exceed $1500 if the sales for the day exceeds 150 units. Since

$$P(X \geq 150) = .8413,$$

the probability that the revenue will exceed $1500 is also .8413.

7.6 Normally Distributed Random Variable

We can analyze the question in another way. Since the unit price for the item is $10, if we let

$$Y = 10X,$$

then Y represents the random variable associated with the total revenue from the sales of this item. According to Theorem 7.6.1, Y is also a normally distributed random variable, and

and
$$E(Y) = 10E(X) = 10 \times 200 = 2000$$
$$V(Y) = 10^2 V(X) = 100 \times 2500 = 250{,}000$$

and, in turn, $\sigma_y = 500$. Thus,

$$P(Y \geq 1500) = .8413.$$

Theorem 7.6.2
Let $X_1, \ldots, X_n$ be statistically independent, normally distributed random variables. If we let

$$Y = X_1 + \cdots + X_n,$$

then Y is also a normally distributed random variable.

This theorem proposes that the sum of statistically independent, normally distributed random variables is also a normally distributed random variable.

EXAMPLE 7.6.6
Let X_1 and X_2 be the random variables associated with the daily sales in dollars for two different items. Assume that X_1 and X_2 are statistically independent, normally distributed, and that

and
$$E(X_1) = 300 \quad V(X_1) = 900$$
$$E(X_2) = 400 \quad V(X_2) = 1600.$$

If we let

$$Y = X_1 + X_2,$$

then Y depicts the random variable associated with the total daily sales in dollars for both items. Further,

and
$$E(Y) = E(X_1) + E(X_2) = 700$$
$$V(Y) = V(X_1) + V(X_2) = 2500,$$

and, in turn, $\sigma_y = 50$. Theorem 7.6.2 proposes that Y is also a normally distributed random variable.

Thus, for example, we have

$$P(Y \geq 600) = .9772,$$

which indicates that the probability that the total daily sales for both items will exceed $600 is .9772.

EXERCISES

41. Assume a service station has been selling an average of 800 gallons of gasoline per day with a standard deviation of 200 gallons. Suppose we also know that the daily sales are normally distributed. What is the probability that the station will sell in any day: (a) between 400 and 1200 gallons, (b) between 600 and 1200 gallons, (c) between 1000 and 1200 gallons, and (d) more than 1200 gallons?

*42. Nylon cords are used in manufacturing a certain item. The breaking strength of the cords should be at least 30 pounds. Suppose now that you have tested the cords supplied by two different firms and have found that the cords supplied by A have a mean breaking strength of 50 pounds with a standard deviation of 20 pounds, and those supplied by B have a mean breaking strength of 45 pounds with a standard deviation of 10 pounds. The breaking strengths of the cords supplied by A and B are normally distributed. Which supplier would you choose? Explain why.

*43. Suppose the college entrance examination board scores are normally distributed with a mean of 500 and a standard deviation of 100. Suppose also that you are the admissions director of a university and you wish to grant admission only to those students in the top 10 percent. What should be the cut-off point below which admission will not be granted?

44. Refer to Exercise 41 and suppose the station manager wishes to have enough gasoline at the beginning of each day to satisfy the estimated demand for the day. However, he does not wish to have more than is needed. In this respect, he is willing to accept 1-percent risk of not having enough gasoline for the day. Then, how many gallons should be stocked at the beginning of each day?

45. A survey indicates that the average waistline of adult males in the United States is 36 inches with a standard deviation of 2 inches. It also indicates that the waistlines are normally distributed. If you are the owner of a clothing-manufacturing firm, what proportion of adult trousers produced should have waist sizes: (a) between 34 and 38 inches, (b) between 32 and 40 inches, and (c) between 40 and 44 inches?

46. Refer to Exercise 41 and assume that the price of gasoline is 40 cents per gallon. What is the probability that the daily revenue for the service station will: (a) exceed $200, (b) be between $200 and $400, and (c) be more than $480?

*47. The following activities for building a house must be completed in the sequence given in Table 7.6.1. The length of time required to complete each of the activities is a variable quantity and is normally distributed with an expected value and standard deviation as indicated in Table 7.6.1.

What is the probability that the siding will be completed (a) less than 22 days after the date that the foundation starts, (b) more than 49 days after the date that the foundation starts?

TABLE 7.6.1

	Expected Completion Time	Standard Deviation
1. foundation	10 days	4 days
2. frame	20 days	7 days
3. siding	10 days	4 days

Chapter 8
Central Limit Theorem

We have pointed out that if $X_1, \ldots, X_n$ are statistically independent, normally distributed random variables, then,

$$Y = X_1 + \cdots + X_n$$

is also a normally distributed random variable. Is Y also a normally distributed random variable if $X_1, \ldots, X_n$ are statistically independent but not normally distributed? The answer is "no."

However, there is a theorem, called the *central limit theorem*, which proposes that we can approximate the probability function of Y with a normal probability function if a number of assumptions are satisfied for the set of random variables $X_1, \ldots, X_n$.

The *central limit theorem* is probably one of the most useful statistical concepts. The theorem was originally proved in 1773 by deMoivre for the sum of Bernoulli-distributed random variables when $\pi = \frac{1}{2}$ and, subsequently, by Laplace in 1812 for the sum of Bernoulli-distributed random variables with arbitrary values of π. The extension of the theorem to the sum of any independent variables did not exist until the twentieth century.

8.1 Central Limit Theorem: Bernoulli Case

We will present in this section the *central limit theorem* as it applies to the sum of Bernoulli-distributed random variables, and we will extend the theorem to a more general case in the following section.

Theorem 8.1.1

Let $X_1, \ldots, X_n$ be a set of Bernoulli-distributed random variables which possess the following characteristics:

1. $X_1, \ldots, X_n$ are statistically independent.
2. $X_1, \ldots, X_n$ have identical probability functions.

Now let $Y = X_1 + \cdots + X_n$. The probability function of Y approaches that of a normal probability function as n becomes larger and larger.

Analysis will reveal that Y in the preceding theorem must be a binomially distributed random variable. Recall Definition 7.2.1 in Chapter 7, which states that the probability function of a binomially distributed random variable is determined by π and n. Thus, the preceding theorem may be restated as:

Theorem 8.1.2

Let Y be a binomially distributed random variable with a given π and n. The probability function of Y is approximately normal if n is relatively large.

The implications of this theorem may be illustrated by Figure 8.1.1. Each of the diagrams in this figure depicts the probability function of a binomially distributed random variable, where π equals $\frac{1}{2}$ and n is a specified value. The four diagrams indicate that as n becomes larger, the probability function of the binomially distributed random variable becomes smoother and apparently approaches a continuous probability function. Theorem 8.1.2 implies that the continuous probability function, to which the probability function of a binomially distributed random variable approaches, is a normal probability function. A normal probability function is superimposed on the binomial probability function for $n = 100$, as shown in (d) of Figure 8.1.1. We observe that the shapes of the two probability functions are quite similar to each other.

EXAMPLE 8.1.1

Assume that we are to toss 100 coins and win $1 for each "heads" obtained. Suppose we wish to find the probability that the total winnings will be between $40 and $60. If we let X_i represent the random variables associated with the individual coins, then the random variable associated with the total winnings is given as $Y = X_1 + \cdots + X_{100}$. Y, then, must be a binomially distributed random variable if we assume that the tossing of each coin is fair. This also means that $E(X_i) = .5$ and $V(X_i) = .25$ and, in turn, that $E(Y) = 50$, $V(Y) = 25$, and $\sigma_y = 5$.

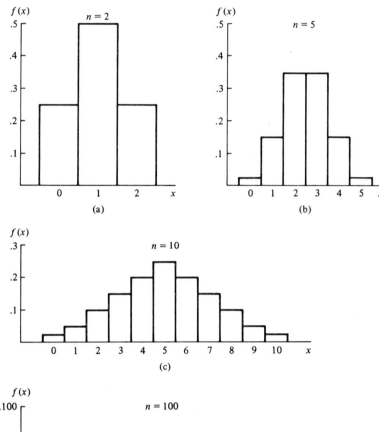

FIGURE 8.1.1 Binomial probability functions for $\pi = .5$.

According to the central limit theorem, we know that Y is approximately normally distributed. We observe that 40 and 60 are two standard deviations away from the mean value of 50, as illustrated in Figure 8.1.2. The shaded area under the normal curve is .9544. Thus, we estimate that the probability that the total amount of winnings will be between $40 and $60 is approximately .9544.

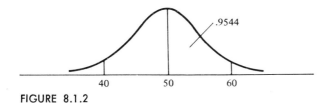

FIGURE 8.1.2

If we examine the binomial probability function, we find that the exact probability that the amount of winnings will be between $40 and $60 is .936. Thus, the difference between the estimated figure and the actual figure is very small.

EXAMPLE 8.1.2

Assume that a manufacturing process is producing 20 percent defectives, on the average. We are to take a sample of 100 items from the process.

Let X_i depict the random variable associated with the result of the ith sample where X_i assumes the value 1 if the item selected is defective and 0 if it is not. Then, $Y = X_1 + \cdots + X_{100}$ is a binomially distributed random variable, if we assume that the process is stable and that there is no pattern in the production of defective and nondefective items. Since $\pi = .2$ by assumption, $E(Y) = n\pi = 20$ and $V(Y) = n\pi(1 - \pi) = 16$. Thus, $\sigma_y = 4$.

Suppose now we wish to find the probability that the sample might contain between 12 and 28 defectives. We observe that 12 and 28 are two standard deviations away from the mean value of 20. Thus, we estimate the probability to be approximately .9544. Our examination of the binomial probability function for $\pi = .2$ and $n = 100$ reveals that the exact probability measure is .968. Again, the difference between the estimated figure and the actual figure is quite small.

EXAMPLE 8.1.3

Refer to Example 8.1.2 and suppose we wish to find the probability that the sample may contain more than 28 defectives. This probability measure is approximately .0228 if we again use the normal probability function to calculate, as shown in Figure 8.1.3. The actual probability obtained from the corresponding binomial probability function is .021.

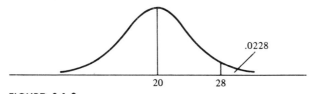

FIGURE 8.1.3

We might ask, "How large does n have to be before the probability function of a binomially distributed random variable becomes approximately normal?" There really is no definite answer to this question. Given the value of π and an increasing value of n, the resulting probability function will be closer to that of a normal probability function, but such a comparison is generally not possible for different sets of π and n. It may be stated, however, that given the value of n, the resulting probability function must be closer to that of a normal probability function as π approaches .5. This means that if π is very close to either 0 or 1, only a very large n may make the resulting probability function approximately normal.

Professor W. G. Cochran has provided the following table (Table 8.1.1), which we can use as a guide in deciding whether or not to use a normal probability function as an approximation for a binomial probability function for various assumed values of π. We feel, however, that a normal approximation may be justified for a smaller n than is suggested by this table.

TABLE 8.1.1

If the Assumed Value of π Is	Use Normal Approximation Only if n Is At Least Equal to
.5	30
.4 or .6	50
.3 or .7	80
.2 or .8	200
.1 or .9	600
.05 or .95	1400

SOURCE: Cochran, W. G., *Sampling Techniques*. New York, John Wiley & Sons, Inc., 1953.

Each of the diagrams in Figure 8.1.4 depicts the probability function of a binomially distributed random variable, where $\pi = .2$ and n is a specified value. The diagram depicting $n = 40$, for example, indicates that when $\pi = .2$, the probability function of a binomially distributed random variable appears to be approximately normal, even though n is only 40.

EXERCISES

1. Suppose you toss 100 coins and will win $1 for each "heads" obtained. Approximate the probability that your total winnings will be (a) between $40 and $60, (b) between $30 and $70, (c) between $35 and $70, and (d) more than $70.

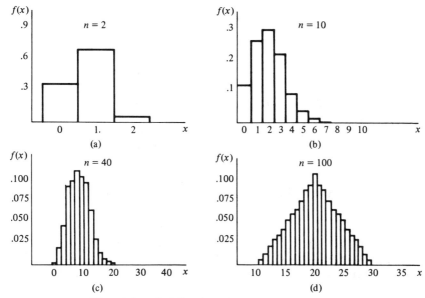

FIGURE 8.1.4 Binomial probability functions for $\pi = .2$.

*2. A manufacturing process is known to produce 10 percent defectives. Suppose you select 100 sample items from the process at random. Approximate the probability that your sample will contain (a) between 4 and 16 defectives, (b) more than 16 defectives, and (c) between 7 and 13 defectives.

3. A producer of a television program claims that exactly 20 percent of the television audience watches his program. You have randomly sampled 10,000 television sets. Approximate the probability that (a) between 1900 and 2100 sets would be turned to that program, and (b) between 1960 and 2060 sets would be turned to that program, assuming the claim of the producer is true.

*4. Assume that 50 percent of eligible California voters are registered Democrats. You have taken a random sample of 10,000 eligible voters in California. Approximate the probability that your sample of 10,000 voters will contain (a) between 4500 and 5500 registered Democrats, (b) between 4900 and 5100 registered Democrats, (c) between 4950 and 5050 registered Democrats, and (d) more than 5100 registered Democrats.

5. Let $X_1, X_2, \ldots, X_{100}$ be a random variable whose values are determined by spinning Roulette Wheel 1 in Figure 8.1.5. The probability function for this roulette wheel is given below.

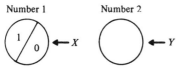

FIGURE 8.1.5 Roulette wheels.

156 Central Limit Theorem

x	$f(x)$
0	.5
1	.5

Let $Y = X_1 + X_2 + \cdots + X_{100}$. Assume that Roulette Wheel 2 reflects the probability function of Y. Suppose you spin Roulette Wheel 2 and win the amount in dollars indicated by the arrow when the wheel stops spinning.
a. How much do you expect to win?
b. What is the probability that you might win more than $60?
c. What is the probability that you might win less than $35?

8.2 Central Limit Theorem: General Case

In this section we will present a more general version of the central limit theorem than we have given in the preceding section. There are more general versions of the central limit theorem than those we will describe here, however, consideration of such theorems is beyond the scope of this textbook.

Theorem 8.2.1
Let $X_1, X_2, \ldots, X_n$ be a set of n random variables possessing the following characteristics:

1. $X_1, X_2, \ldots, X_n$ are statistically independent.
2. $X_1, X_2, \ldots, X_n$ have identical probability functions.
3. $E(X_1), E(X_2), \ldots, E(X_n)$ and $V(X_1), V(X_2), \ldots, V(X_n)$ have finite values.

Now let $Y = X_1 + \cdots + X_n$. Then, the probability function of Y approaches that of a normal probability function as n increases in value.

We will illustrate the implications of this theorem by the following examples. Consider this situation: We are to toss a die and win in dollars the number which appears on top. For each number which appears, the probability function of the random variable will be, then:

x	$f(x)$
1	⅙
2	⅙
3	⅙
4	⅙
5	⅙
6	⅙
	1.0

The probability function is also given in (a) of Figure 8.2.1.

8.2 Central Limit Theorem: General Case

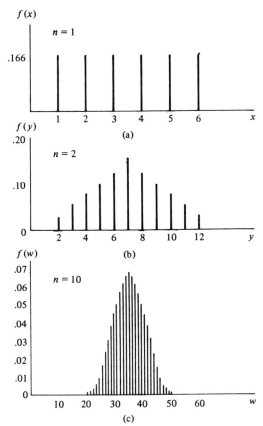

FIGURE 8.2.1 Probability function for the sum of the numbers obtained by tossing n dice.

Assume now that we are to toss two dice and win in dollars the sum of the two numbers which appear on top. Let Y be the random variable associated with the total amount of winnings. Then, the sample space of Y will contain 36 elements. From this sample space we can construct the probability function of Y, as shown in Table 8.2.1. The probability function of Y is also given in (b) in Figure 8.2.1.

Suppose we are to toss 10 dice and win in dollars the sum of the numbers which appear on top. Let W be the random variable associated with the total amount of winnings. Then, the sample space for W will contain $6^{10} = 60,466,176$ elements. Thus, obtaining the probability function by enumerating the sample space would be almost an impossible task. Yet, if we need to find the probability function, it can readily be obtained by an alternate method. Although a discussion of this method is beyond the scope of this textbook, the resulting probability function is given in (c) of Figure 8.2.1.

Let us now compare the three diagrams in Figure 8.2.1. We find that as

TABLE 8.2.1

y	Frequency in Sample Space	$f(y)$
2	1	1/36
3	2	2/36
4	3	3/36
5	4	4/36
6	5	5/36
7	6	6/36
8	5	5/36
9	4	4/36
10	3	3/36
11	2	2/36
12	1	1/36
	36	1.0

the number of dice increases, the probability function corresponding to the random variables associated with the total amount of winnings becomes smoother. In addition, we observe that the shape of the curve for the probability function of W is quite similar to that of a normal probability function. These observations, however, should not be strange to us if we evaluate them in light of Theorem 8.2.1. If we let $X_1, \ldots, X_{10}$ be the random variables associated with the amount of winnings as a result of tossing the individual dice, then, $X_1, \ldots, X_{10}$ are statistically independent, identically distributed random variables having finite means and variances. This means that if we let

$$W = X_1 + \cdots + X_{10},$$

the probability function of W may approximate a normal probability function, according to the central limit theorem.

Suppose now we ask the following question, "What is the probability that the amount of total winnings will be between \$25 and \$45?" Our calculations from finding the exact probability function of W indicate that the answer is .9506.

We can also approximate this probability by using a normal probability function. If we calculate the expected value and the variance of a single die, we find that $E(X_i) = 3.5$ and $V(X_i) \cong 2.91$. This means that

$$E(W) = E(X_1) + \cdots + E(X_{10}) = 35$$

and

$$V(W) = V(X_1) + \cdots + V(X_{10}) \cong 29.1.$$

In turn, $\sigma_w = 5.4$. Then, 25 and 45 are 1.85 standard deviations away from the mean value of 35. Since $P(-1.85 \leq Z \leq 1.85) = .9346$, we estimate the probability that the amount of winnings will be between $25 and $45 to be approximately .9346.

Suppose we wish to find the probability that the amount of winnings will be between $20 and $50. The probability function of W reveals that the exact probability is .9989, whereas our estimate, based on the central limit theorem, reveals that the probability is approximately .9868.

These estimates may be improved by using a so-called *continuity-correction factor*. However, even if we do not utilize a correction technique, we can see that our estimates are very close to the actual figures.

We will now illustrate some simple applications of the central limit theorem.

EXAMPLE 8.2.1

Assume that we are to take a multiple-choice examination, and that each question has six choices, of which only one is correct. We will be given 5 points if our choice is correct, but we will be penalized 1 point if our choice is incorrect. There are 80 questions on the examination, and we will pass it if we can score 40 points or more.

Suppose we have decided to answer the questions by tossing a die. The probability that we will answer any given question correctly is $1/6$ and that we will answer it incorrectly is $5/6$. Thus, each trial corresponding to each question may be described by the probability function in Table 8.2.2. Our calculations reveal that $E(X) = 0$ and $V(X) = 5$. If we now let

$$Y = X_1 + \cdots + X_{80},$$

where each X_i has a probability function identical to that of X in Table 8.2.2, then Y depicts the random variable associated with our total score. Since each X_i is a statistically independent, identically distributed random variable, we have

$$E(Y) = E(X_1) + \cdots + E(X_{80}) = 0$$

and

$$V(Y) = V(X_1) + \cdots + V(X_{80}) = 400,$$

as well as $\sigma_y = 20$. The central limit theorem assures that the probability function of Y is approximately normal. Then, the probability of our passing the examination by resorting to tossing a die is approximately .0228, since

TABLE 8.2.2

x	$f(x)$	$xf(x)$	$[x - E(X)]^2 f(x)$
5	$1/6$	$5/6$	$25/6$
−1	$5/6$	$-5/6$	$5/6$
		0	5

the passing score of 40 is 2 standard deviations away from the expected score, which, of course, is 0.

EXAMPLE 8.2.2

Consider the following situation: We toss a die and move 3 yards forward if 1, 2, 3, or 4 appears and 3 yards backward if either 5 or 6 appears. Assume that we are going to toss the die 200 times. Where will we be after 200 throws?

Consider now any single move. We might represent the outcome of this move by the random variable given in Table 8.2.3. Thus, $E(X) = 1$ and $V(X) = 8$. If we let $Y = X_1 + \cdots + X_{200}$ where each X_i has a probability function identical to that of X in Table 8.2.3, then Y depicts the random variable associated with our position in reference to the starting point after 200 moves. We next calculate that

$$E(Y) = E(X_1) + \cdots + E(X_{200}) = 200$$
and
$$V(Y) = V(X_1) + \cdots + V(X_{200}) = 1600,$$

and, in turn, that $\sigma_y = 40$. $E(Y) = 200$ implies that our expected position after 200 moves is 200 yards ahead of the starting position.

TABLE 8.2.3

x	$f(x)$	$xf(x)$	$[x - E(X)]^2 f(x)$
3	2/3	6/3	8/3
-3	1/3	-3/3	16/3
		1	8

Suppose we wish to know what the probability is that we might be behind the starting position after 200 moves? How does the central limit theorem help us to answer this question? It tells us that the probability function of Y is approximately normal. Since the starting point is 5 standard deviations away from the expected position, the probability in question must be practically 0.

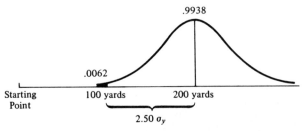

FIGURE 8.2.2 Probability function of position after 200 moves.

What is the probability that we would have advanced more than 100 yards? Figure 8.2.2 shows that the probability is approximately .9938.

Although our analysis, illustrated by Figure 8.2.2, appears to have no useful practical application, we might point out that Albert Einstein used a similar analysis to explain the Brownian motions of elementary particles. We might also point out that financial analysts have applied the underlying concepts of this illustration to explain the behavior of stock prices.

EXAMPLE 8.2.3

Assume that the average demand for a certain item is 30 units with a standard deviation of 10. Suppose we wish to forecast the sales for the next 100 days. Let X_i depict the random variable associated with the sales for the ith day. Suppose we assume that X_i are identically distributed and, furthermore, statistically independent. Then, $Y = X_1 + \cdots + X_{100}$ would depict the random variable associated with the total sales for the 100 days.

Since $E(X_i) = 30$ and $V(X_i) = 100$, we have

$$E(Y) = 3000,$$
$$V(Y) = 10,000,$$

and
$$\sigma_y = 100.$$

Furthermore, Y is approximately normal, according to the central limit theorem.

What is the probability that the sales during this period may exceed 3200 units? This probability is approximately .0228, as shown in Figure 8.2.3.

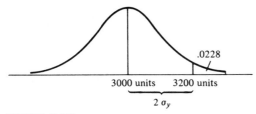

FIGURE 8.2.3

In applying the central limit theorem we have assumed that X_i are statistically independent. Of course, this assumption may not be justifiable in many practical situations.

EXERCISES

6. Suppose you are to toss 100 dice and will win in dollars the sum of the numbers which appear on top. What is the probability that your total amount of winnings will (a) exceed $400, (b) be between $300 and $400, and (c) be less than $250?

*7. Assume you are given 100 true-false questions in an examination. For each answer which is correct you will be given 1 point, but for each incorrect answer you will lose 1 point. Suppose you have decided to answer the questions by tossing a coin. What is the probability that you will score (a) more than 20 points, (b) less than 10 points?

8. Suppose that the nylon cords produced by a firm have a mean breaking strength of 10 pounds and a standard deviation of 2 pounds. Suppose a rope is made with one hundred of these cords. If you assume that the breaking strength of the rope is the sum of the strength of the cords, what is the probability that the rope will support a weight of (a) more than 1100 pounds, (b) more than 1040 pounds?

*9. The daily demand for a certain item is 80 units, on the average, with a standard deviation of 20 units. What is the probability that during the next 100 days the total demand will be (a) between 7500 and 8500 units, (b) between 7800 and 8200 units, and (c) more than 8400 units?

*10. Refer to Exercise 9. Suppose you wish to have enough units of the item at the beginning of the 100-day period, so that the probability that you would be out of stock would be not more than .001. How many units should you have at the beginning of the period?

8.3 Applications

In this section we will apply some of the statistical concepts we have learned thus far to simple inventory-management problems. The two examples now presented characterize the two basic approaches in modern inventory-management theories.

EXAMPLE 8.3.1

Let us return to Example 8.2.3 and assume that the average daily demand for a certain item is 30 units with a standard deviation of 10 units. Assume also that the company selling the item orders it once every 100 days from its supplier. Each order placed will arrive 21 days later. The duration of the time span between the date of the order and of arrival is called the *lead time*. Suppose today is the date that the company places an order. Examining its stockroom, the company finds that there are still 1000 units of this item on hand. How many units should the company order?

The company can resolve the problem in the following manner. The order placed today will arrive 21 days later; the order placed on the next order date will arrive 121 days from today. This means that the quantity of inventory on hand plus the quantity which arrives 21 days from today will have to meet the demand for the next 121 days.

Let Y be the random variable associated with the total demand for the next 121 days. Then,

$$E(Y) = E(X_1) + \cdots + E(X_{121}) = 3630,$$
$$V(Y) = V(X_1) + \cdots + V(X_{121}) = 12{,}100,$$

and, in turn,

$$\sigma_y = 110.$$

This means, for example, that

$$P(Y \geq 3810) \cong .05;$$

that is, the probability that the total demand will exceed 3810 units during the next 121 days is about .05.

Suppose now that the company, for example, wants to order in such quantity that the probability of running out of stock for this item during the period is .05. Then, the company should order 2810 units today, since the present inventory of 1000 units plus the quantity which will arrive 21 days later constitutes 3810 units.

EXAMPLE 8.3.2

Let us assume again that the average daily demand for a certain item is 30 units with a standard deviation of 10 units. Assume that the company uses the following ordering procedure. At the beginning of each day, the company examines the number of units of the item on hand. If it feels that the inventory is sufficient to cover the demand for the lead time, it places no order at all. On the other hand, if it feels that the inventory is not sufficient to cover the demand for the lead time, it places an order of 3000 units.

Suppose the lead time is 25 days. How low would the inventory decline before the company places an order, assuming, for example, that it wishes to accept only a .05 probability that it could be out of stock during the lead-time period?

The company may resolve the problem of deciding when to order in the following manner. Let Y be the random variable associated with the total demand during the lead time. Then, we can express that

$$Y = X_1 + \cdots + X_{25}$$

and, in turn, that

$$E(Y) = E(X_1) + \cdots + E(X_{25}) = 750,$$
$$V(Y) = V(X_1) + \cdots + V(X_{25}) = 2500,$$

and
$$\sigma_y = 50.$$

Thus, for example,

$$P(Y \geq 832) = .05;$$

that is, the probability that the total demand would exceed 832 units during the lead time is .05.

Suppose now the company places an order whenever the inventory is as low as 832 units. Then, the probability that it will run out of stock during the lead time is also .05.

EXERCISES

*11. Suppose a company sells 80 units of an item, on the average, per day, and the standard deviation of daily sales is 20 units. The company orders more units every 60 days; the lead time for the order is 21 days. The company wishes to order in such quantity that the probability of running out of stock before the arrival of the subsequent order would be .01. If there are 2500 units of this item in stock now, how many units should the company order?

*12. Refer to Exercise 11 and suppose that the company orders every 100 days and that the lead time is 44 days. The company is willing to accept a .001 probability that it will run out of stock. If there are 6000 units of the item on hand, how many units should it order?

*13. Suppose a company sells 80 units of an item, on the average, per day with a standard deviation of 20 units. The company uses the following ordering policy: It continuously monitors its inventory holdings and places an order whenever the inventory holding is low, taking into account the lead-time factor. The lead time for an order placed is 36 days. Suppose the company is willing to accept a .01 probability that it will be out of stock during the lead-time period. Then, how low would the inventory have to be before the company places an order?

Chapter 9
Sampling Techniques

In the preceding chapters we have implicitly assumed that the relevant measures pertaining to a probability function of a random variable, such as its mean and its variance, are known. However, this is rarely true in practice; we often face the problem of estimating these values, which we can do by taking a sample from the probability function for the random variable. We will explore in this chapter some of the problems connected with taking a sample from a probability function of a random variable.

9.1 Simple Random Samples

The notion of taking a sample from a probability function is difficult to conceptualize. The sampling, for example, implies that visible objects are selected from a physical process, although the probability function is not something from which we can obtain visible objects. We will, therefore, present some physical counterparts to illustrate the notion of taking a sample from a probability function. We will confine these illustrations to a *simple random sample*.

Definition 9.1.1

Let $X_1, \ldots, X_n$ be the random variables associated with n repeated measurements obtained from the probability function of a random variable X. If the measurements are obtained in a manner to assure that $X_1, \ldots, X_n$ are statistically independent random variables whose probability functions are identical to that of X, then the set $\{X_1, \ldots, X_n\}$ is said to constitute a *simple random sample* of size n from the given probability function.

EXAMPLE 9.1.1

Assume that the probability function from which we wish to take a sample is the one given in Table 9.1.1. A physical counterpart of this probability function is depicted in Figure 9.1.1, which illustrates that whenever the roulette wheel stops spinning, the probability that the arrow will point toward 1 is 10 percent and that it will point toward 0 is 90 percent.

TABLE 9.1.1

x	$f(x)$
0	.9
1	.1
	1.0

Suppose we are to spin the roulette wheel n times to obtain a sequence of n numbers. We can assign a random variable to the outcome of each spin. This means that the outcomes for the n spins may be depicted by a set of random variables $\{X_1, \ldots, X_n\}$. A little reflection will reveal that the random variables are identically distributed to X and are also statistically independent.

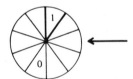

FIGURE 9.1.1   Roulette wheel.

Since, by assumption, the roulette wheel is a physical counterpart to the probability function of X, we may consider that $X_1, \ldots, X_n$ are n repeated measurements obtained from the probability function of X. Thus, the set $\{X_1, \ldots, X_n\}$ may be considered to be a simple random sample of size n from the probability function of X.

EXAMPLE 9.1.2

Assume we know that a manufacturing process produces 10 percent defectives, on the average. Assume, further, that the process is stable and that there is no pattern in the production of defective and nondefective items.

9.1 Simple Random Samples

This assumption implies that the probability that the next item produced will be defective is always 10 percent, regardless of what has been produced previously.

Let X be the random variable associated with selecting an item from the process so that X assumes the value 1 if the item chosen is defective and X assumes the value 0 if it is not defective. Then, the probability function of X is given as

x	$f(x)$
0	.9
1	.1
	1.0

Suppose we are to select n items from the process. The set of the outcomes may be depicted by a set of random variables $\{X_1, \ldots, X_n\}$, where X_i assumes the value 0 if the ith item selected from the process is nondefective, and assumes the value 1 if it is defective. Then, each random variable has the probability function given in Table 9.1.2. We observe that $X_1, \ldots, X_n$ are identically distributed to the probability function X and, at the same time, are statistically independent. We can, therefore, consider the random variables $X_1, \ldots, X_n$ to be associated with the outcomes of n repeated measurements taken from the probability function of X. Thus, we may call the set $\{X_1, \ldots, X_n\}$ a simple random sample of size n from the probability function of the random variable X.

TABLE 9.1.2

x_i	$f(x_i)$
0	.9
1	.1
	1.0

EXAMPLE 9.1.3

Assume that a box contains 10 defective items and 90 nondefective items. Let X depict the random variable associated with selecting just one item from the box. The probability function of X may be given as

x	$f(x)$
0	.9
1	.1
	1.0

Although the probability function of X is related to the outcome of selecting just one item from the box, it also reflects the composition of the entire box. In other words, if we knew the probability function of X, then we would also know how many defectives are in the box.

Assume now that we are to select one item from this box n times, and are to replace each item selected before we select the next item. The set of outcomes to be obtained by n different drawings from the box may be depicted by a set of random variables $\{X_1, \ldots, X_n\}$. Since the item drawn is replaced each time, these random variables must be statistically independent and identically distributed to the probability function of X. We can, therefore, consider the random variables $X_1, \ldots, X_n$ as those associated with the outcomes of n repeated measurements taken from the probability function of X. Thus, the set $\{X_1, \ldots, X_n\}$ may be said to constitute a simple random sample of size n from the probability function of X.

EXAMPLE 9.1.4

Assume that a box contains the following 10 numbers: 1, 1, 1, 2, 2, 2, 2, 3, 3, 3. Let X depict the random variable associated with selecting just one number from the box. The probability function of X may be given as

x	$f(x)$
1	.3
2	.4
3	.3
	1.0

Although the probability function of X is related to the outcome of selecting just one number from the box, it also reflects the composition of the box. If we know that the box contains 10 numbers, then we can construct the composition of the box based on the probability function of X, as shown in Table 9.1.3. Thus, the number of elements in the box and the probability function of X provide complete information with regard to the composition of the box.

Assume now that we are to select one item from this box n times, and are to replace each item selected before we select the next item. The set of outcomes obtained from n different drawings may again be depicted by a set of random variables $\{X_1, \ldots, X_n\}$. The random variables in the set must be statistically independent and identically distributed to the probability

TABLE 9.1.3

Number	$10 f(x)$
1	$10 \times .3 = 3$
2	$10 \times .4 = 4$
3	$10 \times .3 = 3$

function of X. We can, therefore, consider the random variables $X_1, \ldots, X_n$ to be associated with the outcomes of n repeated measurements obtained from the probability function of X. Thus, the set $\{X_1, \ldots, X_n\}$ constitutes a simple random sample of size n from the probability function of X.

EXERCISES

1. Suppose a box contains the following 5 numbers: 0, 0, 1, 1, 1. You are to take a simple random sample of 2 numbers from the box. Describe the probability function from which the sample is to be taken.
*2. Suppose a box contains 20 defectives and 80 nondefectives. You are to select a simple random sample of 5 items from the box. Describe the probability function from which the sample is to be taken.
3. Suppose a process generates the numbers 0 and 1, which have the associated probabilities .1 and .9, respectively. You are to take a simple random sample of 10 items. Describe the probability function from which the sample is to be taken.
4. Suppose a manufacturing process is known to produce 10 percent defectives. You are to select a simple random sample of size n. Describe the probability function from which the sample is to be taken.
5. Suppose a box contains the following three numbers: 2, 3, 3. You are to select a simple random sample of 5 numbers from the box. Describe the probability function from which the sample is to be selected.
*6. Suppose a process generates the numbers 1, 2, and 3 with equal probabilities. You are to select a simple random sample of 5 numbers from the process. Describe the probability function from which the sample is to be selected.

9.2 Techniques for Selecting a Simple Random Sample

Although it is easy to illustrate the definition of a simple random sample, the task of actually taking a simple random sample from the physical counterpart of a probability function may be difficult. We will discuss in the following examples how some of these difficulties may be resolved. In each example we will first introduce a hypothetical problem, which may not have a practical application, and then describe how to take a simple random sample. We will introduce some practical sampling problems and illustrate how the structures of these real problems are similar to the structure of the hypothetical one.

EXAMPLE 9.2.1

Assume that a box contains 80 zeroes and 20 ones. Our task is to select a simple random sample of two numbers from the box. The probability function from which the sample is to be taken may be given as

x	$f(x)$
0	.8
1	.2
	1.0

One safe approach to taking a simple random sample is to select one number from the box, examine whether it is a one or a zero, replace the number into the box, mix the numbers in the box thoroughly, and then select the second number to be examined. If we let X_1 and X_2 be the random variables associated with the two sample trials, then X_1 and X_2 are identically distributed, statistically independent random variables. Thus, the sample satisfies Definition 9.1.1.

The preceding process is cumbersome, because we have to replace the number after each drawing. Suppose we take two numbers from the box without replacing the first one. Would the sampling procedure assure that the resulting sample is still a simple random sample? The answer is "no," for the following reason. Since the first number drawn is not replaced, X_1 and X_2 are statistically dependent random variables. Thus, the sample does not satisfy all the requirements for Definition 9.1.1.

Consider the following real sampling situation. A box contains N items. We wish to take a simple random sample of n items from the box, where $n < N$, and estimate the proportion of defectives in the box. One way to assure that n items selected constitute a simple random sample is to select an item, examine it, and replace it in the box, select the next item, and so on, n times. This procedure is cumbersome, and for some cases, we may not even be able to apply it. For example, suppose an item selected must be destroyed in the process of examining it. Replacement becomes impossible. Thus, it appears that we may run into a sampling problem which is unsolvable, but this is not true for the following reasons. First, we can solve the problem by assuming that the sample is not a simple random sample and by applying the appropriate statistical theory for the given situation. Such a theory is not difficult to develop. On the other hand, when n is smaller than 5 or 10 percent of N, an analysis based on an appropriate theory would not be materially different from one based on the assumption that a simple random sample of n items has been taken. In other words, if n items are selected from the box without replacement, where n is less than 5 or 10 percent of N, a statistical proposition based on an assumption that the sample constitutes a simple random sample may be accurate enough for all practical purposes.

EXAMPLE 9.2.2

Assume that a box contains 100 numbers, whose composition is shown in Figure 9.2.1. Let X be the random variable associated with selecting

9.2 Selecting a Simple Random Sample

```
┌─────────────┐
│  20 Ones    │
│             │
│  80 Zeroes  │
└─────────────┘
```

FIGURE 9.2.1

a number from the box. Then, the probability function of X is

x	$f(x)$
0	.8
1	.2
	1.0

Assume now that the box containing 100 numbers has been divided into two smaller boxes, so that each of the smaller boxes contains 50 numbers. We wish to devise a sampling procedure which will yield a simple random sample of 2 from the original box. We can solve the problem by combining the two smaller boxes into a larger one and selecting the simple random sample from the latter.

There are many real sampling problems whose structures resemble this hypothetical problem, however, it would not always be possible to solve a problem in this way. For example, consider the problem of a company auditor who is trying to estimate the proportion of delinquent customers in the accounts-receivable files. Assume that the files are scattered among several different locations. This is similar to an assumption that the larger box containing all accounts-receivable has been divided into several smaller boxes. If the auditor wishes to select a simple random sample from the larger box, then he must devise a way that will avoid the necessity of combining the smaller boxes.

Let us return to our hypothetical problem. Suppose now we select one number from each of the smaller boxes. Would the sampling procedure assure a simple random sample? The answer is "not likely," for the following reason. Suppose the contents of the two smaller boxes are those shown in Figure 9.2.2. Let X_1 and X_2 be the random variables associated with the sample results

```
     BOX 1                BOX 2
┌─────────────┐      ┌─────────────┐
│  15 Ones    │      │  5 Ones     │
│             │      │             │
│  35 Zeroes  │      │  45 Zeroes  │
└─────────────┘      └─────────────┘
```

FIGURE 9.2.2

172 Sampling Techniques

obtained from Box 1 and Box 2, respectively. Then, X_1 and X_2 are statistically independent random variables. On the other hand, the probability functions of X_1 and X_2 are not identical to that of X. Therefore, the set $\{X_1, X_2\}$ does not constitute a simple random sample.

Suppose, however, that the contents of the two boxes are those shown in Figure 9.2.3. Then, the sampling procedure in question would yield a simple random sample, since X_1 and X_2 are not only statistically independent, but also identically distributed to the probability function of X.

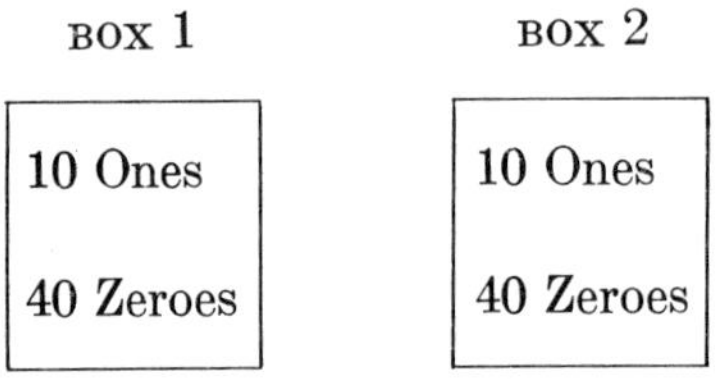

BOX 1

10 Ones

40 Zeroes

BOX 2

10 Ones

40 Zeroes

FIGURE 9.2.3

Intuitively, we can see that if the box containing 100 numbers were divided arbitrarily into two smaller boxes, each containing 50 numbers, then it would be very unlikely that the resulting boxes would each contain 10 ones. This is the reason that the sampling procedure in question would not likely yield a simple random sample. Does this mean that it is impossible to obtain a simple random sample without physically combining the contents of the two smaller boxes into the larger one? The answer is "no," which we will now illustrate.

Let the two smaller boxes be labelled Box 1 and Box 2. Consider next the following sampling scheme: we are to toss a coin and if "heads" appears, select a number from Box 1; but if "tails" appears, we are to select a number from Box 2. We must replace the number drawn into the box from which it was drawn. Then, we must repeat the experiment for the second sample number. The experiment would be repeated as many times as necessary to obtain the required sample size.

To illustrate how this procedure yields a simple random sample, let us arbitrarily assume that the composition of each of the two smaller boxes is illustrated in Figure 9.2.4. Let X_1 be the random variable associated with the

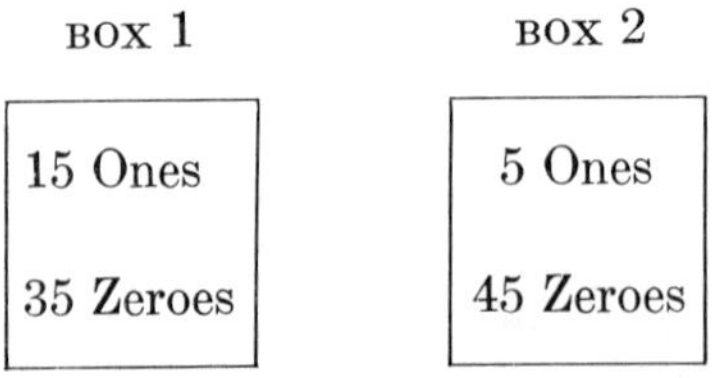

BOX 1

15 Ones

35 Zeroes

BOX 2

5 Ones

45 Zeroes

FIGURE 9.2.4

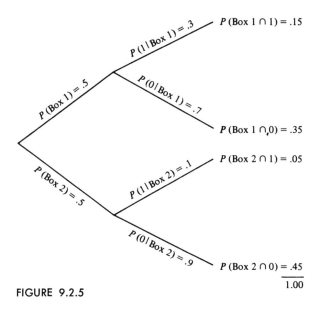

FIGURE 9.2.5

first item sampled. Then, the various possible sample outcomes and their likelihoods are illustrated in Figure 9.2.5. In this figure we find that the probability that the procedure will lead us to select a one is .2 and that it will lead us to select a zero is .8. This means that the probability function of X_1 may be given as

x_1	$f(x_1)$
0	.8
1	.2
	1.0

which is identical to the probability function of the combined box. Let X_2 be the random variable associated with the second number to be sampled. Since we must replace the first number drawn into the box from which we have drawn it, X_2 must be statistically independent of X_1. Furthermore, the probability functions for the two random variables must be identical to each other, as well as to that of X. Thus, the set $\{X_1, X_2\}$ constitutes a simple random sample from the combined large box, even though physically the two smaller boxes remain separate.

Suppose now the percentages of ones and zeroes are different from those illustrated. The procedure would still yield a simple random sample. In other words, the procedure would yield a simple random sample regardless of the compositions of the smaller boxes, assuming each box contains 50 numbers.

On the other hand, suppose Box 1 contains 25 numbers and Box 2 contains 75 numbers. Then, we have to modify the procedure in order to obtain a simple random sample. The modified procedure would be: we are to toss two coins. If two "heads" appear, we must select a number from Box 1; otherwise, we must select a number from Box 2. We must replace the number drawn into the box from which we have drawn it and, then, repeat the experiment for the next number to be sampled.

To illustrate why this modification is necessary, let us arbitrarily assume that the compositions of the smaller boxes are those shown in Figure 9.2.6.

BOX 1	BOX 2
10 Ones	10 Ones
15 Zeroes	65 Zeroes

FIGURE 9.2.6

Various possible sample outcomes and their likelihoods are given in Figure 9.2.7. From Figure 9.2.7 we obtain

x_1	$f(x_1)$
0	.8
1	.2
	1.0

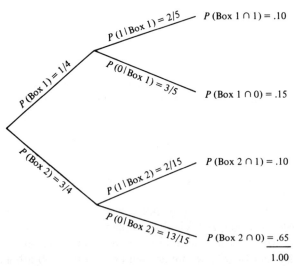

FIGURE 9.2.7

which is identical to the probability function of the combined box. If we let X_2 be the random variable associated with the second number to be sampled, then X_2 must be statistically independent from X_1 and must also have a probability function identical to that of X. Thus, the set $\{X_1, X_2\}$ constitutes a simple random sample from the combined box.

Let us again return to the auditor's sampling problem. Assume that the files are separated into k different locations, where $N_1, \ldots, N_k$ depicts the number of accounts at these locations and $N = N_1 + \cdots + N_k$. If the auditor wishes to select a simple random sample of size n from the combined file containing N accounts, he must devise a scheme similar to the one we have devised for the hypothetical problem. For each sample trial the probability that an account would actually be selected from the ith location is given by N_i/N.

A natural question to raise at this point is, "Why can the auditor not select $(N_1/N)n$ accounts from the first location, $(N_2/N)n$ accounts from the second location, and so on?" Such a procedure may, in fact, be more efficient than the one which yields a simple random sample. An evaluation of such a procedure would, however, require us to discuss a number of additional statistical concepts, in addition to those which we will present in this textbook.

EXAMPLE 9.2.3

Assume that we are to select a simple random sample of 6 numbers from a box containing the following numbers: 1, 1, 1, 1, 2, 2, 2, 2, 2, 2, 3, 3. The probability function from which the sample is to be selected is given as

x	$f(x)$
1	$2/6$
2	$3/6$
3	$1/6$
	1.0

One way to take such a simple random sample is to select a number from the box, examine it, replace it into the box, and select a second number. Repeat this experiment six times.

Suppose we are to select 6 numbers from the box without replacing the numbers drawn. Let $X_1, X_2, \ldots, X_6$ be the random variables associated with the sample trials. Then, they would be statistically dependent random variables. Thus, the set $\{X_1, \ldots, X_6\}$ would not constitute a simple random sample.

Assume next that the box containing 12 numbers has been divided into three smaller boxes, so that one box contains 2 numbers, another contains 4 numbers, and a third contains 6 numbers. It would appear that a simple random sample may be obtained by selecting 1 number from the first box,

176 Sampling Techniques

2 numbers from the second box, and 3 numbers from the third box. As indicated in Example 9.2.2, this procedure is unlikely to yield a simple random sample. To illustrate that this procedure would not yield a simple random sample, suppose the compositions of the boxes are those shown in Figure 9.2.8.

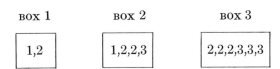

FIGURE 9.2.8

Let X_1 be the random variable associated with the sample drawn from Box 1; let X_2 and X_3 be the random variables associated with the samples drawn from Box 2, where the first number drawn is replaced into the box before the second number is drawn; finally, let X_4, X_5, and X_6 be the random variables associated with the samples from Box 3. Then, the probability functions of the random variables are those given in Table 9.2.1. We find that X_1, X_2, X_3, X_4, X_5, and X_6 are not identically distributed random variables. The sampling procedure would, therefore, not yield a simple random sample.

TABLE 9.2.1

x_1	$f(x_1)$	x_2, x_3	$f(x_2, x_3)$	x_4, x_5, x_6	$f(x_4, x_5, x_6)$
1	1/2	1	1/4	2	1/2
2	1/2	2	2/4	3	1/2
		3	1/4		
	1.0		1.0		1.0

Suppose now that we use the following procedure to select the sample number. We toss a die and choose

        Box 1 if 1 appears
        Box 2 if 2 or 3 appears
        Box 3 if 4, 5, or 6 appears.

Then, we pick a number from the box selected and replace the number into the box from which we have drawn it and repeat the procedure for the second number to be sampled. We repeat the procedure until the sixth number has been observed. Let X_1 be the random variable associated with the first number sampled. Then, various sample outcomes and their likelihoods are shown in Figure 9.2.9.

9.2 Selecting a Simple Random Sample

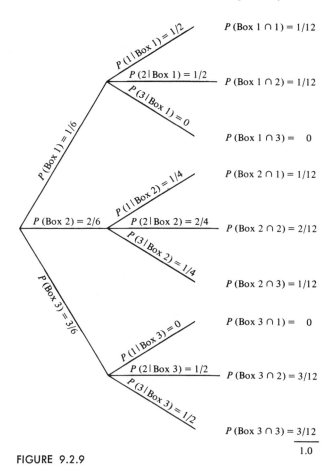

FIGURE 9.2.9

From this illustration we find that the probability function of X_1 is

x_1	$f(x_1)$
1	2/6
2	3/6
3	1/6
	1.0

which is identical to the probability function of X. Since the same procedure is repeated for the successive trials, the probability functions of $X_2, \ldots, X_6$ would be identical to that of X_1 and, in turn, to that of X. Furthermore, the random variables $X_1, \ldots, X_6$ would be statistically independent, since each number drawn from a box is replaced into it before the next number is drawn.

178 Sampling Techniques

Consequently, the set $\{X_1, \ldots, X_6\}$ would constitute a simple random sample of size 6 from the combined box.

We will now describe a real sampling problem, which has a similar structure to that of the hypothetical one. Suppose the auditor described in Example 9.2.2 is planning to select a simple random sample of size n from the company's accounts-receivable files in order to estimate their mean value. The first decision that he must make is whether or not to replace an account he has selected in the files so that it can be redrawn. Our evaluation of the hypothetical problem shows that the resulting sample will not be a simple random sample unless the auditor replaces the selected account in the files before he selects the next account. On the other hand, we can show that if n is less than 5 or 10 percent of N, then, an analysis based on the assumption that a simple random sample is taken will result in a fairly accurate conclusion, even though the accounts are not, in fact, replaced in the files.

Suppose now that the files are separated at k different locations, where $N_1, \ldots, N_k$ depict the number of accounts at these locations. Then, the auditor must devise a scheme similar to the one we have devised for the hypothetical problem so that for each sample trial the probability that an account would be selected from the ith location is N_i/N.

EXERCISES

7. Suppose a box contains 100 items, some of which are defective. Suppose you wish to select a simple random sample of 10 items from the box. How would you obtain such a sample?

*8. Assume that the box given in Exercise 7 is divided into two smaller boxes, one of which contains 20 items and the other of which contains 80 items. Suppose you still wish to select a simple random sample from the combined box, but you do not want to combine the smaller boxes. How would you obtain a simple random sample? Illustrate why your procedure would actually yield a simple random sample.

9.3 Use of Random Numbers

Frequently, an actual sample may have to be taken from a list which contains the elements of the sample space associated with a probability function. Consider, for example, the problem of an instructor who wishes to estimate the average amount of time spent by his students to complete a term project. Assume that the list in Table 9.3.1 contains the students in his class. He wishes to select a simple random sample of four students from the list.

One way in which the instructor can accomplish this objective is to write the name of each student on a separate piece of paper, put the eight

TABLE 9.3.1

Identifying Number	Student's Name
0	James Adams
1	Mary Brown
2	John Cook
3	Robert Davis
4	Frank Edwards
5	Ann Foster
6	Thomas Grey
7	Walter Hanson

pieces of paper in a hat, and draw a name four times. Each name drawn is to be replaced in the hat before the next one is drawn. The four names drawn in succession will then constitute a simple random sample of four.

There is another way that he can select a simple random sample. Instead of replacing the names in the hat, he might put eight pieces of paper containing the numbers 0, 1, . . . , 7 in a hat. Then he would draw a number from the hat four times, replacing each number drawn before drawing the next number. Assume now that the four numbers observed during the sampling process are

$$1, 4, 3, 1.$$

Then, the equivalent simple random sample would be given as

{Mary Brown, Frank Edward, Robert Davis, Mary Brown}

(Note that Mary Brown must be counted twice if we are to satisfy the requirements for a simple random sample.) This set of four numbers generated is called *the random numbers*.

Suppose now that we need to generate a set of random numbers which are frequently associated with sampling problems. Instead of generating a set every time we need one, we could generate a set containing a fairly large number of random numbers and use them as the situation demands. The process of generating random numbers can be simple if we do not need too many of them. For example, we could put the following numbers in a hat: 0, 1, 2, 3, 4, 5, 6, 7, 8, 9; then, we could draw a number from the hat and record it, replace the number in the hat, draw another number, and record it. By repeating this procedure, we would obtain a set of one-digit, non-negative integers. These integers in the set, then, may be considered as one-digit random numbers.

If we need to generate a set of two-digit random numbers, then, we can generate them from the set of one-digit random numbers. Assume that the one-digit random numbers generated are

6, 1, 7, 0, 6, 5, 8, 1, 3, 3, 9, 8, 8, 5, 4, 2, ...
↑

First, we must choose one number in the preceding list by a random design; for example, we might close our eyes and point a finger at a number. Assume that the number chosen is 6, as indicated by the arrow. Then, the sequence of every two numbers starting with 6 may be considered to be a random number. Thus, the random numbers generated are

65, 81, 33, 98, ...

Obviously, it would be cumbersome to generate many one-digit random numbers. However, there is really no need for us to generate even a small set of such numbers, since the Rand Corporation has prepared a table containing one million random digits. An abstract of the table is presented in Appendix B (Table B.9).

Chapter 10
Sampling
Distribution

We have pointed out that the purpose of taking a sample from a probability function is to estimate, for example, the mean or the variance of the probability function. Such a measure is called the *parameter* of the probability function. For example, we would take a sample from a Bernoulli probability function to estimate its parameter π; we would take a sample from an exponential probability function to estimate its parameter λ; or, we would take a sample from a normal probability function to estimate its parameters μ and σ.

Before we proceed to evaluate the problems connected with estimating a parameter of a probability function, however, we need to discuss the concept of *sampling distribution*. The *sampling distribution theory* explores various properties of samples obtained from a probability function whose parameters are assumed to be known. Later we will discover that an understanding of these properties will help us in making some inductive propositions about a probability function whose parameters are *not* known.

10.1 Probability Function of a Sample Proportion

Assume that we are to take a simple random sample of size n from the probability function of a Bernoulli-distributed random variable with a given value of π. We are then to obtain a sequence of n numbers containing zeroes

182 Sampling Distribution

and ones. Let P denote the variable whose value is determined by adding the numbers in the sequence and then by dividing the sum by n. Then, P must be a random variable. We will examine the properties of the probability function for P.

Definition 10.1.1

Let $\{X_1, \ldots, X_n\}$ be a simple random sample of size n from the probability function of a Bernoulli-distributed random variable. Let

$$P = \frac{1}{n}(X_1 + \cdots + X_n).$$

Then P is called the random variable associated with the *sample proportion*.

The reason for the term *proportion* may be stated as follows: The value of P will always be between 0 and 1. Thus, the value of P may be expressed as a percentage, but instead of calling it a percentage, the statisticians have decided to call it a *proportion*.

EXAMPLE 10.1.1

Assume that we are to select a simple random sample of two items from the probability function of a Bernoulli-distributed random variable with $\pi = .5$. The probability function from which we are to select the sample is then given as

x	$f(x)$
0	.5
1	.5
	1.0

Let $Y = X_1 + X_2$ and $P = (1/n)Y$. Then, as Table 10.1.1 illustrates, various outcomes are obtainable from the sampling process. In this table the elements under the column "Likelihood" indicate the probability of obtaining a specific sample outcome; for example, it indicates that the probability of obtaining

TABLE 10.1.1 Outcomes of a Sampling Process

Sample	Likelihood	Value of Y	Value of P
(0,0)	¼	0	0
(0,1)	¼	1	.5
(1,0)	¼	1	.5
(1,1)	¼	2	1.0

(0,0) is $\frac{1}{4}$. The right-hand column shows that P can assume the value 0, .5, or 1.0. We can also find the probabilities that P will assume these values. Thus, we have the function

p	$f(p)$
0	.25
.5	.50
1.0	.25
	1.00

which is, then, an example of the probability function for a sample proportion.

EXAMPLE 10.1.2

Assume that we are to take a simple random sample of two items from the probability function of a Bernoulli-distributed random variable with $\pi = .2$. The probability function from which we are to select the sample is given as

x	$f(x)$
0	.8
1	.2
	1.0

Let $Y = X_1 + X_2$ and $P = (1/n)Y$. The various outcomes obtainable from the sample and their likelihoods are shown in Table 10.1.2.

TABLE 10.1.2 Outcomes of a Sampling Process

Sample	Likelihood	Value of Y	Value of P
(0,0)	.64	0	0
(1,0)	.16	1	.5
(0,1)	.16	1	.5
(1,1)	.04	2	1.0

In turn, we find that the probability function of P is that shown in Table 10.1.3. This function is another probability function of a sample proportion.

Theorem 10.1.1

Let P be a random variable associated with the sample proportion of a simple random sample of size n from a Bernoulli probability function with a given value of π. Then,

and
$$E(P) = \pi$$
$$V(P) = \frac{\pi(1-\pi)}{n}.$$

TABLE 10.1.3

p	$f(p)$
0	.64
.5	.32
1.0	.04
	1.00

EXAMPLE 10.1.3

Let us return to the probability function of P in Example 10.1.1. We can calculate the expected value and the variance of P as shown in Table 10.1.4. Thus, $E(P) = .5$ and $V(P) = \frac{1}{8}$.

TABLE 10.1.4 Expected Value and Variance of P

p	$f(p)$	$pf(p)$	$[p - E(P)]^2 f(p)$
0	.25	0	$\frac{1}{16}$
.5	.50	.25	0
1.0	.25	.25	$\frac{1}{16}$
	1.00	.50	$\frac{1}{8}$

On the other hand, according to Theorem 10.1.1, we have
$$E(P) = \pi = .5$$
and
$$V(P) = \frac{\pi(1-\pi)}{n} = \frac{1}{8},$$

which yields the same results as those obtained in Table 10.1.4.

EXAMPLE 10.1.4

Let us now return to the probability function of P given in Example 10.1.2. We can calculate the expected value and the variance of P as shown in Table 10.1.5. Thus, $E(P) = .2$ and $V(P) = .08$.

TABLE 10.1.5 Expected Value and Variance of P

p	$f(p)$	$pf(p)$	$[p - E(P)]^2 f(p)$
0	.64	0	.0256
.5	.32	.16	.0288
1.0	.04	.04	.0256
	1.00	.20	.0800

10.1 Probability Function of a Sample Proportion

On the other hand, according to Theorem 10.1.1, we have

$$E(P) = \pi = .2$$

and

$$V(P) = \frac{\pi(1-\pi)}{n} = .08,$$

which yields the same results as those obtained by our tabular calculations.

Definition 10.1.2

Let $\sigma_p = \sqrt{\pi(1-\pi)/n}$. Then, σ_p is called the *standard error of proportion*.

The *standard error of proportion* is simply the standard deviation for the probability function of P, and its value is obtained by finding the square root of $V(P)$.

Theorem 10.1.2

Let P be the random variable associated with the sample proportion of a simple random sample of size n from a Bernoulli probability function with a given value of π. The probability function of P approaches a normal probability function as n becomes larger and larger.

The validity of this theorem may be explained as follows. If we let $Y = X_1 + \cdots + X_n$, then Theorem 8.1.1 assures that the probability function of Y approaches that of a normal probability function as n becomes larger and larger. If the probability function approaches a normal probability function, then the probability function of $P = (1/n)Y$ must also approach a normal probability function, according to Theorem 7.6.1, since $1/n$ is a constant.

We will now illustrate some applications of Theorem 10.1.2 with the following examples.

EXAMPLE 10.1.5

Assume that a manufacturing process produces a given percentage of defectives, on the average. Assume further that the actual proportion of defectives produced by the process is 10 percent.

Suppose we take a simple random sample of 100 items from the process and obtain a value of P from the sample. What can we say about P? According to Theorem 10.1.1, we can say that

$$E(P) = .10$$

and

$$V(P) = \frac{(.1)(.9)}{100} = .0009.$$

In turn, $\sigma_p = .03$. According to Theorem 10.1.2, we can state that the probability function of P must be approximately normal, which means, for example, that the probability that the value of P will be between .04 and .16 is

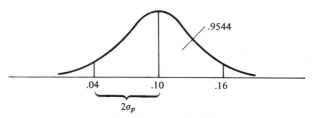

FIGURE 10.1.1 Probability that P will be between .04 and .16.

approximately .9544, as illustrated in Figure 10.1.1. In words this means that if the proportion of defectives produced by the process is really 10 percent, then there is approximately a 95 percent chance that our estimation based on a simple random sample of 100 items from the process will be between 4 and 16 percent.

EXAMPLE 10.1.6

Suppose that we have taken a simple random sample of 2500 voters from a cross-section of the country in order to find out the proportion of the nation's voters who are in favor of a candidate for a national political office. Such a sample may then be considered to have been drawn from a Bernoulli probability function.

Assume, for example, that 50 percent of the voters actually favor the candidate. Then, by assumption, $\pi = .5$ and

$$E(P) = .5$$
and
$$V(P) = .0001,$$

and, in turn, $\sigma_p = .01$. Theorem 10.1.2 further implies that the probability function of P is approximately normal. Then, the probability that the value of P will be between .48 and .52 is approximately .9544, as illustrated in Figure 10.1.2. If the voters are, in fact, equally divided for and against the candidate, there is approximately a 95 percent chance that the proportion of those in a sample of 2500 voters favoring him will be between 48 percent and 52 percent.

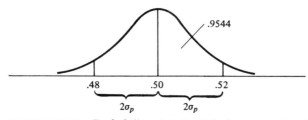

FIGURE 10.1.2 Probability that P will be between .48 and .52.

EXERCISES

1. Suppose a process generates the numbers 0 and 1 with equal probabilities. You are to select a simple random sample of two from the process. Describe the probability function for P. Then, find $E(P)$ and $V(P)$.

*2. A simple random sample of two is to be selected from the following probability function:

x	$f(x)$
0	.8
1	.2
	1.0

Describe the probability function for P. Then, find $E(P)$ and $V(P)$.

3. Suppose a box contains the following five numbers: 0, 0, 1, 1, 1. You are to select a simple random sample of three. Describe the probability function for P. Then, find $E(P)$ and $V(P)$.

4. Assume that you are to take a simple random sample of 100 from the process given in Exercise 1. Find $E(P)$ and $V(P)$. Then, calculate: (a) $P(|P - \pi| \geq .05)$ and (b) $P(|P - \pi| \geq .10)$.

*5. Assume that you are to take a simple random sample of 100 from the probability function given in Exercise 2. Find $E(P)$ and $V(P)$. Then, calculate: (a) $P(|P - \pi| \geq .04)$ and (b) $P(|P - \pi| \geq .08)$.

6. Assume that you are to take a simple random sample of 100 from the box given in Exercise 3. Find $E(P)$ and $V(P)$. Then, calculate: (a) $P(|P - \pi| \geq .08)$ and (b) $P(|P - \pi| < .04)$.

*7. Assume that a manufacturing process produces 10 percent defectives. If you take a simple random sample of 100 from the process, what is the probability that the proportion of defectives in your sample might exceed 16 percent?

8. A Sunday issue of the *San Jose Mercury News* (November 23, 1969) had an article written by George Gallup, the famous political pollster. He stated among other things:

> Of the 1,465 adults interviewed in 300 localities across the nation, 68 percent say they approve of the way he (President Nixon) is handling his overall job. . . .

Suppose that the majority of the adult population in the nation do not really approve of the way Nixon is handling his job. (That is, less than 50 percent of the adult population actually approves of him.) Then, what is the probability that one might obtain the findings reported by Mr. Gallup? (Assume that Mr. Gallup has taken a simple random sample.)

10.2 Probability Function of a Sample Mean

In the preceding section, we have confined our discussion to sampling from a Bernoulli probability function. We will extend this discussion to sampling from *any* probability function.

Assume that we are to take a simple random sample of size n from any probability function with given values for μ and σ. We will then obtain a sequence of n numbers. Let $\bar{X}$ denote the variable whose value will be determined by adding the numbers in the sequence and dividing the sum by n. Then, $\bar{X}$ must be a random variable. The value of $\bar{X}$ is called the *sample mean*. We will examine the properties of the probability function for $\bar{X}$, but first we will define $\bar{X}$.

Definition 10.2.1
Let $\{X_1, \ldots, X_n\}$ be the simple random sample of size n from the probability function of a random variable. Let

$$\bar{X} = \left(\frac{1}{n}\right)(X_1 + \cdots + X_n).$$

Then, $\bar{X}$ is called the random variable associated with the *sample mean*.

If the sample is taken from a Bernoulli probability function, then $\bar{X}$ is equivalent to P. Our discussion in the preceding section would apply to the given situation.

EXAMPLE 10.2.1
Assume that a box contains the following three numbers: 1, 2, and 3. We are to select a sample of two numbers from the box, and replace the first number drawn in the box before we draw the second number. The probability function from which we are to select the sample is equivalent to

x	$f(x)$
1	$\frac{1}{3}$
2	$\frac{1}{3}$
3	$\frac{1}{3}$
	1.0

A simple calculation will reveal that $E(X) = 2$ and $V(X) = \frac{2}{3}$.

Let $Y = X_1 + X_2$ and $\bar{X} = (1/n)Y$. Then, the different possible outcomes obtained from the sample and their likelihoods are those shown in

10.2 Probability Function of a Sample Mean

TABLE 10.2.1 Possible Outcomes from a Sample of Two Numbers

Sample	Likelihood	Value of Y	Value of $\bar{X}$
(1.1)	1/9	2	1.0
(1.2)	1/9	3	1.5
(1.3)	1/9	4	2.0
(2.1)	1/9	3	1.5
(2.2)	1/9	4	2.0
(2.3)	1/9	5	2.5
(3.1)	1/9	4	2.0
(3.2)	1/9	5	2.5
(3.3)	1/9	6	3.0
	1.0		

Table 10.2.1. Summarizing the information provided in this table, we obtain

$\bar{x}$	$f(\bar{x})$
1.0	1/9
1.5	2/9
2.0	3/9
2.5	2/9
3.0	1/9
	1.0

which is the probability function of $\bar{X}$.

EXAMPLE 10.2.2

Assume that we are to select a simple random sample of three items from the following probability function.

x	$f(x)$
1	.6
4	.4
	1.0

Let $Y = X_1 + X_2 + X_3$ and $\bar{X} = (1/3)Y$. Then, various outcomes obtainable from the samples and their likelihoods are shown in Table 10.2.2. Summarizing

TABLE 10.2.2 Possible Outcomes from a Sample of Three Items

Sample	Likelihood	Value of Y	Value of $\bar{X}$
(1,1,1)	.216	3	1
(1,1,4)	.144	6	2
(1,4,1)	.144	6	2
(1,4,4)	.096	9	3
(4,1,1)	.144	6	2
(4,1,4)	.096	9	3
(4,4,1)	.096	9	3
(4,4,4)	.064	12	4

the information in this table, we obtain

$\bar{x}$	$f(\bar{x})$
1	.216
2	.432
3	.288
4	.064
	1.000

which is the probability function of the sample mean.

The preceding two examples show that if a random sample is taken from the probability function of a discrete random variable, then the probability function of the sample mean is also discrete. Thus, we are able to list various values that the sample mean can assume in addition to the associated probabilities. On the other hand, if a sample is to be taken from the probability function of a continuous random variable, then we cannot illustrate the probability function by listing the values that the sample mean can assume. Rather, we must use another approach to illustrate this type of probability function.

EXAMPLE 10.2.3

Assume that we are to take a simple random sample of two items from the probability function depicted by Figure 10.2.1. Simple calculus manipulations will reveal that $E(X) = 5$ and $V(X) = 25/3$. Let $Y = X_1 + X_2$ and $\bar{X} = (1/2)Y$. Then, the density function of Y is that shown in Figure 10.2.2, and, in turn, the density function of $\bar{X}$ is that shown in Figure 10.2.3. Figure 10.2.3, for example, indicates that the probability that the value of $\bar{X}$ will be between 4 and 6 is .36. The graph in Figure 10.2.3 depicts the probability density function of $\bar{X}$.

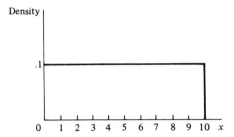
FIGURE 10.2.1

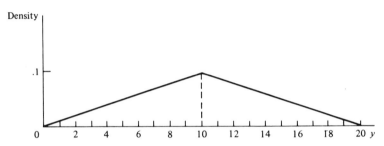
FIGURE 10.2.2

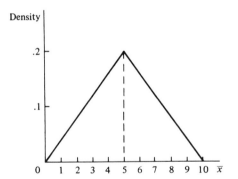
FIGURE 10.2.3

Theorem 10.2.1

Let $\bar{X}$ be a random variable associated with the sample mean of a simple random sample of size n from a probability function with given values of μ and σ. Then,

$$E(\bar{X}) = \mu$$

and
$$V(\bar{X}) = \frac{\sigma^2}{n}.$$

EXAMPLE 10.2.4

Let us return to the probability function of $\bar{X}$ in Example 10.2.1. We can calculate the expected value and the variance of $\bar{X}$ as shown in Table 10.2.3. Thus, $E(\bar{X}) = 2$ and $V(\bar{X}) = 1/3$.

TABLE 10.2.3 Expected Value and Variance of $\bar{X}$

$\bar{x}$	$f(\bar{x})$	$\bar{x}f(\bar{x})$	$[\bar{x} - E(\bar{X})]^2 f(\bar{x})$
1.0	1/9	1/9	2/18
1.5	2/9	3/9	1/18
2.0	3/9	6/9	0
2.5	2/9	5/9	1/18
3.0	1/9	3/9	2/18
		2.0	1/3

We can also calculate these values in another way. Since $\mu = 2$ and $\sigma^2 = 2/3$, according to Theorem 10.2.1, we have

$$E(\bar{X}) = 2$$

and

$$V(\bar{X}) = \frac{(2/3)}{2} = \frac{1}{3},$$

which yields the same results as those obtained in Table 10.2.3.

EXAMPLE 10.2.5

Let us return to the probability function of $\bar{X}$ in Example 10.2.3. If we wish to calculate $E(\bar{X})$ and $V(\bar{X})$, we cannot do so in tabular format, as illustrated in Example 10.2.4, since X in Example 10.2.3 is a continuous random variable, whereas X in Example 10.2.1 is a discrete random variable. On the other hand, since $\mu = 5$ and $\sigma^2 = 25/3$, we can obtain

$$E(\bar{X}) = 5$$

and

$$V(\bar{X}) = \frac{(25/3)}{2} = \frac{25}{6},$$

according to Theorem 10.2.1.

Definition 10.2.2

Let $\sigma_{\bar{x}} = \sigma/\sqrt{n}$. Then, $\sigma_{\bar{x}}$ is called the *standard error of mean*.

The *standard error of mean* is simply the standard deviation of the random variable $\bar{X}$, and its value is obtained by finding the square root of $V(\bar{X})$.

Theorem 10.2.2

Let $\bar{X}$ be the random variable associated with the sample mean of a simple random sample of size n from the probability function of a random variable. Then, the probability function of $\bar{X}$ approaches a normal probability function as n becomes larger and larger.

This theorem is the central limit theorem applied to the probability function of $\bar{X}$. A practical application of the theorem is illustrated by the following example.

EXAMPLE 10.2.6

Assume that we are to take a simple random sample of 100 items from a process producing ball bearings in order to estimate their average diameter. Let $\bar{X}$ denote the random variable associated with the sample mean. Then, we know that $E(\bar{X}) = \mu$ and $V(\bar{X}) = \sigma^2/n$. Furthermore, $\bar{X}$ must be approximately a normal distribution, according to Theorem 10.2.2.

Suppose, for example, $\mu = .5$ inches and $\sigma = .01$ inches. Then, $E(\bar{X}) = .5$ and $\sigma_{\bar{x}} = .01/\sqrt{100} = .001$. This means that the probability that the value of $\bar{X}$ will be, for example, between .498 and .502 inches is approximately .9544, as Figure 10.2.4 illustrates. If the average diameter of the ball bearings produced by the process is, in fact, .500 inches, then there is approximately a 95 percent chance that the average diameter for a simple random sample of 100 ball bearings will be between .498 and .502 inches.

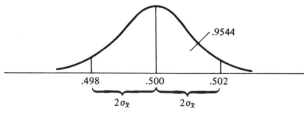

FIGURE 10.2.4

EXERCISES

*9. Suppose a process generates the numbers 1, 2, and 3 with equal probabilities. You are to take a simple random sample of two numbers from the process. Describe the probability function of $\bar{X}$. Then, find $E(\bar{X})$ and $V(\bar{X})$.

*10. Assume that you are to take a simple random sample of 600 from the process given in Exercise 9. Find $E(\bar{X})$ and $V(\bar{X})$. Then, calculate: (a) $P(|\bar{X} - \mu| \leq 1/15)$, (b) $P(-1/15 \leq \bar{X} - \mu \leq 1/15)$, and (c) $P(-1/30 \leq \bar{X} - \mu \leq 1/15)$.

11. Assume a process generates the numbers 1, 2, and 4 with associated probabilities .50, .25, and .25, respectively. You are to take a simple random sample of two items from the process. Describe the probability function of $\bar{X}$. Then, find $E(\bar{X})$ and $V(\bar{X})$.

12. Assume that you are to take a simple random sample of 600 from the process given in Exercise 11. Find $E(\bar{X})$ and $V(\bar{X})$. Then, calculate: (a) $P(|\bar{X} - \mu| \leq .3)$, (b) $P(-.3 \leq \bar{X} - \mu \leq .3)$, and (c) $P(-.45 \leq \bar{X} - \mu \leq .15)$.

*13. The average diameter of the ball bearings produced by a machine is known to be .5 inches, and the standard deviation for the diameters of the ball bearings is .01 inches. Suppose that a simple random sample of 100 is selected from the process. Suppose you are to determine the average diameter for the sample. What is the probability that the value of the sample mean will be: (a) between .497 and .503 inches, (b) less than .498 inches, and (c) greater than .501 inches?

14. Assume that at a given point in history, the average household income in the United States is $10,000 with a standard deviation of $4000. Suppose a simple random sample of 10,000 households is selected from the entire United States. What is the probability that the value of the mean calculated for the sample will: (a) lie between $9900 and $10,100 or (b) differ from the true mean by more than $100?

10.3 Sampling Errors

In evaluating the probability functions of P and $\bar{X}$, we have assumed that the values of π and μ are known, whereas, in practice, this is seldom true. Suppose, therefore, that the underlying reason for obtaining the values of P and $\bar{X}$ from specific samples is to estimate the values of π and μ, respectively. Then, we would naturally want the values of P and $\bar{X}$ to be close approximations of the unknown values of π and μ; that is, we want the difference between the values of P and π and the values of $\bar{X}$ and μ to be as small as possible. The differences between the estimated values, P and $\bar{X}$, and their respective, unknown counterparts, π and μ, are called *sampling errors*.

We will discuss the statistical properties of sampling errors next.

Definition 10.3.1

Let $\boldsymbol{\varepsilon}_p{}^* = P - \pi$, where P is the random variable associated with the sample proportion of a Bernoulli probability function with a given value of π. Then, $\boldsymbol{\varepsilon}_p$ is the random variable associated with the sampling errors in estimating π.

Since $\boldsymbol{\varepsilon}_p$ is a random variable, it has a probability function, which is illustrated in Examples 10.3.1 and 10.3.2.

*From this point on, when a Greek letter depicts a random variable, the Greek letter will be printed in boldface. The value that the random variable can assume, however, will be depicted by the same Greek letter in lightface.

10.3 Sampling Errors

TABLE 10.3.1

p	$f(p)$	ϵ_p	$f(\epsilon_p)$
0	1/4	−.5	1/4
.5	2/4	0	2/4
1.0	1/4	.5	1/4
	1.0		1.0

EXAMPLE 10.3.1

Let us return to Example 10.1.1. The probability function of P for that example is shown in the left-hand column of Table 10.3.1. We recall that the probability function of P was based on the assumption that $\pi = .5$. This means that if the value of P is 0, then the value of ϵ_p must be $-.5$. Furthermore, the probability that ϵ_p will assume the value $-.5$ is $1/4$, since the probability that P will assume the value 0 is $1/4$. The probability that ϵ_p will assume another value can be calculated in a similar manner. The probability function of ϵ_p thus obtained is shown in the right-hand column of Table 10.3.1. The relationship between the probability function of P and of ϵ_p is illustrated in Figure 10.3.1.

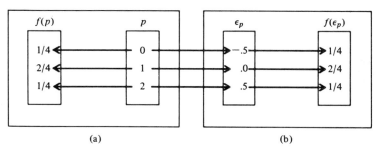

FIGURE 10.3.1 (a) Probability function of P and (b) probability function of ϵ_p.

EXAMPLE 10.3.2

Let us return to Example 10.1.2. The probability function of P for that example is given in Table 10.3.2. We recall that the probability function of P was based on the assumption that $\pi = .2$. Thus, the probability function of ϵ_p may be described as we have done in Table 10.3.2.

Definition 10.3.2

Let $\epsilon_{\bar{x}} = \bar{X} - \mu$, where $\bar{X}$ is the random variable associated with the sample mean taken from a probability function with a given value of μ.

TABLE 10.3.2

p	$f(p)$	ϵ_p	$f(\epsilon_p)$
0	.64	−.2	.64
.5	.32	.3	.32
1.0	.04	.8	.02
	1.00		1.00

Then, $\varepsilon_{\bar{x}}$ is the random variable associated with the sampling errors in estimating μ.

Since $\varepsilon_{\bar{x}}$ is a random variable, it has a probability function, which Table 10.3.3 illustrates.

TABLE 10.3.3 Probability Functions of $\bar{X}$ and $\varepsilon_{\bar{x}}$

$\bar{x}$	$f(\bar{x})$	$\epsilon_{\bar{x}}$	$f(\epsilon_{\bar{x}})$
1.0	1/9	−1.0	1/9
1.5	2/9	−.5	2/9
2.0	3/9	0	3/9
2.5	2/9	.5	2/9
3.0	1/9	1.0	1/9
	1.0		1.0

EXAMPLE 10.3.3

Let us return to Example 10.2.1. The probability function of $\bar{X}$ is also shown in Table 10.3.3. We recall that the probability function of $\bar{X}$ was based on the assumption that $\mu = 2$. Then, the value of $\varepsilon_{\bar{x}}$ is -1.0 if the value of $\bar{X}$ is 1.0. Furthermore, the probability that $\varepsilon_{\bar{x}}$ will assume the value -1.0 is 1/9, since the probability that $\bar{X}$ will assume the value 1.0 is 1/9. In a similar man-

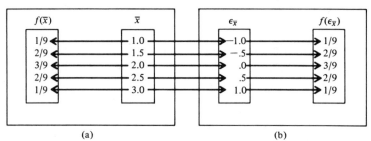

FIGURE 10.3.2 (a) Probability function of $\bar{X}$ and (b) probability function of $\varepsilon_{\bar{x}}$.

ner, we can obtain other values that $\varepsilon_{\bar{x}}$ will assume, as well as their associated probabilities. The probability function of $\varepsilon_{\bar{x}}$ thus obtained, as well as the relationship between the probability function of $\bar{X}$ and that of $\varepsilon_{\bar{x}}$, are illustrated in Figure 10.3.2.

10.4 Expectation and Variance of Sampling Errors

Since ε_p and $\varepsilon_{\bar{x}}$ are random variables, they must each have an expected value and a variance, which obviously can be obtained from the probability functions of these random variables. However, the following two theorems offer simpler ways of obtaining these values.

We will first introduce a theorem concerning the expectation and the variance of ε_p.

Theorem 10.4.1

Let $\varepsilon_p = P - \pi$. Then,

$$E(\varepsilon_p) = 0$$

and

$$V(\varepsilon_p) = \frac{\pi(1-\pi)}{n},$$

where n depicts the sample size.

The validity of this theorem may be stated as follows. We recall that if $Y = a + X$ where a is a constant and X is a random variable, then $E(Y) = a + E(X)$ and $V(Y) = V(X)$. Since $\varepsilon_p = P - \pi$ where P is a random variable, and $E(P) = \pi$, $V(P) = [\pi(1-\pi)]/n$, and π is an unknown constant, then,

$$E(\varepsilon_p) = E(P) - \pi = \pi - \pi = 0$$

and

$$V(\varepsilon_p) = V(P) = \frac{\pi(1-\pi)}{n}.$$

We will also illustrate the theorem by means of the following example.

EXAMPLE 10.4.1

Consider the probability function of ε_p given in Example 10.3.1. From the values given in Table 10.4.1, we find that $E(\varepsilon_p) = 0$ and that $V(\varepsilon_p) = \frac{1}{8}$. On the other hand, according to Theorem 10.4.1, we obtain

$$E(\varepsilon_p) = 0$$

and

$$V(\varepsilon_p) = \frac{(.5)(.5)}{2} = \frac{1}{8},$$

Sampling Distribution

TABLE 10.4.1

ϵ_p	$f(\epsilon_p)$	$\epsilon_p f(\epsilon_p)$	$[\epsilon_p - E(\epsilon_p)]^2 f(\epsilon_p)$
−.5	1/4	−1/8	1/16
0	2/4	0	0
.5	1/4	1/8	1/16
	1.0	0	1/8

which yield the same results as those obtained in Table 10.4.1. If we review Example 10.1.3, we find that $V(P)$ is equal to $V(\epsilon_p)$. This also means that the standard deviation of ϵ_p must also be the same as that of P.

For the sake of completeness, we offer the following definition.

Definition 10.4.1

Let $\sigma_{\epsilon_p} = \sqrt{\pi(1-\pi)/n}$. Then, σ_{ϵ_p} is called the *standard error of proportion*.

Thus, we have given the same name for two different notations, σ_p and σ_{ϵ_p}, since their values must always be the same.

We will now introduce a theorem concerning the expectation and the variance for $\bar{X}$.

Theorem 10.4.2

Let $\varepsilon_{\bar{x}} = \bar{X} - \mu$. Then,

$$E(\varepsilon_{\bar{x}}) = 0$$

and

$$V(\varepsilon_{\bar{x}}) = \frac{\sigma^2}{n}$$

where n depicts the sample size.

The validity of this theorem may be stated as follows: Since $E(\bar{X}) = \mu$ and $V(\bar{X}) = \sigma^2/n$,

$$E(\varepsilon_{\bar{x}}) = E(\bar{X}) - \mu = \mu - \mu = 0$$

and

$$V(\varepsilon_{\bar{x}}) = V(\bar{X}) = \frac{\sigma^2}{n}.$$

We will also illustrate the theorem in the following example.

EXAMPLE 10.4.2

Consider the probability function of $\varepsilon_{\bar{x}}$ given in Example 10.3.3. We make the calculations given in Table 10.4.2. Thus, we find that $E(\varepsilon_{\bar{x}}) = 0$ and $V(\varepsilon_{\bar{x}}) = 1/3$.

TABLE 10.4.2

$\epsilon_{\bar{x}}$	$f(\epsilon_{\bar{x}})$	$\epsilon_{\bar{x}} f(\epsilon_{\bar{x}})$	$[\epsilon_{\bar{x}} - E(\epsilon_{\bar{x}})]^2 f(\epsilon_{\bar{x}})$
-1.0	$1/9$	$-2/18$	$2/18$
$-.5$	$2/9$	$-2/18$	$1/18$
0	$3/9$	0	0
$.5$	$2/9$	$2/18$	$1/18$
1.0	$1/9$	$2/18$	$2/18$
	1.0	0	$1/3$

If we review Example 10.2.4, we will find that $V(\boldsymbol{\epsilon}_{\bar{x}})$ is equal to $V(\bar{X})$. In turn, the standard deviation of $\boldsymbol{\epsilon}_{\bar{x}}$ must also be the same as that of $\bar{X}$. Thus,

Definition 10.4.2

Let $\sigma_{\epsilon_{\bar{x}}} = \sigma/\sqrt{n}$. Then, $\sigma_{\epsilon_{\bar{x}}}$ is called the *standard error of mean*.

Thus, we have given the same name for the two different notations $\sigma_{\bar{x}}$ and $\sigma_{\epsilon_{\bar{x}}}$, since the values corresponding to these notations must always be the same.

EXERCISES

15. A process generates the numbers 0 and 1 with equal probabilities. You are to select a simple random sample of two from the process. Describe the probability functions for $\boldsymbol{\epsilon}_p$. Then, find $E(\boldsymbol{\epsilon}_p)$ and $V(\boldsymbol{\epsilon}_p)$.

*16. A simple random sample of two is to be selected from the following probability function:

x	$f(x)$
0	.8
1	.2
	1.0

Describe the probability function for $\boldsymbol{\epsilon}_p$. Then, find $E(\boldsymbol{\epsilon}_p)$ and $V(\boldsymbol{\epsilon}_p)$.

17. A box contains the following five numbers: 0, 0, 1, 1, 1. You are to select a simple random sample of four. Describe the probability function for $\boldsymbol{\epsilon}_p$. Then, find $E(\boldsymbol{\epsilon}_p)$ and $V(\boldsymbol{\epsilon}_p)$.

*18. A process generates the numbers 1, 2, and 3 with equal probabilities. You are to take a random sample of two numbers from the process. Describe the probability function for $\boldsymbol{\epsilon}_{\bar{x}}$. Then, calculate $E(\boldsymbol{\epsilon}_{\bar{x}})$ and $V(\boldsymbol{\epsilon}_{\bar{x}})$.

19. A process generates the numbers 1, 2, and 4 with the associated probabilities .50, .25, and .25, respectively. Describe the probability function for $\varepsilon_{\bar{x}}$. Then, calculate $E(\varepsilon_{\bar{x}})$ and $V(\varepsilon_{\bar{x}})$.

10.5 Law of Large Numbers

Intuitively, it must be apparent that if two different sizes of simple random samples are taken from a given probability function, the larger sample will more likely yield a smaller sampling error than the smaller sample. This intuitive proposition is not without theoretical justification, as we will illustrate in this section.

Theorem 10.5.1

Let $\varepsilon_p = P - \pi$ for any sample size n. Then, σ_{ε_p} approaches zero as n becomes larger and larger.

The validity of this theorem should be obvious. Since

$$\sigma_{\varepsilon_p} = \frac{\pi(1-\pi)}{n}$$

as n becomes larger and larger, σ_{ε_p} will become smaller and smaller, since $\pi(1-\pi)$ is constant.

We will now extend Theorem 10.5.1 to a more general case.

Theorem 10.5.2

Let $\varepsilon_{\bar{x}} = \bar{X} - \mu$ for any sample size n. Then, $\sigma_{\varepsilon_{\bar{x}}}$ approaches zero as n becomes larger and larger.

The validity of this theorem may be justified by the fact that, since σ is a constant for $\sigma_{\varepsilon_{\bar{x}}} = \sigma/\sqrt{n}$, the value of $\sigma_{\varepsilon_{\bar{x}}}$ will become smaller and smaller as n becomes larger and larger.

Let us now evaluate Theorems 10.5.1 and 10.5.2 in conjunction with the Tchebycheff inequality. Together, these two theorems suggest that the probability that the sampling error will exceed any arbitrary positive value must approach zero as n becomes larger and larger. This proposition is called the *law of large numbers*. We will present this law more formally in the following two theorems.

Theorem 10.5.3

Let $\varepsilon_p = P - \pi$, where P is a random variable associated with the sample proportion to be obtained by a simple random sample of size n from a Bernoulli probability function. Then, for any arbitrary positive number δ, $P(|\varepsilon_p| \geq \delta)$ approaches zero as n becomes larger and larger.

EXAMPLE 10.5.1

Let us examine a sampling problem from the Bernoulli probability function with $\pi = .5$. Assume that even though the value of π is, in fact, .5, we do not know that this is the case. We will, therefore, take a simple random sample of size n from the probability function and calculate the value of P, which will serve as our estimate of π. Suppose we now ask, "What is the probability that the error committed in estimating the value of π might exceed .1?" The answer to this question depends on the size of the sample to be taken, as we will now show.

Let us first assume that $n = 100$. Then,

$$E(\varepsilon_p) = 0$$
and
$$\sigma_{\varepsilon_p} = .05.$$

According to the Tchebycheff inequality, we have

$$P(|\varepsilon_p| \geq .1) = P(|\varepsilon_p| \geq 2\sigma_{\varepsilon_p}) \leq \frac{1}{4}.$$

Thus, we conclude that the probability in question must be less than .25.

Suppose now that $n = 2500$. Then,

$$E(\varepsilon_p) = 0$$
and
$$\sigma_{\varepsilon_p} = .01.$$

According to the Tchebycheff inequality, we have

$$P(|\varepsilon_p| \geq .1) = P(|\varepsilon_p| \geq 10\sigma_{\varepsilon_p}) \leq = \frac{1}{100}.$$

Thus, the probability in question must be less than .01.

It must be apparent now that as the sample size increases, the upper limit for the probability must approach zero. This also suggests that the actual probability must approach zero.

Theorem 10.5.4

Let $\varepsilon_{\bar{x}} = \bar{X} - \mu$ where $\bar{X}$ is the random variable associated with the sample mean to be obtained from a simple random sample of size n from a probability function. Then, for any arbitrary positive number δ, $P(|\varepsilon_{\bar{x}}| \geq \delta)$ approaches zero as n becomes larger and larger.

EXAMPLE 10.5.2

Let us assume that a simple random sample of size n is to be selected from a probability function in order to estimate the value of μ. Let $\bar{X}$ denote the random variable associated with the estimated value of the mean. Then, $\varepsilon_{\bar{x}} = \bar{X} - \mu$ is the random variable associated with the sampling error.

Suppose we ask, "What is the probability that the value of the sampling error will exceed 2 units?" (Note that the numerical value "2" is chosen arbitrarily.) To answer the question in specific, numerical terms, we need first to know the value of the standard deviation of the probability function from which the sample is to be taken. Assume, therefore, that the standard deviation in question, denoted by σ, is 10 units. Then, we can answer our question if we know the sample size.

Let us first assume that $n = 100$. Then,

$$E(\varepsilon_{\bar{x}}) = 0$$

and

$$\sigma_{\varepsilon_{\bar{x}}} = 1.$$

According to the Tchebycheff inequality, we have

$$P(\varepsilon_{\bar{x}}| \geq 2) = P(|\varepsilon_{\bar{x}}| \geq 2\sigma_{\varepsilon_{\bar{x}}}) \leq \frac{1}{4}.$$

Thus, we conclude that the probability in question must be less than .25.

Suppose now that $n = 10{,}000$. Then,

$$E(\varepsilon_{\bar{x}}) = 0$$

and

$$\sigma_{\varepsilon_{\bar{x}}} = .01.$$

Then, according to the Tchebycheff inequality, we have

$$P(|\varepsilon_{\bar{x}}| \geq 2) = P(|\varepsilon_{\bar{x}}| \geq 20\sigma_{\varepsilon_{\bar{x}}}) = \frac{1}{400}.$$

Thus, the probability in question must be less than .0025.

Again, it must be apparent that as the sample size increases, the upper limit for the probability in question must approach zero; then, the actual probability also must approach zero. This is what is implied by Theorem 10.5.4.

EXERCISES

*20. Let π denote the probability that a coin will land on "heads." Let P denote the estimate of π to be obtained from tossing a coin n times. Show that P approaches π as n becomes larger and larger.

21. A die is to be tossed n times. You are interested in the number of times 3 will appear. Let k depict the number of times 3 will appear in n tosses. What does the law of large numbers say about the relationship between k and n? Show the validity of your answer by numerical illustrations.

10.6 Central Limit Theorem for Sampling Errors

One implication which follows from Theorems 10.5.3 and 10.5.4 is that by taking a large enough sample we can be sure that the probability that the

10.6 Central Limit Theorem for Sampling Errors

sampling errors will exceed any specified value will be made as small as we wish. The next question is, "What size sample is needed in order to insure that the probability of sampling errors exceeding the given value will be kept to a specified level or less?" Such a question may be answered by extending the Tchebycheff inequality to the underlying problem. However, a more accurate answer may be obtained by applying the central limit theorem. We will, therefore, extend the central limit theorem to the probability function of sampling errors. In the following section we will discuss the question of sample size.

Theorem 10.6.1

Let $\varepsilon_p = P - \pi$, where P is a random variable associated with the sample proportion to be obtained by a simple random sample of size n from a Bernoulli probability function. Then, the probability function of ε_p approaches a normal probability function as n becomes larger and larger.

The validity of this theorem results from the fact that π is a constant, and the probability function of P approaches a normal probability function as n becomes larger and larger.

The ramifications of this theorem for sampling problems will be illustrated in the following example.

EXAMPLE 10.6.1

Assume that we are to select a simple random sample of 100 items from a Bernoulli probability function in order to estimate the value of π. Theorem 10.5.3 indicates that

$$E(\varepsilon_p) = 0$$

and

$$V(\varepsilon_p) = \frac{\pi(1-\pi)}{n},$$

and, in turn, $\sigma_{\varepsilon_p} = \sqrt{\pi(1-\pi)/n}$. Theorem 10.6.1 indicates that the probability function of ε_p is approximately normal.

Suppose we wish to calculate the probability that the sampling error will exceed δ, where δ is an arbitrarily chosen value. We can proceed to obtain this probability in the following manner. Let

$$h = \frac{\delta}{\sqrt{\pi(1-\pi)/n}}.$$

Then, δ is equal to h times the standard error of proportion. This means that

$$P(|\varepsilon_p| \geq \delta) = P(|\varepsilon_p| \geq h\sigma_{\varepsilon_p}) \doteq P(|Z| \geq h),$$

since ε_p is approximately normally distributed. Thus, once the value of h is determined, finding the probability in question is a straightforward procedure. Using the table of standard normal variables (Table B.1 in Appendix B), we simply calculate $P(|Z| \geq h)$.

There is one problem in obtaining the value of h. In order to obtain the value of h, we need to know the value of π. If we knew the value of π, we would not be trying to estimate it in the first place. Thus, it appears that our discussion has come to a useless end. But we will point out that the situation is not really this bad. Assume, for example, that we wish to calculate the probability that $P(|\varepsilon_p| \geq .1)$. Thus, we arbitrarily set the value of δ to be .1. We might first evaluate $P(|\varepsilon_p| \geq .1)$ by making various assumptions with regard to the value of π, as illustrated in Table 10.6.1.

TABLE 10.6.1

Assumed Value of π	σ_{ε_p}	h	$P(|Z| \geq h)$
.90	.030	3.33	.0008
.80	.040	2.50	.0124
.70	.046	2.17	.0300
.60	.049	2.04	.0414
.50	.050	2.00	.0456
.40	.049	2.04	.0414
.30	.046	2.17	.0300
.20	.040	2.50	.0124
.10	.030	3.33	.0008

Table 10.6.1 shows that if we have reason to believe that the value of π is .2, then $P(|\varepsilon_p| \geq .1)$ is approximately .0124, as indicated in the extreme right-hand column. On the other hand, suppose we have absolutely no information pertaining to the value of π. The table shows that regardless of the value of π, $P(|\varepsilon_p| \geq .1)$ cannot be greater than .0456. Thus, we may conclude in such an instance that, at worst, $P(|\varepsilon_p| \geq .1)$ should be approximately .0456 or less.

Theorem 10.6.2

Let $\varepsilon_{\bar{x}} = \bar{X} - \mu$, where $\bar{X}$ is the random variable associated with the sample mean to be obtained from a simple random sample of size n from a probability function. Then, the probability function of $\varepsilon_{\bar{x}}$ approaches a normal probability function as n becomes larger and larger.

The validity of this theorem results from the fact that μ is a constant and the probability function of $\bar{X}$ approaches a normal probability function as n becomes larger and larger.

The significance of this theorem for sampling problems will be illustrated by the following example.

EXAMPLE 10.6.2

Assume that we are to select a simple random sample of 100 items from a probability function in order to estimate the value of μ. Then, according to Theorem 10.5.4, we know that

$$E(\varepsilon_{\bar{x}}) = 0$$

and

$$V(\varepsilon_{\bar{x}}) = \frac{\sigma^2}{n},$$

and, in turn, $\sigma_{\varepsilon_{\bar{x}}} = \sigma/\sqrt{n}$. Theorem 10.6.2 has shown that the probability function of $\varepsilon_{\bar{x}}$ is approximately normal.

Suppose we wish to calculate the probability that the sampling error will exceed an arbitrarily chosen value, denoted by δ. We can calculate the probability in the following manner. Let

$$h = \frac{\delta}{\sigma/\sqrt{n}}.$$

Then, δ is equal to h times the standard error of mean. This means that

$$P(|\varepsilon_{\bar{x}}| \geq \delta) = P(|\varepsilon_{\bar{x}}| \geq h\sigma_{\varepsilon_{\bar{x}}}) \doteq P(|Z| \geq h),$$

since $\varepsilon_{\bar{x}}$ is approximately normally distributed. Thus, once the value of h is determined, we can calculate the probability by finding $P(|Z| \geq h)$ from a table of standard normal variables, given in Table B.1 of Appendix B.

As was the case when we attempted sampling from a Bernoulli probability function, there is a problem in calculating the value of h, which arises because in order to calculate the value of h, we need to know the value of σ. We would know the value of σ only if we knew the probability function from which the sample is to be taken. Of course, if we knew the probability function in question, then there would be no reason to try to estimate the value of μ by sampling, since the value of μ could be calculated without having to take a sample. Thus, it appears that we are again trapped.

One pragmatic way to resolve the problem is to approach it as if we really knew the value of σ. We might, for example, assume that the value of σ is 10. Suppose we assigned the value of 2 to δ. Then,

$$h = \frac{\delta}{\sigma/\sqrt{n}} = \frac{2}{10/\sqrt{100}} = 1.$$

Thus,

$$P(|\varepsilon_{\bar{x}}| \geq 2) = P(|\varepsilon_{\bar{x}}| \geq 2\sigma_{\varepsilon_{\bar{x}}}) \doteq P(|Z| \geq 2) = .0456.$$

How accurate would be the conclusion attained by assuming a value of σ? Obviously, it would be accurate if the assumed value of σ is fairly close to the true value of σ. Although the preceding statement appears to be merely a tautology, it does offer important advice to us. The statement suggests that if we follow the approach prescribed for resolving the problem connected with the value of σ, then we should make use of whatever information is available to us pertaining to the probability function to be sampled, so that our assumed value of σ will be as close as possible to the true value of σ.

EXERCISES

22. Assume that you are to select a simple random sample of 100 from the process given in Exercise 15. Find $E(\varepsilon_p)$ and $V(\varepsilon_p)$. Then, calculate (a) $P(|\varepsilon_p| \geq .05)$ and (b) $P(|\varepsilon_p| \geq .10)$.

*23. Assume that you are to select a simple random sample of 100 from the probability function given in Exercise 16. Find $E(\varepsilon_p)$ and $V(\varepsilon_p)$. Then, calculate (a) $P(|\varepsilon_p| \geq .08)$ and (b) $P(|\varepsilon_p| \leq .08)$.

24. Assume that you are to take a simple random sample of 100 from a process in order to estimate the proportion of defectives produced by the process. Suppose the process is, in fact, producing 10 percent defectives, even though you do not know that this is the case. Then, what is the probability that your estimating error will exceed 6 percent?

*25. It is known that a process generates 20 percent defectives. Suppose you are to take a sample of 100 items from the process. What is the probability that the proportion of defectives for your sample may differ from the proportion of defectives produced by the process by more than 5 percent?

26. Refer to Exercise 25 and suppose you are to take a sample of 1600 items. What is the probability that the proportion of defectives in your sample may differ from the proportion of defectives produced by the process by more than 3 percent?

*27. Assume that you are to take a simple random sample of 600 items from the process given in Exercise 18. Find $E(\varepsilon_{\bar{x}})$ and $V(\varepsilon_{\bar{x}})$. Then, calculate (a) $P(|\varepsilon_{\bar{x}}| \leq 1/15)$ and (b) $P(-1/30 \leq \varepsilon_{\bar{x}} \leq 1/15)$.

28. Assume that you are to take a simple random sample of 600 from the process given in Exercise 18. Find $E(\varepsilon_{\bar{x}})$ and $V(\varepsilon_{\bar{x}})$. Then, calculate (a) $P(|\varepsilon_{\bar{x}}| \leq .3)$ and (b) $P(-.45 \leq \varepsilon_{\bar{x}} \leq .15)$.

29. Suppose a process generates numbers. The mean of the process is not known, however, the standard deviation of the process is known to be 20. You plan to estimate the process mean by taking a random sample of 100 numbers generated by the process. What is the probability that your estimate, based on a sample of 100, will deviate from the true process mean by more than 4?

30. Refer to the problem given in Exercise 29 and suppose you are to take a random sample of 10,000. What is the probability that your estimate will deviate from the true process mean by more than .5?

31. Assume you are a producer of a certain brand of detergent and you wish to estimate the average weight for the boxes of detergent. Even though the exact mean weight for the manufacturing process is not known, the standard deviation of the process is known to be .1 ounce. Suppose that you take a random sample of

100 boxes of detergent and find the mean weight for your sample. What is the probability that the sample mean thus obtained might be different from the process mean by more than .03 ounce?

*32. A manufacturing process produces ball bearings. You plan to take a simple random sample of 100 ball bearings in order to estimate the average diameter for the ball bearings produced by this process. Suppose the standard deviation for the diameters of the ball bearings is .01 inches. Then, what is the probability that your estimating error will be within .003 inches?

33. You plan to take a simple random sample of 10,000 from the household population in the United States, in order to estimate the average household income. Suppose the standard deviation of household incomes in the United States is $4000. What is the probability that the estimate to be obtained from the sample might differ from the true figure by more than $100?

10.7 Sample Size and Sampling Errors

Frequently, someone will relay his sampling problem to us and then ask, "How large a sample will I need?" Invariably, we have to ask him in return, "How accurate do you want the results from your sample to be?" since he usually fails to discuss this point before he asks about the necessary sample size. Obviously, we cannot answer the question with regard to the sample without knowing what degree of accuracy is desired from the sample.

There is another problem which usually confronts us when we try to answer the question of how large a size is needed for a sample. When we ask about the degree of accuracy desired from the sample, we are frequently told, "I would like very accurate results," "I would like fairly accurate results," or other similar statements. One should take a very large sample if he desires very accurate results from his sample, and, conversely, one should take a relatively small sample if he can accept results which are not very accurate. The preceding statement is a truism. It does not provide a meaningful answer to the question of the sample size, but on the other hand, it suggests that the desired accuracy from a sample should be expressed more specifically than those given in the preceding quotations. There are ways to express the degree of accuracy desired from a sample so that the necessary sample size which will satisfy the desired accuracy can always be calculated.

Consider that we are to take a simple random sample either from a Bernoulli probability function in order to estimate π, or from a non-Bernoulli probability function in order to estimate μ. To say that we would like the sample to give us a very accurate result is equivalent to saying that we would like the sampling error to be very small, which is still ambiguous. However, this ambiguity may be clarified by stating that we would like to take a sample large enough so that the probability of the sampling error exceeding δ will be kept as small as, or less than, α. In notation the statement may be expressed as

$$P(|\varepsilon_p| \geq \delta) \leq \alpha,$$

if the sample is to be taken from a Bernoulli probability function, and

$$P(|\varepsilon_{\bar{x}}| \geq \delta) \leq \alpha,$$

if the sample is to be taken from a non-Bernoulli probability function. It should be understood that the values of δ and α are chosen in such a way that they reflect the accuracy desired from the sample.

Theorem 10.7.1

Let $P(|\varepsilon_p| \geq \delta) \leq \alpha$ reflect the accuracy desired from a simple random sample to be taken from a Bernoulli probability function. Then, the necessary sample size n, which will satisfy the desired accuracy, may be determined as

$$n = \frac{h^2 \pi (1 - \pi)}{\delta^2},$$

where the value of h is such that $P(|Z| \geq h) \leq \alpha$, provided that n is relatively large.

The validity of this theorem may be explained in the following way. Suppose the value of h is such that $P(|Z| \geq h) \leq \alpha$. If the sample size, which is yet to be determined, is large, ε_p will be approximately normally distributed. Then,

$$P(|\varepsilon_p| \geq \delta) = P(|\varepsilon_p| \geq h\sigma_{\varepsilon_p}) = P(|Z| \geq h) \leq \alpha.$$

Thus, $\delta = h\sigma_{\varepsilon_p}$, which, in turn, implies that

$$\delta = h \sqrt{\frac{\pi(1-\pi)}{n}}.$$

Solving the preceding equation for n, we obtain

$$n = \frac{h^2 \pi (1 - \pi)}{\delta^2}.$$

EXAMPLE 10.7.1

Assume that we are to select a simple random sample from a Bernoulli probability function in order to estimate π. We would like to take a sample large enough so that the probability of a sampling error exceeding .1 will be kept as small as or less than .0456. In notation, the degree of accuracy that we desire from the sample may be expressed as

$$P(|\varepsilon_p| \geq .1) \leq .0456.$$

First we must determine the values of h, so that $P(|Z| \geq h) \leq .0456$. From the table of normal probability functions (Table B.1 in Appendix B), we find that $P(|Z| \geq 2) \leq .0456$. Thus, $h = 2$. The necessary sample size may then be determined by finding

$$n = \frac{h^2 \pi (1 - \pi)}{\delta^2} = \frac{4\pi(1 - \pi)}{.01} = 400\pi(1 - \pi).$$

This equation indicates that the sample size in question may be determined only if we know the value of π. Thus, the problem mentioned in the preceding section re-occurs. One way to resolve the problem is to approach it as if we knew the value of π. Table 10.7.1 has been constructed to show the various assumed values of π.

TABLE 10.7.1

Assumed Value of π	$\pi(1 - \pi)$	Sample Size $n = 400\pi(1 - \pi)$
.9	.09	36
.8	.16	66
.7	.21	84
.6	.24	96
.5	.25	100
.4	.24	96
.3	.21	84
.2	.16	66
.1	.09	36

We can use the table in the following way. Suppose, on the basis of past experience, we have good reason to believe that the value of π should be approximately .2. Then, the sample size which will satisfy our desired accuracy is about 66. On the other hand, suppose we have absolutely no knowledge with regard to the value of π. Table 10.7.1 indicates that the largest sample size needed to satisfy the desired accuracy is 100. In other words, if a sample of 100 is taken from the probability function, then the accuracy that we desire from the sample is bound to be satisfied, regardless of the actual value of π.

In this respect, some statisticians have suggested that we might take a small pilot sample and estimate the value of π from it and, then, use the estimated value of π as if it were the true value, and calculate the sample size by using the formula given in Theorem 10.7.1.

Theorem 10.7.2

Let $P(|\varepsilon_{\bar{x}}| \geq \delta) \leq \alpha$ reflect the desired accuracy from a simple random sample to be taken from a non-Bernoulli probability function. Then, the necessary sample size which will satisfy the desired accuracy may be determined as

$$n = \frac{h^2\sigma^2}{\delta^2},$$

where the value of h is such that $P(|Z| \geq h) \leq \alpha$, provided that n turns out to be fairly large.

The validity of this theorem may be shown in the following way. Assume that h is determined so that $P(|Z| \geq h) \leq \alpha$. If the sample size, yet to be determined, turns out to be fairly large, then $\varepsilon_{\bar{x}}$ will be approximately normally distributed. Thus,

$$P(|\varepsilon_{\bar{x}}| \geq \delta) = P(|\varepsilon_{\bar{x}}| \geq h\sigma_{\varepsilon_{\bar{x}}}) = P(|Z| \geq h) \leq \alpha.$$

From this equation we find that $\delta = h\sigma_{\varepsilon_{\bar{x}}}$, which, in turn, implies that

$$\delta = h\frac{\sigma}{\sqrt{n}}.$$

Solving the above equation for n, we obtain

$$n = \frac{h^2\sigma^2}{\delta^2}.$$

EXAMPLE 10.7.2

Assume that we are to select a simple random sample from a probability function in order to estimate μ. We would like to take a sample large enough so that the probability of a sampling error exceeding 10 units will be kept as small as or less than .0456. In notation, the degree of accuracy desired may be expressed as

$$P(|\varepsilon_{\bar{x}}| \geq 10) \leq .0456.$$

First, we must determine the value of h, so that $P(|Z| > h) \leq .0456$. We find that $h = 2$. The necessary sample size may then be determined by finding

$$n = \frac{h^2\sigma^2}{\delta^2} = \frac{4\sigma^2}{100} = .04\sigma^2.$$

This equation indicates that the sample size in question may be determined only if we know the value of σ.

As we have suggested for the sampling problem from a Bernoulli probability function, one pragmatic way to resolve the problem here is to approach it as if we knew the value of σ. Suppose, for example, our experience in connection with the probability function indicates that the value of σ should be about 100. Then, we might assume that the true value of σ is, in fact, 100. The necessary sample size is calculated as

$$n = .04(100)^2 = 400.$$

Thus, we may conclude that the sample size, which will satisfy the desired accuracy, is about 400.

In the event that we cannot guess the value of σ, we can resort to a pilot sample. If this approach is used, then we should take a small pilot sample and estimate the value of σ from the information obtained from this sample. Then, we should use the estimated value of σ as if it were the true value of σ and calculate the sample size by means of the formula given in Theorem 10.7.2. We will describe in detail a method of estimating σ in the following chapter.

EXERCISES

*34. Suppose you wish to estimate the proportion of college students today who support a certain political philosophy. You wish to be at least 95 percent sure that the estimating error does not exceed 1 percent. How large a sample should you take?

35. Refer to the problem given in Exercise 34 and suppose you believe that probably 20 percent of the students support that certain political philosophy. How large a sample should you take in order to satisfy the objective given in Exercise 34?

36. Refer to the problem given in Exercise 29 and suppose you wish to take a sample large enough so that the probability of your sampling error exceeding 2 would be 5 percent, at the most. How large a sample should you take?

*37. Refer to the problem given in Exercise 31 and suppose you wish to take a sample large enough so that the probability of your sampling error exceeding .01 ounces would be 5 percent, at the most. How large a sample should you take?

Chapter 11
Statistical
Estimation

Our discussion in the preceding chapter might have conveyed the impression that the only problem that exists in estimating a parameter of a probability function is the question of sample size. For example, it might appear that once a sample of a given size is taken, the process of obtaining the desired estimate for the parameter from this sample is straightforward. However, there are a number of factors which need to be evaluated in order to determine the manner in which the desired parameter for the probability function is to be estimated from the sample data. We will examine a number of these factors in this chapter.

In this respect, we might point out now that two types of estimates may be obtained from a sample. The first type, a *point estimate*, refers to a single numerical figure given as the estimate for the given parameter. The second type, an *interval estimate* refers to a range between two numerical figures believed to contain the parameter in question. These two types of estimates, however, are not independent of each other since, as we will show later, an interval estimate is usually based on a point estimate.

We will first examine some of the problems connected with making a point estimate and, subsequently, will discuss the problems associated with making an interval estimate.

11.1 Criteria of a Good Estimator

Given a sample, there may be more than one way to utilize the data given in the sample to estimate a particular parameter of a probability function. For example, suppose that we are to estimate the mean value of a probability function from sample data. In the examples presented in the preceding chapters, we invariably used the sample mean as our estimate, which appears to be an intuitively sound procedure. There is also a good theoretical basis for using this procedure. However, use of the sample mean is not the only way we can estimate the mean of the probability function; we can use the sample median as an estimate, or we can use the mid-range value between the largest and the smallest figures in the sample. There are also other approaches that we can use to estimate the parameter in question.

Each of the procedures that we utilize in estimating the parameter is called an *estimator*. A specific numerical value which an estimator yields from a given sample is called an *estimate*. An estimator may yield a very bad estimate at times, but we may still prefer to utilize that estimator if we know that it will yield a fairly accurate estimate most of the time.

Since there are many estimators which can be utilized, some of them may be better than others in some way. Statisticians have devised a number of criteria which may be used to evaluate the usefulness of an estimator. We will consider two criteria in this section.

Definition 11.1.1

Let $\hat{\theta}$ be the random variable associated with a particular estimator for θ, where θ denotes the parameter of a probability function. The estimator is said to be *unbiased* if

$$E(\hat{\theta}) = \theta.$$

EXAMPLE 11.1.1

Consider the probability function given in Table 11.1.1. Our calculations will reveal that $\mu = 2.2$.

Let us assume that we are to select a sample of size 3 in order to estimate μ of the probability function. All possible samples of size 3 and their likelihoods are shown in Table 11.1.2.

TABLE 11.1.1

x	$f(x)$
1	.6
4	.4
	1.0

TABLE 11.1.2

Sample	Likelihood
(1,1,1)	.216
(1,1,4)	.144
(1,4,1)	.144
(1,4,4)	.096
(4,1,1)	.144
(4,1,4)	.096
(4,4,1)	.096
(4,4,4)	.064
	1.000

Suppose now we are to use the sample mean $\bar{X}$ to estimate μ of the probability function. In Table 11.1.3 we indicate the likelihood as well as the mean value for each possible sample. From this table we can derive the probability function of the sample mean $\bar{X}$, which is given in Table 11.1.4. The calculations in the right-hand column of the table show that $E(\bar{X}) = 2.2$.

Now suppose we are to use the sample mean $\bar{X}$ as our estimator for μ of the probability function. This is equivalent to saying that the random variable $\bar{X}$ is to play the role of $\hat{\theta}$ and that μ is to play the role of θ. Since $\mu = 2.2$

TABLE 11.1.3

Sample	Likelihood	Sample Mean
(1,1,1)	.216	1
(1,1,4)	.144	2
(1,4,1)	.144	2
(1,4,4)	.096	3
(4,1,1)	.144	2
(4,1,4)	.096	3
(4,4,1)	.096	3
(4,4,4)	.064	4
	1.000	

TABLE 11.1.4

$\bar{x}$	$f(\bar{x})$	$\bar{x}f(\bar{x})$
1	.216	.216
2	.432	.864
3	.288	.864
4	.064	.256
	1.000	2.200

and $E(\bar{X}) = 2.2$, we have $E(\bar{X}) = \mu$. In alternate notation, we have $E(\hat{\theta}) = \theta$. Thus, the sample mean $\bar{X}$ is an unbiased estimator for μ of the probability function.

Suppose that, instead of using the sample mean, we use the sample median to estimate μ of the probability function. In Table 11.1.5 we indicate the likelihood and median value for each possible sample. From this table we can derive the probability function of the sample median, which is shown in Table 11.1.6. We denote the random variable associated with the sample median by $\hat{X}$. The expected value of $\hat{X}$ is also shown in the bottom of the right-hand column of Table 11.1.6.

TABLE 11.1.5

Sample	Likelihood	Sample Median
(1,1,1)	.216	1
(1,1,4)	.144	1
(1,4,1)	.144	1
(1,4,4)	.096	4
(4,1,1)	.144	1
(4,1,4)	.096	4
(4,4,1)	.096	4
(4,4,4)	.064	4

TABLE 11.1.6

$\hat{x}$	$f(\hat{x})$	$\hat{x}f(\hat{x})$
1	.648	.648
4	.352	1.408
	1.000	2.056

We observe in the table that $E(\hat{X}) = 2.056$, which is different from $\mu = 2.2$. Thus, if we let $\hat{X}$ play the role of $\hat{\theta}$ and let μ play the role of θ, we note that $\hat{\theta}$ is not equal to θ. Therefore, the sample median is not an unbiased estimator for μ of the probability function.

Suppose we use the mid-range of the sample to estimate μ of the probability function. If we denote L_1 and L_2 as the smallest and the largest number in the sample, respectively, the mid-range is defined as

$$\text{mid-range} = \frac{L_1 + L_2}{2}.$$

In Table 11.1.7 we indicate the likelihood and mid-range for each possible sample.

TABLE 11.1.7

Sample	Likelihood	Mid-range
(1,1,1)	.216	1.0
(1,1,4)	.144	2.5
(1,4,1)	.144	2.5
(1,4,4)	.096	2.5
(4,1,1)	.144	2.5
(4,1,4)	.096	2.5
(4,4,1)	.096	2.5
(4,4,4)	.064	4.0
	1.000	

Let the random variable associated with the mid-range value be denoted by $\tilde{X}$. The probability function for $\tilde{X}$ is given in Table 11.1.8. Our calculations reveal that $E(\tilde{X}) = 2.272$. Thus, $E(\tilde{X})$ is not equal to μ and the estimator in question is biased.

TABLE 11.1.8

$\tilde{x}$	$f(\tilde{x})$	$\tilde{x}f(\tilde{x})$
1.0	.216	.216
2.5	.720	1.800
4.0	.064	.256
	1.000	2.272

In summary, among the three estimators that we have considered, only the one associated with the sample mean is unbiased.

We might describe the notion of *bias* in another context. Suppose that $\hat{\theta}_1$, $\hat{\theta}_2$, and $\hat{\theta}_3$ are random variables associated with three different estimators for θ. Assume that Figure 11.1.1 depicts the probability functions of $\hat{\theta}_1$, $\hat{\theta}_2$, and $\hat{\theta}_3$. Then, $\hat{\theta}_2$ is an unbiased estimator, since $E(\hat{\theta}_2) = \theta$, but $\hat{\theta}_1$ and $\hat{\theta}_3$ are biased estimators, since $E(\hat{\theta}_1) < \theta$ and $E(\hat{\theta}_3) > \theta$. An evaluation of Figure 11.1.1 will reveal that $\hat{\theta}_1$ tends to underestimate θ and $\hat{\theta}_3$ tends to overestimate

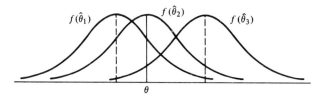

FIGURE 11.1.1 Probability functions of three estimators.

11.1 Criteria of a Good Estimator

θ. Thus, there is a good reason for us to select $\hat{\theta}_2$, an unbiased estimator, over $\hat{\theta}_1$ and $\hat{\theta}_3$.

So far we have considered only one criterion for selecting an estimator: namely, the unbiasedness of the estimator. We will now consider the second criterion for selecting an estimator: the *efficiency* of the estimator.

Definition 11.1.2

Let $\hat{\theta}_1$ and $\hat{\theta}_2$ be the two random variables associated with two different estimators for θ. The estimator associated with $\hat{\theta}_1$ is said to be *efficient* relative to the estimator associated with $\hat{\theta}_2$ if

$$V(\hat{\theta}_1) < V(\hat{\theta}_2).$$

EXAMPLE 11.1.2

Let us return to the three estimators discussed in Example 11.1.1. The variances associated with these estimators for the given sample size are calculated in Tables 11.1.9, 11.1.10, and 11.1.11.

TABLE 11.1.9

$\bar{x}$	$f(\bar{x})$	$[\bar{x} - E(\bar{X})]^2 f(\bar{x})$
1	.216	$(1 - 2.2)^2(.216) = $ .31104
2	.432	$(2 - 2.2)^2(.432) = $ .01728
3	.288	$(3 - 2.2)^2(.288) = $ .18432
4	.064	$(4 - 2.2)^2(.064) = $ .20736
	1.000	.72000

TABLE 11.1.10

$\hat{x}$	$f(\hat{x})$	$[\hat{x} - E(\hat{X})]^2 f(\hat{x})$
1	.648	$(1 - 2.056)^2(.648) = $ .72261
4	.352	$(4 - 2.056)^2(.352) = 1.3\overline{}021$
	1.000	2.05282

TABLE 11.1.11

$\tilde{x}$	$f(\tilde{x})$	$[\tilde{x} - E(\tilde{X})]^2 f(\tilde{x})$
1.0	.216	$(1.0 - 2.272)^2(.216) = $ .3495
2.5	.720	$(2.5 - 2.272)^2(.720) = $ .0374
4.0	.064	$(4.0 - 2.272)^2(.064) = $ .1911
	1.000	.5780

Comparing the variances associated with the three estimators, we find that the sample mid-range has the smallest variance. Thus, it is the *most efficient estimator* among the three estimators compared for the given sample size. However, the mid-range may not be the most efficient estimator for a different sample size.

Such a comparison can be made for some estimators for some sample size. Recall that the variance for the sample mean given in Theorem 10.2.1 is

$$V(\bar{X}) = \frac{\sigma^2}{n}.$$

Without providing a formal theorem, we propose here that, for a large sample size, the variance for the random variable associated with the median is given by

$$V(\hat{X}) = \frac{\pi \sigma^2}{2n}.$$

Since $\pi/2 = 1.57 > 1$, we conclude that the mean is a more efficient estimator than the median for a large sample.

We might describe the notion of efficiency in another context. Assume that $\hat{\theta}_1$ and $\hat{\theta}_2$ are the random variables associated with two different estimators for θ. Assume that Figure 11.1.2 depicts the probability functions of $\hat{\theta}_1$ and $\hat{\theta}_2$. The estimator depicted by $\hat{\theta}_2$ is relatively more efficient than that depicted by $\hat{\theta}_1$, since $V(\hat{\theta}_2) < V(\hat{\theta}_1)$. An evaluation of Figure 11.1.2 will reveal that for a given δ, $P(|\hat{\theta}_2 - \theta| > \delta)$ is likely to be smaller than that of $P(|\hat{\theta}_1 - \theta| > \delta)$; that is to say, $\hat{\theta}_2$ is likely to yield a smaller sampling error than that of $\hat{\theta}_1$. Thus, there is a good reason for selecting $\hat{\theta}_2$ over $\hat{\theta}_1$. Note, however, that we have drawn Figure 11.1.2 so that $E(\hat{\theta}_1) = E(\hat{\theta}_2) = \theta$. Thus, the diagram illustrates that, given two unbiased estimators, we should prefer the one which is more efficient.

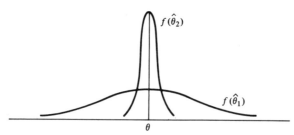

FIGURE 11.1.2 Probability functions of two estimators.

Assume now that Figure 11.1.3 depicts the probability functions of three different estimators, $\hat{\theta}_1$, $\hat{\theta}_2$, and $\hat{\theta}_3$ for θ. The diagram is drawn so that $\hat{\theta}_1$ is unbiased and both $\hat{\theta}_2$ and $\hat{\theta}_3$ are biased. On the other hand, both $\hat{\theta}_2$ and $\hat{\theta}_3$ are more efficient than $\hat{\theta}_1$. Comparing the three probability functions, we observe that we might prefer $\hat{\theta}_1$ over $\hat{\theta}_3$, even though the latter is more efficient,

11.1 Criteria of a Good Estimator

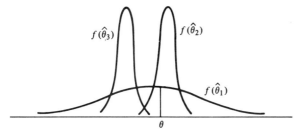

FIGURE 11.1.3 Probability functions of three estimators.

since, at the same time, it is extremely biased. On the other hand, we might prefer $\hat{\theta}_2$ over $\hat{\theta}_1$ even though $\hat{\theta}_2$ is slightly biased, since, at the same time, $\hat{\theta}_2$ is substantially more efficient than $\hat{\theta}_1$.

Even though, in some estimating situations, a biased estimator may conceivably be preferred over an unbiased one, we usually prefer to confine our search for a good estimator to one which is unbiased. Then, we would be making the best possible choice if the estimator selected happens to be the most efficient among all possible unbiased estimators. Therefore, we propose:

Definition 11.1.3

Let $\hat{\theta}$ be the random variable associated with a particular estimator for θ. The estimator is said to be the minimum-variance unbiased estimator for θ if it is unbiased and it is the most efficient among all unbiased estimators for θ. The estimator is also sometimes called the *best unbiased estimator* for θ.

We have already illustrated how to evaluate whether or not an estimator is unbiased. We have also illustrated how to evaluate whether or not an estimator is more efficient than another. However, it is not easy to evaluate whether an unbiased estimator is also the minimum-variance unbiased estimator. Obviously, we cannot compare the variance of this estimator with that of all other unbiased estimators, since there are likely to be an infinite number of them.

There is, however, a method of evaluating whether or not an unbiased estimator is also the minimum-variance unbiased estimator. To discuss this method, however, we need to consider the notion of *sufficient estimators*. A precise mathematical definition of a *sufficient estimator* is quite involved. We will not, therefore, present such a definition. Loosely speaking, however, an estimator is said to be a *sufficient estimator* for a parameter if it contains all relevant information that can be obtained from a sample pertaining to the parameter.

Professors Rao and Blackwell have discovered that an unbiased estimator which is a function of a *sufficient estimator*, has a smaller variance than that of any other estimator which is not a function of a *sufficient estimator*.

Statistical Estimation

The implication of this discovery is that, in trying to find a minimum-variance unbiased estimator, we would need to confine our search only to those estimators which are based on *sufficient estimators*. Of course, there may be more than one unbiased estimator which are based on *sufficient estimators* in a given estimating situation. In that case, we would still have the problem of finding an unbiased estimator with the minimum variance among those unbiased estimators which are based on *sufficient estimators*. However, it can be shown that in many estimating situations there is only one estimator that is based on a *sufficient estimator*. If we find this one estimator, it must be the minimum-variance unbiased estimator.

Our discussion in the preceding paragraph has been a bit vague. We could clarify some of the vagueness by discussing some underlying mathematical logic, but we will not attempt to do so, since it would be far beyond the scope of this textbook. We will, however, indicate for each of the estimators that we evaluate in the following section whether or not it is a minimum-variance unbiased estimator.

EXERCISES

*1. Let $\{X_1, X_2\}$ constitute a simple random sample of two from the probability function given in Table 11.1.12. Illustrate whether or not

$$\hat{\theta} = .5X_1 + .5X_2$$

is an unbiased estimator for π.

TABLE 11.1.12

x	$f(x)$
0	.8
1	.2
	1.0

*2. Refer to Exercise 1 and illustrate whether or not

$$\hat{\theta} = .75X_1 + .25X_2$$

is an unbiased estimator for π.

*3. Let $\{X_1, X_2\}$ constitute a simple random sample of two from the probability function given in Table 11.1.13. Let

$$\hat{\theta}_1 = .5X_1 + .5X_2,$$
$$\hat{\theta}_2 = .75X_1 + .25X_2,$$
and $$\hat{\theta}_3 = .4X_1 + .6X_2.$$

Which of the three is the most efficient estimator for π?

TABLE 11.1.13

x	$f(x)$
0	.8
1	.2
	1.0

4. Let $\{X_1, X_2, X_3\}$ constitute a simple random sample of three from the probability function given in Table 11.1.14. Let $\hat{\theta}_1$, $\hat{\theta}_2$, and $\hat{\theta}_3$ be the random variables associated with the sample mean, sample median, and sample mid-range. For each of these random variables evaluate whether or not it is an unbiased estimator for μ of the probability function for X.

TABLE 11.1.14

x	$f(x)$
1	$\frac{1}{3}$
2	$\frac{1}{3}$
3	$\frac{1}{3}$
	1.0

5. Among the three estimators given in Exercise 4, find the most efficient estimator by comparing the variances of the three estimators.

6. Let $\{X_1, X_2, X_3\}$ constitute a simple random sample of three from the probability function given in Table 11.1.15. Let $\hat{\theta}_1$, $\hat{\theta}_2$, and $\hat{\theta}_3$ be the random variables associated with the sample mean, sample median, and sample mid-range. For each of these random variables evaluate whether or not it is an unbiased estimator for μ of the probability function for X.

TABLE 11.1.15

x	$f(x)$
10	.4
20	.6
	1.0

7. Among the three estimators given in Exercise 6, find the most efficient estimator by comparing the variances of these estimators.

*8. Let $\{X_1, X_2\}$ constitute a simple random sample from the probability function given in Table 11.1.16. Let

$$\hat{\theta}_1 = .5X_1 + .5X_2$$
and
$$\hat{\theta}_2 = .6X_1 + .4X_2$$

be two estimators for μ of the probability function for X. For each of these estimators evaluate whether or not it is unbiased.

TABLE 11.1.16

x	$f(x)$
10	.8
20	.2
	1.0

*9. Of the two estimators given in Exercise 8, which one is relatively more efficient? Answer the question by comparing their variances.

11.2 Point Estimate

In the preceding section we have pointed out that, in many estimating situations, we usually prefer to use an estimator which is the minimum-variance unbiased estimator. We will now point out some of these minimum-variance unbiased estimators.

Theorem 11.2.1

Let $\{X_1, \ldots, X_n\}$ be a simple random sample of size n from a Bernoulli probability function with a parameter π. Then,

$$P = \frac{X_1 + \cdots + X_n}{n}$$

is the minimum-variance unbiased estimator for π.

That P is unbiased is not really a new observation. We have pointed out in Chapter 10 that $E(P) = \pi$. On the other hand, we must ask the reader to accept on faith our proposition that the variance of P is smaller than any other unbiased estimator which we may utilize to estimate π.

An important implication of Theorem 11.2.1 may be stated as follows. Suppose our task is to estimate π of a Bernoulli probability function. Then, we have a very good reason to use P as the estimator for π since, in the first place, it is unbiased and, furthermore, it is the most efficient among all unbiased estimators for π.

We will now illustrate a simple application of Theorem 11.2.1.

EXAMPLE 11.2.1

Assume that a simple random sample of 10 observations from a Bernoulli probability function have revealed the following:

$$\{1,0,0,1,0,1,0,0,0,1\}.$$

If the objective underlying the sampling were to estimate π, then, the estimate is given as

$$P = \frac{4}{10} = .4.$$

In our preceding discussion we have not stated what π and P represent physically. Had the sample been taken from a manufacturing process, π could conceivably represent the proportion of defectives produced by that process and P could represent an estimate of that proportion. Had the sample been taken from a voting public, π could represent the percentage of those voters who are in favor of a given candidate and P could represent an estimate of that percentage.

Theorem 11.2.2

Let $\{X_1, \ldots, X_n\}$ be a simple random sample of size n from a non-Bernoulli probability function. Then,

$$\bar{X} = \frac{X_1 + \cdots + X_n}{n}$$

is an unbiased estimator of μ. Furthermore, if the probability function sampled is normally distributed, $\bar{X}$ is also the minimum-variance unbiased estimator for μ.

That $\bar{X}$ is unbiased, again, is not a new observation. We have already shown in the preceding chapter that $E(\bar{X}) = \mu$. We must again ask the reader, however, to accept on faith the fact that $\bar{X}$ will be the minimum-variance unbiased estimator when the probability function sampled is normal.

These discussions show that, if our task is to estimate μ of a probability function, one very good approach is to use the sample mean as if it were μ. This approach satisfies at least the criterion of being unbiased and, under the normality assumption, it also satisfies the criterion of being a minimum variance. We have also observed in the preceding section that the procedure of using the sample mean is relatively more efficient than that of using the sample median for a large sample size, regardless of the probability function from which the sample is taken.

We will now illustrate a simple application of Theorem 11.2.2.

EXAMPLE 11.2.2

Assume that we plan to take a simple random sample of five observations from a probability function in order to estimate μ. One question confronting us may be whether or not to use, for example, the mean or the median of the sample as the estimated value of μ. Theorem 11.2.2 explains certain desirable features of using the sample mean. Assume now that we have decided to use the sample mean, and that the actual sample has revealed the following five observations:

$$\{22, 28, 21, 26, 23\}.$$

Then, the actual value of our estimate is given as

$$\bar{X} = \frac{22 + 28 + 21 + 26 + 23}{5} = 24,$$

whereas, if we decided to use the sample median, then our estimate would be 23.

When a sample is taken from a probability function, usually we would like to estimate not only the mean of the probability function but also its variance. For this, we now propose:

Theorem 11.2.3

Let $\{X_1, \ldots, X_n\}$ be a simple random sample of size n from a non-Bernoulli probability function. Then,

$$\mathbf{s}^{2*} = \frac{(X_1 - \bar{X})^2 + \cdots + (X_n - \bar{X})^2}{n - 1}$$

is an unbiased estimator of σ^2. Furthermore, if the probability function sampled is normally distributed, $\mathbf{s}^2$ is also the minimum-variance unbiased estimator for σ^2.

We again must ask the reader to accept the validity of this theorem without going through the proof.

EXAMPLE 11.2.3

Refer to Example 11.2.2 and assume that we have decided to use the procedure given in Theorem 11.2.3 to estimate the variance of the probability function in question. Then, the actual estimate of σ^2, based on the sample data given in Example 11.2.2 would be calculated as

$$\mathbf{s}^2 = \frac{(22 - 24)^2 + (28 - 24)^2 + (21 - 24)^2 + (26 - 24)^2 + (23 - 24)^2}{5 - 1}$$

$$= \frac{4 + 16 + 9 + 4 + 1}{4} = 8.5.$$

In turn, we would estimate the standard deviation of the probability function as

$$\mathbf{s} = \sqrt{8.5}.$$

EXERCISES

*10. Assume that you want to estimate the proportion of defectives produced by a manufacturing process. Your sample of 10 yields the information given in Table 11.2.1.

* $\mathbf{s}^2$ is in boldface to indicate that it is a random variable.

TABLE 11.2.1

Observation	Condition of Item Selected
1	good
2	good
3	bad
4	good
5	good
6	good
7	bad
8	good
9	bad
10	good

What would be the point estimate for the proportion of defectives produced by the machine if you were to utilize the minimum-variance unbiased estimator for this purpose?

11. Suppose a large bank considers that a checking account is a desirable account for the bank to offer if the person holding it maintains an average balance of $100 or more. The bank wishes to estimate the proportion of checking accounts which would be classified as desirable accounts. A simple random sample of 20 accounts has revealed the data given in Table 11.2.2. What would be the point estimate for the proportion of desirable accounts for the bank if you were to utilize the minimum-variance unbiased estimator for this purpose?

TABLE 11.2.2

Observation	Account Classification	Observation	Account Classification
1	D	11	ND
2	D	12	D
3	ND	13	D
4	D	14	D
5	ND	15	ND
6	D	16	D
7	D	17	D
8	ND	18	ND
9	D	19	D
10	D	20	D

D = desirable
ND = not desirable

*12. Suppose a cannery receives a truckload of peaches. It wants to estimate the mean and the variance of diameters of the peaches in the truckload. A simple random sample of 26 peaches has revealed the following figures (in inches):

2.625	2.750	3.000	
2.625	2.750	2.625	
2.875	2.625	2.750	
3.000	3.000	2.875	
2.625	3.125	2.750	
3.000	2.875	2.750	
2.750	2.625	2.750	
2.875	2.875	2.875	
2.750	2.875		

What would be the point estimates for the mean and the variance of the diameters for the peaches in the truckload if you were to utilize the minimum-variance unbiased estimators for this purpose?

13. The following 20 sample observations depict monthly water consumption of households in a city (measured in units of 100 cubic feet). A simple random sample of 20 have been selected from the record books of the water company.

13	12	6	8
6	21	10	12
19	5	5	16
4	11	18	10
11	10	9	14

Suppose you wish to estimate the mean and the variance of the amounts of water used by the households. What would be the point estimates if you were to utilize the minimum-variance unbiased estimators for this purpose?

11.3 Maximum Likelihood Estimate

In Section 11.2, we have pointed out that lack of bias and efficiency are two desirable properties that any good estimator should possess. Statisticians, however, have devised a number of other properties that an ideal estimator should possess. We might raise the question, "How do the statisticians go about searching for an estimator which possesses some of these desirable properties?" One procedure they use is called the *maximum likelihood estimate*.

We will now describe the concept of the maximum likelihood estimate. To do so, we will define a term which we have used loosely up to this point.

Definition 11.3.1

Let $f(x|\theta)$ be a probability function, where θ is a parameter of that probability function. Assume that a simple random sample of size n has yielded $X_1 = x_1, \ldots, X_n = x_n$. Then, a function whose domain contains a set of values for θ and whose image set contains the set of values $L(X_1 = x_1, \ldots, X_n = x_n|\theta)$, where

$$L(X_1 = x_1, \ldots, X_n = x_n|\theta)$$
$$= f(X_1 = x_1|\theta)f(X_2 = x_2|\theta) \ldots f(X_n = x_n|\theta)$$

is called the *likelihood function* for the given sample.

If a function is a likelihood function for a given sample outcome, then an element of the domain depicts an assumed value of θ for the probability function from which the sample is taken, and the corresponding element in the image set depicts the likelihood of obtaining the given sample from the probability function with that particular assumed value of θ. We will explain what we mean by the preceding propositions in the next example. At times we will also denote a likelihood function by a simpler notation:

$$L(x_1, \ldots, x_n|\theta) = f(x_1|\theta)f(x_2|\theta) \ldots f(x_n|\theta).$$

EXAMPLE 11.3.1

Assume that a simple random sample of two from a Bernoulli probability function has yielded

$$X_1 = 0 \quad \text{and} \quad X_2 = 1.$$

Suppose we assume further that this sample has come from the Bernoulli probability function with $\pi = .4$. The probability function in question is then given as

x	$f(x)$
0	.6
1	.4
	1.0

or, alternatively, as

$$f(X = 0|\pi = .4) = .6$$
and
$$f(X = 1|\pi = .4) = .4.$$

Let X_1 and X_2 depict the random variables associated with the sample outcomes. Then,

$$f(X_1 = 0|\pi = .4) = .6$$
$$f(X_1 = 1|\pi = .6) = .4,$$

and

$$f(X_2 = 0|\pi = .4) = .6$$
$$f(X_2 = 1|\pi = .4) = .4.$$

Then, we say that the likelihood of obtaining that particular sample outcome is given by the value of its likelihood function; that is,

$$L(X_1 = 0, X_2 = 1|\pi = .4) = f(X_1 = 0|\pi = .4)f(X_2 = 1|\pi = .4)$$
$$= (.6)(.4) = .24.$$

On the other hand, suppose we assume that the sample outcome, $X_1 = 0$ and $X_2 = 1$, comes from the Bernoulli probability function with $\pi = .8$. The value of the likelihood function for the given sample may be calculated as

$$L(X_1 = 0, X_2 = 1 | \pi = .8) = f(X_1 = 0 | \pi = .8) f(X_2 = 1 | \pi = .8)$$
$$= (.2)(.8) = .16.$$

Assume now that the domain of the likelihood function contains only two assumed values of π; namely, $\pi = .4$ and $\pi = .8$. Then, the likelihood function for the given sample is illustrated by Table 11.3.1.

TABLE 11.3.1

Domain π	Image Set: $L(X_1 = 0, X_2 = 1 \| \pi)$
.4	.24
.8	.16

We will now proceed to define the concept of a *maximum likelihood estimate*.

Definition 11.3.2
Let $\hat{\theta}$ depict an estimator for θ. Suppose $\hat{\theta}$ is devised so that the estimated value of θ always maximizes the value of the likelihood function as a function of θ. Then, $\hat{\theta}$ is said to be the *maximum likelihood estimator* for θ.

This definition appears to be complicated, but it really is not, as we will illustrate next.

EXAMPLE 11.3.2
Assume that we have taken a simple random sample of two from a Bernoulli probability function whose π is not known to us. Assume, further, that the particular sample yields

$$X_1 = 0, \quad X_2 = 1.$$

Let us now ask, "What assumed value of π will maximize the likelihood function $L(X_1 = 0, X_2 = 1 | \pi)$?" To answer this question, we might assume several arbitrary values of π and, then, calculate the resulting values for the likelihood function. The assumed values of π and the calculations for the likelihood function are shown in Table 11.3.2.

Table 11.3.2 indicates that the value of π, which maximizes $L(X_1 = 0, X_2 = 1 | \pi)$, is .5. This means that if $\hat{\theta}$ is to be the maximum likelihood estimator for π, then its value should also be .5. How can we devise $\hat{\theta}$ so that its value

TABLE 11.3.2

π	$f(X_1 = 0\|\pi)$	$f(X_2 = 1\|\pi)$	$L(X_1 = 0, X_2 = 1\|\pi)$
.0	(1.0)	(.0)	.00
.1	(.9)	(.1)	.09
.2	(.8)	(.2)	.16
.3	(.7)	(.3)	.21
.4	(.6)	(.4)	.24
.5	(.5)	(.5)	.25
.6	(.4)	(.6)	.24
.7	(.3)	(.7)	.21
.8	(.2)	(.8)	.16
.9	(.1)	(.9)	.09
1.0	(.0)	(1.0)	.00

is .5 if the sample yields $X_1 = 0$, $X_2 = 1$? This can be done by letting $\hat{\theta} = P$, since

$$P = \frac{X_1 + X_2}{2} = \frac{0 + 1}{2} = .5.$$

According to Definition 11.3.2, P is the maximum likelihood estimator for π. We now propose:

Theorem 11.3.1
Let $\{X_1, \ldots, X_n\}$ be a simple random sample of size n from a Bernoulli probability function with a parameter π. Then,

$$P = \frac{X_1 + \cdots + X_n}{n}$$

is the *maximum likelihood estimator* for π.

We have already illustrated the validity of this theorem in Example 11.3.2. However, we will give another illustration.

EXAMPLE 11.3.3
Assume that we have selected a simple random sample of three from a Bernoulli probability function with an unknown π and have obtained the following sample outcome: $X_1 = 0$, $X_2 = 1$, and $X_3 = 0$. Assume arbitrarily that $\pi = \frac{1}{2}$, $\pi = \frac{1}{3}$, and $\pi = \frac{1}{4}$ and calculate $L(X_1 = 0, X_2 = 1, X_3 = 0|\pi)$. The calculations are shown in Table 11.3.3. The table shows that $L(X_1 = 0, X_2 = 1, X_3 = 0|\pi)$ is larger when $\pi = \frac{1}{3}$ than when $\pi = \frac{1}{2}$ or when $\pi = \frac{1}{4}$. It can be shown, in fact, that $\pi = \frac{1}{3}$ maximizes the likelihood function when the sample yields $X_1 = 0$, $X_2 = 1$, and $X_3 = 0$.

Statistical Estimation

TABLE 11.3.3

π	$f(X_1 = 0\|\pi)$	$f(X_2 = 1\|\pi)$	$f(X_3 = 0\|\pi)$	$L(X_1 = 0, X_2 = 1, X_3 = 0\|\pi)$
$\tfrac{1}{2}$	$(\tfrac{1}{2})$	$(\tfrac{1}{2})$	$(\tfrac{1}{2})$	$\tfrac{1}{8} = .125$
$\tfrac{1}{3}$	$(\tfrac{2}{3})$	$(\tfrac{1}{3})$	$(\tfrac{2}{3})$	$\tfrac{4}{27} \cong .148$
$\tfrac{1}{4}$	$(\tfrac{3}{4})$	$(\tfrac{1}{4})$	$(\tfrac{3}{4})$	$\tfrac{9}{64} \cong .141$

Thus, if $\hat{\theta}$ is the maximum likelihood estimator for π, then the value of $\hat{\theta}$ should be $\tfrac{1}{3}$ when the sample yields $X_1 = 0$, $X_2 = 1$, and $X_3 = 0$. This can be done by letting $\hat{\theta} = P$, since

$$P = \frac{X_1 + X_2 + X_3}{3} = \frac{0 + 1 + 0}{3} = \frac{1}{3}.$$

This is precisely what is implied by Theorem 11.3.1.

We have illustrated in the preceding two examples that P satisfies the definition of a maximum likelihood estimator for π. Suppose we do not know that P is the maximum likelihood estimator for π. How would we go about finding the maximum likelihood estimator in question? We will illustrate in the following example the process of ascertaining the maximum likelihood estimator for those who are familiar with some calculus; those who are not familiar with calculus can skip the next example without any loss of continuity.

EXAMPLE 11.3.4

Assume that a simple random sample of two is to be selected from a Bernoulli probability function in order to estimate π. Then, the likelihood function for the sample is

$$L(x_1, x_2|\pi) = f(x_1|\pi)f(x_2|\pi),$$

where x_1 and x_2 depict the values of X_1 and X_2. Since, for example,

$$f(x_1|\pi) = \pi^{x_1}(1 - \pi)^{1-x_1},$$

we can express the following:

$$L(x_1, x_2|\pi) = [\pi^{x_1}(1 - \pi)^{1-x_1}][\pi^{x_2}(1 - \pi)^{1-x_2}]$$
$$= \pi^{(x_1+x_2)}(1 - \pi)^{2-(x_1+x_2)}.$$

The next task is to find the value of π which will maximize the preceding likelihood function. By using calculus, we can find such a value of π by differentiating the likelihood function with respect to π, setting the resulting equation equal to 0, and then finding the value of π which satisfies the equation.

Let

$$L = \log L(x_1, x_2|\pi).$$

Then,
$$L = (x_1 + x_2) \log \pi + [2 - (x_1 + x_2)] \log (1 - \pi)].$$

Differentiating L with respect to π, we have
$$\frac{dL}{d\pi} = \frac{x_1 + x_2}{\pi} - \frac{2 - (x_1 + x_2)}{1 - \pi}.$$

Now we set
$$\frac{x_1 + x_2}{\pi} - \frac{2 - (x_1 + x_2)}{1 - \pi} = 0$$

and solving for π, we find that
$$\pi = \frac{x_1 + x_2}{2},$$

which is really the value of P. Thus, we conclude that P must be the maximum likelihood estimator for π.

By using a procedure similar to the one we have described in Example 11.3.4, we can derive the maximum likelihood estimators for the parameters of other probability functions. One result of such derivation is the following theorem:

Theorem 11.3.2
Let $X_1, \ldots, X_n$ be a simple random sample of size n from a normal probability function. Then,
$$\bar{X} = \frac{X_1 + \cdots + X_n}{n}$$
and
$$\hat{s}^2 = \frac{(X_1 - \bar{X})^2 + \cdots + (X_n - \bar{X})^2}{n}$$

are the maximum likelihood estimators for μ and σ^2, respectively.

Even though we will not prove the theorem, it should be obvious that $\bar{X}$ must be the maximum likelihood estimator for μ. On the other hand, that $\hat{s}^2$ is the maximum likelihood estimator for σ^2 can only be illustrated by calculus, with which we will not burden the reader. We ask the reader to note that although the maximum likelihood estimator for μ is also the minimum-variance unbiased estimator for μ, the maximum likelihood estimator for σ^2 is not the minimum-variance unbiased estimator for σ^2.

EXERCISES

*14. Suppose you wish to estimate π for a Bernoulli-distributed random variable. A simple random sample of four has yielded

$$X_1 = 0,$$
$$X_2 = 1,$$
$$X_3 = 0,$$
$$X_4 = 0.$$

Illustrate that the maximum likelihood estimate for π is .25.

*15. Suppose you have selected a simple random sample of 10 from a normally distributed probability function. The sample has yielded the following information:

−700	−3400
580	1860
−420	660
3160	190
120	−50

What are the point estimates for the mean and the variance of the probability function, if the point estimates are based on the maximum likelihood estimators?

11.4 Interval Estimates: Mean

It is intuitively apparent that a point estimate for a parameter of a probability function will seldom equal the parameter. For this reason, once we have actually obtained a point estimate, we may still be very skeptical that it is, in fact, the same as the parameter, even though the value of the parameter is not known to us. However, if we provide an interval centered around the point estimate as one which contains the parameter, then the possibility that the interval might actually contain the unknown parameter may be greater than that the point estimate is, in fact, equal to the unknown parameter. In this respect, given the sample size, the wider the interval is, the greater the possibility is that the interval contains the unknown parameter. In another respect, other things being the same, if we are given an interval, the possibility that the interval contains the unknown parameter should increase with the sample size. We will now describe the manner by which we can establish such intervals and, in turn, evaluate their usefulness.

We will first assume that the sample is to be taken from a normal probability function. Let

$$Y = \frac{\bar{X} - \mu}{\sigma_{\bar{x}}} = \frac{\bar{X} - \mu}{\sigma/\sqrt{n}}.$$

Then, Y is the standard, normally distributed random variable. Thus, for example, a probabilistic statement that

$$P\{-1.96 \leq Y \leq 1.96\} = .95$$

holds true, regardless of the value of μ, σ, or n.

Assume now that the value of σ is known to us. To be more specific, let us assume that $\sigma = 20$ and $n = 100$. Then,

11.4 Interval Estimates: Mean

$$Y = \frac{\bar{X} - \mu}{\sigma/\sqrt{n}} = \frac{\bar{X} - \mu}{2}.$$

The fact that $-1.96 < Y$ may now be modified to

$$\mu \leq 3.92 + \bar{X},$$

and the fact that $Y < 1.96$ may now be modified to

$$\mu \geq \bar{X} - 3.92.$$

Thus,

$$P\{-1.96 \leq Y \leq 1.96\} = P\{\bar{X} - 3.92 \leq \mu \leq \bar{X} + 3.92\} = .95.$$

The meaning of this statement may be stated as follows. Suppose we are to select a sample of 100 from a normal probability function with $\sigma = 20$ and to establish an interval $\bar{X} \pm 3.92$. Then, the probability that the interval will contain μ is .95.

We may also illustrate the validity of this proposition graphically, as in Figure 11.4.1. Let $k_1 = \mu - 3.92$ and $k_2 = \mu + 3.92$, where μ is still assumed to be unknown. The positions of k_1 and k_2 relative to μ are shown in this figure, where the curve depicts the probability function of $\bar{X}$.

If we take a simple random sample of 100, 95 percent of the sample means will fall between k_1 and k_2. Suppose we establish an interval $\bar{X} \pm 3.92$ around each sample mean. Then, the interval will contain μ if $\bar{X}$ is between k_1

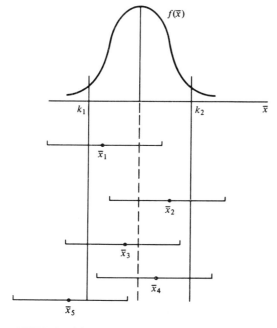

FIGURE 11.4.1

and k_2 and it will not contain μ if $\bar{X}$ is not between k_1 and k_2. Since 95 percent of the sample means are expected to fall between k_1 and k_2, we may also infer that 95 percent of the intervals thus established will contain μ. Thus, we may state that there is a .95 probability that the interval established will contain μ. (Note that this probability statement pertains to an interval which is yet to be established.)

Let us now assume that a sample is actually taken and we have obtained the specific value of $\bar{X}$ from that sample. For example, let us arbitrarily assume that $\bar{X} = 24$. Then, substituting this value in the formula just given, we have

$$P\{20.02 \leq \mu \leq 27.92\} = .95.$$

The interval between 20.02 and 27.92 is called a *confidence interval*, and the coefficient (.95) in this equation is called a *confidence coefficient*. The reason it is called a confidence coefficient and not a *probability coefficient* is that once an actual interval such as the one given is established, then μ must be either within the interval or outside of it. Consequently, according to the classical approach to probability, there is really no probability associated with the proposition that the interval contains μ. Since the proposition is either true or false, it is incorrect to infer, for example, that there is a 95 percent probability that the proposition is correct.

If we cannot make any probabilistic statement about the interval, then why should we trouble ourselves to establish such an interval? Returning to our earlier discussion, we know that if we were to establish many such intervals, then, 95 percent of such intervals would contain μ. Consequently, 95 percent of the time we would be correct in stating that such intervals contain μ. Thus, even though we do not know whether a given interval, in fact, contains μ, we do know the degree of reliability we can attach to the procedure which led to establishing that interval. In turn, our knowledge about the reliability of the procedure used to establish an interval should provide us with a degree of confidence commensurate with the reliability measure for the procedure. This is the reason we establish such an interval and call it a confidence interval.

Although we have considered the confidence coefficient of .95, other numerical values can be assigned as a confidence coefficient. The procedure of establishing the interval will have to be modified somewhat. We will illustrate now how to establish confidence intervals with different confidence coefficients.

EXAMPLE 11.4.1

Let us assume that a simple random sample of 100 observations from a normal probability function with $\sigma = 50$ has yielded a sample mean of 220.

Suppose we wish to establish a 95 percent confidence interval around this sample mean. First, we calculate $\sigma_{\bar{x}} = 50/\sqrt{100} = 5$. Next, we look at the table of normal probabilities (Table B.1 in Appendix B) and find that $P(|Z| \leq 1.96) = .95$. Thus, the interval is given as

$$\bar{X} \pm 1.96\sigma_{\bar{x}} = 220 \pm 1.96(5).$$

Suppose we wish to establish a 99 percent confidence interval. Since we have already calculated that $\sigma_{\bar{x}} = 5$, we now ascertain from Table B.1 that $P\{|Z| \leq 2.58\} = .99$. Thus, the interval to be established is given as

$$\bar{X} \pm 2.58\sigma_{\bar{x}} = 220 \pm 2.58(5).$$

In establishing the confidence interval, we have assumed that the value of σ for the probability function is known, but this is not likely to be the case in practice. There is theoretically a correct way to establish a confidence interval, even if the value of σ is not known. We will discuss the procedure in Chapter 14. We will, however, suggest here one pragmatic way to solve the problem. When a sample is taken, we can obtain the values of $\bar{X}$ as well as **s**. We could, then, use the value of **s** as if it were, in fact, σ. We will now illustrate an application of the procedure just described.

EXAMPLE 11.4.2

Assume that we are interested in finding the mean breaking strength of nylon cords produced by a manufacturing process. Assume, further, that the individual breaking strengths are normally distributed.

Suppose we have randomly selected 100 pieces of cord from the machine's output and have tested each piece for its breaking strength. From this experiment, assume that we have obtained

$$\bar{X} = 8.4 \text{ pounds}$$
and
$$\mathbf{s} = 2.5 \text{ pounds}.$$

Now assume that σ for the machine process is, in fact, 2.5 pounds. Thus, $\sigma_{\bar{x}} = \sigma/\sqrt{n} = 2.5/\sqrt{100} = .25$. Then, a 95 percent confidence interval, for example, would be given as

$$\bar{X} \pm 1.96\sigma_{\bar{x}} = 8.4 \pm 1.96(.25).$$

Thus, the interval is between 7.91 and 8.89 pounds.

So far we have assumed that the sample has been taken from a normal probability function. Suppose now that the sample is to be taken from a probability function which is not normally distributed. Then, the probability function of Y where

$$Y = \frac{\bar{X} - \mu}{\sigma_{\bar{x}}}$$

is not likely to be normally distributed. However, the central limit theorem assures that the probability function of Y becomes approximately normal for a fairly large sample size. Thus, if we utilize the procedure just described and establish a confidence interval, then, the confidence coefficient that we attach to the interval is expected to be approximately close to its true value.

EXAMPLE 11.4.3

Suppose a large grocery store is interested in finding out the average length of time required to service a customer at a check-out counter. The actual length of time in any given situation depends on the efficiency of the clerk, as well as the amount of food bought by the customer.

Assume that the store manager has directed an assistant to select 400 observations of actual service times. Assume, further, that the manner in which these observations are obtained satisfies the criteria for a simple random sample. From the sample, the following figures have been obtained:

$$\bar{X} = 8$$
and
$$s = 5.$$

Thus, the point estimate for the mean service time is 8 minutes.

Suppose now we wish to establish a 95 percent confidence interval centered around this point estimate. We might ask whether or not we can reasonably assume that the probability function of the random variable associated with the service time is normally distributed. In the queuing theory which involves the mathematical study of waiting lines where this sort of question is asked, other assumptions, in addition to that of normality with regard to the probability function of service times, are made. Therefore, assume that the probability function of the service time is *not* normal. Can we still utilize the procedure which we used in Example 11.4.2 in establishing a confidence interval?

The answer is "yes," since when the sample size is as large as 400, the probability function of Y is likely to be very close to that of a normal probability function. Thus, we first calculate $\sigma_{\bar{x}} = 5/\sqrt{400} = .25$ and, in turn,

$$\bar{X} \pm 1.96\sigma_{\bar{x}} = 8 \pm 1.96(.25).$$

Thus, the interval in question is between 8.51 and 9.49 minutes.

We must realize, however, that the confidence interval established is only an approximation, because Y is not exactly a normally distributed random variable, even for the sample size of 400.

EXERCISES

*16. A simple random sample of 20 castings has been selected from a manufacturing process, and the diameters for their entry throats have been measured. The following data (in inches) have been found.

2.989	3.030	2.986	2.997
3.029	3.039	3.040	3.031
3.026	3.000	3.008	3.027
3.010	3.023	2.993	2.998
3.002	2.999	3.004	3.020

Assume that the standard deviation for the diameters in question is .002 inches. Establish a 95 percent confidence interval for the mean diameter for the castings produced by the process. (Assume that the diameters in question are normally distributed.)

17. Refer to Exercise 16 and establish a 99 percent confidence interval for the mean diameter in question, assuming again that the standard deviation for the diameters is .002 inches.

*18. Refer again to Exercise 16 and assume that the standard deviation for the diameters is not known. Then, how would you establish, for example, a 95 percent confidence interval? Establish the interval in question.

*19. Assume a university administrator wants to estimate the average amount of time per week that the students at his university spend in studying outside the classrooms. A simple random sample of 20 students has been selected. The administrator has secretly observed the study patterns for these students for 1 week without telling them that they are under observation. The sample results are given below, and each figure depicts the number of hours spent by the student in studying.

8	12	5	10
10	16	8	18
6	0	12	7
12	5	20	2
4	14	1	10

Establish a 95 percent confidence interval for the average number of hours spent by the student at the university during the week. Make any statistical assumptions that you feel are warranted.

11.5 Interval Estimates: Proportions

In this section we will discuss the problem of establishing an interval estimate when the sample is taken from a Bernoulli probability function. In this respect, we will also assume that the sample size is fairly large. Procedures for establishing an interval estimate for a small size sample does exist, but we will not consider them here.

Recall our proposition in Chapter 10 that when n is large, the probability function of P becomes approximately normal, with a standard deviation of $\sigma_p = \sqrt{\pi(1-\pi)/n}$. Now let

$$Y = \frac{P - \pi}{\sqrt{\pi(1-\pi)/n}}.$$

Then, the probability function of Y becomes very close to that of the standard normal probability function. This means that, if the value of k is such that $P\{|Z| \leq k\} = 1 - \alpha$, then, we have

$$P\left\{-k \leq \frac{P - \pi}{\sqrt{\pi(1 - \pi)/n}} \leq k\right\} \cong 1 - \alpha.$$

Solving for π, we find

$$P\left\{\frac{2nP + k^2 - k\sqrt{4nP + k^2 - 4nP^2}}{2(n + k^2)} \leq \pi \right.$$
$$\left. \leq \frac{2nP + k^2 + k\sqrt{4nP + k^2 - 4nP^2}}{2(n + k^2)}\right\} \cong 1 - \alpha.$$

It can be shown that if n is large, we can ignore k^2 in the preceding equation. Then,

$$P\left\{P - k\sqrt{\frac{P(1 - P)}{n}} \leq \pi \leq P + k\sqrt{\frac{P(1 - P)}{n}}\right\} \cong 1 - \alpha.$$

If we let $\alpha = .05$, then,

$$P\left\{P - 1.96\sqrt{\frac{P(1 - P)}{n}} \leq \pi \leq P + 1.96\sqrt{\frac{P(1 - P)}{n}}\right\} \cong .95.$$

The significance of this equation may be stated as follows. Suppose we are to take a simple random sample of size n from a Bernoulli probability function and we are to establish an interval around the sample P as

$$P \pm 1.96\sqrt{\frac{P(1 - P)}{n}}.$$

Then, the probability that such an interval will contain the unknown π is approximately .95, if the size of the sample to be taken is large. Again, we must emphasize that this interpretation of the probabilistic statement is correct only if we are concerned with an interval yet to be established.

On the other hand, suppose now we have taken the sample and have obtained a specific value for P. For example, assume that the sample of 100 has yielded that $P = .2$. Substituting this value in the preceding equation, we have

$$P\left\{.2 - 1.96\sqrt{\frac{(.2)(.8)}{100}} \leq \pi \leq .2 + 1.96\sqrt{\frac{(.2)(.8)}{100}}\right\}$$
$$= \{.1224 \leq \pi \leq .2776\} \cong .95.$$

This equation, however, no longer depicts a probabilistic statement, since the unknown value of π is either within the interval .1224 and .2776 or it is outside the interval. Again, we should consider the equation as one which depicts a confidence statement, where .95 is the confidence coefficient appropriate to the interval.

Since we have already explored the significance of the term *confidence*, we will not elaborate on its meaning again here. We will, however, illustrate a number of applications of the statistical procedure discussed in this section.

EXAMPLE 11.5.1

Assume that we wish to estimate the proportion of defectives produced by a given manufacturing process. We have taken a simple random sample of 100 items and have found that the sample contains 10 defectives. This means that the point estimate of the proportion of defectives for the process may be given as

$$P = \frac{10}{100} = .1.$$

Suppose now we wish to establish a 95 percent confidence interval centered around this point estimate. To do so, we first calculate

$$\sqrt{\frac{P(1-P)}{n}} = \sqrt{\frac{(.1)(.9)}{100}} = .03$$

and, in turn, the value of k such that $P\{|Z| \leq k\} = .95$. Thus, $k = 1.96$. The confidence interval in question is, then,

$$P \pm 1.96 \sqrt{\frac{P(1-P)}{n}} = .1 \pm 1.96(.03).$$

Thus, the desired interval is between .0412 and .1588.

EXAMPLE 11.5.2

Assume that we wish to estimate the proportion of voters who are in favor of a certain candidate. Assume that we have selected a simple random sample of 2400 voters from the voting population. Suppose that in our sample, 960 of the voters are actually in favor of the candidate. Then, the point estimate for the proportion in question would be

$$P = \frac{960}{2400} = .4.$$

Suppose now we wish to establish a 99 percent confidence interval centered around this point estimate. We first calculate

$$\sqrt{\frac{P(1-P)}{n}} = \sqrt{\frac{(.4)(.6)}{2400}} = .01$$

and, in turn, find the value of k such that $P\{|Z| \leq k\} = .99$. Then, $k = 2.58$. The desired interval is given as

$$P \pm 2.58 \sqrt{\frac{P(1-P)}{n}} = .4 \pm 2.58(.01).$$

Thus, the interval is between .3742 and .4258.

We have stated that the procedure just described for establishing a confidence interval for π is valid only for a fairly large sample size. This results from the fact that P becomes approximately normal only for a large sample size. How large, then, should the size be before we can assume P to be approximately normal? We have already elaborated on this question in Chapter 8 in connection with the central limit theorem.

EXERCISES

*20. A political candidate for state-wide office wants to estimate the percentage of registered voters in the state who favor a new bond issue. A simple random sample of 30 registered voters has revealed the results given in Table 11.5.1. Establish a 95 percent confidence interval for the percentage of voters in the state who favor the bond issue.

TABLE 11.5.1

Voter	Preference	Voter	Preference
1	favor	16	favor
2	not favor	17	not favor
3	favor	18	favor
4	favor	19	favor
5	favor	20	favor
6	not favor	21	not favor
7	not favor	22	not favor
8	not favor	23	favor
9	favor	24	not favor
10	favor	25	not favor
11	not favor	26	not favor
12	favor	27	favor
13	favor	28	not favor
14	not favor	29	favor
15	favor	30	not favor

21. Suppose an advertising company has taken a simple random sample of 1500 households with television sets and has found that 300 of them are turned on to a certain program. Establish a 95 percent confidence interval for the percentage of television sets which are turned on to that program.

*22. The credit manager of a department store wants to estimate the percentage of charge accounts which are at least 2 months overdue. A simple random sample of 1600 accounts has been selected from the charge-account files. The results are shown in the following chart. Establish a 95 percent confidence interval for the percentage of the overdue accounts for the store.

Account Status	Number of Accounts
overdue	400
not overdue	1200
	1600

Chapter 12
Testing
a Hypothesis

12.1 Introduction

In making either the point estimate or the interval estimate for a parameter of a probability function, we did not make any supposition as to what its value may be before we actually select a sample. However, certain statistical problems call for us to specifically make such a supposition and then, to select a sample to see if the supposition can be supported in light of the sample information. Such an approach to statistical induction is called *testing a hypothesis*.

In short, the process of testing a hypothesis may be described as follows. Assume that we are interested in estimating a parameter of a probability function. Let the unknown parameter in question be denoted by θ. Since we do not know the value of θ, we make an assumption that it is θ_1, where θ_1 associates a specific value of θ. Such an assumption pertaining to θ may be called a *null hypothesis*. In notation, the null hypothesis is

$$\theta = \theta_1.$$

After establishing such a null hypothesis, we make a point estimate of θ by taking a sample. Let the point estimator be denoted by $\hat{\theta}$. If the value of $\hat{\theta}$ is very close to that of θ_1, we have good reason to accept the null hypothesis; but if the difference between the value of $\hat{\theta}$ and θ_1 is too large, we have good reason to reject the null hypothesis.

When we reject the null hypothesis, we are, in a sense, making an assumption that the value of θ is other than θ_1. It may be useful, therefore, to specify in advance what alternate assumption we will make if we reject the null hypothesis. Such an alternate assumption, when it is specified in advance, is called the *alternate hypothesis*. An alternate hypothesis may be, for example,

$$\theta \neq \theta_1$$

or it may be

$$\theta = \theta_2$$

where θ_2 is another value besides θ_1.

When a hypothesis specifies a single value for the parameter, it is said to be a *simple hypothesis*. However, when a hypothesis specifies more than one value for the parameter, it is said to be a *composite hypothesis*. For example, if we formulate

$$\text{Null Hypothesis:} \quad \theta = \theta_1$$
$$\text{Alternate Hypothesis:} \quad \theta = \theta_2,$$

then, both the null and the alternate hypotheses are simple hypotheses. On the other hand, if we formulate

$$\text{Null Hypothesis:} \quad \theta = \theta_1$$
$$\text{Alternate Hypothesis:} \quad \theta \neq \theta_1,$$

then, the null hypothesis is simple and the alternate hypothesis is composite. In this chapter we will be concerned only with a simple hypothesis.

Before involving ourselves with the detailed mechanics of hypothesis testing, we might describe a situation in which the approach would be put to use. Assume that a certain manufacturing process is known to produce π_1 percent defectives. A new, supposedly improved process has been invented, but use of this new process is considered to be economically justifiable only if the proportion of defectives produced by the new process is π_2 or less, where $\pi_2 < \pi_1$.

One way to resolve the problem of deciding whether or not to use this new process is to initially establish the following null and alternate hypotheses pertaining to the new process:

$$\text{Null Hypothesis:} \quad \pi = \pi_1$$
$$\text{Alternate Hypothesis:} \quad \pi = \pi_2,$$

where π depicts the percentage of defectives produced by the new process. (Note that both of these hypotheses are simple hypotheses.) We then take a random sample from the new process. If our sample evidence leads us to accept the null hypothesis, then we would conclude that the new process is not better than the old one. On the other hand, if our sample evidence leads us to reject the null hypothesis and to accept the alternate hypothesis, we would conclude that the new process is clearly better than the old one, and we would switch to the new process.

There are two pitfalls in resorting to the hypothesis testing approach in order to resolve the given problem: first, the approach could lead us to keep the old process, even if the new process is significantly better; second, the approach could lead us to switch to the new process, even if it is not really any better than the old one. We should, naturally, wish to avoid committing either type of error.

Since the decision will have to be made based on the information provided in a sample, it is not possible to rule out completely the likelihood of our incurring both types of errors at the same time. On the other hand, as we will illustrate subsequently, the likelihood of our incurring these errors will depend to some extent on how effectively we utilize the information contained in the sample data. We will, therefore, explore some of the problems connected with effective utilization of sample data.

12.2 A Simple Bernoulli Hypothesis

To illustrate some of the problems connected with testing a hypothesis, we will describe in this section a very simple hypothesis testing situation and then we will examine the structure of the given problem.

EXAMPLE 12.2.1

Consider the following situation. We are in possession of two black boxes. The contents of these boxes are given in Figure 12.2.1. Suppose we are to choose one of these boxes and then guess the number of red marbles in the box selected. Assume that we are told that the box contains either one or two red marbles out of four in the box. There are four possible outcomes for this game:

1. We have chosen the box containing one red marble and have guessed that it contains one red marble.
2. We have chosen the box containing one red marble and have guessed that it contains two red marbles.
3. We have chosen the box containing two red marbles and have guessed that it contains one red marble.
4. We have chosen the box containing two red marbles and have guessed that it contains two red marbles.

These four possible outcomes are classified in Table 12.2.1.

1 Red Marble	2 Red Marbles
3 White Marbles	2 White Marbles

FIGURE 12.2.1

12.2 A Simple Bernoulli Hypothesis

TABLE 12.2.1

		Selection of Box	
		ONE RED	TWO REDS
Guess	ONE RED	correct	erroneous
	TWO REDS	erroneous	correct

Instead of describing the problem in terms of guessing the number of red marbles in the selected box, we can also describe the problem in terms of guessing the percentage of red marbles in the box. For example, if the box contains one red marble, then we would depict this fact by letting $\pi = 1/4$. If the box contains two red marbles, then we would depict this situation by letting $\pi = 2/4$.

Suppose we establish a null hypothesis that $\pi_1 = 1/4$ and an alternate hypothesis that $\pi_2 = 2/4$. Table 12.2.1 may be modified as Table 12.2.2. Thus, we have been able to describe the problem in the context of hypothesis testing.

We now define two terms which we will use frequently.

TABLE 12.2.2

		Selection of Box	
		$\pi = 1/4 = \pi_1$	$\pi = 2/4 = \pi_2$
Guess	ACCEPT NULL HYPOTHESIS	correct action	erroneous action
	REJECT NULL HYPOTHESIS	erroneous action	correct action

Definition 12.2.1

A *Type I error* is said to have been committed if the null hypothesis is rejected when it is, in fact, true.

Definition 12.2.2

A *Type II error* is said to have been committed if the null hypothesis is accepted when it is, in fact, false.

Returning to the guessing problem, we commit a Type I error if we guess that the box contains two red marbles when, in fact, it contains one red marble; we commit a Type II error if we guess that the box contains one red marble when, in fact, it contains two red marbles. These propositions are illustrated in Table 12.2.3.

TABLE 12.2.3

		Selection of Box	
		$\pi = 1/4 = \pi_1$	$\pi = 2/4 = \pi_2$
Guess	ACCEPT NULL HYPOTHESIS	no error	Type II error
	REJECT NULL HYPOTHESIS	Type I error	no error

Assume now that there is a penalty associated with each type of error and that this penalty may be described numerically. However, it is not necessary that we do so. We can simply assume now that there are good reasons for trying to avoid incurring either of the two penalties.

Suppose now that after we select a box, we will be allowed to select a marble once or twice from the box, but we must replace the first marble drawn before we draw the second and then make a guess. Our problem is to decide how many times to select a marble and, in turn, how to utilize the information thus obtained in making our guess. The problem really amounts to deciding how large a sample should be taken and how to utilize the information contained in a given size sample. We will put aside the problem of sample size for the moment and consider only the problem of utilizing information contained in a given size sample. The manner in which we utilize the information contained in a sample may be described in terms of a so-called *decision function*, which we formulate.

Definition 12.2.3

A *decision function* is a function whose domain is a sample space and whose image set is a set containing decisions to be made with regard to the null hypothesis. The set containing the decisions is called the *decision space*.

EXAMPLE 12.2.2

Refer to the guessing problem we have been discussing and suppose we have decided to select a marble once before making our guess. Then, there are four different decision functions which we can formulate, as Figure 12.2.2 illustrates. We may note that, according to Decision Function 1, we would accept the null hypothesis, regardless of the sample outcome and, according to Decision Function 4, we would reject the null hypothesis, regardless of the sample outcome.

If we are to select one decision function among the four, there must be some criteria by which we can evaluate these decision functions. The following definitions provide these criteria.

12.2 A Simple Bernoulli Hypothesis

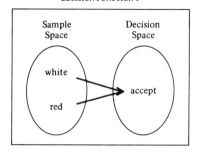
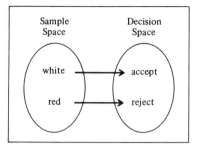
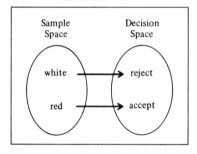
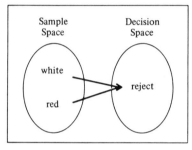

FIGURE 12.2.2

Definition 12.2.4
The probability that a given decision function will lead the decision maker to commit a Type I error is called the α *risk* associated with the decision function.

Definition 12.2.5
The probability that a given decision function will lead the decision maker to commit a Type II error is called the β *risk* associated with the decision function.

These two types of probabilities are called the *error probabilities* associated with a decision function.

We will now illustrate how to calculate the error probabilities in reference to Decision Function 3. Suppose that the box selected contains one red marble; that is, $\pi = 1/4 = \pi_1$. Then, the probability of our picking a red marble is $1/4$; that is,

$$P\left(R|\pi = \frac{1}{4} = \pi_1\right) = \frac{1}{4}$$

and, in turn, the probability of our picking a white marble is $3/4$; that is,

$$P\left(W|\pi = \frac{1}{4} = \pi_1\right) = \frac{3}{4}.$$

Since Decision Function 3 directs us to reject the null hypothesis in the event that we pick a white marble, the probability of rejecting the null hypothesis is $3/4$ if the box selected happens to contain one red marble. However, the null hypothesis is true, since the box contains one red marble. This means that $3/4$ is the probability of committing a Type I error, and, therefore, it is the α risk associated with the decision function.

Suppose now the box selected contains two red marbles and, thus, that the null hypothesis is not true. The probability of our picking a red marble is now $2/4$; that is,

$$P\left(R \mid \pi = \frac{2}{4} = \pi_2\right) = \frac{2}{4}$$

and, in turn, the probability of our picking a white marble is $2/4$; that is,

$$P\left(W \mid \pi = \frac{2}{4} = \pi_2\right) = \frac{2}{4}.$$

This means that $P(R \mid \pi = 2/4 = \pi_2) = 2/4$ now becomes the β risk associated with the decision function, since it represents the probability of accepting the null hypothesis, in spite of the fact that it is not true.

In a similar manner, the α and β risks may be ascertained for the other three decision functions and are given in Table 12.2.4. Before we proceed further, the reader should understand how these error probabilities have been ascertained.

TABLE 12.2.4 Error Probabilities for Decision Functions

Error Probability	Decision Function			
	D_1	D_2	D_3	D_4
α risk	0	$1/4$	$3/4$	1
β risk	1	$2/4$	$2/4$	0

Let us now compare Decision Functions 2 and 3. We observe that Decision Function 2, while having the same β risk as Decision Function 3, has a lower α risk than Decision Function 3. Thus, Decision Function 2 is clearly superior to Decision Function 3. Decision Function 3 is then said to be *dominated* by Decision Function 2. Whenever one decision function is dominated by another, we should rule it out as a consideration.

EXERCISES

*1. Suppose each of two marble boxes contains 10 marbles. One of them contains 2 red marbles and 8 white marbles, and the other contains 5 red marbles and 5

white marbles. You are to choose one of these boxes and then guess the percentage of red marbles in the box selected.
   a. Establish a null hypothesis to deal with your guessing problem.
   b. When would you commit a Type I error?
   c. When would you commit a Type II error?

2. Assume that you will be allowed to draw one marble from the box selected in Exercise 1. Enumerate all the decision functions which may be defined in the sample space.

3. For each decision function which you have enumerated in Exercise 2, calculate the error probabilities associated with each decision function.

4. Compare the error probabilities associated with the decision functions in Exercise 3. Is any decision function dominated by another?

*5. Refer to Exercise 1 and assume that you will be allowed to draw a marble twice before you make a guess. You will replace the first marble drawn, however, in the box before you draw the second. Enumerate all the decision functions which may be defined in the sample space.

*6. For each decision function which you have enumerated for Exercise 5, calculate the error probabilities associated with each decision function.

*7. Compare the error probabilities for the decision functions in Exercise 6. Is any decision function dominated by another?

12.3 Randomization of a Decision Function

The decision functions which we have considered in the preceding section are called *pure decision functions*. On the other hand, we can devise so-called *randomized decision functions*. We will explain the distinction between these two types of functions in this section. To illustrate this distinction, we will again resort to an example.

EXAMPLE 12.3.1

Refer to the guessing problem given in Example 12.2.1 and suppose we are to select a marble twice before we make a guess. Assume also that we will replace the first marble drawn in the box before we select the second marble.

Instead of specifying the sample space in terms of the color of the marble, let us specify it in terms of P, which depicts the percentage of red marbles observed in two independent draws. Then, a conceivable decision function, for example, may be the one given in Figure 12.3.1. This decision function indicates that we are to accept the null hypothesis if we observe either a white marble twice, in which case $P = 0$, or a white marble once, in which case $P = .5$. On the other hand, we are to reject the null hypothesis if we observe a red marble twice, in which case $P = 1.0$.

All conceivable decision functions which may be defined in the sample space in terms of the value of P are given in Table 12.3.1. Again, all of these

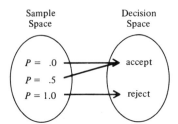

FIGURE 12.3.1 Decision function.

TABLE 12.3.1

Sample Space	Decision Function							
p	D_1	D_2	D_3	D_4	D_5	D_6	D_7	D_8
.0	a	a	a	a	r	r	r	r
.5	a	a	r	r	a	a	r	r
1.0	a	r	a	r	a	r	a	r

a = accept null hypothesis
r = reject null hypothesis

decision functions are pure decision functions. For each of these pure decision functions we have calculated the α and β risks, which are shown in Table 12.3.2.

TABLE 12.3.2 Error Probabilities for Pure Decision Functions

Error Probabilities	Pure Decision Function							
	D_1	D_2	D_3	D_4	D_5	D_6	D_7	D_8
α risk	0	$1/16$	$6/16$	$7/16$	$9/16$	$10/16$	$15/16$	1
β risk	1	$12/16$	$8/16$	$4/16$	$12/16$	$8/16$	$4/16$	0

We will illustrate how the error probabilities have been calculated for one of these decision functions. Let us, for example, choose Decision Function 2. Suppose the box selected contains one red marble; thus, $\pi = 1/4$. Then, the probability of drawing a red marble twice is $1/16$; that is,

$$P\left(P = 1 | \pi = \frac{1}{4} = \pi_1\right) = \frac{1}{16}.$$

Since we would be rejecting the null hypothesis if the observed value of P is 1,

12.3 Randomization of a Decision Function

and since the null hypothesis is, in fact, true when $\pi = \frac{1}{4}$,

$$P\left(P = 1 \mid \pi = \frac{1}{4} = \pi_1\right) = \frac{1}{16}$$

is the probability of rejecting the null hypothesis when it is, in fact, true. Thus, it is the α risk associated with the decision function. Suppose now the box selected contains two red marbles; thus, $\pi = \frac{2}{4}$, and the null hypothesis is not, in fact, true. However, the probability of accepting the null hypothesis is $\frac{12}{16}$, since

$$P\left(P = 0 \text{ or } P = .5 \mid \pi = \frac{2}{4}\right) = \frac{12}{16}$$

and, in turn, we would accept the null hypothesis when the observed value of P is either 0 or .5. Thus, the β risk associated with the decision function is $\frac{12}{16}$.

It may be instructive for the reader to verify the error probabilities of the other decision functions before he proceeds further.

Let us now compare the error probabilities of the decision functions given in Table 12.3.2. We find, for example, that Decision Functions 6 and 7 are dominated by Decision Function 4; Decision Function 5 is dominated by Decision Function 2. It appears that Decision Function 3 is not dominated by any other decision functions, but it can be made to be dominated if we resort to a *randomized decision function*.

We will now offer a definition of a *randomized decision function*.

Definition 12.3.1

A *randomized decision function* is a function whose domain is a sample space and whose image set is a set of pure decision functions.

A randomized decision function more frequently is called a *randomized strategy*, but we prefer to call it a randomized decision function. We will now illustrate what we mean by a randomized decision function.

EXAMPLE 12.3.2

Let us return to Example 12.3.1. Suppose we decide to toss a coin. If "heads" appears, we will choose Decision Function 2, and if "tails" appears, we will choose Decision Function 4. Such a strategy is a randomized decision function and can be depicted, as shown in Figure 12.3.2.

The significance of randomizing two decision functions is that, by doing so, we can obtain a set of error probabilities which are not obtainable without such randomization. The α risk for the preceding randomized decision function may be obtained as follows. Since the probability of choosing Decision Function 2 is $\frac{1}{2}$, and of choosing Decision Function 4 is also $\frac{1}{2}$, the α risk for the randomized decision function is the weighted average of the α risks associated with the two decision functions; that is,

$$\begin{aligned}
\text{α risk for randomized decision function} &= \frac{1}{2}(\alpha \text{ for } D_2) + \frac{1}{2}(\alpha \text{ for } D_4) \\
&= \frac{1}{2}\left(\frac{1}{16}\right) + \frac{1}{2}\left(\frac{7}{16}\right) \\
&= \frac{4}{16},
\end{aligned}$$

where D_2, for example, denotes Decision Function 2. Similarly, we calculate the β risk for the randomized decision function as

$$\begin{aligned}
\text{β risk for randomized decision function} &= \frac{1}{2}(\beta \text{ risk for } D_2) + \frac{1}{2}(\beta \text{ risk for } D_4) \\
&= \frac{1}{2}\left(\frac{12}{16}\right) + \frac{1}{2}\left(\frac{4}{16}\right) \\
&= \frac{8}{16}.
\end{aligned}$$

Thus, for the randomized decision function, we have $\alpha = 4/16$ and $\beta = 8/16$.

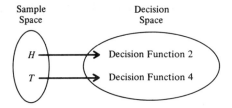

FIGURE 12.3.2 Randomized decision function.

Let us now compare the error probabilities of this randomized decision function and of Decision Function 3. We find that Decision Function 3 is dominated by the randomized decision function. This illustration shows that, although a pure decision function is not dominated by any other pure decision function, it can be dominated by a randomized decision function.

The significance of randomizing the decision function may be illustrated in another context. Consider Figure 12.3.3. Each numbered dot depicts a pure decision function. The dot on the line segment between D_2 and D_4 depicts the randomized decision function described above.

We have considered one possible randomized decision function. However, we can devise another randomized decision function. For example, if we toss a die, we can devise a randomized decision function such as the one given in Figure 12.3.4. The error probabilities for the randomized decision

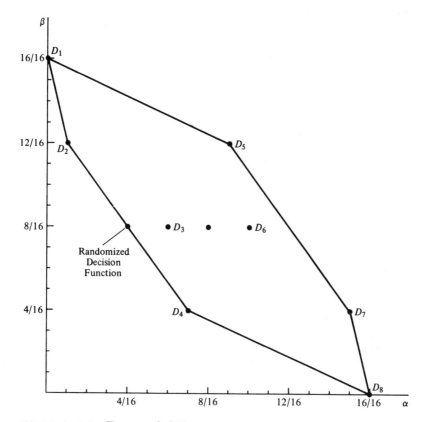

FIGURE 12.3.3 Error probabilities for decision functions.

function given in Figure 12.3.4 are calculated as

$$\alpha = \left(\frac{1}{6}\right)\left(\frac{0}{16}\right) + \left(\frac{1}{6}\right)\left(\frac{9}{16}\right) + \left(\frac{1}{6}\right)\left(\frac{15}{16}\right) + \left(\frac{1}{6}\right)\left(\frac{16}{16}\right) + \left(\frac{1}{6}\right)\left(\frac{1}{16}\right)$$
$$+ \left(\frac{1}{6}\right)\left(\frac{7}{16}\right) = \frac{1}{2}$$

$$\beta = \left(\frac{1}{16}\right)\left(\frac{16}{16}\right) + \left(\frac{1}{6}\right)\left(\frac{12}{16}\right) + \left(\frac{1}{6}\right)\left(\frac{4}{16}\right) + \left(\frac{1}{6}\right)\left(\frac{0}{16}\right) + \left(\frac{1}{6}\right)\left(\frac{12}{16}\right)$$
$$+ \left(\frac{1}{6}\right)\left(\frac{4}{16}\right) = \frac{1}{2}.$$

These error probabilities are depicted by the unnumbered dot in the interior of the polygon in Figure 12.3.3.

Even though we will not prove the proposition, it can be shown that any point in the shaded polygon may be obtained by properly randomizing the pure decision functions. At the same time, it can be shown also that any point outside of the polygon cannot be obtained by randomizing the pure decision functions.

Testing a Hypothesis

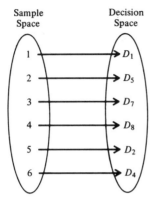

FIGURE 12.3.4 Randomized decision function.

EXERCISES

8. Select any two decision functions among those you have enumerated for Exercise 5 and randomize them in such a way that the probability of choosing one is .5. Then, calculate the error probabilities for the randomized decision function.

*9. For the two decision functions selected for Exercise 8, let the probability of selecting one be .2 and of selecting the other be 8. Then, calculate the error probabilities for each decision function.

10. Select any three decision functions among those you have enumerated for Exercise 5. Then, randomize them in such a way that each has an equal probability of being selected. Calculate the error probabilities for each decision function.

12.4 Admissible Decision Function

Consider now a set of points depicting all conceivable randomized decision functions of those pure decision functions shown in Figure 12.3.3. It can be shown that the set of such points must correspond to the shaded polygon in Figure 12.3.3. The significance of this proposition may be stated: Although there are only a finite number of pure decision functions from which we might choose one, if we allow a randomizing strategy, then there are really an infinite number of decision functions from which we can choose.

Obviously, we cannot evaluate each decision function individually. On the other hand, we do not have to evaluate each decision function in the shaded polygon to come up with a reasonably good one. If we examine the decision functions in the polygon in Figure 12.3.3 carefully, we will observe that only those on the line segments joining D_1, D_2, D_4, and D_8 are not dominated by other decision functions. The set of decision functions on the line segments joining D_1, D_2, D_3, and D_4, none of which is dominated by other decision functions in the shaded polygon in Figure 12.3.3, is called an *efficient set* or a *minimal complete class*, and a decision function in the efficient set is called an *admissible decision function*. Or, alternatively we might define:

Definition 12.4.1

A decision function is said to be *admissible* if it is impossible to reduce one type of risk by choosing another decision function without, at the same time, increasing the other type of risk.

This definition implies that a reasonably good decision function must be an admissible decision function.

Which decision function should be chosen as the single best one among those admissible decision functions in the efficient set? The answer to this question is difficult. If we compare any two among those admissible decision functions, we will find that one of them will have a smaller α risk relative to the other and, at the same time, will have a larger β risk relative to the other. This means that there is a need to trade off one type of risk against another. Two factors must be considered in trading off these errors. The likelihood that the null hypothesis is true is the first factor. For the marble-box guessing problem, then, we must consider the likelihood that we will choose one box as opposed to the other. The second factor to be considered is the relative severity of the penalties associated with the two types of errors. We will discuss the second of these two factors in the next section.

EXERCISES

11. Graph the error probabilities of the pure decision functions which you have enumerated in Exercise 2. Then, obtain a polygon which contains all pure and randomized decision functions.

12. Indicate on the graph you have drawn for Exercise 11 the set of points depicting the admissible decision functions.

13. Select any two pure admissible decision functions which are adjacent to each other from Exercise 12. Randomize the two decision functions so that the probability of selecting one of them is γ, where $0 \leq \gamma \leq 1$ is an arbitrary number which you have assigned. Then, calculate the error probabilities for the resulting decision function. Indicate the point corresponding to the randomized decision function on the graph.

*14. Graph the error probabilities of the decision functions which you have enumerated for Exercise 5. Then, obtain a polygon which contains all pure and randomized decision functions.

*15. Indicate on the graph you have drawn for Exercise 14 the set of points depicting the admissible decision functions.

256   Testing a Hypothesis

*16. Refer to Exercises 14 and 15 and select any two pure admissible decision functions which are adjacent to each other. Randomize the two decision functions so that the probability of selecting one of them is γ, where $0 \leq \gamma \leq 1$ is an arbitrary number which you have assigned. Then, calculate the error probabilities for the resulting decision function. Indicate the point corresponding to the randomized decision function on the graph.

12.5 Loss Function

So far we have refrained from specifying the numerical values associated with each type of error. For the marble-box guessing problem, we can, for example, arbitrarily assign numerical values for the penalties. On the other hand, for many hypothesis-testing problems of practical significance, quantifying the values of the penalties may be quite difficult. For example, what is the value of the penalty associated with the failure of a key component used in a manned spacecraft designed to go to the moon and back? Quantification of that value would indeed be difficult.

Nevertheless, some sort of quantification of the values of the penalties associated with each type of error obviously would be useful in deciding upon the decision function. One way to quantify the values of the penalties associated with the errors is to design a so-called *loss function*.

Definition 12.5.1

Let the domain of a function contain the types of errors that may be committed in testing a null hypothesis, and let the image set contain a set of real numbers associated with the penalties for the errors. The function is called a *loss function* associated with testing the hypothesis.

We might also point out that the numbers in the image set usually represent either monetary values or utilities.

EXAMPLE 12.5.1

Let us return to the marble-box guessing problem described in Example 12.3.1. Suppose the penalties associated with the wrong guesses are those given in Table 12.5.1. Thus, the penalty associated with the Type II error is, by assumption, more serious. The loss function is given in Figure 12.5.1.

For each type of error we have so far specified two numerical quantities: the error probability and the value of the loss associated with the error. We can now weigh the amount of the loss by its probability. The weighed loss is called the *conditional expected loss*.

12.5 Loss Function

TABLE 12.5.1

		Selection of Box	
		$\pi = \frac{1}{4} = \pi_1$	$\pi = \frac{2}{4} = \pi_2$
Guess	ACCEPT NULL HYPOTHESIS	no penalty	100
	REJECT NULL HYPOTHESIS	80	no penalty

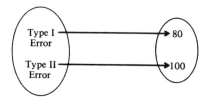

FIGURE 12.5.1 Loss function.

Definition 12.5.2

Let $L(I)$ and $L(II)$ denote the amount of loss associated with the Type I and Type II errors, respectively, and let D denote the decision function selected. Then,

$$R(D|H_1) = \alpha L(I)$$

is called the *conditional expected loss* associated with the decision function D, given that the null hypothesis is true, and

$$R(D|H_2) = \beta L(II)$$

is called the *conditional expected loss* associated with the decision function D, given that the null hypothesis is false.

These conditional expected losses may also be considered the risks that we face by choosing a particular decision function. In fact, in many standard treatises on this subject matter, they are properly called the *risks*. However, we will not call them risks here, since we have already used the term to define the two types of error probabilities: alpha and beta risks.

EXAMPLE 12.5.2

Let us again return to the marble-box guessing problem in Example 12.3.1. Suppose we combine the loss function given in Example 12.5.1 with the pure decision functions given in Example 12.3.1. Then, the conditional expected losses found in Table 12.5.2 may be obtained.

258 Testing a Hypothesis

TABLE 12.5.2 Conditional Expected Losses for Decision Functions

Conditional Expected Loss	Decision Function							
	D_1	D_2	D_3	D_4	D_5	D_6	D_7	D_8
$R(D\|H_1)$	0	5	30	35	45	50	75	80
$R(D\|H_2)$	100	75	50	25	75	50	25	0

The conditional expected losses corresponding to each pure decision function is depicted by a numbered dot in Figure 12.5.2. The entire shaded polygon in the figure depicts the set of conditional expected losses associated with all possible randomized decision functions.

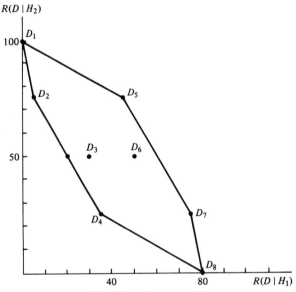

FIGURE 12.5.2 Conditional expected losses.

Consider now any decision function on the line segments for $\overline{D_1D_2}$, $\overline{D_2D_4}$, and $\overline{D_4D_8}$. We observe that it is not dominated by any other decision functions depicted by points in the polygon. It is apparent that it is reasonable to consider only those decision functions on these lower boundary line segments. Thus, we may provide an alternate definition of an *admissible decision function;* that is,

Definition 12.5.3
A decision function is said to be *admissible* if it is impossible to reduce one type of conditional expected loss by choosing another decision func-

tion without, at the same time, increasing the other type of conditional expected loss.

Thus, only those decision functions depicted by the points on the lower boundary line segments will satisfy this definition.

Let us now compare Figures 12.3.3 and 12.5.2. Both shaded polygons depict the set containing all possible pure and randomized decision functions. On the other hand, we find that, for both of these polygons, the boundary segments $\overline{D_1D_2}$, $\overline{D_2D_4}$, and $\overline{D_4D_8}$ constitute the efficient set containing all admissible decision functions.

It is clear, then, that the set of admissible decision functions may be described either in terms of error probabilities or in terms of conditional expected losses. Since the two sets are, in fact, equivalent, it appears that we have gained nothing by describing the set in terms of the conditional expected losses; however, this is not true. We have pointed out that, in trading off the error probabilities, we need to know the likelihood that the null hypothesis is true and the relative severity of the penalties associated with the errors. Since the relative severity of the penalties is reflected by their conditional expected losses, we only have to trade off these conditional expected losses. To do so, however, we still need to know the likelihood that the null hypothesis is true.

EXERCISES

*17. Refer to Exercise 1 and assume that the penalty associated with a Type I error is $200 and the penalty associated with a Type II error is $100. Describe the loss function for the problem.

18. For each pure decision function which you have enumerated for Exercise 2, calculate the conditional expected losses, using the loss function described for Exercise 17.

19. Graph the conditional expected losses found for Exercise 18. Then, find a polygon containing the conditional expected losses for all pure and randomized decision functions.

20. In the polygon obtained for Exercise 19, indicate the set of points corresponding to all admissible decision functions.

*21. For each pure decision function which you have enumerated for Exercise 5, calculate the conditional expected losses, using the loss function you have described for Exercise 17.

*22. Graph the conditional expected losses found for Exercise 21. Then, find a polygon containing the conditional expected losses for all pure and randomized decision functions.

*23. In the polygon obtained for Exercise 22, indicate the set of points corresponding to all admissible decision functions.

12.6 Minimax Decision Function

We have stated that in trading off between the two types of conditional expected losses, we need to have some idea as to the likelihood that the null hypothesis is true. Suppose we have no idea as to the likelihood that the null hypothesis is true. Then, we might wish to obtain a decision function such that the worst possible consequence of utilizing that decision function is not very bad. One such decision function is called a *minimax decision function;* that is,

Definition 12.6.1

Let D be the decision function which yields the smallest possible, maximum conditional expected loss. The decision function is called a *minimax decision function*.

EXAMPLE 12.6.1

Let us return to Example 12.5.2 and compare the two decision functions D_1 and D_8. The conditional expected loss for D_1 is 0 if the null hypothesis is true; that is,

$$R(D_1|H_1) = 0,$$

but the conditional expected loss for D_1 is \$100 if the null hypothesis is false; that is,

$$R(D_1|H_2) = 100.$$

Thus, the maximum conditional expected loss for D_1 is \$100. Similarly, we observe that the maximum conditional expected loss for D_8 is \$80. Thus, if we were to choose a minimax decision function between D_1 and D_8, it would have to be D_8.

The maximum conditional expected losses for all pure admissible decision functions are given in Table 12.6.1. The minimax decision function among the pure admissible decision functions is D_4, for which the maximum conditional expected loss associated is \$35.

TABLE 12.6.1

Decision Function	Maximum Conditional Expected Loss
D_1	100
D_2	75
D_4	35
D_8	80

12.6 Minimax Decision Function

Could we now obtain a decision function whose maximum conditional expected loss is smaller than $35 by randomizing the pure decision functions? The answer is "yes" for the given example.

Suppose, for example, we randomize D_2 and D_4 in such a way that the probability of selecting D_2 is .1 and of selecting D_4 is .9. Then, the conditional expected losses for the resulting randomized decision function are those given in Table 12.6.2. Thus, the maximum conditional expected loss is $32, which clearly is smaller than that for D_4.

TABLE 12.6.2

States with Regard to Hypothesis	Conditional Expected Loss
$\pi = 1/4 = \pi_1 = H_1$	$.1(5) + .9(35) = \$32$
$\pi = 2/4 = \pi_2 = H_2$	$.1(75) + .9(25) = \$30$

Can we reduce the maximum conditional expected loss from $32 by increasing the probability of selecting D_2? The answer is "yes." On the other hand, there is a limit to which the probability of selecting D_2 can be increased without, at the same time, increasing the maximum conditional expected loss. For example, suppose we let the probability of selecting D_2 be increased to .2. Then, the conditional expected losses are those given in Table 12.6.3. Thus, the maximum conditional expected loss is $35, which is the same as that for D_4.

TABLE 12.6.3

States with Regard to Hypothesis	Conditional Expected Loss
$\pi = 1/4 = \pi_1 = H_1$	$.2(5) + .8(35) = \$29$
$\pi = 2/4 = \pi_2 = H_2$	$.2(75) + .8(25) = \$35$

Let γ depict the probability of selecting D_2 and let $1 - \gamma$ depict the probability of selecting D_4. Let us now define the *optimal value* of γ to be that value of γ which yields the minimax decision function among all randomized decision functions between D_2 and D_4. The two preceding illustrations suggest that the optimal value of γ must be somewhere between .1 and .2. Can we, however, find this optimal value for γ without resorting to a trial-and-error procedure? The answer is "yes." We can find the optimal value for γ in one trial. To illustrate how this can be done, let us first define that

$$R[D(\gamma)|H_1] = R(D_2|H_1)\gamma + R(D_4|H_1)(1 - \gamma)$$
$$= 5\gamma + 35(1 - \gamma) = 35 - 30\gamma$$

and

$$R[D(\gamma)|H_2] = R(D_2|H_2)\gamma + R(D_4|H_2)(1 - \gamma)$$
$$= 75\gamma + 25(1 - \gamma) = 25 + 50\gamma.$$

262 Testing a Hypothesis

Then, $R[D(\gamma)|H_1]$ and $R[D(\gamma)|H_2]$ are the two conditional expected losses for the randomized decision function between D_2 and D_4 for a given value of γ. These conditional expected losses are depicted by Figure 12.6.1 for the various values of γ.

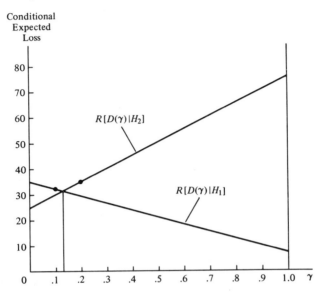

FIGURE 12.6.1

Now let

$$R[D(\gamma)] = \text{Max.} \{R[D(\gamma)|H_1], R[D(\gamma)|H_2]\}.$$

We observe that the values of $R[D(\gamma)]$ for all possible values of γ are depicted by the shaded area in Figure 12.6.1. We have, for example, indicated $R[D(\gamma = .1)]$ and $R[D(\gamma = .2)]$ by the two dots on the graph. We can now locate the optimal value of γ for the minimax decision function. Since the optimal value of γ must be that which minimizes the value of $R[D(\gamma)]$, it must correspond to the location indicated by the vertical line on the graph. Thus, we have ascertained the optimal value of γ graphically.

Note however that when the optimal value of γ is chosen for the minimax decision function, the two conditional expected losses are the same. This equality between the two types of conditional expected losses suggests that it may be ascertained algebraically. The fact that the two conditional expected losses must be the same for the optimal value of γ can now be expressed as

$$R(D_2|H_1)\gamma + R(D_4|H_1)(1 - \gamma) = R(D_2|H_2)\gamma + R(D_4|H_2)(1 - \gamma)$$

and, in turn, as

$$5\gamma + 35(1 - \gamma) = 75\gamma + 25(1 - \gamma).$$

When we solve this equation for γ, we find that the optimal value of γ is $\frac{1}{8}$. Calculating now the conditional expected losses, we find the values given in Table 12.6.4. Thus, the minimax decision function is the randomization between D_2 and D_4 such that the probability of selecting D_2 is $\frac{1}{8}$ and of selecting D_4 is $\frac{7}{8}$. If we re-examine Table 12.6.4 we will find that our algebraic solution is, in fact, equivalent to the one we obtained graphically.

TABLE 12.6.4

States with Regard to Hypothesis	Conditional Expected Loss
$\pi = \frac{1}{4} = \pi_1 = H_1$	$\frac{1}{8}(5) + \frac{7}{8}(35) = \$31.25$
$\pi = \frac{2}{4} = \pi_2 = H_2$	$\frac{1}{8}(75) + \frac{7}{8}(25) = \$31.25$

We now propose that the minimax decision function among the randomized decision functions between D_2 and D_4 is also the minimax decision function among all pure and randomized decision functions given in Figure 12.5.2. This proposition implies that if we wish to obtain the minimax decision function among all decision functions given in Figure 12.5.2, then we only need to choose D_2 and D_4 and randomize the two decision functions. "How do we know that D_2 and D_4 are the two pure decision functions which we need to randomize in order to obtain the minimax decision function?" The reason for choosing D_4 is obvious enough: It is the minimax pure decision function. How can we now justify our choice of D_2 as the other decision function needed for the randomization? We observe that $R(D_4|H_1) = 35$ is greater than $R(D_4|H_2) = 25$. Since $R[D(\gamma)]$ for the randomized minimax decision function must be smaller than 35 and, at the same time, larger than 25, we must choose another decision function to randomize in such a way that $R(D|H_1)$ will be smaller than 35 and, at the same time, that $R(D|H_2)$ will be larger than 25. The best minimax decision function among the remaining pure admissible decision functions, which also satisfies the preceding constraints, is D_2. Thus, we should randomize D_2 and D_4 in order to obtain the randomized minimax decision function.

EXERCISES

24. Among the pure admissible decision functions given in Exercise 20, find the decision function which satisfies the minimax criterion.
25. Among all the admissible decision functions given in Exercise 20, find the decision function which satisfies the minimax criterion.

*26. Among the pure admissible decision functions given in Exercise 23, find the decision function which satisfies the minimax criterion.

*27. Among all the admissible decision functions given in Exercise 23, find the decision function which satisfies the minimax criterion.

12.7 Bayes Decision Function

In discussing the minimax decision function, we have assumed that we have absolutely no knowledge of the likelihood that the null hypothesis is true. However, this may not be the case in many hypothesis-testing situations.

If the decision maker has some knowledge of the likelihood that the null hypothesis is true, he may not wish to use the minimax decision function. Refer once again to the marble-box guessing problem given in Example 12.3.1 and suppose that we believe that the probability that the null hypothesis is true is 1, which is to say that we are absolutely certain that we will choose the box containing one red marble. Then, there is really no reason that we should be concerned that the conditional expected loss associated with the null hypothesis might be false, because by assumption the null hypothesis cannot be false. Consequently, we will choose the decision function which will minimize the conditional expected loss when the null hypothesis is true. This decision function is D_1. On the other hand, if we believe that the probability that the null hypothesis is true is 0, then we will obviously choose the decision function which will minimize the conditional expected loss when the null hypothesis is false. This decision function is D_8.

The situations described perhaps are too extreme. If we are to assign a probability for the null hypothesis being true, we are more likely to assign a value greater than 0 but less than 1. Let ψ be our assessment of the probability that the null hypothesis is true. Such a probability is called a *prior probability*. We will explore later the nature of a prior probability as well as the controversies which surround it. We will for the moment assume, however, that ψ is a valid measure of the probability that the null hypothesis is true. Then, ψ and $1 - \psi$ can be used to weigh the two types of conditional expected losses. A decision function based on this procedure is called a *Bayes decision function*.

Definition 12.7.1

Let ψ be the prior probability that the null hypothesis is true. Let $\hat{D}$ be a decision function which minimizes

$$R(D) = \psi R(D|H_1) + (1 - \psi)R(D|H_2).$$

The decision function $\hat{D}$ is called a Bayes decision function, where $R(D)$ is called the *unconditional expected loss* associated with the decision function D.

We will, however, call $R(D)$ simply the expected loss associated with the decision function D.

EXAMPLE 12.7.1

Let us return again to the marble-box guessing problem. Suppose, for some reason, we believe that the probability that the null hypothesis is true is .5; that is, we believe that there is a 50 percent chance that we will select one type of marble box as opposed to another. Then, the expected losses associated with all pure decision functions are those shown in Table 12.7.1. Thus, D_4 is the Bayes decision function among the pure decision functions for $\psi = .5$.

TABLE 12.7.1

Decision Function	$R(D)$: Expected Loss
D_1	.5(0) + .5(100) = 50
D_2	.5(5) + .5(75) = 40
D_3	.5(30) + .5(50) = 40
D_4	.5(35) + .5(25) = 30
D_5	.5(45) + .5(75) = 60
D_6	.5(50) + .5(50) = 50
D_7	.5(75) + .5(25) = 50
D_8	.5(80) + .5(0) = 40

It must be intuitively apparent that a Bayes decision function must be among the admissible decision functions. The question that may arise is, "How should we ascertain a Bayes decision function among all admissible decision functions?" However, there is really no reason to seek a Bayes decision function among the randomized decision functions, for the following reason.

Theorem 12.7.1

Let D be a Bayes decision function which is a randomized decision function. Then, there exists a pure decision function which is an equivalent Bayes decision function.

Theorem 12.7.1 does not imply that if a decision function is a Bayes decision function, then it must be a pure decision function; it merely states that if there is a Bayes decision function that is a randomized decision function, then we can be sure that there must exist a pure decision function that is also a Bayes decision function. We will illustrate this proposition in the next example.

EXAMPLE 12.7.2

Refer to Example 12.7.1 and assume that $\psi = 5/8$. Then, the expected losses for the admissible pure decision functions are those given in Table 12.7.2. Thus, both D_2 and D_4 are the Bayes decision functions among the pure decision functions.

TABLE 12.7.2

Decision Function	$R(D)$: Expected Loss
D_1	$5/8(0) + 3/8(100) = \$37.50$
D_2	$5/8(5) + 3/8(75) = \$31.25$
D_4	$5/8(35) + 3/8(25) = \$31.25$
D_8	$5/8(80) + 3/8(0) = \$50.00$

Now, suppose we randomize D_2 and D_4 so that the probability of selecting D_2 is $1/2$. The randomized decision function is shown in Figure 12.5.2. We note that this randomized decision function is on the line segment joining D_2 and D_4. Let us now calculate the expected loss for this randomized decision function as

$$R(D) = \gamma R(D_2) + (1 - \gamma)R(D_4)$$
$$= .5(31.25) + .5(31.25) = 31.25,$$

where $R(D_2)$, for example, denotes the expected loss for the pure decision function D_2. Thus, we find that the randomized decision function is also a Bayes decision function. In fact, regardless of how we randomize D_2 and D_4, the resulting decision function will be a Bayes decision function. The validity of this proposition is shown by the fact that for any arbitrarily chosen value of γ, we have

$$R(D) = \gamma(31.25) + (1 - \gamma)(31.25)$$
$$= \gamma(31.25) + (31.25) - \gamma(31.25)$$
$$= 31.25.$$

It can be shown that the only time an admissible randomized decision function becomes a Bayes decision function is when it is on the line segment joining two pure decision functions which are also Bayes decision functions. It is clear, then, that whenever a randomized Bayes decision function exists, there must also exist a pure decision function which is a Bayes decision function. This is what Theorem 12.7.1 implies.

We can now state the significance of the theorem. In searching for a Bayes decision function, there really is no need to evaluate the randomized decision functions, since, as we have pointed out, if there is a randomized Bayes decision function, then there must also exist a pure decision function which is a Bayes decision function.

12.7 Bayes Decision Function

The computational significance of the theorem is clear: in order to find a Bayes decision function, we need only to examine the pure decision functions which are admissible. The student who is familiar with linear programming might readily see the apparent similarity between the search for a Bayes decision function and the search for an optimal solution in linear programming. One fundamental theorem in linear programming states that, if there is a nonextreme optimal point-solution, there must exist an equivalent optimal extreme point-solution. In searching for an optimal solution for a linear programming problem, we need to examine only those solutions corresponding to the extreme points in the efficient set. Thus, a pure Bayes decision function, in a sense, corresponds to an extreme point-solution for a linear programming problem.

We have pointed out that a Bayes decision function must be an admissible decision function. On the other hand, the following theorem also is true, and includes an important ramification for our discussion in the next chapter.

Theorem 12.7.2

Every admissible decision function is a Bayes decision function for an appropriate value of ψ.

Instead of proving this theorem, we will again illustrate its validity with an example.

EXAMPLE 12.7.3

Let us return to the situation given in Example 12.7.1. We have already observed that D_1, D_2, D_4, and D_8 together constitute the set of all pure decision functions which are also admissible.

Suppose now we let $\psi = 5/6$. The calculations shown in Table 12.7.3 indicate that both D_1 and D_2 are Bayes decision functions. This also means that every randomized decision function between D_1 and D_2 is a Bayes decision function when $\psi = 5/6$.

Suppose we let $\psi = 5/8$. Then, our calculations shown in Table 12.7.2 indicate that both D_2 and D_4 are Bayes decision functions, which also means that every randomized decision function between D_2 and D_4 is a Bayes decision function when $\psi = 5/8$.

TABLE 12.7.3

Decision Function	$R(D)$: Expected Loss
D_1	$(5/6)(0) + (1/6)(100) = 16.67$
D_2	$(5/6)(5) + (1/6)(75) = 16.67$
D_4	$(5/6)(35) + (1/6)(25) = 33.33$
D_8	$(5/6)(80) + (1/6)(0) = 66.67$

268 Testing a Hypothesis

Suppose now we let $\psi = 5/14$. Then, the calculations shown in Table 12.7.4 indicate that both D_4 and D_8 are Bayes decision functions, which also means that every randomized decision function between D_4 and D_8 is a Bayes decision function when $\psi = 5/14$.

TABLE 12.7.4

Decision Function	$R(D)$: Expected Loss
D_1	$(5/14)(0) + (9/14)(100) = 64.29$
D_2	$(5/14)(5) + (9/14)(75) = 50.00$
D_4	$(5/14)(35) + (9/14)(25) = 28.57$
D_8	$(5/14)(80) + (9/14)(0) = 28.57$

If we return to Figure 12.5.2, we find that the pure decision functions D_1, D_2, D_4, and D_8, as well as every randomized decision function between D_1 and D_2, D_2 and D_4, and D_4 and D_8 together constitute the entire set of all admissible decision functions. Thus, we have illustrated that every admissible decision function is a Bayes decision function for an appropriate value of ψ.

If we re-examine now the calculations shown in Tables 12.7.2, 12.7.3, and 12.7.4, we will find that the expected loss $R(D)$ associated with a Bayes decision function depends on the value of ψ. For example, when $\psi = 5/6$, Table 12.7.3 shows that $R(D) = 16.67$. But, when $\psi = 5/14$, Table 12.7.4 shows that $R(D) = 28.57$. It should be apparent, then, that there must exist a value of ψ whose corresponding Bayes decision function yields the largest possible value of $R(D)$. Since $R(D)$ depicts expected loss, we will call the value of ψ which maximizes $R(D)$ the *least favorable prior probability* that the null hypothesis is true.

Theorem 12.7.3

Let ψ_0 be the least favorable prior probability that the null hypothesis is true. Then, the Bayes decision function for $\psi = \psi_0$ is equivalent to the minimax decision function.

EXAMPLE 12.7.4

Returning to Example 12.7.3, suppose we let $\psi = 0$. Then, we will find that D_8 is the Bayes decision function and $R(D) = 0$ for this decision function. On the other hand, suppose we let $\psi = 1$. Then, we will find that D_1 is the Bayes decision function and $R(D) = 0$ for this decision function. The values of $R(D)$ for five different values of ψ are summarized in Table 12.7.5.

We observe in Table 12.7.5 that, for the five different values of ψ, the value of $R(D)$ is largest when $\psi = 5/8$. Of course, we do not know yet whether some other value of ψ which is not given in Table 12.7.5 might yield a value

12.7 Bayes Decision Function

TABLE 12.7.5

ψ	$R(D)$
0	0
5/14	28.57
5/8	31.25
5/6	16.67
1	0

of $R(D)$ larger than 31.25. Figure 12.7.1, however, depicts the values of $R(D)$ for all values of ψ between 0 and 1. The figure indicates that the largest value of $R(D)$ is 31.25, which is attained when $\psi = 5/8$.

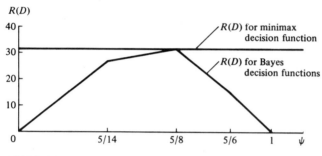

FIGURE 12.7.1

Thus, that $\psi = 5/8$ is really the least favorable prior probability that the null hypothesis is true, and, in turn, the least favorable apriori probability in question yields $R(D) = 31.25$.

Let us now return to Example 12.6.1, in which we have illustrated a minimax decision function. We have shown in Table 12.6.4 that, if we resort to the minimax decision function for the given problem, then the maximum expected loss is 31.25. We observe, then, that the expected loss for the Bayes decision function corresponding to the least favorable apriori probability for the null hypothesis and the maximum expected loss for the minimax decision function are, in fact, the same. This is what is implied by Theorem 12.7.3.

What is the significance of Theorem 12.7.3? It indicates that, if we know the value of ψ, then, there is really no reason for us to prefer the minimax decision function over a Bayes decision function. The theorem also shows that, if the value of ψ is known, the expected loss for any Bayes decision function we might obtain will be either equal to or less than that of the minimax decision function.

Does this mean that we should prefer a Bayes decision function over the minimax decision function, even though we do not know the value of ψ? The answer is "not necessarily so," as we will illustrate.

EXAMPLE 12.7.5

Let us return to our marble-box guessing problem and assume that the prior probability of the null hypothesis being true is $5/14$; that is, $\psi = 5/14$. Then, according to our calculations in Table 12.7.4, both D_4 and D_8 are Bayes decision functions with expected losses of 28.57.

Let us assume now that, even though the true value of ψ is $5/14$, we have mistakenly assigned $5/6$ as the value of ψ. Then, according to our calculations in Table 12.7.3, we would choose either D_1 or D_2, erroneously thinking that the expected loss for D_1 or D_2 is 16.67.

Suppose now that we have actually selected D_2. Since the true value of ψ is really $5/14$ and not $5/6$, the true expected loss for D_2 is not 16.67 but 50.00 as we calculate

$$R(D_2) = \left(\frac{5}{14}\right)(5) + \left(\frac{9}{14}\right)(75) = 50.00.$$

Thus, the expected loss for the decision function D_2 is greater than that of the minimax decision function.

On the other hand, suppose now that the true value of ψ is $5/6$. Then, both D_1 and D_2 are Bayes decision functions with expected losses of 16.67. Assume now that we have mistakenly assigned $5/14$ as the value of ψ. Then, we would choose either D_4 or D_8 as a Bayes decision function, erroneously thinking that the expected losses are 28.57. Assume now that we have selected D_4. Then, since the true value of ψ is $5/6$ and not $5/14$, the true expected loss is calculated as

$$R(D_4) = \left(\frac{5}{6}\right)(35) + \left(\frac{1}{6}\right)(25) = 33.33,$$

which is, again, larger than that of the minimax decision function.

A lesson to be learned from our preceding illustrations may be stated as follows. In many practical applications of statistical hypothesis testing, the exact value of ψ is likely to be unknown and, furthermore, possibly very difficult to estimate. Suppose, however, we have obtained a Bayes decision function by using a guessed value of ψ. Then, the decision function thus obtained will turn out to be better than the minimax decision function if our guessed value of ψ is very close to the unknown but true value of ψ. On the other hand, if the guessed value of ψ is substantially different from the true value of ψ, then, we might have been better off with the minimax decision function or some other type of decision function.

EXERCISES

28. Refer to Exercise 1 and assume that you will choose the box by tossing a coin. Find a Bayes decision function among the admissible decision functions given in Exercise 20.

29. Refer to Exercise 1 and assume that you will use the following scheme to choose a box: toss a die and if the number 1 or 2 appears, choose the box containing 2 red marbles; otherwise, choose the box containing 5 red marbles. Find a Bayes decision function among the admissible decision functions given in Exercise 20.

*30. Refer to Exercise 1 and assume that you will choose a box by means of the following method: toss 2 coins and if you obtain 2 tails, choose the box containing 5 red marbles; otherwise, choose the box containing 2 red marbles. Find a Bayes decision function among the admissible decision functions given in Exercise 23.

12.8 Illustrative Applications

In illustrating the elements of statistical decision theory, we have resorted to the marble-box guessing problem.

We will now show the relevance of the statistical decision theory for simple but practical problems.

EXAMPLE 12.8.1

Assume that a manufacturing process is in one of two states: It produces either 10 percent defectives or 20 percent defectives. If the process produces 10 percent defectives, we assume that it is working satisfactorily; if it produces 20 percent defectives, then we will stop the process for corrective adjustment.

Since we do not know whether the process is producing 10 percent defectives or 20 percent defectives, we have decided to guess the proportion in question by selecting a very small simple random sample from the process. Further, we have decided to select three sample items from the process and then make a decision as to whether to let the machine keep working or to stop it for corrective adjustment.

The similarity between this problem and the marble-box guessing problem should be apparent. We may, for example, convert this problem into an equivalent marble-box guessing problem by assuming that one of the two boxes contains 10 percent defective items and that the other box contains 20 percent defective items.

Let us now proceed to establish a null hypothesis. To do so, we can let either $\pi_1 = .1$ or $\pi_1 = .2$. If we let $\pi_1 = .1$, the null hypothesis is really a supposition that the machine process is working satisfactorily. If we let $\pi_1 = .2$, the null hypothesis states that the machine process is not working satisfactorily and, therefore, should be corrected. The choice between the two is arbitrary. Thus, we will let $\pi_1 = .1$. This means, in turn, that $\pi_2 = .2$ (the alternate hypothesis). Various outcomes and consequences of making a decision are classified in Table 12.8.1.

TABLE 12.8.1

Decision	Status of Machine Process	
	$\pi = .1 = \pi_1$ (NULL HYPOTHESIS TRUE)	$\pi = .2 = \pi_2$ (ALTERNATE HYPOTHESIS TRUE)
ACCEPT NULL HYPOTHESIS: LET MACHINE RUN	correct action	Type II error: let machine run even though not working satisfactorily
REJECT NULL HYPOTHESIS: STOP MACHINE	Type I error: stop machine when working satisfactorily	correct action

Let us now formulate some of the decision functions that we may use. All possible pure decision functions in terms of P are given in Table 12.8.2. There are altogether 16 possible pure decision functions in terms of P. The error probabilities corresponding to these decision functions are given in Table 12.8.3.

TABLE 12.8.2

Sample Space	Decision Function															
p	D_1	D_2	D_3	D_4	D_5	D_6	D_7	D_8	D_9	D_{10}	D_{11}	D_{12}	D_{13}	D_{14}	D_{15}	D_{16}
0	a	a	a	a	a	a	a	a	r	r	r	r	r	r	r	r
1/3	a	a	a	a	r	r	r	r	a	a	a	a	r	r	r	r
2/3	a	a	r	r	a	a	r	r	a	a	r	r	a	a	r	r
1	a	r	a	r	a	r	a	r	a	r	a	r	a	r	a	r

a = accept null hypothesis
r = reject null hypothesis

We will show how to calculate the error probabilities for one of these decision functions. Let us select arbitrarily Decision Function 6. This decision function is also depicted in Figure 12.8.1.

Suppose the null hypothesis is, in fact, true. Then, from Table B.6 in Appendix B (Binomial Probabilities), we find the probability function shown

12.8 Illustrative Applications

TABLE 12.8.3

Error Probability	Decision Function							
	D_1	D_2	D_3	D_4	D_5	D_6	D_7	D_8
α risk	0	.001	.027	.028	.243	.244	.270	.271
β risk	1	.992	.904	.896	.616	.608	.520	.512

Error Probability	Decision Function							
	D_9	D_{10}	D_{11}	D_{12}	D_{13}	D_{14}	D_{15}	D_{16}
α risk	.729	.730	.756	.757	.972	.973	.999	1
β risk	.488	.480	.392	.384	.104	.096	.008	0

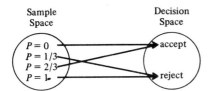

FIGURE 12.8.1 Decision Function 6.

in Table 12.8.4. Since we will reject the null hypothesis if we observe that P is either $\tfrac{1}{3}$ or 1, the probability of rejecting the null hypothesis is

$$P\left(P = \frac{1}{3} \,\bigg|\, \pi = .1\right) + P(P = 1 | \pi = .1) = .243 + .001 = .244,$$

which corresponds to the α risk.

Suppose that the alternate hypothesis is, in fact, true. Then we find the probability function given in Table 12.8.5. Since we will still accept the null

TABLE 12.8.4

p	$P(P = p | \pi = .1)$
0	.729
$\tfrac{1}{3}$	.243
$\tfrac{2}{3}$	.027
1	.001
	1.000

TABLE 12.8.5

p	$P(P = p \mid \pi = .2)$
0	.512
1/3	.384
2/3	.096
1	.008
	1.000

hypothesis if we observe that P is either 0 or $\tfrac{2}{3}$, the probability of accepting the null hypothesis even though the alternate hypothesis is true is given as

$$P(P = 0 \mid \pi = .2) + P\left(P = \frac{2}{3} \;\middle|\; \pi = .2\right) = .512 + .096 = .608,$$

which corresponds to the β risk.

Suppose now we wish to choose an admissible decision function among the pure decision functions given in Table 12.8.3. Our examination of the error probabilities for these decision functions reveals, however, that none of them is dominated by another pure decision function. Does this imply that all the pure decision functions given are admissible? The answer is "no," as we will illustrate, since some of them are dominated by a randomized decision function.

The error probabilities of these pure decision functions are indicated by the numbered dots in Figure 12.8.2. The shaded area in the figure depicts the set of all randomized decision functions. The figure shows that D_1, D_2, D_4, D_8, and D_{16} are pure decision functions, which are also admissible. The set of all admissible decision functions is given by the lower boundary segments $\overline{D_1 D_2}$, $\overline{D_2 D_4}$, $\overline{D_4 D_8}$, and $\overline{D_8 D_{16}}$.

So far we have not been concerned with the numerical values associated with each type of error for the manufacturing process. Let us now assume that the following chart depicts the loss function for the respective errors.

Type of Error	Amount of Loss
Type I error	$2000
Type II error	$1000

The loss functions indicate that the monetary value of the loss associated with erroneously stopping the manufacturing process is twice that associated with failing to stop the process when it is unsatisfactory.

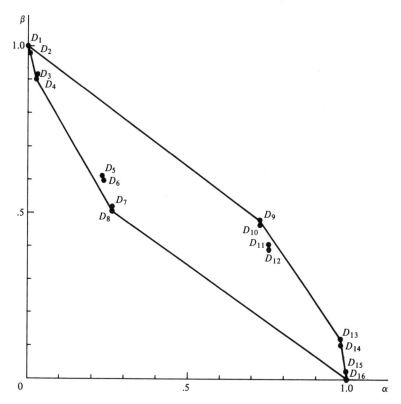

FIGURE 12.8.2   Error probabilities of decision functions.

The conditional expected losses associated with the pure decision functions are shown in Table 12.8.6, as well as graphically by numbered dots in Figure 12.8.3. The shaded area in the figure depicts the conditional expected losses for the set of all randomized decision functions.

TABLE 12.8.6

Conditional Expected Loss	Decision Function							
	D_1	D_2	D_3	D_4	D_5	D_6	D_7	D_8
$R(D\|H_1)$	0	2	54	58	486	488	540	548
$R(D\|H_2)$	1000	992	904	896	616	608	520	512

Conditional Expected Loss	Decision Function							
	D_9	D_{10}	D_{11}	D_{12}	D_{13}	D_{14}	D_{15}	D_{16}
$R(D\|H_1)$	1458	1460	1512	1514	1944	1946	1998	2000
$R(D\|H_2)$	488	480	392	384	104	96	8	0

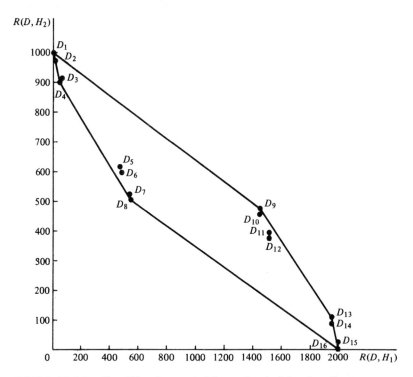

FIGURE 12.8.3 Conditional expected losses for decision function.

Suppose now that we wish to find the decision function which satisfies the criterion for a minimax decision function. To do so, we first find the minimax decision function among those which are pure and admissible, which is D_8. We now observe that $R(D_8|H_1) = 548$ is greater than $R(D_8|H_2) = 512$. This means, then, that another decision function for the randomization should be chosen so that $R(D|H_1)$ will be smaller than 548 and, at the same time, $R(D|H_2)$ will be greater than 512. The best minimax decision function among the remaining pure admissible decision functions, which also satisfies the necessary constraints, is D_4. Thus, we randomize D_4 and D_8 in order to obtain the randomized minimax decision function. Note that $R(D_4|H_1) = 58$ and $R(D_4|H_2) = 896$.

The value of γ may, therefore, be obtained by solving the following equation,

$$58\gamma + 548(1 - \gamma) = 896\gamma + 512(1 - \gamma),$$

for γ. Thus, we find that $\gamma = .041$. This implies that the decision function should be devised so that the probability of selecting D_4 is approximately .04 and of selecting D_8 is approximately .96, in which case the maximum conditional expected loss is given as

$$\text{Max.}[58\gamma + 548(1 - \gamma) = 527;\ 896\gamma - 512(1 - \gamma) = 527] = 527.$$

Up to this point we have assumed that we have no idea of the likelihood that the null hypothesis is true. Suppose now that our past experience with the machine process reveals that approximately 60 percent of the time the process produces 10 percent defectives and 40 percent of the time it produces 20 percent defectives. Then, we might devise the following prior probability function for π.

π	$f(\pi)$
.1	.60
.2	.40
	1.00

This means that for the given null hypothesis, we can let $\psi = .6$ and, in turn, proceed to obtain a Bayes decision function.

Recall Theorem 12.7.1 in Section 12.7 which states that we can obtain a Bayes decision function by evaluating the expected losses corresponding to those pure decision functions which are admissible. Our calculations of these expected losses are shown in Table 12.8.7. Thus, D_4 is a Bayes decision function, and the value of the expected loss associated with the decision function is $392.40, the smallest loss associated with any decision function we can devise.

TABLE 12.8.7

Decision Function	$R(D)$: Expected Loss
D_1	.6(0) + .4(1000) = $400.00
D_2	.6(2) + .4(992) = $398.00
D_4	.6(58) + .4(896) = $393.20
D_8	.6(548) + .4(512) = $533.60
D_{16}	.6(2000) + .4(0) = $1200.00

EXERCISES

*31. Assume that a manufacturing process produces either 10 percent defectives or 20 percent defectives. If it produces 10 percent defectives, it is assumed to be operating satisfactorily; if it produces 20 percent defectives, it is assumed to be not operating satisfactorily. The following null hypothesis is established: The proportion of defectives produced by the process is 10 percent.
 a. What is the significance of committing a Type I error?
 b. What is the significance of committing a Type II error?

*32. Referring to Exercise 31, assume that you will draw a simple random sample of 2 items from the process to make the decision. Enumerate all the pure decision functions which might be defined in the sample space.

*33. For each decision function in Exercise 32 calculate the error probabilities.

*34. Refer to Exercise 33 and graph the set containing the error probabilities for all conceivable decision functions, both pure and randomized.

*35. For the set graphed in Exercise 34, indicate the set associated with all admissible decision functions.

*36. Refer to Exercise 31 and assume that the cost of erroneously stopping the process is $1000, whereas the cost of failing to stop the process when it is not working satisfactorily is $2000. Find the conditional expected losses for the pure decision functions enumerated for Exercise 32.

*37. Graph the set containing the conditional expected losses for all conceivable decision functions in Exercise 36, both pure and randomized.

*38. For the set obtained in Exercise 37, indicate the set associated with all admissible decision functions.

*39. Among the admissible decision functions in Exercise 38, find the decision function which satisfies the minimax criterion.

*40. Refer to Exercise 31. Suppose it is known that about 80 percent of the time the process produces 10 percent defectives and 20 percent of the time it produces 20 percent defectives. Find a Bayes decision function.

41. Refer to Exercise 31. Suppose it is known that approximately 20 percent of the time the machine process produces 10 percent defectives and 80 percent of the time it produces 20 percent defectives. Find a Bayes decision function.

Chapter 13
Classical
Decision
Functions

Recall our discussion that, in order to devise a Bayes decision function, we must explicitly formulate a loss function associated with the two types of errors and decide upon a prior probability that the null hypothesis is true. In practice, however, formulating a realistic loss function or deciding upon a prior probability can be quite difficult.

One school of thought, known as the *Bayesian statisticians*, advocates that, regardless of the difficulties encountered, we should always formulate a loss function explicitly and decide upon a prior probability before deciding upon a decision function.

On the other hand, suppose we neither wish to explicitly formulate a loss function nor determine explicitly a prior probability that the null hypothesis is true. Can we still devise a reasonable decision function? The answer is "yes," according to another school of thought, known as the *classical statisticians*.

We will elaborate in this chapter on the manner in which classical statisticians derive such reasonable decision functions.

13.1 Likelihood-Ratio Test

In the event that we can neither formulate a loss function nor decide upon the prior probabilities associated with the null and alternate hypotheses, the only reasonable alternative open to us is to choose a decision function among the set of all admissible decision functions. The reader may recall that we do not have to explicitly formulate a loss function in order to ascertain a set of all admissible decision functions. When the sample size is very small, we can ascertain a set of all admissible decision functions by first enumerating all possible pure decision functions. On the other hand, when the sample size is very large, it would be very cumbersome to enumerate all such pure decision functions.

However, the following theorem provides a method for finding admissible decision functions without having to enumerate all possible decision functions.

Theorem 13.1.1

In testing a simple null hypothesis against a simple alternate hypothesis, every Bayes decision function is a *likelihood-ratio test*.

Although we will not prove this theorem, the significance of the theorem may be stated as follows. Recall our proposition in the preceding chapter that every admissible decision function is a Bayes decision function. Since every Bayes decision function is a *likelihood-ratio test*, it follows that every admissible decision function must be a *likelihood-ratio test*. Thus, if it is not very difficult to devise a *likelihood-ratio test*, then devising a *likelihood-ratio test* may be one way to obtain an admissible decision function. It so happens that devising a *likelihood-ratio test* is often not very difficult.

Although we have elaborated on the usefulness of a *likelihood-ratio test*, we have not yet defined what it is. We offer, then, the following definition.

Definition 13.1.1

Let the null hypothesis be $\pi = \pi_1$ and the alternate hypothesis be $\pi = \pi_2$, where $\pi_1 \neq \pi_2$. A decision function is said to be a *likelihood-ratio test* if for a given value k, the null hypothesis is accepted if $\lambda > k$, rejected if $\lambda < k$, and either action is taken if $\lambda = k$, where

$$\lambda = \frac{L(P = p | \pi = \pi_1)}{L(P = p | \pi = \pi_2)}.$$

In this definition $L(P = p | \pi = \pi_1)$, for example, denotes the likelihood that P assumes a specific value p, given that the null hypothesis is true, and $L(P = p | \pi = \pi_2)$ denotes the likelihood that P assumes the same value p, given that the null hypothesis is false.

EXAMPLE 13.1.1

Let us return to the marble-box guessing problem discussed in the preceding chapter. Assume that a decision function is to be based on a simple random sample of 2 marbles from the box selected. Then, the possible sample

13.1 Likelihood-Ratio Test

TABLE 13.1.1

Likelihood	Sample Space		
	$P = 0$	$P = .5$	$P = 1.0$
$L(P = p\|\pi = \frac{1}{4})$	$\frac{9}{16}$	$\frac{6}{16}$	$\frac{1}{16}$
$L(P = p\|\pi = \frac{2}{4})$	$\frac{4}{16}$	$\frac{8}{16}$	$\frac{4}{16}$
λ	$\frac{9}{4}$	$\frac{3}{4}$	$\frac{1}{4}$

outcomes, their likelihoods, and the likelihood-ratios are those given in Table 13.1.1.

Suppose we now devise a decision function so that we

Accept the null hypothesis if $\lambda > \frac{1}{2}$ and reject the null hypothesis otherwise.

Graphically, we have Figure 13.1.1. Similarly, we list the likelihood-ratio tests given in Table 13.1.2 and their corresponding decision functions in terms of the observed value of P. Thus, the corresponding decision functions are D_1, D_2, D_4, and D_8. If we now examine Figure 12.5.2, we find that these decision functions constitute the set of all admissible pure decision functions that we can devise for the sample size of 2.

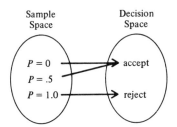

FIGURE 13.1.1 Decision function.

TABLE 13.1.2

Sample Space p	k			
	$\frac{1}{5}$	$\frac{1}{2}$	1	3
.0	accept	accept	accept	reject
.5	accept	accept	reject	reject
1.0	accept	reject	reject	reject
Decision Function	D_1	D_2	D_4	D_8

282   Classical Decision Functions

Note that the values of k selected for the likelihood-ratio tests do not correspond to any value that λ can actually assume. Suppose now we select a value that λ can actually assume; for example, let $k = \frac{1}{4}$. Then, two possible likelihood-ratio tests emerge:

1. Accept the null hypothesis if $\lambda \geq \frac{1}{4}$; otherwise, reject it.
2. Accept the null hypothesis if $\lambda > \frac{1}{4}$; otherwise, reject it.

Translating these two likelihood-ratio tests, we have the decision functions given in Table 13.1.3.

TABLE 13.1.3

Sample Space	Decision Function	
p	D_1	D_2
.0	accept	accept
.5	accept	accept
1.0	accept	reject
Likelihood-Ratio test	accept if $\lambda \geq \frac{1}{4}$	accept if $\lambda > \frac{1}{4}$

Thus, D_1 and D_2 are equivalent decision functions when $k = \frac{1}{4}$, which, in turn, implies that all possible randomized decision functions between D_1 and D_2 must also be equivalent to either D_1 or D_2. Thus, any point on the line segment joining D_1 and D_2 must be a likelihood-ratio test. Similarly, it can be shown that any decision function depicted by a point on the line segment joining D_2 and D_4 is a likelihood-ratio test for $k = \frac{3}{4}$ and any decision function depicted by a point on the line segment joining D_4 and D_8 is a likelihood-ratio test for $k = \frac{9}{4}$.

We have illustrated what is implied by Theorem 13.1.1 and Definition 13.1.1; that is, if we devise a likelihood-ratio test, then we can be sure that the resulting decision function will be an admissible decision function.

Let us now re-examine the likelihood-ratio tests given for $k = \frac{1}{5}$, $k = \frac{1}{2}$, $k = 1$, and $k = 3$. The equivalent decision functions expressed in terms of p are shown in Table 13.1.4. This table indicates that a certain pattern exists among the given decision functions. This apparent pattern is incorporated in the following theorem.

Theorem 13.1.2

Let a null hypothesis be $\pi = \pi_1$ and an alternate hypothesis be $\pi = \pi_2$, where $\pi_1 < \pi_2$. Then, a decision function is a likelihood-ratio test if for a given number ℓ the null hypothesis is accepted if $P < \ell$, rejected if $P > \ell$, and either action is taken if $P = \ell$.

TABLE 13.1.4

Sample Space p	k			
	$\frac{1}{5}$	$\frac{1}{2}$	1	3
.0	a	a	a	r
.5	a	a	r	r
1.0	a	r	r	r

a = accept null hypothesis
r = reject null hypothesis

Although we will not provide a formal proof of this theorem, the validity of the theorem must be intuitively obvious. If $\pi_1 < \pi_2$ by assumption, then it is more rational to conclude that $\pi = \pi_1$ if $P < \ell$ and that $\pi = \pi_2$ if $P > \ell$ than to conclude that the opposite is true. We will, however, illustrate this theorem with an example.

EXAMPLE 13.1.2

Let us return to Example 13.1.1, in which we have illustrated that a likelihood-ratio test is an admissible decision function. We will now show that any decision function based on Theorem 13.1.2 is also a likelihood-ratio test and, in turn, an admissible decision function.

Consider the decision functions given in Table 13.1.5. We find that these decision functions are identical to the likelihood-ratio tests for $k = \frac{1}{5}$, $k = \frac{1}{2}$, $k = 1$, and $k = 3$, respectively. Thus, they must be admissible decision functions.

Note that the values of ℓ selected for the decision functions in Table 13.1.5 do not correspond to any value that P can assume. Suppose now that we let ℓ be one of the values that P can actually assume. Then, the decision

TABLE 13.1.5

Sample Space p	Value of ℓ			
	$1 < \ell$	$.5 < \ell < 1$	$0 < \ell < .5$	$\ell < 0$
.0	a	a	a	r
.5	a	a	r	r
1.0	a	r	r	r
Decision Function	D_1	D_2	D_4	D_8

a = accept null hypothesis
r = reject null hypothesis

284 Classical Decision Functions

TABLE 13.1.6

Sample Space p	Value of ℓ					
	$\ell = 1$		$\ell = .5$		$\ell = 0$	
.0	a	a	a	a	a	r
.5	a	a	a	r	r	r
1.0	a	r	r	r	r	r
Decision Function	D_1	D_2	D_2	D_4	D_4	D_8

a = accept null hypothesis
r = reject null hypothesis

functions which emerge are those shown in Table 13.1.6. Thus, when $\ell = .5$, for example, the corresponding decision function is either D_2 or D_4 or any randomized decision function between D_2 and D_4.

We have now shown that, regardless of the value of ℓ, a decision function based on Theorem 13.1.2 will be an admissible decision function.

We will now evaluate the usefulness of the ideas discussed so far.

EXAMPLE 13.1.3

Let us return to Example 12.8.1, in which we assumed that a manufacturing process produces either 10 percent defectives or 20 percent defectives. The null hypothesis which we formulated was $\pi = .1$, and the alternate hypothesis which we formulated was $\pi = .2$.

TABLE 13.1.7 Admissible Decision Functions

Sample Space p	Decision Function											
	D_1	D_2	D_3	D_4	D_5	D_6	D_7	D_8	D_9	D_{10}	D_{11}	D_{12}
.0	a	a	a	a	a	a	a	a	a	a	a	r
.1	a	a	a	a	a	a	a	a	a	a	r	r
.2	a	a	a	a	a	a	a	a	a	r	r	r
.3	a	a	a	a	a	a	a	a	r	r	r	r
.4	a	a	a	a	a	a	a	r	r	r	r	r
.5	a	a	a	a	a	a	r	r	r	r	r	r
.6	a	a	a	a	a	r	r	r	r	r	r	r
.7	a	a	a	a	r	r	r	r	r	r	r	r
.8	a	a	a	r	r	r	r	r	r	r	r	r
.9	a	a	r	r	r	r	r	r	r	r	r	r
1.0	a	r	r	r	r	r	r	r	r	r	r	r

a = accept null hypothesis
r = reject null hypothesis

Assume now that we are to test the hypotheses by selecting a simple random sample of 10 items from the process. Obviously, there are a fairly large number of pure decision functions which we can devise. On the other hand, Theorem 13.1.2 indicates that there are only 12 pure decision functions which are admissible. These decision functions are shown in Table 13.1.7. We can, therefore, rule out from consideration any pure decision functions which are not listed in this table. Which decision function among the 12 should we choose? We will attempt to answer this question in Section 13.3.

EXERCISES

*1. Assume that a manufacturing process produces 10 or 20 percent defectives. The null hypothesis pertaining to the process has been formulated as $\pi = .1$, where π depicts the proportion of defectives produced. A simple random sample of 2 is to be selected to test the hypothesis. Let P depict the percentage of defectives in the sample. Calculate the likelihood-ratio λ for all possible values of P.

*2. Refer to Exercise 1 and construct a decision function based on the likelihood-ratio test in terms of λ, which will yield a unique decision function when expressed in terms of P. Then, transform the decision function in terms of P.

*3. Refer to Exercise 1 and construct a decision function based on the likelihood-ratio test in terms of λ, which will yield two different decision functions in terms of P. Then, transform the decision function in terms of P.

4. Refer again to Exercise 1 and suppose the sample size is increased to 5. Describe the class of pure decision functions in terms of P which are admissible.

*5. A company has been sponsoring a certain television program. The company believes that either 20 or 40 percent of the audience watches the program, on the average, and establishes a null hypothesis that $\pi = .4$. Suppose the company decides to test the hypothesis by selecting a simple random sample of 10. Construct a decision function based on the likelihood-ratio test in terms of λ, which will yield a unique decision function when expressed in terms of P, where P depicts the percentage of those in the sample who watch the program. Then, transform the decision function in terms of P.

6. Refer to Exercise 5 and construct a decision function based on the likelihood-ratio test in terms of λ, which when expressed in terms of P will yield two different decision functions. Then, transform the decision function in terms of P.

*7. Suppose the sample size for the problem given in Exercise 5 is increased to 100. How would you characterize the class of decision functions in terms of P which are admissible?

13.2 Simple Hypothesis: Continuous Case

Up to this point we have been concerned with the testing of hypotheses pertaining to Bernoulli probability functions. We will now extend our analysis to the testing of hypotheses pertaining to continuous probability functions. First, however, we will describe the nature of the difficulty which we will encounter in devising a decision function if the probability function sampled is assumed to be continuous.

Assume that the null hypothesis pertaining to a continuous probability function is $\mu = \mu_1$ and the alternate hypothesis is $\mu = \mu_2$, where $\mu_1 \neq \mu_2$. Then, all possible outcomes of the decision-making process are those given in Table 13.2.1.

TABLE 13.2.1

		State of Probability Function	
		$\mu = \mu_1$	$\mu = \mu_2$
Decision	ACCEPT NULL HYPOTHESIS	correct decision	Type II error
	REJECT NULL HYPOTHESIS	Type I error	correct decision

Let us now assume that we will draw a simple random sample of size n from this continuous probability function and, then, will decide whether to accept or reject the null hypothesis. A decision function must divide the sample space into two mutually exclusive subsets S_1 and S_2 and, in turn, associate every element of one subset with accepting the null hypothesis and every element of the other set with rejecting the null hypothesis. If the sample size is 2, then such a decision function may be illustrated graphically, as in Figure 13.2.1. The region in the sample space which corresponds to "accept" in the decision space is called the *acceptance region*, and the region which corresponds to "reject" is called the *rejection region* (also known as the *critical region*). Note that the rejection region in the sample space is drawn arbitrarily, which means that we can draw another rejection region, again, arbitrarily. In fact, we can draw an infinite number of such rejection regions.

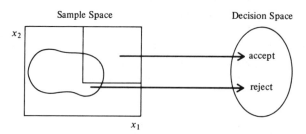

FIGURE 13.2.1 Decision function.

Once a decision function is selected, we can evaluate the α and β risks associated with the decision function. The error probabilities are defined as:

Definition 13.2.1

Let the domain of the decision function be divided into S_1 and S_2, where S_1 denotes the acceptance region and S_2, the rejection region. If the outcome of the random sample is given as $\{X_1, \ldots, X_n\}$, then,

$$\alpha = P(\{X_1, \ldots, X_n\} E S_2 | \mu = \mu_1)$$
and
$$\beta = P(\{X_1, \ldots, X_n\} E S_1 | \mu = \mu_2).$$

This definition is a natural extension of Definitions 12.2.4 and 12.2.5.

In testing a Bernoulli hypothesis, we have observed that there are only a finite number of pure decision functions to evaluate, even though that number could be very large. Now we observe that when the domain of the decision function is a continuous space, there are an infinite number of pure decision functions which we can devise.

Is there, then, an efficient set of admissible decision functions? The answer to this question is "yes." However, it must be obvious that we cannot expect to ascertain such a set by enumerating the pure decision functions, since there are an infinite number of them. Is it, then, possible to characterize the class of decision functions which are admissible? The following theorem is a characterization of such a class.

Theorem 13.2.1

Let a null hypothesis be $\mu = \mu_1$ and an alternate hypothesis be $\mu = \mu_2$. A decision function is admissible if it is devised so that, for a given number k, the null hypothesis is accepted if $\lambda > k$, rejected if $\lambda < k$, and either action is taken if $\lambda = k$, where

$$\lambda = \frac{L(x_1, \ldots, x_n | \mu = \mu_1)}{L(x_1, \ldots, x_n | \mu = \mu_2)}.$$

This theorem indicates that if the decision function is a likelihood-ratio test, then it must be an admissible decision function.

In testing a Bernoulli hypothesis, we have discovered that a likelihood-ratio test expressed in terms of λ can be cumbersome to devise. However, we have pointed out that an equivalent likelihood-ratio test can be devised in terms of P. We now confront a similar situation: In testing a null hypothesis, $\mu = \mu_1$, against an alternate hypothesis, $\mu = \mu_2$, the likelihood-ratio test expressed in terms of λ is cumbersome to devise. We will, therefore, not attempt to devise one here. However, it can be shown that under certain assumptions an equivalent likelihood-ratio test can be devised in terms of the sample mean $\bar{X}$.

Theorem 13.2.2

Let a simple null hypothesis pertaining to a normal probability function be $\mu = \mu_1$ and an alternate hypothesis be $\mu = \mu_2$, where $\mu_1 < \mu_2$. Assume that the standard deviation of the probability function is known and is given as σ. Then a decision function is a likelihood-ratio test if, for a given number ℓ, the null hypothesis is accepted if $\bar{X} < \ell$, rejected if $\bar{X} > \ell$, and either action is taken if $\bar{X} = \ell$.

Although we will not prove the validity of this proposition, its soundness must be intuitively apparent. First, since the hypothesis really pertains to the mean of the probability function, it would be reasonable to base our decision on the observed value of the sample mean. Second, since $\mu_1 < \mu_2$ by assumption, in the event that the observed value of $\bar{X}$ is less than an arbitrary value ℓ, it is more sound to conclude that $\mu = \mu_1$ than to conclude that $\mu = \mu_2$, for example. These propositions are, in a sense, incorporated in Theorem 13.2.2. We will now illustrate the usefulness of these concepts.

EXAMPLE 13.2.1

Assume that we wish to test a null hypothesis, $\mu = 0$, pertaining to a normal probability function against an alternate hypothesis, $\mu = 2$. Assume that the standard deviation is 1, regardless of the value of μ. Suppose that a random sample of 4 is to be selected to test the hypothesis.

Consider the following three arbitrary decision functions:

D_1: Accept the null hypothesis if $\bar{X} \leq 1$; otherwise, reject it.
D_2: Accept the null hypothesis if $\bar{X} \geq 1$; otherwise, reject it.
D_3: Accept the null hypothesis if $.5 \leq \bar{X} \leq 1.5$; otherwise, reject it.

The error probabilities for the three decision functions are shown in Table 13.2.2.

TABLE 13.2.2

Error Probability	D_1	D_2	D_3
α risk	.0228	.9772	.8426
β risk	.0228	.9772	.1574

Before we proceed further, perhaps we should evaluate how these error probabilities have been ascertained. Let us, for example, consider D_3. According to this decision function, we are to reject the null hypothesis if the value of the sample mean $\bar{X}$ is either less than .5 or greater than 1.5. Thus, the α risk for the decision function is really the probability of obtaining a sample whose mean value is either less than .5 or greater than 1.5, even though the sample actually has been obtained from the probability function with $\mu = 0$. In notation, we have

$$\alpha = P(\bar{X} < .5 \text{ or } \bar{X} > 1.5 | \mu = 0).$$

13.2 Simple Hypothesis: Continuous Case

From our discussion in Chapter 10 we know that if the sample actually is obtained from the normal probability function with $\mu = 0$ and $\sigma = 1$, then $\bar{X}$ must be normally distributed with $E(\bar{X}) = 0$ and $\sigma_{\bar{x}} = 1/\sqrt{4} = .5$. Thus, we may convert the above equation into

$$\alpha = P\left\{\frac{\bar{X} - E(\bar{X})}{\sigma_{\bar{x}}} < .5 \middle| \mu = 0\right\} + P\left\{\frac{\bar{X} - E(\bar{X})}{\sigma_{\bar{x}}} > 1.5 \middle| \mu = 0\right\}$$
$$= P\{\bar{X} < 0 + 1\sigma_{\bar{x}}\} + P\{\bar{X} > 0 + 3\sigma_{\bar{x}}\},$$

and, in turn, we obtain

$$\alpha = P(Z < 1) + P(Z > 3) = .8413 + .0013 = .8426.$$

The α risk obtained is illustrated in Figure 13.2.2.

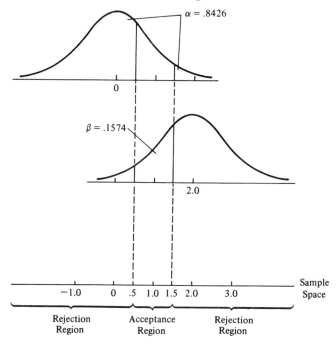

FIGURE 13.2.2 Error probabilities of a decision function.

Let us now calculate the β risk for the decision function, D_3. According to this decision function, we are to accept the null hypothesis if the value of the sample mean $\bar{X}$ is between .5 and 1.5. Thus, the β risk for the decision function is really the probability of obtaining a sample whose mean value is between .5 and 1.5, even though the sample actually has been obtained from the probability function with $\mu = 2$. In notation, then, we have

$$\beta = P(.5 \leq \bar{X} \leq 1.5 | \mu = 2).$$

We know that if the sample actually has been obtained from the normal probability function with $\mu = 2$ and $\sigma = 1$, then $\bar{X}$ must be normally distributed with $E(\bar{X}) = 2$ and $\sigma_{\bar{x}} = .5$. Thus, we may convert the above equation into

$$\beta = P\left\{.5 \leq \frac{\bar{X} - E(\bar{X})}{\sigma_{\bar{x}}} \leq 1.5 \Big| \mu = 2\right\}$$
$$= P(0 - 3\sigma_{\bar{x}} \leq \bar{X} \leq 0 - 1\sigma_{\bar{x}}),$$

and, in turn,

$$\beta = P(-3 \leq Z \leq -1) = .1574.$$

The β risk obtained also is illustrated in Figure 13.2.2.

The error probabilities for the remaining two decision functions (D_1 and D_2) can be ascertained in a similar manner.

Let us now compare the error probabilities for the three decision functions. We find that both D_2 and D_3 are dominated by D_1. On the other hand, Theorem 13.2.2 shows that D_1 is an admissible decision function, whereas D_2 and D_3 are not. Therefore, we should not be surprised by the fact that we are able to devise a decision function, for example, D_1, which dominates D_2 and D_3.

EXERCISES

*8. A process produces nylon fishing lines. It is believed that the average breaking strength of the lines is either 4 or 5 pounds. Further, it is assumed that the breaking strength of the lines is normally distributed with a standard deviation of 1 pound. The null hypothesis is $\mu = 4$, where μ depicts the average weight of the lines. Suppose you wish to test the hypothesis by selecting a simple random sample of 25. Construct an arbitrary decision function in terms of $\bar{X}$ which is also a likelihood-ratio test.

*9. For the decision function that you have devised for Exercise 8, calculate the values of the α and β risks.

*10. Referring to Exercise 8, how would you characterize an admissible decision function?

13.3 The Neyman-Pearson Lemma

We have pointed out in the preceding sections that our reason for devising a likelihood-ratio test for a simple null hypothesis is that the test yields an admissible decision function. Motivation for devising a likelihood-ratio test may be explained in a different context, however.

13.3 The Neyman-Pearson Lemma

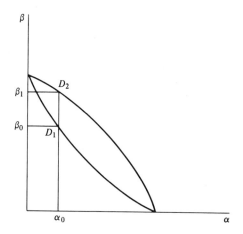

FIGURE 13.3.1

Assume that the shaded portion in Figure 13.3.1 depicts the class of all decision functions for testing a simple null hypothesis against a simple alternate hypothesis.

Suppose that we are willing to accept the α risk of α_0. Then, any decision function depicted by a point on the vertical line segment between D_1 and D_2 in Figure 13.3.1 will yield the α risk equal to α_0. Thus, there are an infinite number of decision functions which will yield this particular value for the α risk. On the other hand, each of these decision functions yields a different value for the β risk.

Obviously, we would like to choose the decision function which will yield the smallest possible β risk. This decision function corresponds to D_1 in Figure 13.3.1 and is said to be the *most powerful decision function* for the given value of α for the following reason. For any decision function, $1 - \beta$ depicts the power to detect the situation when the alternate hypothesis is, in fact, true. However, the largest power is attained for a given α whenever the value of the β risk is smallest for the given α.

Suppose now that we wish to find the most powerful decision function for a given value of α. It is apparent that the decision function must be an admissible decision function, and, in turn, it may be ascertained by devising a likelihood-ratio test. The fact that the most powerful decision function in question can be obtained by devising a likelihood-ratio test was not very obvious before the decision theorists conceived the notion of admissibility during the 1950s. Nevertheless, Neyman and Pearson have shown the validity of the proposition in their historic paper written in 1933, in which they proposed:

Theorem 13.3.1

Let $\{X_1, \ldots, X_n\}$ constitute a random sample of size n for testing a simple null hypothesis against a simple alternate hypothesis, and let S_1 and S_2, respectively, denote the acceptance and rejection regions of a decision function. The most powerful decision function for a given value of α is obtained by devising a likelihood-ratio test in such a way that

$$P(\{X_1, \ldots, X_n\} E S_2 | H_1) = \alpha.$$

This theorem is known as the Neyman-Pearson lemma and states that the most powerful decision function for a given α is a likelihood-ratio test devised in such a way that the probability that the sample outcome will fall in the rejection region is equal to α when the null hypothesis is, in fact, true.

We will now illustrate some applications of the Neyman-Pearson lemma.

EXAMPLE 13.3.1

Suppose we have a null hypothesis that $\mu = 100$ pertaining to a normal probability function and an alternate hypothesis that $\mu = 120$. Assume that the standard deviation pertaining to the probability function is 50. We are to select a random sample of 25 from the probability function to test the hypothesis, and we wish to find the most powerful decision function subject to the constraint that $\alpha = .05$.

The Neyman-Pearson lemma indicates that the decision function in question can be obtained by devising a likelihood-ratio test. We have also shown in Theorem 13.2.2 that when the probability function sampled is assumed to be normal, a likelihood-ratio test may be in the form:

Accept the null hypothesis if $\bar{X} \leq \ell$, and
reject the null hypothesis if $\bar{X} > \ell$.

This means that the most powerful decision function for $\alpha = .05$ is found by determining the value of ℓ such that

$$P(\bar{X} > \ell | \mu = 100) = .05.$$

Since $\sigma_{\bar{x}} = 50/\sqrt{25} = 10$ and $E(\bar{X}) = 100$ when $\mu = 100$, we have

$$P(\bar{X} > 100 + 1.64\sigma_{\bar{x}} | \mu = 100) = .05$$

or

$$P(\bar{X} > 116.4 | \mu = 100) = .05.$$

Thus, the value of ℓ should be 116.4, and the null hypothesis should be accepted if the observed value of $\bar{X}$ is 116.4 or less.

What is the β risk corresponding to this decision function? It is given as

$$\beta = P(\bar{X} \leq 116.4 | \mu = 120).$$

Since $E(\bar{X}) = 120$ when $\mu = 120$, we have

$$\beta = P[\bar{X} \le E(\bar{X}) - 3.6 | \mu = 120]$$
$$= P[\bar{X} \le E(\bar{X}) - .36\sigma_{\bar{x}} | \mu = 120] = .3594.$$

Graphically, the decision function and error probabilities are shown in Figure 13.3.2. The power to reject the null hypothesis if the alternate hypothesis is true is $1 - .3594$ or $.6406$.

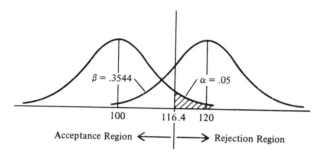

FIGURE 13.3.2

EXAMPLE 13.3.2

Let us return to Example 13.1.3, in which the null hypothesis is $\pi = .1$ and the alternate hypothesis is $\pi = .2$. For the sample size 10, we have already described 12 pure decision functions which are admissible.

Let us consider D_{10}. We calculate that

$$\alpha = P(P > .1 \,\pi = .1) = \sum_{k=2}^{10} b(k, n = 10, \pi = .1) = .2639$$

and $\quad \beta = P(P \le .1 | \pi = .2) = \sum_{k=0}^{1} b(k, n = 10, \pi = .2) = .3758.$

The error probabilities for the other decision functions are calculated and are shown in Table 13.3.1. D_{10} is the most powerful decision function for $\alpha = .2639$. Suppose now that we wish to find a decision function which is most powerful for $\alpha = .15$. There is no pure decision function which will satisfy the given objective. We can, however, obtain the desired decision function by randomizing two pure decision functions. In order to do this, we select the decision functions to randomize so that their α risks are nearest to .15: one of them is above .15, and another is below .15. Thus, D_9 and D_{10} are selected. To find the probabilities of selecting D_9 and D_{10}, we solve the equation,

$$.0702\gamma + (1 - \gamma)(.2639) = .15,$$

for γ. We obtain that $\gamma = .6$.

Thus, if we select D_9 with a probability of .6 and D_{10} with a probability of .4, the resulting decision function is the most powerful for $\alpha = .1478$. How

TABLE 13.3.1 Error Probabilities of Decision Functions

| Error | Decision Function | | | | | |
Probability	D_1	D_2	D_3	D_4	D_5	D_6
α risk	.0000	.0000	.0000	.0000	.0000	.0001
β risk	1.0000	1.0000	1.0000	.9999	.9991	.9936

| Error | Decision Function | | | | | |
Probability	D_7	D_8	D_9	D_{10}	D_{11}	D_{12}
α risk	.0016	.0128	.0702	.2639	.6513	1.0000
β risk	.9672	.8791	.6778	.3758	.1074	.0000

powerful is the decision function in question? To answer this question, we first calculate the β risk associated with the decision function:

$$\beta = .6(.6778) + .4(.3758) = .5724.$$

The power of the decision function is $1 - \beta$ or .4276.

EXERCISES

*11. Refer again to Exercise 8 and suppose that you want to find the decision function which is most powerful for $\alpha = .05$. How should you formulate the decision function?

12. Refer to Exercise 8. Can you find a decision function which yields $\alpha = .05$ but which is less powerful than the one you have devised for Exercise 11? If so, devise such a decision function.

13. In Exercise 1 a sample of 5 was to be selected to test the hypothesis. Find the pure decision function which is most powerful for $\alpha \leq .3$.

14. Refer to Exercise 13 and find the most powerful decision function among either pure or randomized decision functions for $\alpha = .3$.

13.4 Composite Hypothesis: Bernoulli Case

In specifying either the null or the alternate hypothesis, we have assumed that the hypothesized parameter has one value; however, we might hypothesize that the parameter has more than one value. For example, in a Bernoulli case, we might let the null hypothesis be $\pi \leq \pi_0$ and the alternate hypothesis be $\pi > \pi_0$, where π_0 is a specifically given value. Both the null and the alternate hypotheses are then composite.

We will first formulate composite null and alternate hypotheses for the marble-box guessing problem before we explore some of the difficulties associated with devising a suitable decision function for the problem.

EXAMPLE 13.4.1

Assume that there are 11 marble boxes whose outside appearances are identical. The compositions of these boxes are given in Figure 13.4.1. We are to select one of these boxes and then guess the proportion of the marbles in the box which are red.

```
┌──────────┐   ┌──────────┐              ┌──────────┐
│  0 Red   │   │  1 Red   │   . . . . .  │ 10 Red   │
│ 10 White │   │  9 White │              │  0 White │
└──────────┘   └──────────┘              └──────────┘
```

FIGURE 13.4.1

There are two types of penalties incurred by erroneous guessing:

Type I: incurred if we guess erroneously when the box selected contains 1 red marble or less
Type II: incurred if we guess erroneously when the box selected contains 2 red marbles or more.

One convenient way to deal with the problem is to establish a null hypothesis that $\pi \leq .1$ and an alternate hypothesis that $\pi > .1$, where π depicts the proportion of red marbles in the box selected.

The various outcomes of our guessing are shown in Table 13.4.1.

Suppose we are allowed to select a marble twice from the box selected and then make our guess. Then, the pure decision functions which we might devise are those shown in Table 13.4.2. In order to evaluate the desirability of these decision functions, we would need to calculate the error probabilities

TABLE 13.4.1 Various Outcomes of Guessing

		Selection of Box	
		$\pi \leq .1$	$\pi > .1$
Guess	ACCEPT NULL HYPOTHESIS	no error	Type II error
	REJECT NULL HYPOTHESIS	Type I error	no error

TABLE 13.4.2 Pure Decision Functions

Sample Space p	Decision Function							
	D_1	D_2	D_3	D_4	D_5	D_6	D_7	D_8
0	a	a	a	a	r	r	r	r
.5	a	a	r	r	a	a	r	r
1.0	a	r	a	r	a	r	a	r

a = accept null hypothesis
r = reject null hypothesis

associated with these decision functions, which can be ascertained by initially constructing a power function whose definition is given as:

Definition 13.4.1
A function whose domain contains various assumed values of π for the probability function sampled and whose image set contains the associated probability of rejecting the null hypothesis is called the *power function* pertaining to a decision function.

EXAMPLE 13.4.2
For the decision functions given in Table 13.4.2, we will now construct power functions. Let us use Decision Function 2 as an example. Figure 13.4.2 illustrates the power function for this decision function. The power function is obtained in the following manner. Suppose the box selected contains 1 red marble; that is, $\pi = .1$. Then, the probability of drawing a red marble twice would be .01; that is,

$$P(P = 1|\pi = .1) = .01.$$

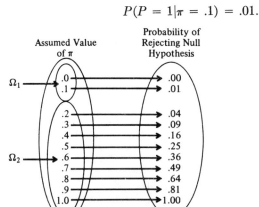

FIGURE 13.4.2 Power function for Decision Function 2.

13.4 Composite Hypothesis: Bernoulli Case

Since Decision Function 2 indicates that we should reject the null hypothesis if the observed value of P is 1, the probability of rejecting the null hypothesis is .01 if $\pi = .1$.

Let us now assume that the box selected contains 5 red marbles; that is, $\pi = .5$. Then, the probability of drawing a red marble twice is .25; that is,

$$P(P = 1 | \pi = .5) = .25.$$

Thus, .25 is the probability of rejecting the null hypothesis when $\pi = .5$. In a similar manner the other elements of the power function can be ascertained.

Let us now examine the power function obtained. We have divided the domain of the power function into two mutually exclusive subsets, Ω_1 and Ω_2. If the box selected corresponds to an element of Ω_1, the null hypothesis is, in fact, true. If the box selected corresponds to an element of Ω_2, then the alternate hypothesis is, in fact, true. This means that the probabilities associated with the elements of Ω_1 in the power function are the probabilities of rejecting the null hypothesis when it is, in fact, true. These probabilities are α risks associated with the decision function. On the other hand, the probabilities associated with the elements of Ω_2 in the power function are the probabilities of rejecting the null hypothesis when it is, in fact, false. These probabilities are the $1 - \beta$ risks associated with the decision function.

Power functions corresponding to all the pure decision functions given in Table 13.4.2 are shown in Table 13.4.3. From these power functions the error probabilities shown in Table 13.4.4 have been ascertained.

In testing a simple null hypothesis against a simple alternate hypothesis, we have confined our search for a suitable decision function to the class of admissible decision functions. We have found that such a decision function is the most powerful for a given value of α. We can now extend this idea to

TABLE 13.4.3 Power Functions

Assumed Value of π	Decision Function							
	D_1	D_2	D_3	D_4	D_5	D_6	D_7	D_8
.0	0	.00	.00	.00	1.00	1.00	1.00	1.00
.1	0	.01	.18	.19	.81	.82	.99	1.00
.2	0	.04	.32	.36	.64	.68	.96	1.00
.3	0	.09	.42	.51	.49	.58	.91	1.00
.4	0	.16	.48	.64	.36	.52	.84	1.00
.5	0	.25	.50	.75	.25	.50	.75	1.00
.6	0	.36	.48	.84	.16	.52	.64	1.00
.7	0	.49	.42	.91	.09	.58	.51	1.00
.8	0	.64	.32	.96	.04	.68	.36	1.00
.9	0	.81	.18	.99	.01	.82	.19	1.00
1.0	0	1.00	.00	1.00	.00	1.00	.00	1.00

TABLE 13.4.4 Error Probabilities

Type of Risk	Assumed Value of π	Decision Function							
		D_1	D_2	D_3	D_4	D_5	D_6	D_7	D_8
α	.0	.0	.00	.00	.00	1.00	1.00	1.00	1.00
	.1	.0	.01	.18	.19	.81	.82	.99	1.00
β	.2	1.0	.96	.68	.64	.36	.32	.04	.00
	.3	1.0	.91	.58	.49	.51	.42	.09	.00
	.4	1.0	.84	.52	.36	.64	.48	.16	.00
	.5	1.0	.75	.50	.25	.75	.50	.25	.00
	.6	1.0	.64	.52	.16	.84	.48	.36	.00
	.7	1.0	.51	.58	.09	.91	.42	.49	.00
	.8	1.0	.36	.68	.04	.96	.32	.64	.00
	.9	1.0	.19	.82	.01	.99	.82	.81	.00
	1.0	1.0	.00	1.00	.00	1.00	1.00	1.00	.00

testing a composite null hypothesis against a composite alternate hypothesis. The class from which we will choose a decision function is called the *class of uniformly most powerful decision functions*.

Definition 13.4.2

Let the null hypothesis be $\pi \varepsilon \Omega_1$ and the alternate hypothesis be $\pi \varepsilon \Omega_2$. A decision function for a given α_0 is said to be *uniformly most powerful* if it is devised so that $\alpha \leq \alpha_0$ for all π in Ω_1 and $1 - \beta$ is maximum for each π in Ω_2.

EXAMPLE 13.4.3

To illustrate Definition 13.4.2, let us return to Example 13.4.2. The null hypothesis is $\pi \leq .1$ and the alternate hypothesis is $\pi > .1$. When the null hypothesis is true, $\pi \varepsilon \Omega_1$, and when the alternate hypothesis is true, $\pi \varepsilon \Omega_2$. This, in turn, implies that if $\pi \leq .1$, then $\pi \varepsilon \Omega_1$, and if $\pi > .1$, then $\pi \varepsilon \Omega_2$.

Let us now randomize D_2 and D_4 so that the probability of selecting D_2 is $1/18$ and of selecting D_4 is $17/18$. The power function for the resulting decision function is given in Table 13.4.5. If we compare the power function for this randomized decision function with that for D_3, we will find the largest value for α is .18 for both decision functions. On the other hand, the value of $1 - \beta$ is higher for the randomized decision function for each π in Ω_2. Thus, the randomized decision function is uniformly more powerful relative to that of D_3.

Suppose now that we are able to devise a decision function with $\alpha \leq .18$ whose $1 - \beta$ is the highest for each π in Ω_2 compared to any other decision function having $\alpha \leq .18$. The decision so devised is then called the uniformly most powerful decision function for $\alpha \leq .18$.

TABLE 13.4.5 Power Function

Assumed Value of π	Probability of Rejectng Null Hypothesis
.0	$\frac{1}{18}(0) + \frac{17}{18}(0) = .00$
.1	$\frac{1}{18}(.01) + \frac{17}{18}(.19) = .18$
.2	$\frac{1}{18}(.04) + \frac{17}{18}(.36) = .34$
.3	$\frac{1}{18}(.09) + \frac{17}{18}(.51) = .48$
.4	$\frac{1}{18}(.16) + \frac{17}{18}(.64) = .61$
.5	$\frac{1}{18}(.25) + \frac{17}{18}(.75) = .72$
.6	$\frac{1}{18}(.36) + \frac{17}{18}(.84) = .81$
.7	$\frac{1}{18}(.49) + \frac{17}{18}(.91) = .88$
.8	$\frac{1}{18}(.64) + \frac{17}{18}(.96) = .95$
.9	$\frac{1}{18}(.81) + \frac{17}{18}(.99) = .97$
1.0	$\frac{1}{18}(1.0) + \frac{17}{18}(1.0) = .99$

How can we find such a uniformly most powerful decision function with a specific value for α? The illustration shown in Table 13.4.5 suggests that such a decision function may well have to be randomized. For the moment, therefore, instead of trying to find the uniformly most powerful decision function for a specific value for α, we will describe a way to find any pure decision function which is uniformly most powerful.

Theorem 13.4.1

Let the null hypothesis be $\pi \leq \pi_0$ and the alternate hypothesis be $\pi > \pi_0$. Let the decision function be such that the null hypothesis is accepted if $P \leq \ell$ and rejected otherwise, where ℓ is a given constant. The decision function, then, is uniformly most powerful for $\alpha \leq \alpha_0$, where

$$\alpha_0 = P(P > \ell | \pi = \pi_0).$$

We have already shown that such a decision function is the most powerful one for testing a simple null hypothesis, $\pi = \pi_1$, against a simple alternate hypothesis, $\pi = \pi_2$, where $\pi_1 < \pi_2$. Theorem 13.4.1 is really an extension of this idea. Although we will not attempt to prove the theorem, its underlying logic should be apparent. Since the null hypothesis is $\pi \leq \pi_0$, whenever we observe that $P < \ell$, it is more reasonable to conclude that $\pi \leq \pi_0$ than, for example, to conclude that $\pi > \pi_0$. We will, however, illustrate the validity of the theorem by using the pure decision functions given in Example 13.4.2.

EXAMPLE 13.4.4

Consider the error probabilities of the pure decision functions given in Example 13.4.2. We have already pointed out in Example 13.4.3 that D_3 cannot be uniformly most powerful. We can also illustrate that D_5, D_6, and D_7 cannot be uniformly most powerful. Suppose we randomize D_4 and D_8 such that the probability of selecting D_4 is $19/81$ and of selecting D_8 is $62/81$. Then, the power function for the resulting decision function is that shown in Table 13.4.6.

TABLE 13.4.6 Power Function

Assumed Value of π	Probability of Rejecting Null Hypothesis
.0	$19/81(0) + 62/81(1.0) = .77$
.1	$19/81(.19) + 62/81(1.0) = .81$
.2	$19/81(.36) + 62/81(1.0) = .85$
.3	$19/81(.51) + 62/81(1.0) = .88$
.4	$19/81(.64) + 62/81(1.0) = .92$
.5	$19/81(.75) + 62/81(1.0) = .94$
.6	$19/81(.84) + 62/81(1.0) = .96$
.7	$19/81(.91) + 62/81(1.0) = .98$
.8	$19/81(.96) + 62/81(1.0) = .99$
.9	$19/81(.99) + 62/81(1.0) = .99$
1.0	$19/81(1.0) + 62/81(1.0) = .99^+$

If we compare D_5 and the randomized decision function, we will find that the largest α is .81 for both decision functions. On the other hand, the randomized decision function is more powerful for each π in Ω_2. Thus, D_5 cannot be uniformly most powerful. In a similar manner, it can be shown that both D_6 and D_7 cannot be uniformly most powerful.

So far, we have shown that D_3, D_5, D_6, and D_7 cannot be uniformly most powerful. What can we say about the remaining four decision functions? We will find that it is impossible to devise a decision function which is uniformly more powerful than any one of these remaining four decision functions while keeping the maximum α risk the same. This discovery is not surprising, however, if we examine these remaining four decision functions with regard to Theorem 13.4.1, since such an examination will reveal that they are, in fact, uniformly most powerful.

EXAMPLE 13.4.5

Assume that a machine process is performing satisfactorily if the machine produces 1 defective or less per 10 items. On the other hand, the machine should be stopped for a corrective adjustment if it produces more than 1 defective per 10 items. Our problem is to guess whether the machine produces 1 or less defectives per 10, which is similar to the marble-box guessing problem described in the beginning of this section.

13.4 Composite Hypothesis: Bernoulli Case

In order to resolve the problem we can establish the null hypothesis, $\pi \leq .1$, and the alternate hypothesis, $\pi > .1$. Naturally, if we accept the null hypothesis, we would let the machine continue production without interruption, whether it is really operating satisfactorily or not. On the other hand, if we reject the null hypothesis, we would stop the machine, regardless of whether it is operating satisfactorily or not. The possible errors which we may commit are given in Table 13.4.7.

TABLE 13.4.7 Possible Errors in Guessing

		State of Process	
		$\pi \leq .1$	$\pi > .1$
Guess	ACCEPT NULL HYPOTHESIS	correct action	Type II error
	REJECT NULL HYPOTHESIS	Type I error	correct action

Assume now that we have decided to select a random sample of 10 items from the process and, then, either to accept or reject the null hypothesis on the basis of the sample information. Let us assume that we wish to consider initially only the pure decision functions. According to Theorem 13.4.1, there are only 12 pure decision functions which are uniformly most powerful, and these are given in Table 13.4.8. If we now return to Example 13.1.3, we will find that this set of uniformly most powerful decision functions is the same as

TABLE 13.4.8

Sample Space p	Decision Function											
	D_1	D_2	D_3	D_4	D_5	D_6	D_7	D_8	D_9	D_{10}	D_{11}	D_{12}
.0	a	a	a	a	a	a	a	a	a	a	a	r
.1	a	a	a	a	a	a	a	a	a	a	r	r
.2	a	a	a	a	a	a	a	a	a	r	r	r
.3	a	a	a	a	a	a	a	a	r	r	r	r
.4	a	a	a	a	a	a	a	r	r	r	r	r
.5	a	a	a	a	a	a	r	r	r	r	r	r
.6	a	a	a	a	a	r	r	r	r	r	r	r
.7	a	a	a	a	r	r	r	r	r	r	r	r
.8	a	a	a	r	r	r	r	r	r	r	r	r
.9	a	a	r	r	r	r	r	r	r	r	r	r
1.0	a	r	r	r	r	r	r	r	r	r	r	r

a = accept null hypothesis
r = reject null hypothesis

the set of admissible decision functions for testing a simple null hypothesis against a simple alternate hypothesis.

Suppose we now select D_{10} in Table 13.4.8. Then, the power function for this decision function is given in Table 13.4.9. D_{10} is thus the uniformly most powerful decision function for $\alpha \leq .2639$.

TABLE 13.4.9 Power Function for D_{10}

Assumed Value of π	Probability of Rejecting Null Hypothesis
.0	.0000
.1	.2639
.2	.6242
.3	.8507
.4	.9536
.5	.9893
.6	.9983
.7	.9999
.8	.9999+
.9	.9999+
1.0	.9999+

Suppose now we wish to devise a uniformly most powerful decision function whose maximum α risk is less than .2639. We might then select D_9. The power function for D_9 is given in Table 13.4.10. Thus, D_9 is the uniformly most powerful decision function for $\alpha \leq .0702$.

Suppose we wish to find the uniformly most powerful decision function for $\alpha \leq .15$. There is no pure decision function which will satisfy this objective.

TABLE 13.4.10 Power Function for D_9

Assumed Value of π	Probability of Rejecting Null Hypothesis
.0	.0000
.1	.0702
.2	.3222
.3	.6172
.4	.8327
.5	.9453
.6	.9877
.7	.9984
.8	.9999
.9	.9999+
1.0	.9999+

However, we can devise the desired decision function by randomizing D_9 and D_{10} so that the probability of selecting D_9 is approximately .6 and of selecting D_{10} is approximately .4. The power function of the resulting decision function shown in Table 13.4.11 indicates that the maximum α risk for the randomized decision function is .1477, or approximately .15.

TABLE 13.4.11 Power Function

Assumed Value of π	Probability of Rejecting Null Hypothesis
.0	.6(0) + .4(0) = .0000
.1	.6(.0702) + .4(.2639) = .1477
.2	.6(.3222) + .4(.6242) = .4430
.3	.6(.6172) + .4(.8507) = .7106
.4	.6(.8327) + .4(.9536) = .8810
.5	.6(.9453) + .4(.9893) = .9629
.6	.6(.9877) + .4(.9983) = .9919
.7	.6(.9984) + .4(.9999) = .9989
.8	.6(.9999) + .4(.9999+) = .9998
.9	.6(.9999+) + .4(.9999+) = .9999+
1.0	.6(.9999+) + .4(.9999+) = .9999+

EXAMPLE 13.4.6

Referring to the problem given in Example 13.4.1, let us assume that the sample size is to be increased to 100. Recall that the null hypothesis is $\pi \leq .1$ and that the alternate hypothesis is $\pi > .1$. Suppose that we wish to obtain the uniformly most powerful decision function for $\alpha_0 = .05$, which means that the value of ℓ should be determined so that

$$\sum_{k=\ell+1}^{100} b(k, n=100, \pi=.1) \leq .05.$$

Even though Table B.6 (binomial probabilities) in Appendix B does not give binomial probabilities for $n = 100$, there exist tables of binomial probabilities for $n = 100$. (See, for example, Schlaifer [1959] cited in the Bibliography.) If we consult such a table, we will find that

$$\sum_{k=15}^{100} b(k, n=100, \pi=.1) = .0726$$

and

$$\sum_{k=16}^{100} b(k, n=100, \pi=.1) = .0399.$$

Now let D_a be acceptance of the null hypothesis if $P \leq .14$ and rejection if $P > .14$, and let D_b be acceptance of the null hypothesis if $P \leq .15$ and rejection if $P > .15$. The largest α risks associated with the two decision functions are .0726 and .0399, respectively. If we randomize D_a and D_b so that the probability of selecting D_a is 101/327 and of selecting D_b is 226/327, then the largest α risk for the resulting decision function is given as

$$.0726 \left(\frac{101}{327}\right) + .0399 \left(\frac{226}{327}\right) = .0500.$$

Thus, the uniformly most powerful decision function for $\alpha = .05$ is the randomized decision function between D_a and D_b so that the probability of selecting D_a is 101/327.

Suppose now that we wish to find the uniformly most powerful decision function among the pure decision functions with the largest α risk of *approximately* .05. Either D_a or D_b would obviously serve the purpose. Given this objective, we can also find a suitable decision function without looking at a table of binomial probabilities. This can be done by applying the central limit theorem. Since P is approximately normally distributed, all we need to do is find the value of ℓ such that

$$P(P > \ell | \pi = .1) \cong .05,$$

since we will incur the largest α risk when $\pi = \pi_0$, if the decision function is of the form given in Theorem 13.4.1. Since $E(P) = .1$ and $\sigma_p = \sqrt{(.1)(.9)/100} = .03$ when $\pi = .1$, the preceding equation may be modified to

$$P[P > \ell = E(P) + 1.65\sigma_p | \pi = .1] = P(P > \ell = .1 + .0495 | \pi = .1) \cong .05.$$

Rounding off the last digit, we obtain the resulting decision function, which is to accept the null hypothesis if $P \leq .1500$ and reject it otherwise.

The power to reject the null hypothesis may be calculated for various values of $\pi > .1$. For example, assume that $\pi = .2$. Then, the alternate hypothesis is true. Thus, the power function is given as

$$P(P > \ell | \pi = .2) = 1 - \beta.$$

Since $E(P) = .2$ and $\sigma_p = \sqrt{(.2)(.8)/100} = .04$ when $\pi = .2$, the preceding equation may be modified to

$$P(P > \ell = E(P) - 1.25\sigma_p | \pi = .2) \cong .8944.$$

The decision function and the error probabilities are shown in Figure 13.4.3.

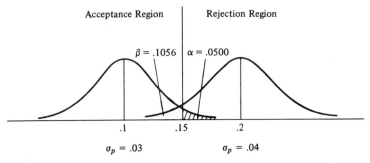

FIGURE 13.4.3

The power to reject the null hypothesis may be calculated in a similar manner for other values of $\pi > .1$.

EXERCISES

15. A company has been sponsoring a certain television program. At the moment, the company officials are debating whether or not to keep the program for another season. They believe that the program should be kept if 20 percent or more of the viewers watch the program. You are a statistical consultant to the firm and, therefore, have established the following hypotheses:

$$\text{Null Hypothesis: } \pi \geq .2$$
$$\text{Alternate Hypothesis: } \pi < .2.$$

What are the implications of committing a Type I and a Type II error?

*16. Refer to Exercise 15 and suppose that you have decided to test the hypotheses with a simple random sample of 2. Describe all pure decision functions which you can devise in terms of P, where P depicts the percentage of those in the sample who watch the program. Then, ascertain the power function for each of the decision functions for some assumed values of π.

*17. From the power functions which you have ascertained for Exercise 16, ascertain the set of error probabilities for some assumed values of π.

*18. Among the decision functions you have found for Exercise 16, find those which are uniformly most powerful for some given value of α.

*19. Refer to Exercise 15 and assume that a sample of 5 is to be selected. Characterize the set of pure decision functions which are uniformly most powerful for some specified values of α.

20. Find a pure decision function from those you have found for Exercise 19 which is uniformly most powerful for an α risk not exceeding .4.

21. Refer to Exercise 19 and find a decision function among either pure or randomized decision functions which will yield $\alpha = .4$ and, at the same time, will be uniformly most powerful.

*22. Refer to Exercise 16 and assume that the sample size is to be increased from 2 to 1600. Find a decision function which is uniformly most powerful for $\alpha \leq .05$.

*23. Suppose you are to utilize the decision function found for Exercise 22. What is the probability that the television program may be continued for another season, even though the proportion of those who watch the program is 15 percent? Does this probability constitute a risk, and if so, what kind of a risk is it?

13.5 Composite Hypothesis: Continuous Case

In Section 13.2 we considered a simple null hypothesis, $\mu = \mu_1$, and an alternate hypothesis, $\mu = \mu_2$. In many cases, however, it is more appropriate to formulate a null hypothesis, $\mu \leq \mu_0$, and an alternate hypothesis, $\mu > \mu_0$;

or, a null hypothesis, $\mu \geq \mu_0$, and an alternate hypothesis, $\mu < \mu_0$. The null and alternate hypotheses thus formulated are composite.

We will illustrate the usefulness of this approach by the following example.

EXAMPLE 13.5.1

Consider a manufacturing process which produces nylon cords. The average breaking strength of the cords should be 100 pounds or more. We wish to stop the process when the average breaking strength is less than 100 pounds.

In order to resolve the problem, we might establish a null hypothesis that $\mu \geq 100$ pounds and an alternate hypothesis that $\mu < 100$ pounds. We would, for example, be stopping the process if we reject the null hypothesis. The possible consequences of our decisions are shown in Table 13.5.1.

TABLE 13.5.1

		State of Process	
		$\mu \geq 100$	$\mu < 100$
Decision	ACCEPT NULL HYPOTHESIS	correct decision	Type II error
	REJECT NULL HYPOTHESIS	Type I error	correct decision

Let us assume that we will select a random sample of size n from the probability function and then we will decide whether to accept or reject the null hypothesis. As was the case when we tested a simple null hypothesis, $\mu = \mu_1$, against a simple alternate hypothesis, $\mu = \mu_2$, there are, in fact, an infinite number of pure decision functions which we can devise.

Can we characterize the class of uniformly most powerful decision functions relatively easily? The answer to this question depends on the shape of the probability function assumed to be sampled. It can be shown that if the probability function sampled is a member of the so-called exponential family, then, it is relatively easy for us to characterize the class. (Proof of this proposition is fairly complex, and we will not be concerned with it here.) What are some members of the exponential family? Among the probability functions we have studied so far, the normal and exponential probability functions would fall into this family.

We will now describe how to characterize the class of uniformly most powerful decision functions, assuming that the sample is taken from a normal probability function.

13.5 Composite Hypothesis: Continuous Case

Theorem 13.5.1
Let a null hypothesis pertaining to a normal probability function be that $\mu \leq \mu_0$ and an alternate hypothesis be that $\mu > \mu_0$. Let a decision function be such that the null hypothesis is accepted if $\bar{X} < \ell$, rejected if $\bar{X} > \ell$, and either action is taken if $\bar{X} = \ell$, where ℓ is a constant. Then, the decision function is uniformly most powerful for $\alpha \leq \alpha_0$, where

$$\alpha_0 = P(\bar{X} > \ell | \mu = \mu_0).$$

We have already shown that such a decision function is the most powerful for testing a simple null hypothesis, $\mu = \mu_1$, against a simple alternate hypothesis, $\mu = \mu_2$, where $\mu_1 < \mu_2$. Theorem 13.5.1 is an extension of this idea. Although we will not prove the theorem, its underlying logic should be intuitively apparent. Since the null hypothesis is that $\mu \leq \mu_0$ if $\bar{X} < \ell$, it appears more reasonable to conclude that $\mu < \mu_0$ than to conclude that $\mu > \mu_0$. We might, however, point out that such intuitive reasoning will be fallacious at times. We will mention without elaboration that the decision function given in Theorem 13.5.1 may not be uniformly most powerful if, for example, the values of σ are different for different values of μ.

As we have done in testing composite Bernoulli hypotheses, we will define a power function for testing the hypothesis $\mu \leq \mu_0$ against $\mu > \mu_0$.

Definition 13.5.1
A function whose domain contains various assumed values of μ for the probability function sampled and whose image set contains the associated probability of rejecting the null hypothesis is called the *power function*.

EXAMPLE 13.5.2
Consider a manufacturing process which produces a certain type of sheet metal. At present, the tensile strength of the sheet metal is 20,000 pounds per square inch, on the average. A modified process is being tested to see if it will increase the average tensile strength of the sheet metal.

Our problem is to guess whether the modified process will actually increase the average tensile strength of the sheet metal. The null hypothesis that we might establish is that $\mu \leq 20{,}000$ pounds, which really is an assertion that the modified process is not any better than the old process. The alternate hypothesis is $\mu > 20{,}000$ pounds, which really is an assertion that the modified process is better than the old process. The consequences of accepting or rejecting the null hypothesis are shown in Table 13.5.2. We see that the consequence of committing a Type I error is the conclusion that the modified process is better than the old one, even though, in fact, it is not. The consequence of committing a Type II error is the conclusion that the modified process is not any better than the old one when, in fact, it is better.

TABLE 13.5.2

		State of Process	
		$\mu \leq 20{,}000$	$\mu > 20{,}000$
Decision	ACCEPT NULL HYPOTHESIS	correct decision	Type II error
	REJECT NULL HYPOTHESIS	Type I error	correct decision

Assume that we plan to test the hypotheses with a simple random sample of 100 items from the process. What type of decision function should we devise to test the hypotheses? Assume that the probability function of the tensile strength is normal and has the same variance, regardless of its mean, and assume, furthermore, that it is 100 pounds. Then, a uniformly most powerful decision function is of the form:

Accept the null hypothesis if $\bar{X} \leq \ell$; otherwise, reject it.

Suppose we arbitrarily let $\ell = 20{,}020$ pounds. Then, the null hypothesis should be accepted if the observed value of the sample mean is 20,020 pounds or less, otherwise, it should be rejected.

The power function for the decision function is shown in Figure 13.5.1 for various discrete values of μ. We will also illustrate how the elements of this power function are calculated. Assume, for example, that $\mu = 20{,}000$. Then, $E(\bar{X}) = 20{,}000$ and $\sigma_{\bar{x}} = 100/\sqrt{100} = 10$. In turn,

$$P(\bar{X} > 20{,}020 | \mu = 20{,}000) = P[\bar{X} > E(\bar{X}) + 2\sigma_{\bar{x}} | \mu = 20{,}000] = .0228.$$

On the other hand, suppose that $\mu = 20{,}030$. Then, $E(\bar{X}) = 20{,}030$ and $\sigma_{\bar{x}} = 100/\sqrt{100} = 10$. Thus,

$$P(\bar{X} > 20{,}020 | \mu = 20{,}030) = P[\bar{X} > E(\bar{X}) - 1\sigma_{\bar{x}} | \mu = 20{,}030] = .8413.$$

The other elements of the power function are calculated in a similar manner.

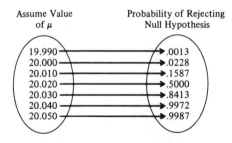

FIGURE 13.5.1 Power function.

13.5 Composite Hypothesis: Continuous Case

Suppose now we are to calculate the power function for all conceivable values of μ. Then, we would obtain a curve such as the one shown in Figure 13.5.2. This curve depicts a power function and is called a *power curve*.

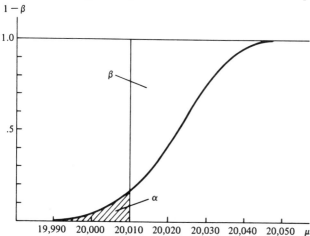

FIGURE 13.5.2 Power curve.

The power function might be used to calculate various error probabilities, such as those in Table 13.5.3. These error probabilities also are calculated for the power curve. Further, we can draw curves depicting the error probabilities, as shown in Figure 13.5.3.

TABLE 13.5.3

Type of Risk	Assumed Value of μ	Error Probability
α	19,990	.0013
	20,000	.0228
β	20,010	.8413
	20,020	.5000
	20,030	.1587
	20,040	.0228
	20,050	.0013

If we utilize the given decision function, what is the largest α risk? We find from the curve in Figure 13.5.3 that the largest α risk is .0228. Thus, the decision function in question is uniformly most powerful for $\alpha \leq .0228$.

Suppose now that we wish to find the uniformly most powerful decision function for $\alpha \leq .05$. Then, the value of ℓ would be such that

$$P(\bar{X} > \ell | \mu = 20{,}000) = .05,$$

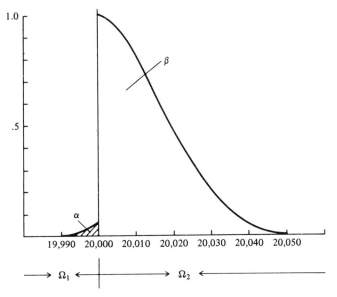

FIGURE 13.5.3   Error probabilities.

since the largest α risk exists when $\mu = \mu_0$. Since $E(\bar{X}) = 20{,}000$ when $\mu = 20{,}000$ and $\sigma_{\bar{x}} = 100/\sqrt{100} = 10$, we have

$$P(\bar{X} > \ell = 20{,}000 + 1.65\sigma_{\bar{x}}|\mu = 20{,}000) = .05$$

or

$$P(\bar{X} > \ell = 20{,}016.5|\mu = 20{,}000) = .05.$$

The resulting decision function is to accept the null hypothesis if $\bar{X} \leq 20{,}016.5$ and to reject it otherwise.

EXAMPLE 13.5.3

Consider a manufacturing process which produces a certain type of light bulb. It is generally known that the probability function pertaining to the lifetime of such bulbs is exponentially distributed; that is, $f(t) = ae^{-at}$ where t depicts the time that the light bulb fails. The density function of the lifetime is illustrated in Figure 13.5.4. Thus, we have $\mu = 1/a$ and $a = 1/\mu$. Thus, if we know μ, then we can determine a.

Assume that if $\mu \geq 1000$ hours, the process is in a satisfactory condition and we will let the process run without interruption. On the other hand, if $\mu < 1000$ hours, then the process is in an unsatisfactory condition and we will stop the process. The null hypothesis which we can establish is $\mu \geq 1000$ hours, and the alternate hypothesis is $\mu < 1000$ hours.

Let us assume that the sample size is 100. Does a uniformly most powerful decision function exist, and, if so, can we characterize it relatively easily? The answer is "yes" to both of these questions.

13.5 Composite Hypothesis: Continuous Case   311

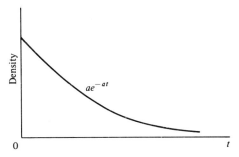

FIGURE 13.5.4

Theorem 13.5.2

Let the null hypothesis pertaining to an exponential probability function be $\mu \geq \mu_0$ and let the alternate hypothesis be $\mu < \mu_0$. Let a decision function exist such that the null hypothesis is accepted if $\bar{X} \geq \ell$ and rejected otherwise, where ℓ is a constant. The decision function is uniformly most powerful for $\alpha \leq \alpha_0$, where

$$\alpha_0 = P(\bar{X} < \ell | \mu = \mu_0).$$

Suppose now that we wish to find the uniformly most powerful decision function for $\alpha \leq .05$. All we need to do is find the value of ℓ such that

$$P(\bar{X} < \ell | \mu = 1000) = .05.$$

Finding the exact value of ℓ is rather difficult. On the other hand, since $\bar{X}$ is normally distributed, according to the central limit theorem, we can find an approximate value of ℓ such that

$$P(\bar{X} < \ell | \mu = 1000) \cong .05.$$

We will now explain how we can obtain such an approximate value of ℓ.

Since the probability function sampled is assumed to be exponentially distributed, and we know that if $\mu = 1000$, then $\sigma_x = 1000$ and, in turn, $\sigma_{\bar{x}} = 1000/\sqrt{100} = 100$. Then,

$$P(\bar{X} < 1000 - 1.64\sigma_{\bar{x}} | \mu = 1000) \cong .05;$$

that is,

$$P(\bar{X} < 836 | \mu = 1000) \cong .05.$$

The decision function to be devised is:

    Accept the null hypothesis if $\bar{X} \geq 836$;
    otherwise, reject it.

How powerful is the decision function? This question can be evaluated by constructing the power function. Instead of constructing the entire power function, we will calculate one element of that function. Suppose $\mu = 800$.

Then, the alternate hypothesis is, in fact, true, since $\mu < 1000$. How powerful is the decision function in detecting this situation? This probability is calculated as follows. Since $\mu = 800$, $E(\bar{X}) = 800$ and $\sigma_{\bar{x}} = 800/\sqrt{100} = 80$, then,

$$P(\bar{X} < 836 | \mu = 800) = P[\bar{X} < E(\bar{X}) + .45\sigma_{\bar{x}} | \mu = 800] \cong .6736,$$

which indicates the power of the decision function to reject the null hypothesis when $\mu = 800$. The β risk that can be taken when $\mu = 800$ is approximately .3264.

EXERCISES

*24. The management of a company which has a large office staff wants to make sure that the average time taken by workers for a morning coffee break does not exceed 30 minutes; if it does exceed 30 minutes, the management plans to announce the fact in the next office bulletin. Suppose the null hypothesis established to deal with the problem is $\mu \leq 30$ minutes and the alternate hypothesis is $\mu > 30$ minutes. Then, what are the implications of making Type I and Type II errors?

*25. Refer to Exercise 24 and suppose a simple random sample of 100 observations is to be made. Assume that the distribution of time for coffee breaks for individual workers is normally distributed with the standard deviation of 5 minutes. Find an arbitrary decision function which is uniformly most powerful. For this decision function, what is the maximum α?

*26. For the decision function found in Exercise 25, ascertain the power function for some assumed values of μ.

*27. For the decision function found in Exercise 25, what is the probability of concluding that the average in question is less than 30 minutes, in spite of the fact that it is 32 minutes? Is this probability a risk, and if so, what kind of a risk is it?

*28. Refer to Exercise 25 and devise a decision function which is uniformly most powerful for $\alpha \leq .05$. Then, ascertain the power function for some assumed values of μ.

*29. For the decision function you have devised for Exercise 28, what is the probability of concluding that the average is less than 30 minutes, in spite of the fact that it is 32 minutes?

30. The lifetime of a certain type of picture tube is known to be exponentially distributed. The producer of the picture tube wants the average life to be at least 30 months. Thus, the null hypothesis is $\mu \geq 30$, and the alternate hypothesis is $\mu < 30$. A simple random sample of 100 picture tubes is to be made to test the hypothesis. Devise a decision function which is uniformly most powerful for $\alpha_0 \cong .05$ where $\alpha \leq \alpha_0$.

31. For the decision function found for Exercise 30, what is the approximate probability that the null hypothesis may be accepted, even though $\mu = 20$? Is this a risk, and if so, what kind of a risk is it?

13.6 Composite Hypothesis: Two-sided

The composite hypothesis which we have considered is of the form:

Null Hypothesis: $\theta \leq \theta_0$ (or $\theta \geq \theta_0$)
Alternate Hypothesis: $\theta > \theta_0$ (or $\theta < \theta_0$).

There are, however, occasions when we might wish to formulate the null and alternate hypotheses as

Null Hypothesis: $\theta = \theta_0$
Alternate Hypothesis: $\theta \neq \theta_0$.

In this case, the null hypothesis is simple, whereas the alternate hypothesis is composite and is said to be two-sided.

Before we proceed to discuss the problems connected with devising decision functions, we will provide examples in which two-sided alternate hypotheses would be appropriate.

EXAMPLE 13.6.1

Suppose we wish to evaluate whether or not a given coin is fair. A given coin is fair if $\pi = .5$, where π depicts the probability that the coin will land on "heads." On the other hand, if the coin is not fair, then $\pi > .5$ or $\pi < .5$. Consequently, the appropriate hypotheses which we might formulate are

Null Hypothesis: $\pi = .5$
Alternate Hypothesis: $\pi \neq .5$.

The alternate hypothesis is two-sided and composite.

The possible outcomes of the decisions are given in Table 13.6.1. A Type I error is committed when we conclude that the coin is not fair even though it is. A Type II error is committed when we conclude that the coin is fair even though it is not.

TABLE 13.6.1

		State of Coin	
		$\pi = .5$	$\pi \neq .5$
Decision	ACCEPT NULL HYPOTHESIS	correct decision	Type II error
	REJECT NULL HYPOTHESIS	Type I error	correct decision

EXAMPLE 13.6.2

The average net weight of the cans from a canning process is assumed to be 80 ounces. We wish to evaluate, by taking a random sample, whether the average net weight for the process is, in fact, 80 ounces. The null and alternate hypotheses which we might formulate are:

$$\text{Null Hypothesis:} \quad \mu = 80 \text{ ounces}$$
$$\text{Alternate Hypothesis:} \quad \mu \neq 80 \text{ ounces.}$$

The possible outcomes of our decisions are given in Table 13.6.2. A Type I error is committed when we conclude that the process mean is not 80 ounces when, in fact, it is 80 ounces. A Type II error is committed when we conclude that the process mean is 80 ounces when, in fact, it is not.

TABLE 13.6.2

		State of Process	
		$\mu = 80$	$\mu \neq 80$
Decision	ACCEPT NULL HYPOTHESIS	correct decision	Type II error
	REJECT NULL HYPOTHESIS	Type I error	correct decision

Having described the type of problems which can be formulated as two-sided alternate hypotheses, we will now explore some of the difficulties connected with devising decision functions for such hypotheses. In testing the composite hypotheses of the form $\theta \leq \theta_0$ against $\theta > \theta_0$, we have pointed out that if the probability function sampled is, for example, a binomial or a normal function, the class of uniformly most powerful decision functions not only exists, but also can be characterized relatively easily. On the other hand, when the hypotheses to be tested are of the form $\theta = \theta_0$ against $\theta \neq \theta_0$, the class of uniformly most powerful decision functions does not even exist. To show intuitively why this is so, we will use an example.

EXAMPLE 13.6.3

Let us return to Example 13.6.1, in which the null hypothesis pertains to the fairness of a coin. Recall that the hypotheses are:

$$\text{Null Hypothesis:} \quad \pi = .5$$
$$\text{Alternate Hypothesis:} \quad \pi \neq .5.$$

Suppose we are to test the hypotheses by tossing the coin twice. Let P depict the percent of times that the coin lands on "heads." Then, P can assume 0,

.5, or 1. The possible pure decision functions are given in Table 13.6.3. The power functions for these decision functions are provided in Table 13.6.4 for some discrete values of π, and the error probabilities for these decision functions are given in Table 13.6.5.

TABLE 13.6.3 Possible Pure Decision Functions

Sample Space p	Decision Function							
	D_1	D_2	D_3	D_4	D_5	D_6	D_7	D_8
.0	a	a	a	a	r	r	r	r
.5	a	a	r	r	a	a	r	r
1.0	a	r	a	r	a	r	a	r

a = accept null hypothesis
r = reject null hypothesis

TABLE 13.6.4 Power Functions for Decision Functions

Assumed Value of π	Decision Function							
	D_1	D_2	D_3	D_4	D_5	D_6	D_7	D_8
.0	0	.00	.00	.00	1.00	1.00	1.00	1.00
.1	0	.01	.18	.19	.81	.82	.99	1.00
.2	0	.04	.32	.36	.64	.68	.96	1.00
.3	0	.09	.42	.51	.49	.58	.91	1.00
.4	0	.16	.48	.64	.36	.52	.84	1.00
.5	0	.25	.50	.75	.25	.50	.75	1.00
.6	0	.36	.48	.84	.16	.52	.64	1.00
.7	0	.49	.42	.91	.09	.58	.51	1.00
.8	0	.64	.32	.96	.04	.68	.36	1.00
.9	0	.81	.18	.99	.01	.82	.19	1.00
1.0	0	1.00	.00	1.00	.00	1.00	.00	1.00

Let us compare D_2 and D_5. Both decision functions have the same α risk. We also observe that D_2 is more powerful relative to D_5 if $\pi > .5$; however, D_5 is more powerful relative to D_2 if $\pi < .5$. Thus, neither decision function is uniformly more powerful relative to the other.

Let us next compare D_3 and D_6. We find that D_6 is uniformly more powerful relative to D_3 for $\alpha = .5$. Is D_6 the uniformly most powerful function for $\alpha = .5$? The answer is "no." Let us randomize D_5 and D_7 so that the probability of selecting D_5 is $\frac{1}{2}$ and of selecting D_7 is $\frac{1}{2}$. The error probabilities for the resulting decision function are given in Table 13.6.6. We find that both D_6 and the randomized decision function have the same α risk. On the other hand, we find that the randomized decision function is more powerful relative to D_6 when $\pi < .5$.

TABLE 13.6.5 Error Probabilities

Type of Risk	Assumed Value of π	Decision Function							
		D_1	D_2	D_3	D_4	D_5	D_6	D_7	D_8
β	.0	1.00	1.00	1.00	1.00	.00	.00	.00	.00
	.1	1.00	.99	.82	.81	.19	.18	.01	.00
	.2	1.00	.96	.68	.64	.36	.32	.04	.00
	.3	1.00	.91	.58	.49	.51	.42	.09	.00
	.4	1.00	.86	.52	.36	.64	.48	.16	.00
α	.5	.00	.25	.50	.75	.25	.50	.75	1.00
β	.6	1.00	.64	.52	.16	.84	.48	.36	.00
	.7	1.00	.51	.58	.09	.91	.42	.49	.00
	.8	1.00	.36	.68	.04	.96	.32	.64	.00
	.9	1.00	.19	.82	.01	.99	.18	.81	.00
	1.0	1.00	.00	1.00	.00	1.00	.00	1.00	.00

TABLE 13.6.6

Type of Risk	Assumed Value of π	Error Probabilities
β	.0	$.5(.00) + .5(.00) = .00$
	.1	$.5(.19) + .5(.01) = .10$
	.2	$.5(.36) + .5(.04) = .20$
	.3	$.5(.51) + .5(.09) = .30$
	.4	$.5(.64) + .5(.16) = .40$
α	.5	$.5(.25) + .5(.75) = .50$
β	.6	$.5(.84) + .5(.36) = .60$
	.7	$.5(.91) + .5(.49) = .70$
	.8	$.5(.96) + .5(.64) = .80$
	.9	$.5(.99) + .5(.81) = .90$
	1.0	$.5(1.0) + .5(1.0) = 1.00$

It can also be shown that if we randomize D_2 and D_4 so that the probability of selecting D_2 is $\frac{1}{2}$ and of selecting D_4 is $\frac{1}{2}$, then the resulting decision function will have the same α risk as D_6 but will be more powerful than D_6 when $\pi > .5$.

These illustrations show that D_6, although it dominates D_3, nevertheless, is not a uniformly most powerful decision function for $\alpha = .5$.

Can we, then, find a decision function which is uniformly most powerful for any given α? The answer is "no." The reader will find it impossible to devise a decision function which is uniformly most powerful for any given α.

What sort of decision function should we devise in the absence of a uniformly most powerful decision function? A class of decision functions which we will devise is based on the so-called *generalized likelihood-ratio test*.

Definition 13.6.1
Let the null hypothesis be $\theta = \theta_0$ and let the alternate hypothesis be $\theta \neq \theta_0$. A decision function is said to be a *generalized likelihood-ratio test* if, for $0 \leq k \leq 1$, the null hypothesis is accepted if $\lambda > k$, rejected if $\lambda < k$, and either action is taken when $\lambda = k$, where

$$\lambda = \frac{L(\hat{\theta}|\theta = \theta_0)}{\text{Max. } L(\hat{\theta}|\theta = \theta_0 \text{ or } \theta \neq \theta_0)},$$

where $L(\hat{\theta}|\theta = \theta_0)$ denotes the likelihood of observing $\hat{\theta}$ in the sample when $\theta = \theta_0$ and Max. $L(\hat{\theta}|\theta = \theta_0 \text{ or } \theta \neq \theta_0)$ denotes the largest possible likelihood of observing $\hat{\theta}$, regardless of whether $\theta = \theta_0$ or $\theta \neq \theta_0$.

To avoid any confusion, we point out that if the hypothesis pertains to π, then $\hat{\theta} = P$, and if it pertains to μ, then $\hat{\theta} = \bar{X}$.

EXAMPLE 13.6.4
Let us now return to Example 13.6.3 and assume again that a sample of size 2 is to be taken. Then, the numerator of the generalized likelihood-ratio test simply is the probability of observing the specific sample result when the null hypothesis is, in fact, true. Thus,

Sample Space p	$L(P = p\|\pi = .5)$
.0	.25
.5	.50
1.0	.25

The denominator of the ratio is calculated as follows. Suppose the observed value of P is 0. Then, P could have been obtained from the probability function with any given value of π. The likelihood of obtaining the given value of P will differ for different values of π, as shown in Table 13.6.7. However, Max. $L(P = 0|\pi = \pi_i) = 1.0$. The maximum values of the likelihood for different observed values of P are shown in Table 13.6.8 and the likelihood ratios for the observed values of P are shown in Table 13.6.9.

Now suppose we devise a decision function such that we

> Accept the null hypothesis if $\lambda > .25$;
> otherwise, we reject it.

TABLE 13.6.7

π_i	$L(P = 0\|\pi = \pi_i)$
.0	1.00
.1	.81
.2	.64
.3	.49
.4	.36
.5	.25
.6	.16
.7	.09
.8	.04
.9	.01
1.0	.00

TABLE 13.6.8

Sample Space p	Max. $(P = p\|0 \leq \pi \leq 1)$
.0	1.0
.5	.5
1.0	1.0

TABLE 13.6.9

Sample Space p	$\dfrac{L(P = p\|\pi = .5)}{\text{Max. } L(P = p\|0 \leq \pi \leq 1)}$
.0	.25/1.0 = .25
.5	.5/.5 = 1.00
1.0	.25/1.0 = .25

Then, we would accept the null hypothesis if $P = .5$ but reject it if P is either 0 or 1. The decision function devised is based on the generalized likelihood-ratio test.

If we carefully re-examine the decision function thus derived, we will find that it has an appealing feature. Recall in Example 13.6.3 that the null hypothesis that $\pi = .5$ is another way of proposing that the given coin is fair. The decision function which we have formulated indicates that if we were to toss the coin twice, then we should conclude that the coin is fair if it lands once on "heads" and once on "tails," but we should reject it if it lands twice on "heads" or twice on "tails." We could certainly have formulated such a decision function without having to devise a likelihood-ratio test. Although the motivation for devising a generalized likelihood-ratio test is intuitively

13.6 Composite Hypothesis: Two-sided

apparent, its actual application can be cumbersome, which we will illustrate in the following example.

EXAMPLE 13.6.5

Consider a Bernoulli null hypothesis that $\pi = .4$ and a Bernoulli alternate hypothesis that $\pi \neq .4$. Assume that the hypothesis is to be tested by taking a random sample of 5. The sample space, $L(P = p|\pi = .4)$, Max. $L(P = p|0 \leq \pi \leq 1)$, and values of λ are given in Table 13.6.10.

TABLE 13.6.10

Sample Space p	$L(P = p\|\pi = .4)$	Max. $L(P = p\|0 \leq \pi \leq 1)$	λ
.0	.078	1.000	.078
.2	.259	.410	.631
.4	.346	.346	1.000
.6	.230	.346	.664
.8	.077	.410	.188
1.0	.010	1.000	.010

Suppose we wish to find a pure decision function based on the generalized likelihood-ratio test so that the α risk associated with the decision function is .2 or less. Then, the value of k should be determined so that

$$P(\lambda < k|\pi = .4) \leq .2.$$

To do so, we must first ascertain the probability function of λ, assuming that the null hypothesis is true. Such a probability function is given in Table 13.6.11.

TABLE 13.6.11

λ	$P(\lambda\|\pi = .4)$
.010	.010
.078	.078
.188	.077
.631	.259
.664	.230
1.000	.346
	1.000

We now find from the probability function that

320 Classical Decision Functions

$$P(\lambda \leq .188 | \pi = .4) = .165$$

and

$$P(\lambda \leq .631 | \pi = .4) = .424.$$

Thus, the desired decision function should be:

Accept the null hypothesis if $\lambda \geq .188$;
otherwise, reject it.

The α risk associated with the decision function is then .165, which is certainly less that .2.

It must be evident that the decision function expressed in terms of λ can be cumbersome: We need to calculate $L(P = p | \pi = .4)$, then Max. $L(P = p | 0 \leq \pi \leq 1)$, and, in turn, the ratio λ. Calculations for finding these values could be very involved if n is fairly large.

Is there an easier way to devise a decision function based on the generalized likelihood-ratio test when we are testing a simple Bernoulli null hypothesis? The answer is "yes." When the sample size becomes larger and larger, a certain function of λ approaches one of the so-called chi-square probability functions. We will, therefore, discuss this subject again in Chapter 15, at which point we will introduce chi-square probability functions. Before we leave the subject, however, we may point out that the decision function based on the generalized likelihood-ratio test may also be expressed in terms of P. Such a decision function is shown in the right-hand square in Figure 13.6.1. The relationship between the two forms of expressing the decision function is also illustrated in this figure.

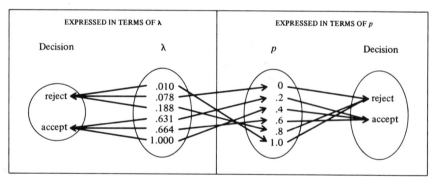

FIGURE 13.6.1 Generalized likelihood-ratio test.

So far we have been concerned with the problem of handling a two-sided Bernoulli alternate hypothesis. We will now turn our attention to the problem of handling a two-sided alternate hypothesis pertaining to a continuous probability function.

Let the null hypothesis pertaining to a continuous probability function be $\mu = \mu_0$ and let the alternate hypothesis be $\mu \neq \mu_0$. Since the alternate hypothesis is composite, we can show again that a uniformly powerful decision function does not exist for any given α. It appears, however, that the decision function should be based on the observed value of $\bar{X}$. If the decision function is based on $\bar{X}$, then one sound way to formulate the decision function appears to be to accept the null hypothesis if the observed value of $\bar{X}$ is very close to μ_0 and to reject the null hypothesis if the observed value of $\bar{X}$ is very far from μ_0. In fact, this is precisely what we would do by using the generalized likelihood-ratio test if the probability function sampled is, for example, assumed to be normally distributed.

Theorem 13.6.1

Let the null hypothesis pertaining to a normal probability function be $\mu = \mu_0$ and let the alternate hypothesis be $\mu \neq \mu_0$. Assume that the standard deviation of the probability function is the same for both hypotheses. Then, the generalized likelihood-ratio test for a given α will result in accepting the null hypothesis if

$$\mu_0 - \ell \leq \bar{X} \leq \mu_0 + \ell$$

and in rejecting the null hypothesis if

$$\mu_0 - \ell > \bar{X} > \mu_0 + \ell,$$

where $\alpha = P(\mu_0 - \ell > \bar{X} > \mu_0 + \ell | \mu = \mu_0)$.

EXAMPLE 13.6.6

Let us return to Example 13.6.2, in which the null hypothesis pertaining to the mean of the canning process is $\mu = 80$ ounces, and the alternate hypothesis is $\mu \neq 80$ ounces.

Assume now that the weights of the cans are normally distributed and the standard deviation for the process is 1 ounce, regardless of the mean. Suppose that we are to select a simple random sample of 25 cans to test the hypothesis. We wish to devise a decision function based on the generalized likelihood-ratio test, which will yield $\alpha = .05$. Theorem 13.6.1 indicates that the structure of the decision function must be of the form:

$\mu_0 - \ell$	μ_0	$\mu_0 + \ell$
Reject null hypothesis if $\bar{X}$ falls here	Accept null hypothesis if $\bar{X}$ falls here	Reject null hypothesis if $\bar{X}$ falls here

The next question is, "How should we determine the value of ℓ?" Theorem 13.6.1 indicates that the value of ℓ should be such that

$$P(\mu_0 - \ell > \bar{X} > \mu_0 + \ell | \mu = \mu_0) = .05.$$

This means that the value of ℓ should be such that the probability of $\bar{X}$ falling between $\mu_0 - \ell$ and $\mu_0 + \ell$ would be .95 when the null hypothesis is true.

When the null hypothesis is true, $E(\bar{X}) = 80$ and $\sigma_{\bar{x}} = 1/\sqrt{25} = .2$, which means that

$$P(80 - 1.96\sigma_{\bar{x}} \leq \bar{X} \leq 80 + 1.96\sigma_{\bar{x}}|\mu = 80) = .95.$$

Thus,

$$\ell = 1.96\sigma_{\bar{x}} = .392$$

and, in turn, we should accept the null hypothesis if $\bar{X}$ falls between 79.602 and 80.392; otherwise, we should reject it.

The power function for the preceding decision function for various assumed values of μ is shown in Figure 13.6.2. The components of the power

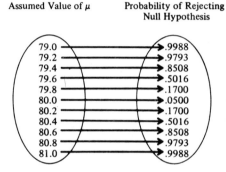

FIGURE 13.6.2 Power function.

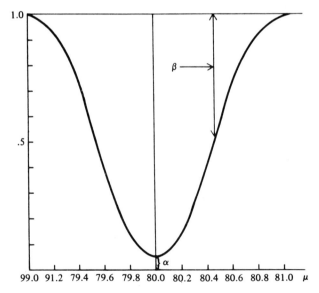

FIGURE 13.6.3 Power curve.

function are calculated as follows. Assume that $\mu = 79.20$. Then, $E(\bar{X}) = 79.20$ and $\sigma_{\bar{x}} = 1/\sqrt{25} = .2$. Thus,

$$P(79.608 > \bar{X} > 80.392) = P[E(\bar{X}) - 2.04\sigma_{\bar{x}} > \bar{X}]$$
$$+ P[\bar{X} > E(\bar{X}) + 5.96\sigma_{\bar{x}}] = .9793 + 0 = .9793.$$

Assume now that $\mu = 80.2$. Then, $E(\bar{X}) = 80.2$ and $\sigma_{\bar{x}} = 1/\sqrt{25} = .2$. Thus,

$$P(79.608 > \bar{X} > 80.392) = P[E(\bar{X}) - 2.96\sigma_{\bar{x}} > \bar{X}]$$
$$+ P[\bar{X} > E(\bar{X}) + .96\sigma_{\bar{x}}] = .0015 + .1685 = .1700.$$

A continuous version of the power function is depicted by the power curve in Figure 13.6.3. From this power curve we find, for example, that $\beta = .4984$ when $\mu = 80.4$. The β risk for any value of $\mu \neq 80$ may be found in a similar way.

EXERCISES

*32. Suppose you wish to test a hypothesis that the public is equally divided for and against a certain political issue. You establish a null hypothesis that $\pi = .5$, where π depicts the percentage of the public for the issue. The alternate hypothesis is $\pi \neq .5$. What, then, are the implications of committing a Type I and a Type II error?

*33. Referring to Exercise 32, suppose you have decided to test the hypotheses with a simple random sample of 3. Describe all pure decision functions that may be defined in the sample space.

*34. Refer to Exercise 32 and find a decision function based on the generalized likelihood-ratio test, so that the null hypothesis is accepted if $\lambda \geq .5$. Describe the decision function in terms of P.

*35. For the decision function found for Exercise 34, ascertain the power function for some assumed values of π.

*36. For the decision function found for Exercise 34, what is the value of the α risk?

*37. For the decision function found for Exercise 34, what is the probability that you may conclude that the public is equally divided for and against the issue, even though 20 percent are, in fact, in favor of the issue and 80 percent are against the issue?

38. Refer to Exercise 32 and assuming that the sample size is still 5, find a decision function based on the generalized likelihood-ratio test so that the null hypothesis is accepted if $\lambda \geq .2$. Describe the decision function in terms of P.

39. For the decision function you have found for Exercise 38, what is the value of the α risk?

40. For the decision function found for Exercise 38, what is the probability that you will conclude that the public is equally divided for and against the issue, even though 20 percent are for and 80 percent are against the issue?

*41. A manufacturing process produces boxes of detergent. The mean weight for the boxes is 50 ounces. It is known that the weights of the individual boxes are normally distributed with a standard deviation of 1 ounce. You now establish a null hypothesis that $\mu = 50$ and an alternate hypothesis that $\mu \neq 50$. What are the implications of committing a Type I and a Type II error?

*42. Assume that you are to test the hypotheses in Exercise 41 by selecting a simple random sample of 100 boxes from the process. Devise an arbitrary decision function which is based on the generalized likelihood-ratio test.

*43. For the decision function you have devised for Exercise 42, what is the value of the α risk?

*44. For the decision function you have devised for Exercise 42, what is the probability that you will conclude that the average weight of the process is 50 ounces, in spite of the fact that it really is 51 ounces?

*45. Refer to Exercise 41 and assuming that a simple random sample of 100 is to be selected, devise a decision function based on the generalized likelihood-ratio test, which will yield $\alpha = .0456$. Then, construct a power function for some assumed values of μ.

*46. For the decision function you have devised for Exercise 45, what is the probability that you will conclude that the average weight is 50 ounces when, in fact, it is (a) 49.6 ounces, (b) 49.8 ounces, (c) 50.1 ounces, and (d) 50.3 ounces?

*47. Are the probabilities found for Exercise 46 risks? If so, what kind of risks are they? If not, then explain why they are not.

Chapter 14
Bayes
Decision
Functions

In this chapter we will examine in detail some of the salient features of a Bayes decision function and explore its uses in solving practical decision problems.

14.1 Prior Probability Function

In Chapters 12 and 13, in which we tested hypotheses pertaining to a parameter θ of a probability function, we did not explicitly state whether θ itself is a random variable. In connection with the Bayes decision function discussed in Chapter 12, however, we implicitly assumed that θ is a random variable; on the other hand, in connection with the classical decision function discussed in Chapter 13, we implicitly assumed that θ is not a random variable.

In one sense, the question of whether or not we should assume θ to be a random variable in a specific hypothesis-testing situation is the heart of the controversy between the Bayesians and the classical statisticians. A classical statistician believes that θ should never be treated as if it were a random variable, whereas a Bayesian believes that θ should always be treated as if it were a random variable.

Instead of arbitrarily resolving the controversy, we will describe some of the problems connected with treating θ as if it were a random variable for a given situation. We hope that these illustrations will help the reader to resolve the controversy for himself in a suitable manner.

EXAMPLE 14.1.1

Let us return to a marble-box guessing problem, in which one box contains 1 red marble out of 4 and the other box contains 2 red marbles out of 4. If the box contains 1 red marble, then we will let $\pi = 1/4$; if it contains 2 red marbles, then we will let $\pi = 2/4$.

Suppose now we wish to treat π as if it were a random variable. (We now denote $\boldsymbol{\pi}$ in boldface, since we assume it to be a random variable.) Then, we would need to determine the probability function of $\boldsymbol{\pi}$. Here the source of the difficulty arises. Suppose we are told that we must choose one of the boxes by tossing a coin. If we were Bayesians, we might derive the desired probability function in the following manner. We first assume that the coin is likely to be fair and, in turn, that each side of the coin is equally obtainable. Since we choose a box by tossing this coin, the desired probability function should be

π	$P(\boldsymbol{\pi} = \pi)$
$1/4$	$1/2$
$2/4$	$1/2$
	1.0

On the other hand, if we were classical statisticians, we could argue that, after all, the coin could be biased and, therefore, it would be impossible to obtain such a probability function without making a long series of coin-tossing experiments.

Suppose now that we have no way of knowing how we will proceed to choose one of the boxes. Then, if we were classical statisticians, we would argue that the desired probability function cannot be obtained, since we could choose a box by tossing a coin, a thumbtack, a die or another device. The probability function of $\boldsymbol{\pi}$ would differ depending on what device we utilized to select the box. On the other hand, if we were Bayesians, we could argue that we should be able to assess the desired probability function based on subjective reasoning.

The probability function of $\boldsymbol{\pi}$, if obtained at all, is called a *prior probability function*. A formal definition is now given as:

Definition 14.1.1

Let $\boldsymbol{\theta}$ be the parameter of a probability function in a hypothesis-testing situation. Suppose we assume that $\boldsymbol{\theta}$ is a random variable. The probability function of $\boldsymbol{\theta}$ ascertained before we select a sample is called the *prior probability function* of $\boldsymbol{\theta}$.

Since the prior probability function is obtained before a sample is taken, Definition 14.1.1 clearly implies that the prior probability function must be obtained subjectively without any objective experimentation. However, the definition does not imply that any objective evidence, such as that obtained from previous samples, should not be taken into account in deriving the probability function; it does imply that no new objective experiments should be conducted specifically for the given hypothesis-testing situation prior to ascertaining the probability function of θ.

14.2 Posterior Probability Function

Once we ascertain the prior probability function of θ, we can obtain a Bayes decision function by applying Definition 12.7.1. However, a Bayes decision function can be ascertained in another way: by first ascertaining the *posterior probability function* of θ for the specifically given sample observation.

Later in this section we will describe the nature of a *posterior probability function*, but our immediate attention will be focused on the Bayes theorem, which we need to utilize in order to develop a *posterior probability function* of θ.

Theorem 14.2.1

Let θ be both a parameter of a probability function and a random variable whose prior probability function is given as $\{[\theta = \theta_1, P(\theta = \theta_1)], \ldots, [\theta = \theta_n, P(\theta = \theta_n)]\}$. Let $P(\hat{\theta} = \hat{\theta}_i | \theta = \theta_j)$ denote the conditional probability that the sample yields $\hat{\theta} = \hat{\theta}_i$ when $\theta = \theta_j$. Then,

$$P(\theta = \theta_k | \hat{\theta} = \hat{\theta}_\ell) = \frac{P(\hat{\theta} = \hat{\theta}_\ell | \theta = \theta_k) P(\theta = \theta_k)}{\sum_{\text{all } j} \sum_{\text{all } i} P(\hat{\theta} = \hat{\theta}_i | \theta = \theta_j) P(\theta = \theta_j)}.$$

This theorem is called the Bayes theorem for a discrete probability function and is an extension of the Bayes theorem for discrete events which we discussed in Chapter 3.

EXAMPLE 14.2.1

Let us now return to the marble-box guessing problem, in which one of the two boxes contains 1 red marble out of 4 and the other box contains 2 red marbles out of 4. Assume that the prior probability of choosing one type of box is $1/2$. Thus, the prior probability function is given as

π	$P(\pi = \pi)$
$1/4$	$1/2$
$2/4$	$1/2$
	1.0

Suppose now that we have selected a box and have drawn 1 marble from this box and found it to be white. Thus, the sample size is 1 and the observed value of P is 0. What is the probability that the box contains 1 red marble? In notation, this probability is depicted as

$$P\left(\pi = \frac{1}{4} \middle| P = 0\right).$$

This question can also be answered by first constructing the probability tree shown in Figure 14.2.1. From this probability tree we find that

$$P\left(\pi = \frac{1}{4} \middle| P = 0\right) = \frac{P(P = 0 \cap \pi = 1/4)}{P(P = 0 \cap \pi = 1/4) + P(P = 0 \cap \pi = 2/4)}$$

$$= \frac{3/8}{3/8 + 2/8} = \frac{3}{5}.$$

However, if we examine the process by which we have obtained this probability, we find that

$$P\left(\pi = \frac{1}{4} \middle| P = 0\right)$$

$$= \frac{P(P = 0|\pi = 1/4)P(\pi = 1/4)}{P(P = 0|\pi = 1/4)P(\pi = 1/4) + P(P = 0|\pi = 2/4)P(\pi = 2/4)},$$

which is an expression of Bayes' theorem, where $\pi = \theta$ and $P = \hat{\theta}$.

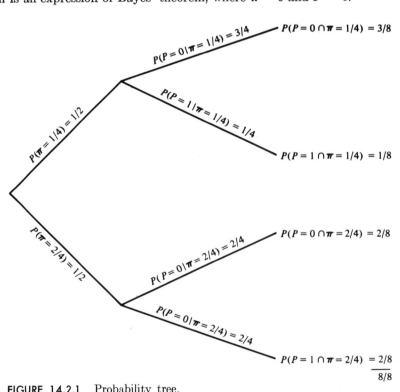

FIGURE 14.2.1 Probability tree.

14.2 Posterior Probability Function

The conditional probability $P(\pi = \frac{1}{4}|P = 0)$ is called the *posterior probability* of π, given that the sample yields $P = 0$. In a similar manner, we can calculate $P(\pi = \frac{2}{4}|P = 0)$. If we calculate the conditional probability of π for all possible values of π, given that $P = 0$, we have, in fact, ascertained the *posterior probability function* of π for the given $P = 0$. The *posterior probability function* can now be defined:

Definition 14.2.1
Let the domain of a function contain the set $\{(\theta = \theta_1|\hat{\theta} = \hat{\theta}_t), \ldots, (\theta = \theta_n|\hat{\theta} = \hat{\theta}_t)\}$ and let the image set contain the associated probabilities $P\{(\theta = \theta_1|\hat{\theta} = \hat{\theta}_t), \ldots, P(\theta = \theta_n|\hat{\theta} = \hat{\theta}_t)\}$. The function is then said to be the *posterior probability function* of θ, given that the sample yields $\hat{\theta} = \hat{\theta}_t$.

EXAMPLE 14.2.2
If we use the preceding marble-box guessing problem, the posterior probability function of π when the sample yields $P = 0$ is that shown in Figure 14.2.2, and the posterior probability function of π when $P = 1$ is that shown in Figure 14.2.3.

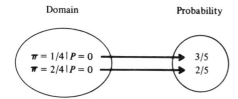

FIGURE 14.2.2 Posterior probability function.

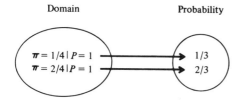

FIGURE 14.2.3 Posterior probability function.

Each probability in these posterior probability functions has been calculated by applying the Bayes theorem. It might be instructive for the reader to calculate these probabilities himself before he proceeds further.

Since θ now has both a prior and a posterior probability function, we need a notational distinction between the two. This distinction is made by denoting a prior probability that θ will assume a specific value θ_k by

$$P_1(\theta = \theta_k)$$

330   Bayes Decision Functions

and a posterior probability that θ will assume θ_k, given that the sample yields $\hat{\theta} = \hat{\theta}_t$, by

$$P_2(\theta = \theta_k | \hat{\theta} = \hat{\theta}_t).$$

EXERCISES

*1. There are two marble boxes, each containing 10 marbles. One of them contains 2 red marbles and 8 white marbles, and the other one contains 5 red marbles and 5 white marbles. You are to choose one of these boxes and guess the percentage of red marbles in the box that you select. Let π depict the proportion of red marbles in the box that you select. Suppose you toss a fair coin and choose one of the boxes as a result of the outcome of the toss. Ascertain a prior probability function for π.

2. Refer to Exercise 1 and assume that you will be allowed to draw 1 marble from the box selected before you make your guess. Let P depict the percentage of red marbles observed; that is, $P = 0$ if the marble observed is white and $P = 1$ if the marble observed is red. On the basis of the prior probability function of π ascertained in Exercise 1, find the posterior probability function of π, given that your observation yields (a) $P = 0$, and (b) $P = 1$.

*3. Using the problem in Exercise 1, assume that you will be allowed to draw a marble twice from the box that you select, and you must replace the first marble drawn before you draw the second one. Let P depict the percentage of red marbles observed in the two draws. Assuming that the prior probability function of π is the same as that given in Exercise 1, ascertain the posterior probability function of π for each possible sample value of P.

4. Using the problem in Exercise 1, assume again that you will be allowed to draw a marble twice, and must replace the first marble drawn before you draw the second. Assume, however, that your assessment of the prior probability that you will choose the box containing 2 red marbles is .25. Ascertain the posterior probability function of π for each possible sample value of P.

5. Refer to Exercise 1 and assume that you will be allowed to draw 4 marbles, replacing each before you draw the next. Assume also that your assessment of the probability that you will choose the box containing 2 red marbles is .6. Ascertain the posterior probability function of π for each possible sample value of P.

*6. Assume that a manufacturing process produces either 10 or 20 percent defectives. Let π depict the percentage of defectives produced. You are to take a simple random sample of 2 from the process. Let P depict the percentage of defectives for your sample. Assume that the process produces 10 percent defectives 80 percent of the time and 20 percent defectives 20 percent of the time. Ascertain the posterior probability function of π for each possible sample value of P.

7. Using the manufacturing process described in Exercise 6, assume that you will take a simple random sample of 4. Assuming the same prior probability function of π as ascertained in Exercise 6, ascertain the posterior probability function of π for each possible sample value of P.

8. Using the manufacturing process described in Exercise 6, assume that you will take a simple random sample of 10. Assuming the same prior probability function of π as ascertained in Exercise 6, ascertain the posterior probability function of π, given that the sample yields (a) $P = .1$, and (b) $P = .4$.

14.3 Finding a Bayes Decision Function: Alternate Method

We now return to the Bayes decision function. It might be worthwhile at this point to review the definition of a Bayes decision function. According to Definition 12.7.1, a Bayes decision function is one which minimizes

$$R(D) = \psi R(D|H_1) + (1 - \psi)R(D|H_2),$$

where ψ is the prior probability that the null hypothesis is true, $R(D|H_1)$ is the conditional expected loss when the null hypothesis is true, and $R(D|H_2)$ is the conditional expected loss when the alternate hypothesis is true.

As we have noted, however, the process of actually finding a Bayes decision function by utilizing this formula can be cumbersome. Even though the search should be confined to pure decision functions in the admissible set, we might still need to evaluate a large number of them before we find a Bayes decision function. We have pointed out, however, that there is an alternate way to find a Bayes decision function, and we will proceed to discuss the nature of this alternate procedure for testing a simple null hypothesis against a simple alternate hypothesis.

Before we proceed, some notational clarification should be helpful for our discussion. We have defined that $L(\mathrm{I})$ depicts the loss associated with rejecting the null hypothesis when it is, in fact, true and that $L(\mathrm{II})$ depicts the loss associated with accepting the null hypothesis when it is, in fact, false. The loss function for the given decision problem can also be depicted by the notations in Table 14.3.1. In this table $D = A_1$ implies that the null hypothesis is accepted, and $D = A_2$ implies that it is rejected. For a given decision function, let $P(A_1|H_1)$ and $P(A_2|H_1)$, respectively, denote the probability of accepting and rejecting the null hypothesis when it is, in fact, true and let $P(A_1|H_2)$ and $P(A_2|H_2)$ denote the probability of accepting and rejecting the null hypothesis when it is, in fact, false. Thus, $P(A_2|H_1) = \alpha$ and $P(A_1|H_2) = \beta$. What really has led us to Definition 12.5.2 are

TABLE 14.3.1   Loss Function

Decision	$H = H_1$	$H = H_2$
$D = A_1$	$L(A_1, H_1) = 0$	$L(A_1, H_2) = L(\mathrm{II})$
$D = A_2$	$L(A_2, H_1) = L(\mathrm{I})$	$L(A_2, H_2) = 0$

$$R(D|H_1) = P(A_1|H_1)L(A_1,H_1) + P(A_2|H_1)L(A_2,H_1) = \alpha L(\text{I})$$
and
$$R(D|H_2) = P(A_1|H_2)L(A_1,H_2) + P(A_2|H_2)L(A_2,H_2) = \beta L(\text{II}).$$

Let $P(H_1)$ depict the probability that the null hypothesis is true and let $P(H_2)$ depict the probability that the alternate hypothesis is true. $P(H_1)$ and $P(H_2)$ may be either prior or posterior probabilities, which, however, is not significant at the moment. We can now give the following definitions:

$$R(H|A_1) = P(H_1)L(A_1,H_1) + P(H_2)L(A_1,H_2) = (1 - \psi)L(\text{II})$$
and
$$R(H|A_2) = P(H_1)L(A_2,H_1) + P(H_2)L(A_2,H_2) = \psi L(\text{I}).$$

$R(H|A_1)$ and $R(H|A_2)$ are called, respectively, the conditional expected loss of accepting and rejecting the null hypothesis. Each of these then depicts the amount of the expected loss associated with taking a particular action with regard to the null hypothesis.

How do then $R(H|A_1)$ and $R(H|A_2)$ differ from $R(D|H_1)$ and $R(D|H_2)$? These two sets of conditional expected losses are based on two fundamentally different assumptions. In defining $R(D|H_1)$ and $R(D|H_2)$, we have assumed that, even though we choose a particular decision function, chance still dictates whether A_1 or A_2 of the decision space for the given decision function will actually be chosen. In a sense, then, we have assumed that the state for H is given but for D is not. On the other hand, in defining $R(H|A_1)$ and $R(H|A_2)$ we have made an opposite assumption; that is, that the state for D is given but for H is not. Thus, we have assumed that although we choose either A_1 or A_2 in the decision space, chance still dictates whether H will be H_1 or H_2.

In defining $R(H|A_1)$ and $R(H|A_2)$, we did not state whether $P(H_1)$ and $P(H_2)$ are prior or posterior probabilities. We will now make that distinction.

Definition 14.3.1

Let

$$R_1(H|A_1) = (1 - \psi_1)L(\text{II})$$
and
$$R_1(H|A_2) = \psi_1 L(\text{I}),$$

where ψ_1 is the prior probability that the null hypothesis is true. Then, $R_1(H|A_1)$ is called the *prior conditional expected loss* for accepting the null hypothesis, and $R_1(H|A_2)$ is called the *prior conditional expected loss* for rejecting the null hypothesis.

The usefulness of this definition is illustrated in the following example.

EXAMPLE 14.3.1

Let us return to a marble-box guessing problem, in which $\pi_1 = 1/4$, $\pi_2 = 3/4$, $\psi_1 = 1/2$, and the consequences of our guessing are given in Table 14.3.2. The numbers in the columns depict the values of losses associated with the different outcomes of our guessing.

14.3 Finding a Bayes Decision Function

TABLE 14.3.2

		State of Box	
		$\pi = 1/4 = \pi_1$	$\pi = 2/4 = \pi_2$
Guess	A_1: ACCEPT NULL HYPOTHESIS	0	100
	A_2: REJECT NULL HYPOTHESIS	80	0

Suppose we were to accept the null hypothesis. Then, the expected loss associated with this action is given by:

$$R(H|A_1) = \psi_1(0) + (1 - \psi_1)L(\text{II})$$
$$= \left(\frac{1}{2}\right)(0) + \left(\frac{1}{2}\right)(100)$$
$$= (1 - \psi_1)L(\text{II})$$
$$= 50.$$

On the other hand, suppose we were to reject the null hypothesis. Then, the expected loss associated with this action is given by

$$R(H|A_2) = \psi_1 L(\text{I}) + (1 - \psi_1)(0)$$
$$= \left(\frac{1}{2}\right)(80) + \left(\frac{1}{2}\right)(0)$$
$$= \psi_1 L(\text{I})$$
$$= 40.$$

Suppose now we wish to accept or reject the null hypothesis without taking a sample from the box. Then, it would be preferable to reject the null hypothesis, since the conditional expected loss of rejecting the null hypothesis is less than that of accepting it. Thus, we propose:

Theorem 14.3.1

Let $R_1(H|A_1)$ and $R_1(H|A_2)$ be the prior conditional expected losses for accepting and rejecting the null hypothesis, respectively. A Bayes decision function without sampling is to choose A_1 or A_2 such that

$$R_1(D) = \text{Min. } \{R_1(H|A_1), R_1(H|A_2)\},$$

where $R(D)$ is the expected loss for the particular chosen action. $R(D)$ will also be called the expected loss associated with the Bayes decision function without sampling.

TABLE 14.3.3 Posterior Probability Functions of $\tilde{\pi}$

Sample Space	$\tilde{\pi}$	$P_1(\tilde{\pi} = \pi)$	$P(P = 0\|\tilde{\pi} = \pi)$	$P(P = 0\|\tilde{\pi} = \pi)P_1(\tilde{\pi} = \pi)$	$P_2(\tilde{\pi} = \pi\|P = 0)$
$P = 0$	1/4	1/2	9/16	9/32	9/13
	2/4	1/2	4/16	4/32	4/13
		1.0	13/16	13/32	13/13

Sample Space	$\tilde{\pi}$	$P_1(\tilde{\pi} = \pi)$	$P(P = .5\|\tilde{\pi} = \pi)$	$P(P = .5\|\tilde{\pi} = \pi)P_1(\tilde{\pi} = \pi)$	$P_2(\tilde{\pi} = \pi\|P = .5)$
$P = .5$	1/4	1/2	6/16	6/32	6/14
	2/4	1/2	8/16	8/32	8/14
			14/16	14/32	14/14

Sample Space	$\tilde{\pi}$	$P_1(\tilde{\pi} = \pi)$	$P(P = 1\|\tilde{\pi} = \pi)$	$P(P = 1\|\tilde{\pi} = \pi)P_1(\tilde{\pi} = \pi)$	$P_2(\tilde{\pi} = \pi\|P = 1)$
$P = 1.0$	1/4	1/2	1/16	1/32	1/5
	2/4	1/2	4/16	4/32	4/5
			5/16	5/32	5/5

EXAMPLE 14.3.2

Returning now to Example 14.3.1, we find that

$$R_1(D) = \text{Min.} \{R_1(H|A_1) = 50, R_1(H|A_2) = 40\} = 40,$$

which indicates that we should reject the null hypothesis and that the expected loss associated with this decision is 40.

We will now proceed to develop a Bayes decision function with sampling. To do so, we will again use an example.

EXAMPLE 14.3.3

Using the marble-box guessing problem in Example 14.3.1, let us assume that a simple random sample of 2 is to be drawn from the box selected. Then, the posterior probability function of π for $P = 0$, $P = .5$, and $P = 1.0$ are shown in Table 14.3.3.

Suppose now that $P = 0$. Then, from the extreme right-hand column of Table 14.3.3 we find that the posterior probability function of π is

π	$P_2(\pi = \pi | P = 0)$
$\frac{1}{4}$	$\frac{9}{13}$
$\frac{3}{4}$	$\frac{4}{13}$
	1.0

Thus, $\psi_2 = \frac{9}{13}$ depicts the posterior probability that the null hypothesis is true in light of the sample observation.

We now propose:

Definition 14.3.2

Let

$$R_2(H|A_1, P = p) = (1 - \psi_2)L(\text{II})$$

and

$$R_2(H|A_2, P = p) = \psi_2 L(\text{I}).$$

Then, $R_2(H|A_1, P = p)$ and $R_2(H|A_2, P = p)$ are called, respectively, the *posterior conditional expected losses* of accepting and rejecting the null hypothesis when the sample observation yields $P = p$.

We might point out that, in the preceding definition, ψ_2 denotes the posterior probability that the null hypothesis is true when the sample yields $P = p$.

EXAMPLE 14.3.4

Let us return to Example 14.3.3. Assume again that the sample yields $P = 0$. Suppose we were to accept the null hypothesis. Then, the conditional expected loss associated with this action is given by

336   Bayes Decision Functions

$$R_2(H|A_1, P = 0) = \psi_2(0) + (1 - \psi_2)L(II)$$
$$= \left(\frac{9}{13}\right)(0) + \left(\frac{4}{13}\right)(100)$$
$$= (1 - \psi_2)L(II)$$
$$= \frac{400}{13}.$$

On the other hand, suppose we were to reject the null hypothesis. Then, the conditional expected loss for this action is given by

$$R_2(H|A_2, P = 0) = \psi_2 L(I) + (1 - \psi_2)(0)$$
$$= \left(\frac{9}{13}\right)(80) + \left(\frac{4}{13}\right)(0)$$
$$= \psi_2 L(I)$$
$$= \frac{720}{13}.$$

What should we do when we observe that $P = 0$? Intuitively, it would be sound for us to take the action which minimizes the posterior conditional expected loss; that is, we should accept the null hypothesis. In fact, this action is the result of utilizing a Bayes decision function, as we will see.

Theorem 14.3.2
Let $R_2(H|A_1, P = p)$ and $R_2(H|A_2, P = p)$ be the posterior conditional expected losses for accepting and rejecting the null hypothesis, respectively, given that $P = p$. Then, a decision function is a Bayes decision function if it is chosen so that

$$R_2(D|P = p) = \text{Min. } \{R_2(H|A_1, P = p), R_2(H|A_2, P = p)\},$$

where $R_2(D|P = p)$ depicts the conditional expected loss for the particular chosen action when $P = p$.

EXAMPLE 14.3.5
In the preceding example we have pointed out that the null hypothesis should be accepted if $P = 0$. Now suppose that $P = .5$. Then the posterior probability function of π is

π	$P_2(\pi = \pi | P = .5)$
1/4	6/14
3/4	8/14
	14/14

which, in turn, implies that $\psi_2 = 6/14$. We next calculate that

14.3 Finding a Bayes Decision Function

$$R_2(H|A_1, P = .5) = (1 - \psi_2)L(II)$$
$$= \left(\frac{8}{14}\right)(100)$$
$$= \frac{800}{14}$$

and
$$R_2(H|A_2, P = .5) = \psi_2 L(I)$$
$$= \left(\frac{6}{14}\right)(80)$$
$$= \frac{480}{14}.$$

Thus, the null hypothesis should be rejected, since

$$(1 - \psi_2)L(II) > \psi_2 L(I).$$

In turn, the conditional expected loss for the Bayes decision function for $P = .5$ is

$$R_2(D|P = .5) = \text{Min.} \left\{\frac{800}{14}, \frac{480}{14}\right\} = \frac{480}{14}.$$

Finally, suppose $P = 1.0$. Then, the posterior probability function of π is

π	$P_2(\pi = \pi|P = 1.0)$
1/4	1/5
3/4	4/5
	1.0

Thus, $\psi_2 = 1/5$. We next calculate that

$$R_2(H|A_1, P = 1.0) = (1 - \psi_2)L(II)$$
$$= \left(\frac{4}{5}\right)(100)$$
$$= \frac{400}{5}$$

and
$$R_2(H|A_2, P = 1.0) = \psi_2 L(I)$$
$$= \left(\frac{1}{5}\right)(80)$$
$$= \frac{80}{5},$$

which indicates that we should reject the null hypothesis, since $(1 - \psi_2)L(II) > \psi_2 L(I)$. In turn,

338 Bayes Decision Functions

$$R_2(D|P = 1.0) = \text{Min.} \left\{ \frac{400}{5}, \frac{80}{5} \right\} = \frac{80}{5}.$$

14.4 Equivalence Relationship between Two Bayesian Procedures

We have maintained in the preceding section that the Bayes decision functions given in Definition 12.7.1 and in Theorem 14.3.2 are equivalent. We will illustrate the validity of this proposition in this section.

Let us return to Examples 14.3.4 and 14.3.5. Figure 14.4.1 summarizes the results from these two examples. If we now return to Example 12.7.1, we will find that the decision function in Figure 14.4.1 corresponds to D_4, which we concluded is a Bayes decision function for the given marble-box guessing problem when the prior probability of the null hypothesis being true is ½.

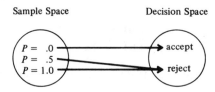

FIGURE 14.4.1

Thus, whether we derive a Bayes decision function according to Definition 12.7.1 or according to Theorem 14.3.2, the resulting decision functions from both are, in fact, equivalent.

Theorem 14.4.1
Let $P(P = p_0), \ldots, P(P = p_k)$ be the probabilities that P will assume the values $p_0, \ldots, p_k$. Then,

$$R_2(D) = R_2(D|P = p_0)P(P = p_0) + \cdots + R_2(D|P = p_k)P(P = p_k)$$

is the expected loss associated with the Bayes decision function.

EXAMPLE 14.4.1
Returning to Examples 14.3.3, 14.3.4, and 14.3.5, we offer the summaries given in Table 14.4.1. According to Theorem 14.4.1,

$$R_2(D) = \left(\frac{400}{13}\right)\left(\frac{13}{32}\right) + \left(\frac{480}{14}\right)\left(\frac{14}{32}\right) + \left(\frac{80}{5}\right)\left(\frac{5}{32}\right) = 30.$$

However, Definition 12.7.1 proposes that

$$R(D) = \psi_1 R(D|H_1) + (1 - \psi_1)R(D|H_2),$$

TABLE 14.4.1

p_i	$R_2(D\|P = p_i)$	$P(P = p_i)$
.0	$400/_{13}$	$13/_{32}$
.5	$480/_{14}$	$14/_{32}$
1.0	$80/_5$	$5/_{32}$
		$32/_{32}$

which, in turn, may be expressed as

$$R(D) = \psi_1 \alpha L(\mathrm{I}) + (1 - \psi_1)\beta L(\mathrm{II}),$$

where α and β depict the error probabilities associated with the Bayes decision function. Thus,

$$R(D) = \left(\frac{1}{2}\right)\left(\frac{7}{16}\right)(80) + \left(\frac{1}{2}\right)\left(\frac{4}{16}\right)(100) = 30.$$

We find that $R_2(D)$ and $R(D)$ calculated according to Theorem 14.4.1 and Definition 12.7.1 are the same. Further, we may note that

$$R_2(D|P = 0)P(P = 0) = (1 - \psi_1)R(D|H_2) = \frac{400}{32}$$

and
$$R_2(D|P = .5)P(P = .5) + R_2(D|P = 1.0)P(P = 1.0)$$
$$= \psi_1 R(D|H_1) = \frac{560}{32}.$$

A closer examination will reveal, however, that $R_2(D|P = 0)P(P = 0)$ and $(1 - \psi_1)R(D|H_2)$ both depict the component of $R_2(D)$ and $R(D)$ associated with accepting the null hypothesis. Thus, their numerical values must be the same. Similarly, $R_2(D|P = .5)P(P = .5) + R_2(D|P = 1.0)P(P = 1.0)$ and $\psi_1 R(D|H_1)$ should have the same values, because they both depict the component of $R_2(D)$ and $R(D)$ associated with rejecting the null hypothesis.

14.5 Value of Information Obtained from a Sample

So far we have been concerned with the problem of devising a Bayes decision function for a given sample size. However, an equally important question that needs to be answered is, "How large a sample should we take?" Two factors influence the answer to this question: (1) the cost of taking the sample of size n and (2) the value of the information provided by the given sample size.

Bayes Decision Functions

We will now describe how to assess the value of the information obtained from the sample.

Definition 14.5.1

Let $E(S_n)$ depict the expected value of information obtainable from a sample of size n. Then,

$$E(S_n) = R_1(D) - R_2(D).$$

The underlying logic of this definition must be fairly obvious. We have defined $R_1(D)$ to be the expected loss for the Bayes decision function without sampling and $R_2(D)$ to be the expected loss for the Bayes decision function with sampling. Then, the difference between $R_1(D)$ and $R_2(D)$ depicts the extent to which the amount of expected loss can be reduced by taking the sample of the given size. Thus, the difference between $R_1(D)$ and $R_2(D)$ can be considered to be the expected value of the information provided by the sample.

EXAMPLE 14.5.1

Let us return to a marble-box guessing problem. In Example 14.3.2, we calculated that

$$R_1(D) = 40,$$

and in Example 14.4.1 we calculated that

$$R_2(D) = 30.$$

So, for the sample of 2, we have

$$E(S_2) = R_1(D) - R_2(D) = 40 - 30 = 10,$$

which depicts the expected value of the information obtainable from a sample of size 2.

In a similar manner, we can calculate $E(S_n)$ for different sample sizes. On the other hand, regardless of the sample size, the value of the information obtained from the sample cannot exceed that which is obtained from the perfect information pertaining to the status of the hypotheses. Thus, we offer:

Definition 14.5.2

Let $R_3(D)$ depict the expected loss associated with a Bayes decision function when perfect information is available. Then,

$$E(PI) = R_1(D) - R_3(D)$$

depicts the expected value of the perfect information.

14.5 Value of Information Obtained from a Sample   341

EXAMPLE 14.5.2

Let us return again to the marble-box guessing problem. What is the meaning of the term *perfect information*? Suppose we select a box and look into the box before we guess. Then, if we select the box containing 1 red marble, we will know that it contains 1 red marble. Similarly, if we select the box containing 2 red marbles, we will know that it contains 2 red marbles.

How much do we expect to lose by playing this game? Obviously, nothing, since we will always guess correctly, regardless of which box we choose. Thus,

$$R_3(D) = 0$$

and, in turn,

$$E(PI) = 40 - 0 = 40.$$

The expected value of perfect information is, then, the expected loss associated with the Bayes decision function without sampling.

What is the practical significance of $E(PI)$? It indicates that, regardless of the sample size, it is not worthwhile for us to spend more than $E(PI)$ for sampling. $E(PI)$, therefore, establishes the upper limit of the amount to be spent in sampling. Again, the underlying logic is straightforward: if one expects to lose $E(PI)$ without any sampling, then there is no point in spending more than $E(PI)$, regardless of how much information can be obtained from the sample.

The next question is, then, "Is there an optimal sample size and, if so, is it calculable?" The answer to these questions is "yes."

Theorem 14.5.1

Let $C(n)$ be the cost of taking a sample of size n. Then, the optimal sample size is that which maximizes

$$G(n) = E(S_n) - C(n),$$

provided that $G(n)$ is positive for at least some values of n. If $G(n)$ is nonpositive for all values of n, then the optimal sample size n is zero.

The underlying logic of this theorem should be apparent. If $G(n)$ is nonpositive for all values of n, then it implies that the expected net gain from a sample is not positive, regardless of the sample size; therefore, it would not be worthwhile to take any sample. On the other hand, if $G(n)$ is positive for some values of n and, at the same time, is different for different values of n, then the optimal sample should be that which maximizes the expected net gain from the sample.

Although the underlying logic of Theorem 14.5.1 is simple, the computations involved in actually finding the optimal n can be cumbersome. We will illustrate in the following example the problem of finding an optimal sample size.

EXAMPLE 14.5.3

Let us return again to the marble-box guessing problem. Again assume that the box contains either 1 or 2 red marbles.

Assume that after we select a box we will be allowed to select a marble as many times as we wish, but we must replace each marble before we select the next. However, each time we select a marble, we must pay $2. We must now decide how many times we should select a marble before we make a guess. In this respect, we still assume that the probability that we will choose one type of box is $\frac{1}{2}$.

We have already calculated in Examples 14.5.2 that

$$E(PI) = 40.$$

Thus, we should not select a marble more than 20 times under any circumstance.

Let us next calculate $G(1)$. We already know that $C(1) = \$2$. But we still need to calculate $E(S_1)$. To do so, we must first calculate the posterior probability function of π for $n = 1$ as shown in Table 14.5.1.

If $P = 0$, then $\psi_2 = \frac{3}{5}$. Thus,

$$R_2(H|A_1, P = 0) = (1 - \psi_2)L(\text{II})$$
$$= \left(\frac{2}{5}\right)(100)$$
$$= 40$$

and
$$R_2(H|A_2, P = 0) = \psi_2 L(\text{I})$$
$$= \left(\frac{3}{5}\right)(80)$$
$$= 48,$$

which indicates that we should accept the null hypothesis that $\pi = \frac{1}{4}$.

If $P = 1$, then $\psi_2 = \frac{1}{3}$. Thus,

$$R_2(H|A_1, P = 1) = (1 - \psi_2)L(\text{II})$$
$$= \left(\frac{2}{3}\right)(100)$$
$$= \frac{200}{3}$$

and
$$R_2(H|A_2, P = 1) = \psi_2 L(\text{I})$$
$$= \left(\frac{1}{3}\right)(80)$$
$$= \frac{80}{3},$$

TABLE 14.5.1

Sample Space $P = 0$	π	$P_1(\mathfrak{k} = \pi)$	$P(P = 0 \mid \mathfrak{k} = \pi)$	$P(P = 0 \mid \mathfrak{k} = \pi) P_1(\mathfrak{k} = \pi)$	$P_2(\mathfrak{k} = \pi \mid P = 0)$
	$1/4$	$1/2$	$3/4$	$3/8$	$3/5$
	$2/4$	$1/2$	$2/4$	$2/8$	$2/5$
		1.0		$5/8$	$5/5$

Sample Space $P = 1$	π	$P_1(\mathfrak{k} = \pi)$	$P(P = 1 \mid \mathfrak{k} = \pi)$	$P(P = 1 \mid \mathfrak{k} = \pi) P_1(\mathfrak{k} = \pi)$	$P_2(\mathfrak{k} = \pi \mid P = 1)$
	$1/4$	$1/2$	$1/4$	$1/8$	$1/3$
	$2/4$	$1/2$	$2/4$	$2/8$	$2/3$
		1.0		$3/8$	$3/3$

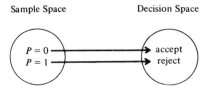

FIGURE 14.5.1

which indicates that we should reject the null hypothesis. The resulting decision function is shown in Figure 14.5.1. In turn,

$$R_2(D) = P(P = 0)R_2(D|P = 0) + P(P = 1)R_2(D|P = 1)$$
$$= \left(\frac{5}{8}\right)(40) + \left(\frac{3}{8}\right)\left(\frac{80}{3}\right)$$
$$= 35$$

and $E(S_1) = R_1(D) - R_2(D) = 40 - 35 = 5.$

Thus,
$$G(1) = E(S_1) - C(1) = 5 - 2 = 3.$$

In words this means that the expected value of the net gain from the sample of size 1 is $3.

Let us now calculate $G(2)$. We have already calculated that, if $n = 2$, then

$$E(S_2) = 10.$$

Thus,
$$G(2) = E(S_2) - C(2) = 10 - 4 = 6.$$

We next calculate $G(3)$ in the following manner. First, we obtain the posterior probability function of π, as shown in Table 14.5.2. Then, we obtain the Bayes decision function, shown in Table 14.5.3 and in Figure 14.5.2. In turn,

$$R_2(D) = P(P = 0)R_2(D|P = 0) + P\left(P = \frac{1}{3}\right)R_2\left(D|P = \frac{1}{3}\right)$$
$$+ P\left(P = \frac{2}{3}\right)R_2\left(D|P = \frac{2}{3}\right) + P(P = 1)R_2(D|P = 1)$$
$$= \left(\frac{35}{128}\right)\left(\frac{800}{35}\right) + \left(\frac{51}{128}\right)\left(\frac{2160}{51}\right) + \left(\frac{33}{128}\right)\left(\frac{720}{33}\right) + \left(\frac{9}{128}\right)\left(\frac{80}{9}\right)$$
$$\cong 29.30.$$

TABLE 14.5.2 Posterior Probability Function of π

Sample Space	π	$P_1(\boldsymbol{\pi} = \pi)$	$P(P = 0 \mid \boldsymbol{\pi} = \pi)$	$P(P = 0 \mid \boldsymbol{\pi} = \pi)P_1(\boldsymbol{\pi} = \pi)$	$P_2(\boldsymbol{\pi} = \pi \mid P = 0)$
$P = 0$	$1/4$	$1/2$	$27/64$	$27/128$	$27/35$
	$2/4$	$1/2$	$8/64$	$8/128$	$8/35$
		1.0		$35/128$	$35/35$

Sample Space	π	$P_1(\boldsymbol{\pi} = \pi)$	$P(P = 1/3 \mid \boldsymbol{\pi} = \pi)$	$P(P = 1/3 \mid \boldsymbol{\pi} = \pi)P_1(\boldsymbol{\pi} = \pi)$	$P_2(\boldsymbol{\pi} = \pi \mid P = 1/3)$
$P = 1/3$	$1/4$	$1/2$	$27/64$	$27/128$	$27/51$
	$2/4$	$1/2$	$24/64$	$24/128$	$24/51$
		1.0		$51/128$	$51/51$

Sample Space	π	$P_1(\boldsymbol{\pi} = \pi)$	$P(P = 2/3 \mid \boldsymbol{\pi} = \pi)$	$P(P = 2/3 \mid \boldsymbol{\pi} = \pi)P_1(\boldsymbol{\pi} = \pi)$	$P_2(\boldsymbol{\pi} = \pi \mid P = 2/3)$
$P = 2/3$	$1/4$	$1/2$	$9/64$	$9/128$	$9/33$
	$2/4$	$1/2$	$24/64$	$24/128$	$24/33$
		1.0		$33/128$	$33/33$

Sample Space	π	$P_1(\boldsymbol{\pi} = \pi)$	$P(P = 1 \mid \boldsymbol{\pi} = \pi)$	$P(P = 1 \mid \boldsymbol{\pi} = \pi)P_1(\boldsymbol{\pi} = \pi)$	$P_2(\boldsymbol{\pi} = \pi \mid P = 1)$
$P = 1.0$	$1/4$	$1/2$	$1/64$	$1/128$	$1/9$
	$2/4$	$1/2$	$8/64$	$8/128$	$8/9$
		1.0		$9/128$	$9/9$

TABLE 14.5.3 Bayes Decision Function

Sample Space	Decision Space	$R_2(H\|A_i, P = p_i)$	Optimal Decision	$R_2(D\|P = p_i)$
$P = 0$	A_1 A_2	$(^{8}\!/_{35})(100) = {}^{800}\!/_{35}$ $(^{27}\!/_{35})(80) = {}^{2160}\!/_{35}$	A_1	${}^{800}\!/_{35}$
$P = \frac{1}{3}$	A_1 A_2	$(^{24}\!/_{51})(100) = {}^{2400}\!/_{51}$ $(^{27}\!/_{51})(80) = {}^{2160}\!/_{51}$	A_2	${}^{2160}\!/_{51}$
$P = \frac{2}{3}$	A_1 A_2	$(^{24}\!/_{33})(100) = {}^{2400}\!/_{33}$ $(^{9}\!/_{33})(80) = {}^{720}\!/_{33}$	A_2	${}^{720}\!/_{33}$
$P = 1.0$	A_1 A_2	$(^{8}\!/_{9})(100) = {}^{800}\!/_{9}$ $(^{1}\!/_{9})(80) = {}^{80}\!/_{9}$	A_2	${}^{80}\!/_{9}$

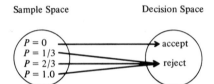

FIGURE 14.5.2 Bayes decision function.

Thus,
$$E(S_3) = 40 - 29.30 = 10.70$$
and, in turn
$$G(3) = E(S_3) - C(3) = 10.70 - 6 = 4.70.$$

The preceding illustrations show that $G(1) < G(2)$ but $G(2) > G(3)$. Our intuition suggests that $G(3) > G(4) > G(5)$, and so on. If we were to actually calculate $G(n)$ for different values of n, then, we would find that our intuition is correct. Thus, $G(n)$ is maximized when $n = 2$ and the optimal sample size is 2.

EXERCISES

*9. Return to the marble-box guessing problem in Exercise 1, in which 1 box contains 2 red marbles out of 10 and the other contains 5 red marbles out of 10. Assume that you are penalized $100 if you guess erroneously when you select the box containing 2 red marbles and $200 if you guess erroneously when you select the box containing 5 red marbles. Assume again that the prior probability of selecting one type of box is .5. Assume also that you have established a null

14.5 Value of Information Obtained from a Sample

hypothesis that $\pi = .2$. Calculate the prior conditional expected losses of accepting and rejecting the null hypothesis. Suppose you have to guess without taking a sample from the box. What is the Bayes decision function without sampling?

*10. Using the given marble-box guessing problem in Exercise 9, assume that you will be allowed to select a marble twice from the box selected, replacing the first before you select the second. Assuming the prior probability function for π is the same as that given in Exercise 9, calculate the posterior conditional expected losses of accepting and rejecting the null hypothesis established in Exercise 9 for each possible sample value of P. Then, indicate the Bayes decision function for each possible sample value of P.

*11. From the Bayes decision functions found for different values of P in Exercise 10, calculate the expected loss associated with the Bayes decision function (given by Theorem 14.4.1).

12. For the marble-box guessing problem described in Exercise 10, ascertain the Bayes decision function according to Definition 12.7.1 in Chapter 12. Calculate the expected loss associated with the Bayes decision function. Then, show that the Bayes decision function ascertained in Exercise 10 and here are equivalent.

*13. For the marble-box guessing problem, on the basis of your analysis in the preceding exercises, what is the expected value of information obtained from a sample of 2?

*14. For the marble-box guessing problem, suppose that each time you draw a marble from the box you must pay $3. Then, what are the optimal number of times that you should draw a marble?

*15. Return to the manufacturing process given in Exercise 6, which might be producing either 10 or 20 percent defectives. Assume that the cost of erroneously stopping the machine is $1000 and of failing to stop the machine when the process is not working satisfactorily is $2000. Assume again that the process produces 10 percent defectives 80 percent of the time and 20 percent defectives 20 percent of the time. Assume also that you have established the null hypothesis that the process is putting out 10 percent defectives. Calculate the prior conditional expected losses of accepting or rejecting the null hypothesis. If you have to make the decision with regard to the process without sampling, what should be your optimal decision?

*16. For the problem in Exercise 15, what is the maximum amount that you should be willing to pay for any information obtainable from sampling?

*17. Using the problem in Exercise 15, assume that you will select a simple random sample of 4 from the process. Assume the same prior probability function for π as that given in Exercise 15. Calculate the posterior conditional expected losses of accepting and rejecting the null hypothesis established in Exercise 15 for each possible sample value of P. Then, indicate the Bayes decision function for each possible sample value P.

18. From the Bayes decision function given in Exercise 17, calculate the expected loss associated with the Bayes decision function, according to Theorem 14.4.1.

19. For the problem described in Exercise 15, ascertain the Bayes decision function according to Definition 12.7.1 in Chapter 12. Calculate the expected loss associated with the Bayes decision function. Then, show that the Bayes decision function ascertained in Exercise 17 and here are, in fact, equivalent.
20. For the problem in Exercise 15, on the basis of your analysis in the preceding problems, what is the expected value of information obtained from the sample of 4?

14.6 Preposterior Analysis

In ascertaining the expected value of the sample information, we had to ascertain first the values $R_1(D)$ and $R_2(D)$. However, the expected value of the sample information may be calculated without explicitly calculating the values of $R_1(D)$ and $R_2(D)$. This alternate method of calculating the expected value of the sample information, however, requires that we first ascertain the *preposterior probability function* of π. This alternate procedure probably does not lessen the computational burden for the kind of problem which we have considered so far, in which the sample is assumed to have been taken from a discrete probability function.

On the other hand, as we will illustrate later, when the sample is assumed to be taken from a continuous probability function, the method which we are about to discuss here is the only practical way to find the expected value of the sample information.

In introducing the procedure, conceivably we can wait until we discuss the problem of sampling from a continuous probability function. However, we believe that the reader may be able to perceive better the logic underlying the procedure if we illustrate it for a problem whose sample is assumed to have been taken from a discrete probability function.

We will now define the notion of a *preposterior probability function* of π.

Definition 14.6.1

Let $\tilde{\pi} = E(\pi|P = p)$. Then, the probability function of $\tilde{\pi}$ is called the *preposterior probability function* of π.

In a sense the preposterior probability function of π is the probability function of the conditional expected value of π, given that $P = p$.

EXAMPLE 14.6.1

Let us now return to a marble-box guessing problem and assume that a sample of 2 is to be selected. Then, the posterior probability functions of π are those given in Table 14.6.1.

14.6 Preposterior Analysis

TABLE 14.6.1 Posterior Probability Function of $\tilde{\pi}$

π	$P_2(\pi = \pi\|P = 0)$	$P_2(\pi = \pi\|P = .5)$	$P_2(\pi = \pi\|P = 1.0)$
$1/4$	$9/13$	$6/14$	$1/5$
$2/4$	$4/13$	$8/14$	$4/5$
	1.0	1.0	1.0

Assume now that the sample yields $P = 0$. Then,

$$\tilde{\pi} = \left(\pi = \frac{1}{4}\right) P\left(\pi = \frac{1}{4}\Big| P = 0\right) + \left(\pi = \frac{2}{4}\right) P\left(\pi = \frac{2}{4}\Big| P = 0\right)$$

$$= \left(\frac{1}{4}\right)\left(\frac{9}{13}\right) + \left(\frac{2}{4}\right)\left(\frac{4}{13}\right)$$

$$= \frac{17}{52}.$$

The probability that $\tilde{\pi}$ assumes $17/52$ is equal to $P(P = 0)$. Thus, $P(\tilde{\pi} = 17/52) = 13/32$. In a similar manner, we derive that $P(\tilde{\pi} = 22/56) = 14/32$ and $P(\tilde{\pi} = 9/20) = 5/32$. In summary, then, the preposterior probability function of π is given as

$\tilde{\pi}$	$P_2(\tilde{\pi} = \tilde{\pi})$
$17/52 = .326$	$13/32$
$22/56 = .392$	$14/32$
$9/20 = .450$	$5/32$
	1.0

We will now illustrate how this preposterior probability function can be used to find the expected value of the sample information. Recall that the loss functions given for the marble-box guessing problem are those shown in Table 14.6.2. We can, for example, assume that these loss functions have come from the following equations:

TABLE 14.6.2 Loss Functions

	State of Box	
Decision	$\pi = .25$	$\pi = .50$
A_1	0	100
A_2	80	0

and
$$L(II) = L(A_1|\tilde{\pi} = \bar{\pi}) = -100 + 400\bar{\pi}$$
$$L(I) = L(A_2|\tilde{\pi} = \bar{\pi}) = 160 - 320\bar{\pi}.$$

Suppose that $\bar{\pi} = 17/52$. Then, $L(A_1|\tilde{\pi} = 17/52) = 30.77$ and $L(A_2|\tilde{\pi} = 17/52) = 55.38$. Thus, A_1, for which we accept the null hypothesis, will entail a smaller loss. Similarly, we ascertain $L(A_1|\tilde{\pi} = \bar{\pi})$ and $L(A_2|\tilde{\pi} = \bar{\pi})$ for $\bar{\pi} = 22/56$ and $\bar{\pi} = 9/20$, as shown in Table 14.6.3. The Bayes decision function is shown in Figure 14.6.1.

TABLE 14.6.3

$\bar{\pi}$	Decision Space	$L(A_i|\tilde{\pi} = \bar{\pi})$	Optimal Decision	$L(\tilde{\pi} = \bar{\pi})$
$17/52$	A_1	$30.77 = 400/13$	A_1	30.40
	A_2	$55.38 = 720/13$		
$22/56$	A_1	$57.14 = 800/14$	A_2	34.28
	A_2	$34.28 = 480/14$		
$9/20$	A_1	$80.00 = 400/5$	A_2	16.00
	A_2	$16.00 = 80/5$		

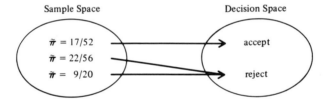

FIGURE 14.6.1 Bayes decision function.

If we now return to Example 14.3.2, we will find that the Bayes decision function without sampling is to reject the null hypothesis. Then, the only time that the Bayes decision function with sampling is different from that without sampling is when $\bar{\pi} = 17/52$, because the former indicates that we should accept the null hypothesis when $\bar{\pi} = 17/52$; this means that the information obtained from the sample is valuable only when the information yields $\bar{\pi} = 17/52$, and it must be given as

$$V(S_2) = L\left(A_2|\tilde{\pi} = \frac{17}{52}\right) - L\left(A_1|\tilde{\pi} = \frac{17}{52}\right)$$
$$= \frac{720}{13} - \frac{400}{13}$$
$$= \frac{320}{13}.$$

The expected value of the sample information is, then,

$$E(S_2) = P_2\left(\tilde{\pi} = \frac{17}{52}\right) V(S_2)$$
$$= \left(\frac{13}{32}\right)\left(\frac{320}{13}\right)$$
$$= 10.$$

The expected value of the sample information thus obtained is the same as that which we obtained in Section 14.5.

EXERCISES

*21. Return to the marble-box guessing problem in Exercises 1 and 9. Assume again that you will be allowed to draw a marble twice, replacing the first before you draw the second. Ascertain the preposterior probability function of π.

*22. Refer to the marble-box guessing problem in Exercise 21 and find the Bayes decision function by carrying out the preposterior analysis. Show that the expected value of the sample information obtained through your preposterior analysis is the same as that obtained in Exercise 13.

23. Return to the process-control problem in Exercises 6 and 15. Assume again that a simple random sample of 4 is to be selected from the process. Ascertain the preposterior probability function of π.

24. Refer to the process-control problem in Exercise 23 and find the Bayes decision function by carrying out the preposterior analysis. Show that the expected value of the sample information obtained through analysis is the same as that obtained in Exercise 20.

14.7 Composite Hypothesis

So far we have worked only with a Bayes decision function for testing a simple null hypothesis against a simple alternate hypothesis. However, a real problem of any significance probably does not lend itself to such a simple formulation. For this reason we have used the trivial problem of guessing the color of marbles in our illustrations.

However, we can devise a Bayes decision function for testing a composite hypothesis against an alternate composite hypothesis. In formulating such a decision function, we will provide a more practical situation for the problem.

EXAMPLE 14.7.1

Assume that a university computer center has received free-of-charge 10 canned programs from the manufacturer of its computer. At the same time, the manufacturer assumes no responsibility if the programs have bugs. Thus, some debugging may be needed before some of the programs will actually run.

Let π depict the percentage of the programs which contain some bugs, that is, which are defective. On the basis of past experience, the following prior probability function is assigned to π.

π	$P_1(\pi = \pi)$
.0	.4
.1	.3
.2	.2
.3	.1
	1.0

A local software firm formed by a group of energetic students has proposed the following options to the computer center:

Option 1: It will debug any program for $200 per program.

Option 2: It will charge $300 to test the entire lot of programs for bugs, whether or not any of the programs need debugging. Also, it will debug any or all programs that need debugging at no additional charge.

The cost associated with the two options for the assumed values of π are shown in Table 14.7.1. This table indicates that if $\pi < .15$, then the first option would entail a lower cost and if $\pi > .15$, the second option would entail a lower cost.

TABLE 14.7.1

Assumed Value of π	Options	
	1	2
.0	0	300
.1	200	300
.2	400	300
.3	600	300

The problem of deciding which option to accept may now be formulated in terms of a set of hypotheses:

$$\text{Null Hypothesis:} \quad \pi \leq .15$$
$$\text{Alternate Hypothesis:} \quad \pi > .15.$$

If the null hypothesis is accepted, then the computer center will accept Option 1, and, conversely, if the null hypothesis is rejected, it will accept Option 2. The types of error that can be committed are shown in Table 14.7.2.

14.7 Composite Hypothesis

TABLE 14.7.2

Assumed Value of π	Decision	
	A_1: ACCEPT NULL HYPOTHESIS	A_2: REJECT NULL HYPOTHESIS
.0	correct decision	Type I error
.1	correct decision	Type I error
.2	Type II error	correct decision
.3	Type II error	correct decision

Thus, for each assumed value of π, the correct decision is one which will yield the smaller cost. The difference between the cost associated with the correct decision and the incorrect decision is the amount of loss associated with making the wrong decision for the given value of π. We can depict the amount of this loss as a function of π and the decision, as it is given in Table 14.7.3.

TABLE 14.7.3 Loss Function for Given Values of π

Assumed Value of π	Decision	
	A_1: ACCEPT NULL HYPOTHESIS	A_2: REJECT NULL HYPOTHESIS
.0	$L(A_1, \pi = 0) = 0$	$L(A_2, \pi = 0) = 300$
.1	$L(A_1, \pi = .1) = 0$	$L(A_2, \pi = .1) = 100$
.2	$L(A_1, \pi = .2) = 100$	$L(A_2, \pi = .2) = 0$
.3	$L(A_1, \pi = .3) = 300$	$L(A_2, \pi = .3) = 0$

The losses in Table 14.7.3 indicate that the actual amount of loss associated with a particular decision is dependent not only on whether or not it is an erroneous decision, but also on how erroneous the decision really is in light of the given value of π. This factor, then, has to be taken into account in calculating the conditional expected losses of accepting and rejecting the null hypothesis. Thus, we modify Definitions 14.3.1 and 14.3.2 as follows:

Definition 14.7.1

Let $R_1(\pi|A_1)$ and $R_1(\pi|A_2)$ be the prior conditional expected losses of accepting and rejecting the null hypothesis. Then,

$$R_1(\pi|A_1) = P_1(\pi = \pi_0)L(A_1|\pi = \pi_0)$$
$$+ \cdots + P_1(\pi = \pi_k)L(A_1|\pi = \pi_k)$$

and $R_1(\pi|A_2) = P_1(\pi = \pi_0)L(A_2|\pi = \pi_0)$
$$+ \cdots + P_1(\pi = \pi_k)L(A_2|\pi = \pi_k).$$

Let $R_2(\pi|A_1, P = p)$ and $R_2(\pi|A_2, P = p)$ be the posterior expected losses of accepting and rejecting the null hypothesis for $P = p$. Then,

$$R_2(\pi|A_1, P = p) = P_2(\pi = \pi_0|P = p)L(A_1|\pi = \pi_0)$$
$$+ \cdots + P_2(\pi = \pi_k|P = p)L(A_1|\pi = \pi_k)$$

and $R_2(\pi|A_2, P = p) = P_2(\pi = \pi_0|P = p)L(A_2|\pi = \pi_0)$
$$+ \cdots + P_2(\pi = \pi_k|P = p)L(A_2|\pi = \pi_k).$$

Note that we have made one slight modification in the notations for the conditional expected losses: Instead of denoting them as $R(H|A_1)$ and $R(H|A_2)$, as we have done previously, we now denote them as $R(\pi|A_1)$ and $R(\pi|A_2)$. The underlying motivation may be stated as follows. Previously, the null hypothesis was either true or false, and we assumed that a chance device selects for us either H_1 or H_2. However, now the hypothesis can be true in two different ways and it can be false in two different ways, depending on the value of π. Therefore, it is more convenient to assume that a chance device selects for us a particular value of π rather than H_1 or H_2.

EXAMPLE 14.7.2

Let us now return to Example 14.7.1. Applying the first part of Definition 14.7.1, we calculate that

$$R_1(\pi|A_1) = (.4)(0) + (.3)(0) + (.2)(100) + (.1)(300) = 50$$
and $R_1(\pi|A_2) = (.4)(300) + (.3)(100) + (.2)(0) + (.1)(0) = 150.$

Suppose the computer center were to make the decision without taking any sample from the lot of programs. Then, according to Theorem 14.3.1, the optimal decision is that which corresponds to

$$R_1(D) = \text{Min.} \{R_1(\pi|A_1) = 50, R_1(\pi|A_2) = 150\} = 50.$$

Thus, the null hypothesis should be accepted, and, in turn, the computer center should choose Option 1. The expected loss associated with this option is then $50, and the value of the perfect information is also $50.

Let us assume now that the computer center can test any program to see whether or not it needs debugging for $8 per program. It might appear that the computer center should test all of the programs in the lot before making the decision, however, then the total cost of examining the 10 programs would exceed the expected value of the perfect information by $30. An alternative approach would be to examine only a portion of the lot and then to make a decision.

Assume now that a sample of 1 program is to be selected from the lot. Then, the posterior probability function of π for $P = 0$ and $P = 1$ are shown in Table 14.7.4.

Suppose the observed value of P is, in fact, 0; that is, the program

TABLE 14.7.4 Posterior Probability Functions of $\tilde{\pi}$

Sample Space	π	$P_1(\tilde{\pi} = \pi)$	$P(P = 0\|\tilde{\pi} = \pi)$	$P(P = 0\|\tilde{\pi} = \pi)P_1(\tilde{\pi} = \pi)$	$P_2(\tilde{\pi} = \pi\|P = 0)$
	.0	.4	1.0	.40	$40/90$
$P = 0$	.1	.3	.9	.27	$27/90$
	.2	.2	.8	.16	$16/90$
	.3	.1	.7	.07	$7/90$
				.90	$90/90$

Sample Space	π	$P_1(\tilde{\pi} = \pi)$	$P(P = 1\|\tilde{\pi} = \pi)$	$P(P = 1\|\tilde{\pi} = \pi)P_1(\tilde{\pi} = \pi)$	$P_2(\tilde{\pi} = \pi\|P = 1)$
	.0	.4	.0	.00	.0
$P = 1$	.1	.3	.1	.03	.3
	.2	.2	.2	.04	.4
	.3	.1	.3	.03	.3
				.10	1.0

356 Bayes Decision Functions

tested does not need debugging. The optimal decision for this observation is then ascertained by calculating:

$$R_2(\pi|A_1, P = 0) = \left(\frac{40}{90}\right)(0) + \left(\frac{27}{90}\right)(0) + \left(\frac{16}{90}\right)(100)$$
$$+ \left(\frac{7}{90}\right)(300) = \frac{3700}{90}$$

$$R_2(\pi|A_2, P = 0) = \left(\frac{40}{90}\right)(300) + \left(\frac{27}{90}\right)(100) + \left(\frac{16}{90}\right)(0)$$
$$+ \left(\frac{7}{90}\right)(0) = \frac{14{,}700}{90},$$

which shows that the null hypothesis should be accepted and Option 1 should be chosen.

Suppose now that the observed value of P is 1; that is, the program tested does need debugging. The optimal decision is ascertained by calculating:

$$R_2(\pi|A_1, P = 1) = (.3)(0) + (.4)(100) + (.3)(300) = 130$$
$$R_2(\pi|A_2, P = 1) = (.3)(100) + (.4)(0) + (.3)(0) = 30,$$

which indicates that Option 2 should be chosen. Thus, the resulting Bayes decision function is given in Figure 14.7.1.

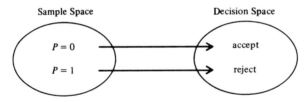

FIGURE 14.7.1 Bayes decision function.

We next calculate:

$$R_2(D) = R_2(D|P = 0)P(P = 0) + R_2(D|P = 1.0)P(P = 1.0)$$
$$= \left(\frac{3700}{90}\right)(.9) + (30)(.1) = 40.$$

Then, according to Definition 14.5.1,

$$E(S_1) = R_1(D) - R_2(D) = 50 - 40 = 10$$

and, in turn,

$$G(1) = E(S_1) - C(1) = 10 - 8 = 2.$$

Thus, the net expected gain from testing one program is $2.

Suppose now a sample of 2 is to be selected from the lot of programs without replacing the sample. Then, the posterior probability function of π is calculated as shown in Table 14.7.5.

TABLE 14.7.5 Posterior Probability Function of $\tilde{\pi}$

Sample Space	π	$P_1(\tilde{\pi} = \pi)$	$P(P=0\|\tilde{\pi}=\pi)$	$P(P=0\|\tilde{\pi}=\pi)P_1(\tilde{\pi}=\pi)$	$P_2(\tilde{\pi}=\pi\|P=0)$
$P=0$	.0	.4	90/90	360/900	360/730
	.1	.3	72/90	216/900	216/730
	.2	.2	56/90	112/900	112/730
	.3	.1	42/90	42/900	42/730
				730/900	730/730

Sample Space	π	$P_1(\tilde{\pi} = \pi)$	$P(P=.5\|\tilde{\pi}=\pi)$	$P(P=.5\|\tilde{\pi}=\pi)P_1(\tilde{\pi}=\pi)$	$P_2(\tilde{\pi}=\pi\|P=.5)$
$P=.5$	.0	.4	0	0	0
	.1	.3	18/90	54/900	54/160
	.2	.2	32/90	64/900	64/160
	.3	.1	42/90	42/900	42/160
				160/900	160/160

Sample Space	π	$P_1(\tilde{\pi} = \pi)$	$P(P=1\|\tilde{\pi}=\pi)$	$P(P=1\|\tilde{\pi}=\pi)P_1(\tilde{\pi}=\pi)$	$P_2(\tilde{\pi}=\pi\|P=1)$
$P=1$	.0	.4	0	0	0
	.1	.3	0	0	0
	.2	.2	2/90	4/900	4/10
	.3	.1	6/90	6/900	6/10
				10/900	10/10

358 Bayes Decision Functions

TABLE 14.7.6

Sample Space	Decision Space	$R_2(\pi\|A_i, P = p_i)$	Optimal Decision	$R_2(D\|P = p)$
$P = 0$	A_1	23,800/730	A_1	23,800/730
	A_2	129,600/730		
$P = .5$	A_1	18,400/160	A_2	5400/160
	A_2	5400/160		
$P = 1.0$	A_1	220	A_2	0
	A_2	0		

The Bayes decision function can be found by calculating the values shown in Table 14.7.6. The resulting Bayes decision function is given in Figure 14.7.2 and, in turn,

$$R_2(D) = \left(\frac{730}{900}\right)\left(\frac{23,800}{730}\right) + \left(\frac{160}{900}\right)\left(\frac{5400}{160}\right)$$
$$\cong 32.44.$$

Thus,
$$E(S_2) = R_1(D) - R_2(D) = 17.56$$

and
$$G(2) = E(S_2) - C(2) = 17.56 - 16 = 1.56.$$

The preceding analysis indicates that $G(1) > G(2)$. If we were to calculate $G(3)$, $G(4)$, and so on, we would find that $G(2) > G(3) > G(4)$, and so on. Thus, the optimal sample size is 1.

Therefore, the computer center should take a sample of 1 program, and if the sampled program does not need debugging, it should accept the null hypothesis and choose Option 1; if the sampled program does need debugging, it should reject the null hypothesis and choose Option 2.

So far we have calculated $E(S_n)$ by explicitly ascertaining the values $R_1(D)$ and $R_2(D)$. We will now illustrate how we can obtain $E(S_n)$ without explicitly ascertaining $R_1(D)$ and $R_2(D)$.

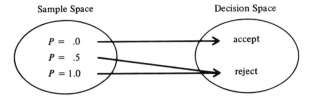

FIGURE 14.7.2 Bayes decision function.

14.7 Composite Hypothesis

Assume that we wish to calculate $E(S_n)$ for $n = 1$. We first must find the preposterior probability function of π in the following manner. Assume that $P = 0$. Then,

$$\tilde{\pi}_0 = (\pi = 0)P_2(\pi = 0|P = 0) + (\pi = .1)P_2(\pi = .1|P = 0)$$
$$+ (\pi = .2)P_2(\pi = .2|P = 0) + (\pi = .3)P_2(\pi = .3|P = 0)$$
$$= (0)\left(\frac{40}{90}\right) + (.1)\left(\frac{27}{90}\right) + (.2)\left(\frac{16}{90}\right) + (.3)\left(\frac{7}{90}\right)$$
$$= \frac{80}{900}$$

and $\tilde{\pi}_1 = (\pi = 0)P_2(\pi = 0|P = 1) + (\pi = .1)P_2(\pi = .1|P = 1)$
$$+ (\pi = .2)P_2(\pi = .2|P = 1) + (\pi = .3)P_2(\pi = .3|P = 1)$$
$$= (0)(0) + (.1)(.3) + (.2)(.4) + (.3)(.3)$$
$$= \frac{180}{900}.$$

The preposterior probability function of π is given as

$\tilde{\pi}$	$P_2(\tilde{\pi} = \tilde{\pi})$
80/900	.9
180/900	.1
	1.0

Let us now re-examine Table 14.7.3, which depicts the losses for making decisions for different values of π. The loss function may then be depicted by:

and
$$L(A_1|\tilde{\pi} > .15) = 2000(\tilde{\pi} - .15)$$
$$L(A_2|\tilde{\pi} \leq .15) = 2000(.15 - \tilde{\pi}).$$

The Bayes decision function which can be determined is shown in Table 14.7.7 and in Figure 14.7.3.

Recall that the Bayes decision function obtained without sampling was to accept the null hypothesis. Thus, the only time that the sample provides

TABLE 14.7.7 Bayes Decision Function

$\tilde{\pi}$	Decision	$L(A_i|\tilde{\pi} = \tilde{\pi})$	Optimal Decision
80/900	A_1 A_2	0 1100/9	A_1
180/900	A_1 A_2	100 0	A_2

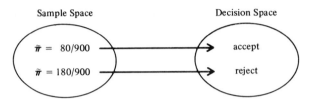

FIGURE 14.7.3   Bayes decision function.

useful information is when it yields $\tilde{\pi} = {}^{180}\!/_{900}$. The value of sample information when it yields $\tilde{\pi} = {}^{180}\!/_{900}$ is then given as

$$V(S_1) = L(A_1|\tilde{\pi} = .2) - L(A_2|\tilde{\pi} = .2) = 100 - 0 = 100,$$

and, the expected value of the sample information is

$$E(S_1) = P\left(\tilde{\pi} = \frac{180}{900}\right) V(S_1) = (.1)(100) = 10,$$

which is the same as that which we obtained previously.

EXERCISES

*25. Suppose you are in charge of controlling a production process. Each 1 percent defective produced by the process means the company loses $100. For example, if the company produces 20 percent defectives, then the cost to the company is $2000. The percentage of defectives can always be decreased to a 10 percent level by a readjustment, but it costs the company $1500 to shut down the process and readjust it. Let π depict the percentage of defectives produced by the process. From the past data, the following prior probability functions for various values of π have been ascertained.

π	$P(\tilde{\pi} = \pi)$
.1	.4
.2	.3
.3	.2
.4	.1

Describe the loss function for the given problem. What null and alternate hypotheses can you establish to deal with the problem? Is it necessary that you establish such hypotheses for the problem?

*26. Return to the process-control problem in Exercise 25 and suppose you were to make the decision with regard to the process without taking any sample from the process. What should be your optimal decision?

*27. You now contemplate taking a sample from the process discussed in Exercise 25 before you make your decision. An item sampled must be destroyed in order to

14.7 Composite Hypothesis

tell whether or not it is defective. Thus, the sampling cost is fairly expensive, even for a small sample. What is the maximum amount that you should be willing to spend in sampling?

*28. Using the process-control problem in Exercise 25, assume that you will select a simple random sample of 2 from the process. Calculate the posterior conditional expected losses for stopping and not stopping the process for each possible value of P. Then, indicate the optimal decision for each observed value of P.

*29. From your decision functions found for Exercise 25, calculate the expected loss associated with the Bayes decision function.

*30. Using the process-control problem in Exercise 29, what is the expected value of the information obtainable from a sample of size 2?

*31. Using the process-control problem in Exercise 29, assume still that a simple random sample of 2 is to be selected from the process. Ascertain the preposterior probability function of π.

*32. Using the process-control problem in Exercise 29, find the Bayes decision function by carrying out the preposterior analysis. Show also that the expected value of the sample information obtained through your preposterior analysis is the same as that obtained in Exercise 30.

*33. Using the process-control problem in Exercise 29, suppose each sample item would cost $100. What is, then, the optimal sample size?

34. A large department-store chain is contemplating making a purchase of 1000 transistor radios. The store will sell these radios for $20 each. There are two different manufacturers from which the store can purchase these radios. Manufacturer A charges $15 per radio but will refund $20 for each defective radio. Manufacturer B only charges $10 per radio but will not give any refund for defective radios. The store will, therefore, have to refund $20 to each customer who returns a defective radio. From past experience with Manufacturer B, the store has been able to formulate the following prior probability functions for π, where π depicts the percentage of defective radios produced by Manufacturer B.

π	$P(\pi = \pi)$
.1	.3
.2	.3
.3	.2
.4	.1
.5	.1

Describe the loss function for the problem. In view of your loss function, what null and alternate hypotheses can you establish to deal with the problem?

35. Using the department store problem in Exercise 34, suppose that the store is to make the decision without any sampling. Which manufacturer should the store choose?

36. Suppose the store in the problem given in Exercise 34 is contemplating taking a sample of some radios produced by Manufacturer B before it makes a decision. What should be the maximum amount spent in sampling?

37. Assume that Manufacturer B will allow the store to take a sample of 5 radios free of charge. Calculate the posterior conditional expected losses for purchasing from Manufacturers A and B for each possible sample value of P. Then, indicate the optimal decision for the store for each observed value of P.

38. Utilizing the information obtained in Exercise 37, calculate the expected loss associated with the Bayes decision function.

39. Return to Exercise 34. What is the expected value of information obtainable from a sample of size 5?

40. Using the problem in Exercise 34, still assume that a simple random sample of 5 is to be selected. Ascertain the preposterior probability function of π.

41. Using the problem in Exercise 34, find the Bayes decision function by carrying out the preposterior analysis. Show also that the expected value of the sample information calculated through your preposterior analysis is the same as the one you calculated in Exercise 39.

14.8 Bayes Decision Function for Normal Probability Function without Sampling

In the preceding sections we have considered a Bayes decision function which has entailed a discrete probability function. We will now extend our analysis to a normal probability function, which is a continuous probability function.

Before we proceed with the details of formulating a Bayes decision function, we will describe a situation which might call for a Bayes decision function.

Let X be a normally distributed random variable with $E(X) = \mu$ and $V(X) = \sigma^2$. Let us next define two linear functions:

and
$$g_1(\mu) = a_1 + b_1\mu$$
$$g_2(\mu) = a_2 + b_2\mu,$$

where the values of a_1, a_2, b_1, and b_2 are assumed to be known and $a_1 < a_2$ but $b_1 > b_2$. The two linear functions we have defined are illustrated in Figure 14.8.1.

We might, for example, assume that $g_1(\mu)$ and $g_2(\mu)$ are two different cost functions, or two different revenue functions associated with two different processes. If $g_1(\mu)$ and $g_2(\mu)$ depict two cost functions, then, we assume that our problem is to choose a process between the two so that

$$g(\mu) = \text{Min.} \{g_1(\mu), g_2(\mu)\},$$

where $g(\mu)$ depicts the cost of operating the chosen process for a specific value

14.8 Normal Probability Function without Sampling

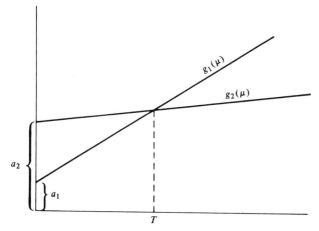

FIGURE 14.8.1 Two linear functions of μ.

of μ. On the other hand, if $g_1(\mu)$ and $g_2(\mu)$ depict two revenue functions, we assume that our problem is to choose a process between the two so that

$$g(\mu) = \text{Max. } \{g_1(\mu), g_2(\mu)\},$$

where $g(\mu)$ depicts the revenue of operating the chosen process for a specific value of μ.

If the value of μ is known, choosing a process which will minimize the cost or choosing the process which will maximize the revenue becomes a trivial problem, since the values of a_1, a_2, b_1, and b_2 are assumed to be known. However, the problem of choosing a process arises precisely because we assume that the value of μ is not known to us.

To illustrate the nature of our problem, let us assume that Figure 14.8.1 depicts two cost functions associated with two different processes. Then, when $\mu < T$, Process 1 will minimize the total cost; when $\mu > T$, Process 2 will minimize the total cost; and when $\mu = T$, the total cost will be the same for the two processes.

The problem confronting us, then, can be formulated in terms of a set of hypotheses. For example,

Null Hypothesis: $\mu \leq T$
Alternate Hypothesis: $\mu > T$.

The errors associated with accepting and rejecting the null hypothesis are given in Table 14.8.1. Assume that we will select Process 1 if the null hypothesis is accepted and Process 2 if the null hypothesis is rejected. Then, the loss functions for accepting and rejecting the null hypothesis are given in Table 14.8.2.

On the other hand, suppose we assume that $g_1(\mu)$ and $g_2(\mu)$ are the revenue functions. Then, Process 1 will yield a higher total expected revenue if

TABLE 14.8.1 Errors Associated with Accepting or Rejecting Null Hypothesis

Decision	State Pertaining to μ	
	$\mu \leq T$	$\mu > T$
A_1: ACCEPT NULL HYPOTHESIS	no error	Type II error
A_2: REJECT NULL HYPOTHESIS	Type I error	no error

TABLE 14.8.2 Loss Functions

Decision	State Pertaining to μ	
	$\mu \leq T$	$\mu > T$
A_1: ACCEPT NULL HYPOTHESIS	0	$g_1(\mu) - g_2(\mu)$
A_2: REJECT NULL HYPOTHESIS	$g_2(\mu) - g_1(\mu)$	0

$\mu > T$; Process 2 will yield a higher total expected revenue if $\mu < T$; and the total expected revenue will be the same if $\mu = T$. Thus, if the null hypothesis is $\mu \leq T$ and the alternate hypothesis is $\mu > T$, then accepting the null hypothesis implies that Process 2 should be chosen as the preferable action and rejecting the null hypothesis implies that Process 1 should be chosen as the preferable action. The loss function may again be given as shown in Table 14.8.2, except that $[g_2(\mu) - g_1(\mu)]$, for example, now depicts the difference between the two revenues.

EXAMPLE 14.8.1

A government agency is in the process of negotiating a service contract for the typists who are working in the agency. Approximately 10,000 typewriters will be covered by the service contract. The typewriter service company with whom the agency is negotiating has proposed the following two options for its consideration:

Option	Total Annual Charge
1	$\$130,000 + \$30,000\mu$
2	$\$220,000 + \$20,000\mu$

14.8 Normal Probability Function without Sampling

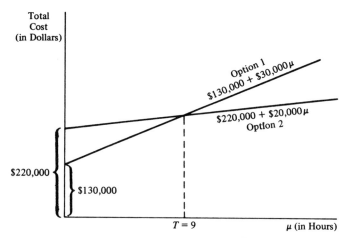

FIGURE 14.8.2 Cost functions for two options.

where μ depicts the average number of man-hours spent to service each typewriter during the year. Figure 14.8.2 shows that $T = 9$ hours, which means that the agency should choose Option 1 if $\mu < 9$ hours; it should choose Option 2 if $\mu > 9$ hours; and it should be indifferent to the two options if $\mu = 9$ hours.

If the agency knew the value of μ, then it would know which option to choose. However, the value of μ is not likely to be known, and, therefore, the problem confronting the agency is to guess the value of μ before it decides which of the two options to choose. This problem confronting the agency may now be formulated in terms of two hypotheses:

$$\text{Null Hypothesis: } \mu \leq 9$$
$$\text{Alternate Hypothesis: } \mu > 9.$$

The errors which may be committed in accepting or rejecting the null hypothesis are those given in Table 14.8.4, and the loss functions are given in Table 14.8.5.

TABLE 14.8.4 Errors Associated with Accepting or Rejecting the Null Hypothesis

Decision	State of Man-Hour Requirement	
	$\mu \leq 9$	$\mu > 9$
A_1: ACCEPT NULL HYPOTHESIS	no error	Type II error
A_2: REJECT NULL HYPOTHESIS	Type I error	no error

TABLE 14.8.5 Loss Functions

	State of Man-Hour Requirement	
Decision	$\mu \leq 9$	$\mu > 9$
A_1: ACCEPT NULL HYPOTHESIS	0	$-90{,}000 + 10{,}000\mu$
A_2: REJECT NULL HYPOTHESIS	$90{,}000 - 10{,}000\mu$	0

Table 14.8.5 indicates that if, for example, $\mu = 5$, then the amount of loss associated with rejecting the null hypothesis is given by $90{,}000 - 10{,}000\mu = 40{,}000$; or, if $\mu = 10$, the amount of loss associated with accepting the null hypothesis is given by $-90{,}000 + 10{,}000\mu = 10{,}000$.

Up to this point we have not stated explicitly whether or not we should assume that μ is a random variable. If we assume that μ is not a random variable, then the only reasonable decision function which we can devise is of the classical type.

We will now assume that μ is a normally distributed random variable with a known prior probability function. Thus, from now on μ will be denoted in boldface ($\boldsymbol{\mu}$). We will denote $E(\boldsymbol{\mu}) = \bar{\mu}_1$ and $V(\boldsymbol{\mu}) = \bar{\sigma}_1{}^2$. We can then proceed to devise a Bayes decision function with and without sampling. We will first consider the Bayes decision function without sampling.

As we have discussed previously, the first step in devising a Bayes decision function is to calculate the conditional expected losses for accepting and rejecting the null hypothesis. We will first define these losses for the prior distribution of $\boldsymbol{\mu}$.

Definition 14.8.1

Let $L(A_1|\mu)$ be the loss associated with accepting the null hypothesis when $\boldsymbol{\mu} = \mu$, and let $L(A_2|\mu)$ be the loss associated with rejecting the null hypothesis when $\boldsymbol{\mu} = \mu$. Then,

$$R_1(\boldsymbol{\mu}|A_1) = \int_T^\infty L(A_1|\mu) f_1(\mu)\, d\mu$$

$$R_1(\boldsymbol{\mu}|A_2) = \int_{-\infty}^T L(A_2|\mu) f_1(\mu)\, d\mu,$$

where $f_1(\mu)$ is the prior density function of $\boldsymbol{\mu}$.

The underlying idea for calculating $R_1(\boldsymbol{\mu}|A_1)$ and $R_1(\boldsymbol{\mu}|A_2)$ is the same as that for calculating $R_1(\pi|A_1)$ and $R_1(\pi|A_2)$ given in Definition 14.7.1. The only difference is that the assumption of a continuous probability function for $\boldsymbol{\mu}$ has forced us to use calculus to define the terms.

14.8 Normal Probability Function without Sampling

According to Theorem 14.3.1, then, the Bayes decision function without sampling is to choose A_1 or A_2 so that the conditional expected loss for the chosen action, $R_1(D)$, is equal to

$$R_1(D) = \text{Min.} \{R_1(\mathbf{\mu}|A_1), R_1(\mathbf{\mu}|A_2)\}.$$

Thus, it appears that we need to calculate both $R_1(\mathbf{\mu}|A_1)$ and $R_1(\mathbf{\mu}|A_2)$ in order to determine the Bayes decision function. However, if we assume that $f_1(\mu)$ is a normal density function, we do not need to calculate both $R_1(\mathbf{\mu}|A_1)$ and $R_1(\mathbf{\mu}|A_2)$, as shown by the following theorem.

Theorem 14.8.1

Let $\bar{\mu}_1$ be the mean of $f_1(\mu)$. Then, $R_1(D) = R_1(\mathbf{\mu}|A_1)$ if $\bar{\mu}_1 < T$, $R_1(D) = R_1(\mathbf{\mu}|A_2)$ if $\bar{\mu}_1 > T$, and $R_1(D) = R_1(\mathbf{\mu}|A_1) = R_1(\mathbf{\mu}|A_2)$ if $\bar{\mu}_1 = T$.

This theorem suggests that we should accept the null hypothesis if $\bar{\mu}_1 \leq T$ and reject it if $\bar{\mu}_1 > T$.

EXAMPLE 14.8.2

Let us return to Example 14.8.1 and assume that $f_1(\mu)$ is a normal density function with $\bar{\mu}_1 = 6$ hours. Then, Theorem 14.8.1 indicates that

$$R_1(D) = R_1(\mathbf{\mu}|A_1)$$

and, in turn, that we should accept the null hypothesis, which means that the government agency should choose Option 1 offered by the service company.

We can justify the validity of this conclusion in another context. Consider again the cost functions for the two options offered by the service company to the government agency. They are shown again here:

$$g_1(\mathbf{\mu}) = 130{,}000 + 30{,}000\mathbf{\mu}$$
$$g_2(\mathbf{\mu}) = 220{,}000 + 20{,}000\mathbf{\mu},$$

where $g_1(\mathbf{\mu})$ depicts the cost function for Option 1 and $g_2(\mathbf{\mu})$ depicts the cost function for Option 2. Then, the expected total cost for the two options is given as

$$E[g_1(\mathbf{\mu})] = E(130{,}000 + 30{,}000\mathbf{\mu}) = 130{,}000 + 30{,}000E(\mathbf{\mu})$$
and $\quad E[g_2(\mathbf{\mu})] = E(220{,}000 + 20{,}000\mathbf{\mu}) = 220{,}000 + 20{,}000E(\mathbf{\mu}).$

However, since by assumption $E(\mathbf{\mu}) = \bar{\mu}_1 = 6$, we have

$$E[g_1(\mathbf{\mu})] = 130{,}000 + (30{,}000 \times 6) = 410{,}000$$
and $\quad E[g_2(\mathbf{\mu})] = 220{,}000 + (20{,}000 \times 6) = 440{,}000.$

This means that by accepting Option 1 the government agency can expect to save \$30,000 a year.

Once the Bayes decision function without sampling is attained, the next step in our analysis is to ascertain the value of having perfect information.

Since the value of perfect information is equal to $R_1(D)$, which is the expected loss for the Bayes decision function without sampling, we can presumably apply Definition 14.8.1 to obtain the value. Strict adherence to these definitions would require using calculus; however, a simpler way of calculating $R_1(D)$ is provided by the following theorem.

Theorem 14.8.2
Let $h = (T - \tilde{\mu}_1)/\tilde{\sigma}_1$. Then,

$$R_1(D) = |b_1 - b_2|\tilde{\sigma}_1 G(|h|),$$

where $G(|h|)$ is the density of a normalized loss function.

Note that in this theorem $\tilde{\mu}_1$ and $\tilde{\sigma}_1$, respectively, denote the mean and the standard deviation of the prior density function $f_1(\mu)$. A table depicting the density $G(|h|)$ is given in Table B.8 in Appendix B.

EXAMPLE 14.8.3
Let us return to Example 14.8.2 and assume that $\tilde{\sigma}_1 = 3$. Then,

$$h = \frac{9 - 6}{3} = 1$$

and
$$G(|1|) = .08332.$$
Thus,
$$R_1(D) = |10,000|(3)(.08332) = 2500.$$

Thus, the expected value of the perfect information is $2500, which means that the sampling cost should never exceed $2500.

14.9 Bayes Decision Function for Normal Probability Function with Sampling

We will now consider the problem of sampling from a normal probability function. In devising a Bayes decision function with sampling for testing a hypothesis pertaining to π, we have found it convenient to devise the decision function by first finding the posterior probability function of π for the given value of P. This is also true when we devise a Bayes decision function with sampling for testing a hypothesis pertaining to μ of a normal probability function.

The process of deriving a posterior probability function of a normal probability function is much more difficult to illustrate than the process of deriving a binomial probability function, for example. We will, therefore, present some of the theorems involved without either proving or illustrating them.

Theorem 14.9.1

Let the prior density function of **µ** be

$$f_1(\mu) = \frac{1}{\sqrt{2\pi}\tilde{\sigma}_1} \exp\left\{-\frac{1}{2}\left(\frac{\mu - \tilde{\mu}_1}{\tilde{\sigma}_1}\right)^2\right\}.$$

Then, the posterior density function of **µ**, given that the sample has yielded $\bar{X} = \bar{x}$ is

$$f_2(\mu|\bar{x}) = \frac{1}{\sqrt{2\pi}\sqrt{\frac{\tilde{\sigma}_1^2 \sigma_{\bar{x}}^2}{\tilde{\sigma}_1^2 + \sigma_{\bar{x}}^2}}} \exp\left\{-\frac{1}{2}\left(\frac{\mu - \frac{\sigma_{\bar{x}}^2 \tilde{\mu}_1 + \tilde{\sigma}_1^2 \bar{x}}{\tilde{\sigma}_1^2 + \sigma_{\bar{x}}^2}}{\sqrt{\frac{\tilde{\sigma}_1^2 \sigma_{\bar{x}}^2}{\tilde{\sigma}_1^2 + \sigma_{\bar{x}}^2}}}\right)^2\right\}.$$

Suppose we let

$$\tilde{\mu}_2 = \frac{\sigma_{\bar{x}}^2 \tilde{\mu}_1 + \tilde{\sigma}_1^2 \bar{x}}{\tilde{\sigma}_1^2 + \sigma_{\bar{x}}^2}, \qquad \tilde{\sigma}_2^2 = \frac{\tilde{\sigma}_1^2 \sigma_{\bar{x}}^2}{\tilde{\sigma}_1^2 + \sigma_{\bar{x}}^2}.$$

Then,

$$f_2(\mu|\bar{x}) = \frac{1}{\sqrt{2\pi}\tilde{\sigma}_2} \exp\left\{-\frac{1}{2}\left(\frac{\mu - \tilde{\mu}_2}{\tilde{\sigma}_2}\right)^2\right\},$$

which is also an expression of a normal density function. Thus, we propose:

Theorem 14.9.2

Let **µ** be a normally distributed random variable with a prior mean of $\tilde{\mu}_1$ and a variance of $\tilde{\sigma}_1^2$. Then, the posterior probability function of **µ**, given $\bar{X} = \bar{x}$, is also normally distributed with a mean and a variance of, respectively,

$$\tilde{\mu}_2 = \frac{\sigma_{\bar{x}}^2 \tilde{\mu}_1 + \tilde{\sigma}_1^2 \bar{x}}{\tilde{\sigma}_1^2 + \sigma_{\bar{x}}^2}; \qquad \tilde{\sigma}_2^2 = \frac{\tilde{\sigma}_1^2 \sigma_{\bar{x}}^2}{\tilde{\sigma}_1^2 + \sigma_{\bar{x}}^2}.$$

EXAMPLE 14.9.1

Let us return to the problem of the government agency which must choose one of two options offered by a typewriter service company. Assume that the agency has decided to select a simple random sample of 9 typewriters and simulate 1 year's use of these typewriters in a short period of time and then determine the man-hours required to service these typewriters during the simulation period. Assume also that the service company will service these typewriters free of charge.

Suppose now that the average time required to service these 9 typewriters is 10 hours. Thus, the sample yields $\bar{X} = 10$ hours. Then, the posterior mean and the variance for the average service time for all 10,000 typewriters can be calculated as

$$\tilde{\mu}_2 = \frac{(1)(6) + (9)(10)}{9+1} = 9.6$$

$$\tilde{\sigma}_2^2 = \frac{(9)(1)}{9+1} = .9,$$

where $\sigma_{\bar{x}}^2 = \tilde{\sigma}_1^2/n = 9/9 = 1$.

Once the posterior probability function of μ has been obtained, as shown in Example 14.9.1, a Bayes decision function can be devised by first calculating the conditional expected losses for accepting and rejecting the null hypothesis. Extending Definition 14.7.1, we have:

Definition 14.9.1

Let $R_2(\mu|A_1,\bar{x})$ and $R_2(\mu|A_2,\bar{x})$ be the conditional expected losses for accepting and rejecting the null hypothesis, given that the sample yields $\bar{X} = \bar{x}$. Then,

$$R_2(\mu|A_1,\bar{x}) = \int_T^\infty L(A_1|\mu) f_2(\mu|\bar{x}) \, d\mu$$

and

$$R_2(\mu|A_2,\bar{x}) = \int_{-\infty}^T L(A_2|\mu) f_2(\mu|\bar{x}) \, d\mu.$$

A Bayes decision function with sampling may now be devised by extending Theorem 14.3.2.

Theorem 14.9.3

Let $R_2(\mu|A_1,\bar{x})$ and $R_2(\mu|A_2,\bar{x})$ be the posterior conditional expected losses for accepting and rejecting the null hypothesis for the given $\bar{X} = \bar{x}$. Then, a decision function is a Bayes decision function if it is chosen so that

$$R_2(D|\bar{x}) = \text{Min.} \{R_2(\mu|A_1,\bar{x}), R_2(\mu|A_2,\bar{x})\},$$

where $R_2(D|\bar{x})$ depicts the conditional expected loss for the particular chosen action, when the sample yields $\bar{X} = \bar{x}$.

Since the underlying logic of Theorem 14.9.3 is the same as that of Theorem 14.3.2, we will not elaborate on it again here. However, there is one difference in applying the two theorems. When we used Theorem 14.3.2 in connection with a discrete probability function, we had to calculate both $R_2(\pi|A_1, P = p)$ and $R_2(\pi|A_2, P = p)$. However, we do not have to calculate both $R_2(\mu|A_1,\bar{x})$ and $R_2(\mu|A_2,\bar{x})$ when we use Theorem 14.9.3, as we will show in the following theorem.

Theorem 14.9.4

Let $\tilde{\mu}_2$ be the posterior mean of $f_2(\mu|\bar{x})$. Then, $R_2(D|\bar{x}) = R_2(\mu|A_1,\bar{x})$ if $\tilde{\mu}_2 < T$; $R_2(D|\bar{x}) = R_2(D|A_2,\bar{x})$ if $\tilde{\mu}_2 > T$; and $R_2(D|\bar{x}) = R_2(\mu|A_1,\bar{x}) = R_2(\mu|A_2,\bar{x})$ if $\tilde{\mu}_2 = T$.

If we were to translate the Bayes decision function given in Theorem 14.9.3 according to Theorem 14.9.4, then,

> The null hypothesis should be accepted if $\tilde{\mu}_2 \leq T$;
> rejected if $\tilde{\mu}_2 > T$.

EXAMPLE 14.9.2

Let us now return to Example 14.9.1, in which we have assumed that a sample of 9 has yielded $\bar{X} = 10$ and, have calculated that $\tilde{\mu}_2 = 9.6$ and $\tilde{\sigma}_2 = \sqrt{.9}$. Recall that the original null hypothesis was that $\mu \leq 9$ and in turn, that $T = 9$. Thus, the null hypothesis should be rejected and Option 2 should be chosen instead of Option 1.

Even though a Bayes decision function can be ascertained without calculating $R_2(D|\bar{x})$ explicitly, if we wish to calculate $R_2(D|\bar{x})$, however, this can be done by applying Definition 14.9.1, which, of course, would require use of calculus. On the other hand, by extending Theorem 14.9.3, we have:

Theorem 14.9.5

Let $h = (T - \tilde{\mu}_2)/\tilde{\sigma}_2$. Then,

$$R_2(D|\bar{x}) = |b_1 - b_2|\tilde{\sigma}_2 G(|h|),$$

where $G(|h|)$ is the density of the normalized loss function.

EXAMPLE 14.9.3

Let us now return to Example 14.9.2, in which $T = 9$, $\tilde{\mu}_2 = 9.6$, and $\tilde{\sigma}_2{}^2 = \sqrt{.9} \cong .949$. Then,

$$R_2(D|\bar{X} = 10) = |10{,}000|(.949)\left[G\left(\left|\frac{9 - 9.6}{.949}\right|\right)\right]$$
$$= (9490 \times .1313)$$
$$= 1227.$$

14.10 Preposterior Analysis of a Normal Probability Function

In the preceding section we have illustrated how we can devise a Bayes decision function for a given sample size. On the other hand, we might at the same time wish to evaluate whether it is worthwhile to take the sample of a given size at all. Such an analysis requires that we calculate $R_1(D)$ and $R_2(D)$, where

$$R_2(D) = \int_{-\infty}^{\infty} R_2(D|\bar{x}) f(\bar{x}) \, d\bar{x},$$

since $E(S_n)$, the expected value of the sample information, is the difference between $R_1(D)$ and $R_2(D)$. The manner of ascertaining $R_2(D)$, shown in the preceding equation, is analogous to that suggested in Theorem 14.4.1 for testing the hypothesis pertaining to π. If we are to calculate $R_2(D)$ in the same manner now, however, it would entail using cumbersome calculus manipulations.

Recall that in testing the hypothesis pertaining to π (Theorem 14.4.1) we were also able to ascertain the expected value of the sample information without explicitly calculating $R_2(D)$. This required us first to derive the preposterior probability function of π. Similarly, we will now illustrate how we can obtain the expected value of the sample information without explicitly calculating $R_2(D)$, but by first ascertaining the preposterior probability function of μ.

In connection with testing the hypothesis pertaining to π, we have defined the preposterior probability function of π as the probability function of $\tilde{\pi}$, where $\tilde{\pi}$ depicts the expected value of the posterior probability function of π, given $P = p$. Thus, the preposterior probability function of π is the probability function of $\tilde{\pi}$, where $\tilde{\pi} = E[f_2(\pi|P=p)]$, and where $f_2(\pi|P=p)$ depicts the posterior probability function of π, given $P = p$. Similarly, we may propose:

Definition 14.10.1

Let $\tilde{\mu} = E[f_2(\mu|\bar{x})]$. Then, the probability function of $\tilde{\mu}$ is called the preposterior probability function of μ.

The preposterior probability function of μ can also be defined in another context. Since

$$\tilde{\mu}_2 = E[f_2(\mu|\bar{x})],$$

we know that $\tilde{\mu} = \tilde{\mu}_2$. Thus,

Definition 14.10.2

The probability function of $\tilde{\mu}_2$ is called the preposterior probability function of μ.

We now propose, without proof:

Theorem 14.10.1

Let μ be a normally distributed random variable with a prior mean and a variance of $\tilde{\mu}_1$ and $\tilde{\sigma}_1^2$, respectively. Then, the preposterior probability function of μ for the given sample size n is also normally distributed with a mean and a variance, respectively, of

$$\tilde{\mu}_3 = \tilde{\mu}_1$$

and

$$\tilde{\sigma}_3^2 = \tilde{\sigma}_1^2 \left(\frac{\tilde{\sigma}_1^2}{\tilde{\sigma}_1^2 + \sigma_{\bar{x}}^2} \right).$$

EXAMPLE 14.10.1

In the preceding example we have assumed that $\tilde{\mu}_1 = 6$, $\tilde{\sigma}_1 = 3$, and we must select a sample of size 9. Then,

$$\tilde{\mu}_3 = 6,$$

$$\tilde{\sigma}_3^2 = 9\left(\frac{9}{9+1}\right) = 8.1,$$

and, in turn, $\tilde{\sigma}_3 = \sqrt{8.1} = 2.846$.

Having ascertained the preposterior probability function of $\mathbf{\mu}$, we can find the expected value of the sample information by applying the following reasoning. Suppose the Bayes decision function without sampling, for example, leads us to accept the null hypothesis. Then, the information obtained from the sample is of value only when it leads us to reject the null hypothesis. This means that if $\tilde{\mu}_1 \leq T$, the sample information is of value if it yields $\tilde{\mu}_2 > T$. Conversely, suppose the Bayes decision function without sampling leads us to reject the null hypothesis. Then, the sample information is of value when it leads us to accept the null hypothesis. Thus, if $\tilde{\mu}_1 > T$, then the sample information is of value if it yields $\tilde{\mu}_2 \leq T$.

Suppose now that $\tilde{\mu}_1 \leq T$, but $\tilde{\mu}_2 > T$. Then, what is the value of the information thus obtained? We argue that it must be equivalent to $L(A_1|\mathbf{\mu} = \tilde{\mu}_2)$, since the sample information will leads us to avoid incurring this particular loss. Conversely, if $\tilde{\mu}_1 > T$, but $\tilde{\mu}_2 \leq T$, then the value of the information is equivalent to $L(A_2|\mathbf{\mu} = \tilde{\mu}_2)$. Thus, we propose:

Theorem 14.10.2

Let $E(S_n)$ depict the expected value of the sample information. Then,

$$E(S_n) = \int_T^\infty L(A_1|\tilde{\mu}_2) f_2(\tilde{\mu}_2)\, d\tilde{\mu}_2$$

whenever $\mu_1 \leq T$, and

$$E(S_n) = \int_{-\infty}^T L(A_2|\tilde{\mu}_2) f_2(\tilde{\mu}_2)\, d\tilde{\mu}_2$$

whenever $\mu_1 > T$.

Although the value of $E(S_n)$ in this theorem can be calculated by means of calculus, the following theorem provides a simpler method.

Theorem 14.10.3

Let $h = (T - \tilde{\mu}_3)/\tilde{\sigma}_3$. Then,

$$E(S_n) = |b_1 - b_2|\tilde{\sigma}_3 G(|h|),$$

where $G(|h|)$ is the density of the normalized loss function.

Bayes Decision Functions

This procedure is quite similar to those shown in Theorems 14.8.2 and 14.9.5.

EXAMPLE 14.10.2

Let us return to Example 14.10.1. Recall that $\mu_1 = 6$, $T = 9$, and $\tilde{\sigma}_3 = 2.846$. Thus,

$$E(S_9) = |10,000|(2.846)G\left(\left|\frac{9-6}{2.846}\right|\right)$$
$$= (2.846)(.08019) = 228.25,$$

which depicts the expected value of the sample information.

What is the significance of the preceding analysis for the government agency in question? It indicates that the agency should not spend more than $228.25 for the information obtainable from the sample size of 9.

EXERCISES

*42. A government agency is in the process of negotiating a service contract for the typewriters being used in the agency. Approximately 10,000 typewriters will be covered by the contract. The typewriter service company with whom the agency is negotiating has proposed the following two options for its consideration: (a) a flat amount of $300,000 per year, (b) $30,000 multiplied by the average number of man-hours spent per typewriter during the year. Formulate the appropriate loss function for the problem facing the agency. If you were to establish a set of hypotheses, how would you formulate them? Can you solve the problem without formulating these hypotheses?

*43. Studying the past records, the agency discussed in Exercise 42 has come to the following conclusions: The mean service time for these typewriters is 8 hours, and the standard deviation is 4 hours. Suppose the agency does not want to obtain any new information through sampling. Then, which of the two options described in Exercise 42 should it accept?

*44. Suppose, however, that the agency discussed in Exercises 42 and 43 wants to obtain some new information through sampling. Then, what is the maximum amount that it should be willing to pay for such new information?

*44. Suppose now that the government agency discussed in Exercise 42 has decided to select a simple random sample of 16 typewriters and to determine what the mean service time is for these 16 typewriters by simulating 1 year's use of them. The sample yields a mean of 15 hours. In the light of this information, how would you revise the mean service time and the standard deviation for the service time for 10,000 typewriters which are being used by the agency?

*45. In the light of the new information obtained in Exercise 44, which of the two options should the government agency accept now?

14.10 Preposterior Analysis

*46. Before it decides to take a sample of 16 typewriters, suppose the agency would like to know how valuable the information obtained from such a sample would be. How would you ascertain such a value? Calculate the value in question.

*47. In the light of your calculations for Exercise 46, would you recommend to the agency that the sample of 16 typewriters is reasonably adequate?

48. A commercial bank in California has 100,000 checking depositors. At present the bank charges 15 cents per check in each monthly statement for its services. If a depositor, for example, writes 10 checks during a month, he will be charged $1.50 during the month. The bank officers are currently debating whether to continue with the present service-charge policy or to adopt a new one. A new policy proposed is that a depositor be charged $1, regardless of the number of checks written during the month, plus 10 cents per check written during the month. The officers of the bank agree that they should choose the policy which will maximize the total amount of monthly service charges for the next twelve months. Formulate the appropriate loss function for the problem facing the bank. If you were to establish a set of hypotheses, how would you formulate them? Can you solve the problem without formulating these hypotheses?

49. Refer to Exercise 48 and assume that the bank officers have come to the following conclusions. The depositors write, on the average, 25 checks per month with a standard deviation of 10 checks. Assume that the bank officers do not want to obtain any new information through sampling. Then, which of the two policies should they adopt; that is, should they keep the present service-charge policy or introduce a new service-charge policy?

50. Referring to Exercise 48, suppose the bank officers want to obtain new information through sampling. Then, what is the maximum amount that it should be willing to pay for such new information?

51. Refer to Exercise 48 and suppose the bank officers have directed the research division of the bank to take a simple random sample of 1000 checking accounts in order to calculate the sample mean and standard deviation. The actual sample has yielded a mean of 15 checks with a standard deviation of 7 checks. In the light of this new information, how should you revise the mean and standard deviation for the number of checks written by 100,000 checking depositors during a month?

52. Referring to Exercise 48, in the light of the new information obtained in Exercise 51, which of the two policies should the bank accept?

53. Referring to Exercise 48, suppose that the bank officers, before deciding upon the sample size of 1000, wanted to know how valuable would be the information obtained from their sample. How would you go about ascertaining such a value? Calculate the value in question.

54. In the light of your calculations for Exercise 53, would you recommend to the officers of the bank that the sample of 1000 is reasonably adequate?

Chapter 15
Chi-Square, Student-t, and F Distributions

In the preceding chapters we have observed that the normal probability function plays a very important role in statistical inference. However, a number of probability functions, derived from a normal probability function, also play very important roles in statistical inference. In this chapter we will consider three such probability functions and, in turn, explore their usefulness.

15.1 Chi-Square Distribution

Let us denote Z as a normally distributed random variable with $E(Z) = 0$ and $V(Z) = 1$. Thus, Z is the standard normal variable. Suppose now we define that

$$Y = Z^2.$$

Then, Y must be a random variable. Y is said to be a *chi-square distributed random variable* with 1 degree of freedom. Thus,

Definition 15.1.1

Let Z be the normally distributed random variable with $E(Z) = 0$ and $V(Z) = 1$. Then, $Y = Z^2$ is the *chi-square distributed random variable* with 1 degree of freedom.

Let us now explore two apparent relationships between Y and Z, the first of which is that Y is a function of Z, and the second of which is that both positive and negative values of Z will correspond to the same positive value of Y. The relationship between Y and Z is illustrated in Figure 15.1.1. Thus, if Z assumes either 2 or -2, for example, Y assumes 4. Perhaps the following illustration will illuminate the relationship between Y and Z further.

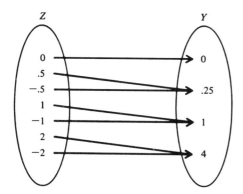

FIGURE 15.1.1 Functional relationship between Y and Z.

EXAMPLE 15.1.1

Let Y be a chi-square distributed random variable with 1 degree of freedom. Suppose we wish to find $P(Y \geq 4)$. We know that $Y \geq 4$ if either $Z \geq 2$ or $Z \leq -2$. Thus,

$$P(Y \geq 4) = P(Z \leq -2) + P(Z \geq 2)$$
$$= .0228 + .0228$$
$$= .0456.$$

Suppose we wish to find $P(1 \leq Y \leq 4)$. This probability can be found by calculating the following:

$$P(1 \leq Y \leq 4) = P(-2 \leq Z \leq -1) + P(1 \leq Z \leq 2)$$
$$= .1359 + .1359$$
$$= .2718.$$

The relationship between Y and Z is also illustrated in Figure 15.1.2, which shows their respective density functions. In this figure $P(Y \geq 4) = P(Z \leq -2) + P(Z \geq 2) = .0456$. It also illustrates that Y cannot be negative, since it is always a squared value.

The chi-square distributed random variable which we have just considered is a special case within the family of chi-square distributed random variables. A more general definition of a chi-square distributed random variable is now offered.

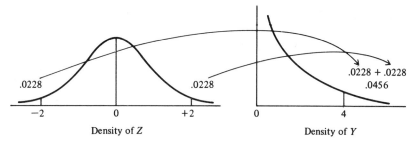

FIGURE 15.1.2 Density function of Y and Z.

Definition 15.1.2

Let Z_i be a statistically independent, normally distributed random variable with $E(Z_i) = 0$ and $V(Z_i) = 1$ for $i = 1, \ldots, n$. Then,

$$Y = Z_1^2 + \cdots + Z_n^2$$

is the chi-square distributed random variable with n degrees of freedom.

Usually, instead of using Y to depict the random variable, we use χ^2. We will follow this convention hereafter.

The density function of χ^2, when it is graphed, will appear as shown in Figure 15.1.3, where n depicts the degrees of freedom. In this illustration, as n becomes larger, the density function of χ^2 approaches that of a normal probability function. This is really not a strange phenomenon if we consider the central limit theorem. We might also point out without providing a proof that the expected value of χ^2 is n.

A table of chi-square values is provided in Appendix B (see Table B.2). We will find in the table, for example, the same values shown in the chart

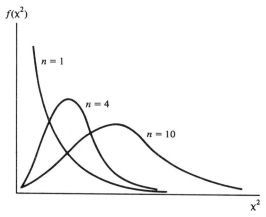

FIGURE 15.1.3 Density function of χ^2.

d.f.	.05
1	3.84
2	5.99
3	7.81
4	9.49
⋮	⋮
10	18.31
⋮	⋮
20	31.41

here. These values indicate, for example, that

$$P(\chi_{10}^2 \geq 18.31) = .05,$$

where the subscript for χ^2 denotes the degree of freedom.

It may appear that the chi-square distributed random variable has a limited usefulness, because not many empirical phenomena can be portrayed by it. However, we can convert a non-chi-square random variable into a chi-square distributed random variable by an appropriate transformation. One such transformation is shown by the following theorem.

Theorem 15.1.1

Let X_i be a statistically independent, normally distributed random variable with $E(X_i) = \mu_i$ and $V(X_i) = \sigma_i^2$. Then,

$$Y = \left(\frac{X_1 - \mu_1}{\sigma_1}\right)^2 + \cdots + \left(\frac{X_n - \mu_n}{\sigma_n}\right)^2$$

is a chi-square distributed random variable with n degrees of freedom.

The validity of this theorem can be shown as follows. Since X_i is a normally distributed random variable, we have

$$\frac{X_i - \mu_i}{\sigma_i} = Z_i.$$

Thus,

$$\left(\frac{X_1 - \mu_1}{\sigma_1}\right)^2 + \cdots + \left(\frac{X_n - \mu_n}{\sigma_n}\right)^2 = Z_1^2 + \cdots + Z_n^2$$

and, since X_i is statistically independent, Z_i must also be statistically independent. Consequently, Y must be the chi-square distributed random variable with n degrees of freedom.

EXAMPLE 15.1.2

Assume that we are given 25 pennies, 25 nickels, 25 dimes, and 25 quarters. We are to use these 100 coins to generate empirically the probability function of χ^2 with 4 degrees of freedom.

We may proceed to accomplish the task in the following manner. Let P_1 be the percentage of "heads" obtained by tossing 25 pennies; P_2 for tossing 25 nickels; and P_3 and P_4 for tossing 25 dimes and quarters, respectively. Then,

$$E(P_1) = E(P_2) = E(P_3) = E(P_4) = .5$$
and
$$V(P_1) = V(P_2) = V(P_3) = V(P_4) = .01,$$

and, in turn, $\sigma_{P_i} = .1$ for $i = 1, \ldots, 4$. Now let

$$U_1 = \frac{P_1 - .5}{.1}, \ldots, U_4 = \frac{P_4 - .5}{.1}.$$

Then, $U_1, \ldots, U_4$ are approximately normally distributed random variables according to the central limit theorem, and, in turn, $E(U_i) = 0$ and $V(U_i) = 1$, which means that

$$W = U_1^2 + U_2^2 + U_3^2 + U_4^2$$

should be approximately the chi-squared distributed random variable with 4 degrees of freedom.

One such experiment was carried out by tossing 100 coins 100 times, thereby obtaining 100 values of W. The cumulative frequency function of W thus generated is shown in Figure 15.1.4. If the reader compares the shape of this empirically generated cumulative frequency function with that of the

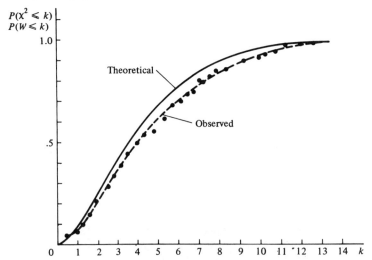

FIGURE 15.1.4 Theoretical and observed cumulative frequency for W.

actual cumulative probability function of χ^2 with 4 degrees of freedom, he will find that the former is a good approximation of the latter.

The following theorem pertaining to a chi-square distribution will be useful for our subsequent discussion.

Theorem 15.1.2

Let χ_m^2 and χ_n^2 be chi-square distributed random variables with m and n degrees of freedom, respectively. Then,

$$Y = \chi_m^2 + \chi_n^2$$

is a chi-square distributed random variable with $m + n$ degrees of freedom.

The validity of this theorem can be shown in the following manner. Suppose we let

$$\chi_m^2 = Z_1^2 + \cdots + Z_m^2$$

and

$$\chi_n^2 = Z_{m+1}^2 + \cdots + Z_{m+n}^2.$$

Then,

$$Y = Z_1^2 + \cdots + Z_m^2 + Z_{m+1}^2 + \cdots + Z_{m+n}^2,$$

which by definition is the chi-square distributed random variable with $m + n$ degrees of freedom.

EXERCISES

*1. Let Y be a chi-square distributed random variable with 1 degree of freedom. Using the table of normal probability function in Appendix B (Table B.1), find (a) $P(Y \geq 2.25)$, (b) $P(1 \leq Y \leq 2.25)$, (c) $P(1 \leq Y \leq 9.00)$, and (d) $P(Y \geq 9.00)$.

*2. Let Y be a chi-square distributed random variable with 1 degree of freedom. Using Table B.1 in Appendix B, find the value of k such that (a) $P(Y \geq k) = .95$, (b) $P(Y \geq k) = .50$, (c) $P(Y \geq k) = .90$, and (d) $P(Y \geq k) = .95$.

*3. Let Y be a chi-square distributed random variable with 1 degree of freedom. Using the table of chi-square probability functions (Table B.2 in Appendix B), find the value of k such that (a) $P(Y \geq k) = .975$, (b) $P(Y \geq k) = .050$, (c) $P(Y \geq k) = .025$, and (d) $P(Y \geq k) = .005$.

.4. Let Y be a chi-square distributed random variable with 10 degrees of freedom. Using Table B.2 in Appendix B, find the value of k such that (a) $P(Y \geq k) = .975$, (b) $P(Y \geq k) = .050$, (c) $P(Y \geq k) = .025$, and (d) $P(Y \geq k) = .001$.

*5. Let Y be a chi-square distributed random variable with 20 degrees of freedom. Using Table B.2 in Appendix B, find the value of k such that (a) $P(Y \geq k) = .975$, (b) $P(Y \geq k) = .050$, (c) $P(Y \geq k) = .025$, and (d) $P(Y \geq k) = .005$.

*6. Assume that X_1 and X_2 are statistically independent, normally distributed random variables with

and
$$E(X_1) = 10, \quad V(X_1) = 4$$
$$E(X_2) = 20, \quad V(X_2) = 9.$$

Suppose we propose that

$$Y = \left(\frac{X_1 - 10}{2}\right)^2 + \left(\frac{X_2 - 20}{3}\right)^2.$$

Then, find the value of k such that (a) $P(Y \geq k) = .05$, and (b) $P(Y \geq k) = .01$.

7. Let $X_1, \ldots, X_{100}$ be statistically independent, normally distributed random variables. Assume also that $X_1, \ldots, X_{100}$ all have an identical probability function with $E(X_i) = 10$ and $V(X_i) = 25$. Let

$$Y = \left(\frac{X_1 - 10}{5}\right)^2 + \cdots + \left(\frac{X_{100} - 10}{5}\right)^2.$$

Then, find the value of k such that (a) $P(Y \geq k) = .05$, and (b) $P(Y \geq k) = .01$.

15.2 Chi-Square and Statistical Inference Pertaining to Variance

In the preceding chapters we have been concerned only with statistical inference pertaining to π or μ of a probability function. We will now illustrate that a statistical inference pertaining to the variance of the probability function can be resolved by utilizing a chi-square probability function. This application is based on the validity of the following theorem.

Theorem 15.2.1
Let $X_1, \ldots, X_n$ be a simple random sample of size n from a normal probability function. Then,

$$Y = \left(\frac{X_1 - \bar{X}}{\sigma}\right)^2 + \cdots + \left(\frac{X_n - \bar{X}}{\sigma}\right)^2$$

is the chi-square distributed random variable with $n - 1$ degrees of freedom.

The validity of this theorem can be illustrated as follows. Let

$$W = \left(\frac{X_1 - \mu}{\sigma}\right)^2 + \cdots + \left(\frac{X_n - \mu}{\sigma}\right)^2.$$

Then, according to Theorem 15.1.1, W is the chi-square distributed random variable with n degrees of freedom. We can, however, decompose W into

$$W = \left(\frac{(X_1 - \bar{X}) + (\bar{X} - \mu)}{\sigma}\right)^2 + \cdots + \left(\frac{(X_n - \bar{X}) + (\bar{X} - \mu)}{\sigma}\right)^2$$
$$= \left(\frac{X_1 - \bar{X}}{\sigma}\right)^2 + \cdots + \left(\frac{X_n - \bar{X}}{\sigma}\right)^2 + \left(\frac{\bar{X} - \mu}{\sigma/\sqrt{n}}\right)^2$$

The last term $[(\bar{X} - \mu)/\sigma/\sqrt{n}]^2$ is obviously Z^2, which is the chi-square distributed random variable with 1 degree of freedom.

Now let

$$W = Y + \left(\frac{\bar{X} - \mu}{\sigma/\sqrt{n}}\right)^2.$$

It can be shown that $[(\bar{X} - \mu)/\sigma/\sqrt{n}]^2$ is statistically independent of Y. Since χ^2 is additive, we have

$$W = \chi^2_{n-1} + \chi_1^2.$$

Thus, Y is the chi-square distributed random variable with $n - 1$ degrees of freedom.

The preceding theorem can be described in another form. Recall that in Chapter 11 we defined:

$$s^2 = \frac{(X_1 - \bar{X})^2 + \cdots + (X_n - \bar{X})^2}{n - 1}.$$

Thus,

$$(X_1 - \bar{X})^2 + \cdots + (X_n - \bar{X})^2 = (n - 1)s^2.$$

If we substitute this equation in Theorem 15.2.1, we have

Theorem 15.2.2

Let $X_1, \ldots, X_n$ be a simple random sample of size n from a normal probability function, and let

$$s^2 = \frac{(X_1 - \bar{X})^2 + \cdots + (X_n - \bar{X})^2}{n - 1}.$$

Then,

$$Y = \frac{(n - 1)s^2}{\sigma^2}$$

is the chi-square distributed random variable with $n - 1$ degrees of freedom.

We will now apply Theorem 15.2.2 to the problem of statistical inference pertaining to the variance of a probability function.

EXAMPLE 15.2.1

Let X be a random variable associated with the diameters of ball bearings produced by a certain process, and assume that X is normally distributed. The process is assumed to be working satisfactorily if the variance of the diameters is .01 inches or less.

Thus, we establish the following hypotheses:

Null Hypothesis: $\sigma^2 \leq .01$ inches
Alternate Hypothesis: $\sigma^2 > .01$ inches.

Suppose now we plan to select a simple random sample of 11 in order to decide whether to accept or reject the null hypothesis. Our immediate task is then to devise a decision function for the sample space before we actually take the sample.

In testing a hypothesis pertaining to the mean of a probability function, the reader may recall that the decision function has been specified in terms of the sample mean. It must be also intuitively apparent that, in testing a hypothesis pertaining to the variance of a probability function, we should specify the decision function in terms of the sample variance. This intuitive reasoning can also be justified on a theoretical ground.

Recall also that in testing the following set of hypotheses,

$$\text{Null Hypothesis:} \quad \mu \leq \mu_0$$
$$\text{Alternate Hypothesis:} \quad \mu > \mu_0,$$

the uniformly most powerful decision function for $\alpha \leq \alpha_0$ has been to accept the null hypothesis if $\bar{X} \leq \ell$ and to reject it otherwise, where the critical value ℓ has been determined such that

$$P(\bar{X} > \ell | \mu = \mu_0) = \alpha_0.$$

Our immediate question is, then, "Does a uniformly most powerful decision function for $\alpha \leq \alpha_0$ exist for the following set of hypotheses,

$$\text{Null Hypothesis:} \quad \sigma^2 \leq \sigma_0^2$$
$$\text{Alternate Hypothesis:} \quad \sigma^2 > \sigma_0^2,$$

and, if so, how can we characterize such a decision function?" We will answer this question first by stating that a uniformly most powerful decision function does exist and, then, by stating that the decision function in question is to accept the null hypothesis if $\mathbf{s}^2 \leq \ell$ and to reject it otherwise, where the critical value ℓ is determined such that

$$P(\mathbf{s}^2 > \ell | \sigma^2 = \sigma_0^2) = \alpha_0.$$

Although we do not plan to give proofs for the preceding propositions, we ask the reader to note the similarity between the decision functions pertaining to σ^2 just discussed and those pertaining to μ, which we have elaborated in Chapter 13.

Let us now return to the illustrative problem and proceed with determining ℓ such that

$$P(\mathbf{s}^2 > \ell | \sigma^2 = .01) = .05.$$

The value of ℓ, then, can be ascertained in the following manner. Let

$$P(\chi_{10}^2 > k) = .05.$$

Since, according to Theorem 15.2.2, $\chi_{10}^2 = 10\mathbf{s}^2/\sigma^2$ and, in turn, we have

$$s^2 = \left(\frac{\sigma^2}{10}\right)\chi_{10}^2,$$

$$P\left[\left(\frac{\sigma^2}{10}\right)\chi_{10}^2 > \left(\frac{\sigma^2}{10}\right)k\middle|\sigma^2 = .01\right] = P(s^2 > \ell|\sigma^2 = .01) = .05.$$

Thus,

$$P(.001\chi_{10}^2 > .001k) = P(s^2 > \ell|\sigma^2 = .01) = .05.$$

From Table B.2 in Appendix B we find that $P(\chi_{10}^2 > 18.31) = .05$. Thus, $k = 18.31$ and, in turn,

$$\ell = .001k = .01831;$$

that is,

$$P(s^2 > .01831|\sigma^2 = .01) = .05.$$

The resulting decision function is to accept the null hypothesis if $s^2 \leq .01831$ and to reject it otherwise. This decision function will limit the α risk to .05, at most.

EXAMPLE 15.2.2

Suppose we wish to establish a confidence interval around the observed value of s^2 instead of testing a hypothesis pertaining to σ^2; for example, we wish to find the values of k_1 and k_2 such that

$$P(k_1 \leq \sigma^2 \leq k_2) = .95,$$

where .95 is the confidence coefficient. Assuming that the probability function sampled is normally distributed, the interval in question may be ascertained in the following manner. Let $\chi^2_{\nu,.975}$ and $\chi^2_{\nu,.025}$ be the constants so that

$$P(\chi_\nu^2 \geq \chi^2_{\nu,.975}) = .975$$
and
$$P(\chi_\nu^2 \geq \chi^2_{\nu,.025}) = .025,$$

where ν denotes the degrees of freedom. Thus,

$$P(\chi^2_{\nu,.975} \leq \chi_\nu^2 \leq \chi^2_{\nu,.025}) = .95.$$

Since $\chi_\nu^2 = (n-1)s^2/\sigma^2$, where $\nu = n - 1$, the preceding equation can be modified to

$$P\left[\chi^2_{n-1,.975} \leq \frac{(n-1)s^2}{\sigma^2} \leq \chi^2_{n-1,.025}\right] = .95.$$

On the other hand,

$$P\left[\frac{(n-1)s^2}{\sigma^2} \geq \chi^2_{n-1,.975}\right] = P\left[\sigma^2 \leq \frac{(n-1)s^2}{\chi^2_{n-1,.975}}\right]$$

and

$$P\left[\frac{(n-1)s^2}{\sigma^2} \leq \chi^2_{n-1,.025}\right] = P\left[\sigma^2 \geq \frac{(n-1)s^2}{\chi^2_{n-1,.025}}\right].$$

Combining the preceding two equations, we obtain

$$P\left[\frac{(n-1)s^2}{\chi^2_{n-1,.025}} \leq \sigma^2 \leq \frac{(n-1)s^2}{\chi^2_{n-1,.975}}\right] = .95,$$

which is an expression of a confidence interval.

To illustrate our procedure, let us assume for the problem given in Example 15.2.1 that a sample of 11 has yielded

$$s^2 = .1.$$

Then, $\nu = 11 - 1 = 10$ and

$$\chi^2_{10,.975} = 3.25$$
$$\chi^2_{10,.025} = 20.50.$$

Thus,

$$P\left[\frac{10s^2}{\chi^2_{10,.025}} \leq \sigma^2 \leq \frac{10s^2}{\chi^2_{10,.975}}\right] = P\left[\frac{(10)(.1)}{20.50} \leq \sigma^2 \leq \frac{(10)(.1)}{3.25}\right]$$
$$= P(.0488 \leq \sigma^2 \leq .308) = .95,$$

and, in turn, $k_1 = .0488$ and $k_2 = .308$.

EXERCISES

*8. Let X be a normally distributed random variable. Suppose you propose two hypotheses pertaining to the variance of X:

$$\text{Null Hypothesis: } \sigma^2 \leq .6$$
$$\text{Alternate Hypothesis: } \sigma^2 > .6.$$

You plan to accept one of the two hypotheses after you take a simple random sample of 16 from the probability function of X. Formulate a decision function which is uniformly most powerful for $\alpha \leq .5$.

*9. Using the situation in Exercise 8, suppose your sample observations yield a sample variance of 1.2. Do you accept or reject the null hypothesis?

*10. Let X be a normally distributed probability function. Suppose you propose two hypotheses pertaining to the variance of X:

$$\text{Null Hypothesis: } \sigma^2 \geq 5$$
$$\text{Alternate Hypothesis: } \sigma^2 < 5.$$

You plan to accept one of the two hypotheses by taking a sample of 26 observations from the probability function of X. Formulate a decision function which is uniformly most powerful for $\alpha = .05$.

*11. Using the situation in Exercise 10, suppose your sample of 26 observations yields a sample variance of 4. Do you accept or reject the null hypothesis?

12. Suppose your company produces a certain type of steel bar. Even though you want all the steel bars to be of the same length, you realize that the production process cannot be perfect. Therefore, you are willing to tolerate a certain amount of variance in the length of these bars; however, you want to stop the process whenever the variance exceeds .01 inches. You have decided to control the process by randomly selecting 16 bars from the process. If you wish to accept no more than a .05 probability of erroneously stopping the process, then, what should be your optimal decision function?

*13. A simple random sample of 16 from a normal probability function has yielded

$$\bar{X} = 20, \quad \text{and } s^2 = 4.$$

Establish a 95 percent confidence interval for the variance of the probability function.

14. A simple random sample of 26 from a normal probability function has yielded

$$\bar{X} = 100, \quad \text{and } s^2 = 40.$$

Establish a 95 percent confidence interval for the variance of the probability function.

*15. The following 20 sample observations depict monthly water consumption of households in a city (in 100 cubic feet). A simple random sample of 20 have been selected from the record books of the water company. (Note that these figures have been given also in Chapter 11, Exercise 13.)

13	12	6	8
6	21	10	12
19	5	5	16
4	11	18	10
11	10	9	14

Assume that the water consumption of households in the city may be approximated by a normal probability function. Establish a 95 percent confidence interval for the variance of water consumption in the city.

16. A cannery has received a truckload of peaches and wants to estimate the variance of the diameters of peaches in the truckload. A simple random sample of 26 peaches has revealed the following figures. (Note that these figures have been given also in Chapter 11, Exercise 12.)

2.625	2.750	3.000
2.625	2.750	2.625
2.875	2.625	2.750
3.000	3.000	2.875
2.625	3.125	2.750
3.000	2.875	2.750
2.750	2.625	2.750
2.875	2.875	2.875
2.750	2.875	

Assume that the diameters of these peaches are normally distributed. Establish a 95 percent confidence interval for the diameters of the peaches in the truckload.

15.3 Student-t Distribution

Let us denote Z_i as statistically independent, standard, normally distributed random variables. Let

$$Y = \frac{Z\sqrt{\nu}}{\sqrt{Z_1^2 + \cdots + Z_\nu^2}}.$$

Then, Y is said to be the *student-t distributed random variable* with ν degrees of freedom. Since

$$Z_1^2 + \cdots + Z_\nu^2$$

is the chi-square distributed random variable with ν degrees of freedom, we can also define the *student-t distributed random variable* in terms of Z and the chi-square distributed random variables; that is:

Definition 15.3.1

Let Z be a normally distributed random variable with $E(Z) = 0$ and $V(Z) = 1$, and let U be the chi-square distributed random variable with ν degrees of freedom. Then,

$$Y = \frac{Z\sqrt{\nu}}{\sqrt{U}} = \frac{Z}{\sqrt{U/\nu}}$$

is said to be the student-t distributed random variable with ν degrees of freedom.

As was the case with a chi-square distributed random variable there is a student-t distributed random variable for each degree of freedom.

The probability density functions for the student-t distributed random variables for different degrees of freedom are shown in Figure 15.3.1.

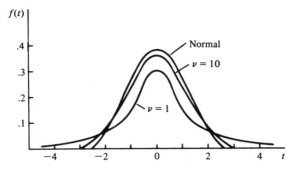

FIGURE 15.3.1 Probability density functions for student-t distributed random variables.

15.3 Student-t Distribution

This illustration also shows that the density function for $\nu = 1$ is relatively flat compared to the normal density function. On the other hand, as ν becomes larger and larger, the shape of the density functions of student-t distributed random variables approaches the shape of the normal density function.

A puzzling property of the student-t distributed random variable occurs when $\nu = 1$. When $\nu = 1$, the student-t distributed random variable can be expressed as

$$Y = \frac{Z}{\sqrt{Z_1^2/1}} = \frac{Z}{Z_1}.$$

Thus, Y is a ratio of two statistically independent, standard normal variables, Z and Z_1. A special name is provided for Y when $k = 1$: It is called the *Cauchy-distributed random variable*. An interesting property of this random variable is that, although it has a median, it has neither a mean nor a variance. Even though a Cauchy-distributed random variable has very few practical applications, it plays an important role in statistical theories. In Chapter 8, we have pointed out that if we let

$$W = Y_1 + \cdots + Y_n,$$

then, W approaches Z as n becomes larger and larger, if Y_i satisfy a set of assumptions, one of which is that Y_i have finite means and variances. On the other hand, if Y_i are Cauchy-distributed random variables, then this assumption is no longer satisfied. Thus, the central limit theorem does not apply if Y_i are Cauchy-distributed random variables. In fact, it can be shown that W is also Cauchy-distributed, regardless of n, if Y_i are Cauchy-distributed random variables.

A table of student-t distributed random variables is provided in Appendix B (see Table B.3). A portion of the table is shown here in Table 15.3.1. The table indicates that if Y is a student-t distributed random variable with 10 degrees of freedom, then, for example,

TABLE 15.3.1

	Percentage Point				
ν	.9	.5	.1	.05	.01
1	.158	1.000	6.314	12.706	63.657
2	.142	.816	2.920	4.303	9.965
.	.	.	.	.	.
.	.	.	.	.	.
10	.129	.700	1.812	2.228	3.169
.	.	.	.	.	.
.	.	.	.	.	.
∞	.126	.674	1.645	1.960	2.576

$$P(-.7 \leq Y \leq .7) = .5,$$
$$P(-1.812 \leq Y \leq 1.812) = .1,$$
and
$$P(-2.228 \leq Y \leq 2.228) = .05.$$

On the other hand, if Y is a student-t distributed random variable with ∞ degrees of freedom, then, for example,

$$P(-1.960 \leq Y \leq 1.960) = .05.$$

The reader will note, however, that

$$P(-1.960 \leq Z \leq 1.960) = .05,$$

where Z is the standard normal variable.

EXERCISES

17. Let Y be a student-t distributed random variable with 5 degrees of freedom. Find the value of k such that (a) $P(-k \geq Y \geq k) = .01$, (b) $P(-k \geq Y \geq k) = .05$, and (c) $P(-k \geq Y \geq k) = .10$.

*18. Let Y be a student-t distributed random variable with 15 degrees of freedom. Find the value of k such that (a) $P(-k \geq Y \geq k) = .01$, (b) $P(-k \geq Y \geq k) = .05$, and (c) $P(-k \geq Y \geq k) = .10$.

19. Let Y be a student-t distributed random variable with 25 degrees of freedom. Find the value of k such that (a) $P(-k \geq Y \geq k) = .01$, (b) $P(-k \geq Y \geq k) = .05$, and (c) $P(-k \geq Y \geq k) = .10$.

*20. Let Y be a student-t distributed random variable with 15 degrees of freedom. Find the value of k such that (a) $P(Y \geq k) = .05$, and (b) $P(Y \geq k) = .025$.

21. Let Y be a student-t distributed random variable with 15 degrees of freedom. Find the value of k such that (a) $P(Y \leq -k) = .05$, and (b) $P(Y \leq -k) = .025$.

15.4 t-Distribution and Statistical Inference for a Small Sample

In Chapter 11, when we attempted to establish a confidence interval for μ of a normal probability function, we encountered a conceptual difficulty when we tried to justify the suggestion that the value of **s** obtained from the sample should be used as if it were really σ. We will now show that this pragmatic detour is not really necessary. First, however, we will present a theorem which we will need to utilize for the given problem.

Theorem 15.4.1

Let $X_1, \ldots, X_n$ be a simple random sample of size n, from a normal probability function. Then,

15.4 t-Distribution and Statistical Inference

$$Y = \frac{\bar{X} - \mu}{s/\sqrt{n}}$$

is the student-t distributed random variable with $n - 1$ degrees of freedom.

The validity of this theorem can be stated as follows. Let

$$Y = \frac{\bar{X} - \mu}{s/\sqrt{n}} = \frac{\bar{X} - \mu}{\sigma/\sqrt{n}} \bigg/ \frac{s}{\sigma}.$$

Then,

$$Y = \frac{Z}{s/\sigma},$$

since

$$\frac{\bar{X} - \mu}{\sigma/\sqrt{n}} = Z.$$

In turn,

$$Y = \frac{Z}{\sqrt{s^2/\sigma^2}} = \frac{Z\sqrt{n-1}}{\sqrt{(n-1)s^2/\sigma^2}}.$$

Since $(n-1)s^2/\sigma^2 = \chi^2_{n-1}$, according to Definition 15.3.1, Y is the student-t distributed random variable with $n - 1$ degrees of freedom.

We will now apply the theorem to an interval estimation problem.

EXAMPLE 15.4.1

Assume that we wish to estimate the average breaking strength of a certain industrial packing tape. Assume also that the breaking strength is normally distributed. Let $\bar{X}$ and s depict the mean and the standard deviations, respectively, to be obtained from a simple random sample of size n. Let us define now

$$Y = \frac{\bar{X} - \mu}{s/\sqrt{n}}.$$

Then, since Y is the student-t distributed random variable with $n - 1$ degrees of freedom, if we let $t_{n-1,.05}$ depict

$$P(-t_{n-1,.05} \leq Y \leq t_{n-1,.05}) = .95,$$

then,

$$P\left(-t_{n-1,.05} \leq \frac{\bar{X} - \mu}{s/\sqrt{n}} \leq t_{n-1,.05}\right) = .95.$$

This equation can be modified to

$$P\left[\bar{X} - (t_{n-1,.05})\left(\frac{s}{\sqrt{n}}\right) \leq \mu \leq \bar{X} + (t_{n-1,.05})\left(\frac{s}{\sqrt{n}}\right)\right] = .95,$$

which is an expression of a confidence interval.

Suppose, for example, a sample of 9 yields

$$\bar{X} = 950 \text{ pounds}$$

and

$$s = 60 \text{ pounds}.$$

Then, the 95 percent confidence interval is established as follows. From Table B.3 in Appendix B we find that

$$t_{8,.05} = 2.306.$$

Thus,

$$P\left[\bar{X} - 2.306\left(\frac{s}{\sqrt{n}}\right) \leq \mu \leq \bar{X} + 2.306\left(\frac{s}{\sqrt{n}}\right)\right]$$
$$= P(903.88 \leq \mu \leq 996.12) = .95.$$

In Chapter 11 we suggested that we establish a confidence interval by assuming that the value of s obtained from the sample is really σ and, at the same time, by utilizing the table of normal probability functions in Appendix B. Had we followed this approach for the preceding interval estimation problem, the interval would have been between 910.80 and 989.20, which is considerably different from the theoretically correct interval established by utilizing the student-t distribution. On the other hand, suppose we assume that

$$\bar{X} = 950$$

and

$$s = 60$$

were obtained from a sample of 100 instead of from a sample of 9. Then, the correct interval established by using the student-t distribution is between

$$950 - t_{99,.05}\left(\frac{60}{\sqrt{100}}\right) = 950 - 1.99(6) = 938.06$$

and

$$950 + t_{99,.05}\left(\frac{60}{\sqrt{100}}\right) = 950 + 1.99(6) = 961.94,$$

and the interval established by using the table of normal probability functions (Table B.1 in Appendix B) is between

$$950 - 1.96\left(\frac{60}{\sqrt{100}}\right) = 938.24$$

and

$$950 + 1.96\left(\frac{60}{\sqrt{100}}\right) = 961.76,$$

which is not materially different from the former.

In devising a classical decision function for testing a set of hypotheses pertaining to μ of a normal probability function, it was assumed in Chapter 13 that σ of the probability function was known. We will now examine the consequence of devising a decision function if we do not assume that σ^2 is known.

Let the hypotheses be:

Null Hypothesis: $\mu = \mu_0$; $0 < \sigma^2 < \infty$
Alternate Hypothesis: $\mu \neq \mu_0$; $0 < \sigma^2 < \infty$.

Assume that a simple random sample of n are to be selected to test the hypotheses. We know, then, that

$$Y = \frac{\bar{X} - \mu_0}{s/\sqrt{n}}$$

is the student-t distributed random variable with $n - 1$ degrees of freedom, if the null hypothesis is true. Suppose now that we devise a decision function which specifies that we

Accept the null hypothesis if $-t_{n-1,.05} \leq Y \leq t_{n-1,.05}$;
Reject the null hypothesis, otherwise.

Then, the α risk associated with the decision function is .05. The decision function may also be modified to satisfy a different value of α.

EXAMPLE 15.4.2

To be more specific, assume that the hypotheses to be tested are

Null Hypothesis: $\mu = 1000$; $0 < \sigma^2 < \infty$
Alternate Hypothesis: $\mu \neq 1000$; $0 < \sigma^2 < \infty$,

where σ^2 is an unspecified value. Assume that we wish to devise a decision function which will yield $\alpha = .05$ where the sample size is 25. Then,

$$Y = \frac{\bar{X} - 1000}{s/\sqrt{25}}$$

is the student-t distributed random variable with 24 degrees of freedom. In turn,

$$t_{24,.05} = 2.06.$$

Thus, the decision function which will yield $\alpha = .05$ can be:

Accept the null hypothesis if $-2.06 \leq Y \leq 2.06$;
Reject it, otherwise.

Suppose now that an actual sample yields

$$\bar{X} = 1060$$
and
$$s = 100.$$

Then,

$$Y = \frac{1060 - 1000}{100/\sqrt{25}} = 3.00.$$

Since the actual value of Y is outside the acceptance region, we should reject the null hypothesis.

The decision function that we have just described is somewhat heuristically derived. It can be shown, however, that the structure of the decision function satisfies the criteria of the generalized likelihood-ratio test.

EXERCISES

22. A simple random sample of 4 from a normal probability function yields
$$\bar{X} = 40, \quad \text{and } s^2 = 16.$$
Establish a 95 percent confidence interval for the mean of the probability function, using Table B.3 in Appendix B.

*23. A simple random sample of 16 from a normal probability function yields
$$\bar{X} = 40, \quad \text{and } s^2 = 16.$$
Establish a 95 percent confidence interval for the mean of the probability function, using Table B.3 in Appendix B.

*24. Suppose you have selected a simple random sample of 5 from a normal probability function. The sample yields 2950, 1000, 1560, 1940, and 3340. Establish a 95 percent confidence interval for the mean of the probability function.

25. A simple random sample of 20 castings has been selected from a manufacturing process, and the diameters of their entry throats have been measured. The sample data are given in inches. (Note that these figures have been given also in Chapter 11, Exercise 16.)

2.989	3.030	2.986	2.997
3.029	3.039	3.040	3.031
3.026	3.000	3.008	3.027
3.010	3.023	2.993	2.998
3.002	2.999	3.004	3.020

Establish a 95 percent confidence interval for the mean diameter of the entry throats of the castings produced by the process. Then, compare the interval with that established for Exercise 18 in Chapter 11, utilizing Table B.1 in Appendix B.

*26. A university administrator wants to estimate the average amount of time per week that the students at his university spend in studying outside of the classroom. A simple random sample of 20 students has yielded the following data. (Note that these figures have been given in Chapter 11, Exercise 19.)

8	12	5	10
10	16	8	18
6	0	12	7
12	5	20	2
4	14	1	10

Assume that the time each student spends in studying is normally distributed.

Then, establish a 95 percent confidence interval for the mean study-time for all students, utilizing the Table B.3 in Appendix B. Then, compare the interval with that established for Exercise 19 of Chapter 11, utilizing Table B.1 in Appendix B.

27. Let X be a normally distributed random variable. The following two hypotheses are proposed with regard to the mean of X:

$$\text{Null Hypothesis: } \mu = 100$$
$$\text{Alternate Hypothesis: } \mu \neq 100.$$

The variance of X, however, is not known. Assume that you will test the hypotheses with a simple random sample of 9 from the probability function. Devise a decision function which you believe to be reasonably good and, at the same time, gives only a 5 percent α risk.

28. Using the situation in Exercise 27, suppose the sample of 9 actually yields

$$\bar{X} = 85, \quad \text{and } s^2 = 225.$$

Should you accept or reject the null hypothesis?

*29. Let X be a normally distributed random variable. The following hypotheses pertaining to the mean of X are proposed:

$$\text{Null Hypothesis: } \mu = 2000$$
$$\text{Alternate Hypothesis: } \mu \neq 2000.$$

The variance of X, however, is not known. You plan to accept or reject the null hypothesis after you take a simple random sample of 10 observations from the probability function. Assume that you are willing to accept a 5 percent α risk. Devise a decision function which you believe to be reasonably good for your purpose.

*30. Using Exercise 29, suppose your sample actually yields the following:

2760	1960
2640	1350
2960	2440
3800	1740
3280	2580

Should you accept or reject the null hypothesis, in light of these data?

15.5 F Distribution

Let us denote U_i and V_i as statistically independent, normally distributed random variables, where $E(U_i) = 0$, $E(V_i) = 0$, $V(U_i) = 1$, and $V(V_i) = 1$. Thus, U_i and V_i are standard normal variables. Let

$$Y = \frac{(U_1^2 + \cdots + U_m^2)/m}{(V_1^2 + \cdots + V_n^2)/n}.$$

Then, Y is said to be the *F distributed random variable* with m degrees of freedom for the numerator and n degrees of freedom for the denominator. Since

$$U_1^2 + \cdots + U_m^2$$

is the chi-square distributed random variable with m degrees of freedom, and

$$V_1^2 + \cdots + V_n^2$$

is the chi-square distributed random variable with n degrees of freedom, we can propose:

Definition 15.5.1

Let χ_m^2 and χ_n^2 be chi-square distributed random variables with m and n degrees of freedom, respectively. Then,

$$Y = \frac{\chi_m^2/m}{\chi_n^2/n}$$

is said to be the *F distributed random variable* with m degrees of freedom for the numerator and n degrees of freedom for the denominator.

Thus, for each ordered pair of integers (m,n) there exists a corresponding F distributed random variable.

Density functions for a number of F distributed random variables are shown in Figure 15.5.1. The first integer within each set of parentheses depicts the degree of freedom for the numerator, and the second integer depicts the degree of freedom for the denominator.

A table of F distributions is provided in Appendix B (see Table B.4). The table is constructed so that

$$F_{(m,n),.05} = P(Y \geq F_{(m,n),.05}) = .05,$$

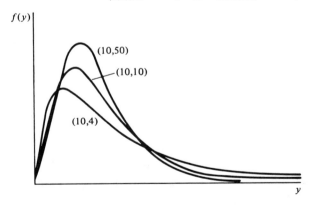

FIGURE 15.5.1 Density functions for several F distributed random variables.

15.5 F Distribution

where Y is the F distributed random variable with (m,n) degrees of freedom. For example,

$$F(10,20) = 2.35$$

implies that

$$P(Y \geq 2.35) = .05,$$

when Y is the F distributed random variable with $(10,20)$ degrees of freedom. Suppose we wish to find

$$F_{(m,n),.95} = P(Y \geq F_{(m,n),.95}) = .95.$$

Such a table is not provided in Appendix B. However, the following theorem will enable us to find such a probability if the need arises.

Theorem 15.5.1

Let Y and W be F distributed random variables with (m,n) and (n,m) degrees of freedom, respectively. If

$$P(W \geq F_{(n,m),.05}) = .05$$

then,

$$P\left(Y \geq \frac{1}{F_{(n,m),.05}}\right) = .95.$$

The validity of this theorem is illustrated as follows. The fact that

$$P\left(W = \frac{\chi_n^2/n}{\chi_m^2/m} \geq F_{(n,m),.05}\right) = .05$$

implies that

$$P\left(1/W = \frac{\chi_m^2/m}{\chi_n^2/n} \leq 1/F_{(n,m),.05}\right) = .05,$$

But $1/W = Y$. Thus,

$$P\left(Y \leq \frac{1}{F_{(n,m),.05}}\right) = .05,$$

which, in turn, implies that

$$P\left(Y \geq \frac{1}{F_{(n,m),.05}}\right) = .95.$$

EXAMPLE 15.5.1

Let Y be the F distributed random variable with $(10,20)$ degrees of freedom, and let W be the F distributed random variable with $(20,10)$ degrees of freedom. Then,

and
$$F_{(10,20),.05} = 2.35$$
$$F_{(20,10),.05} = 2.77.$$

We then find that
$$F_{(10,20),.95} = \frac{1}{F_{(20,10),.05}} = \frac{1}{2.77}$$
and
$$F_{(20,10),.95} = \frac{1}{F_{(10,20),.05}} = \frac{1}{2.35}.$$

Suppose now that Y is the F distributed random variable with $(1,n)$ degrees of freedom. Then we can let
$$Y = \frac{Z^2}{\chi_n^2/n} = \left(\frac{Z}{\sqrt{\chi_n^2/n}}\right)^2.$$

The term in parentheses depicts the student-t distributed random variable with n degrees of freedom. Thus, we propose:

Theorem 15.5.2
Let W depict the student-t distributed random variable with n degrees of freedom. Let
$$Y = W^2.$$
Then, Y is the F distributed random variable with $(1,n)$ degrees of freedom.

This theorem signifies that for some problems we can use either the student-t distribution or the F distribution.

EXERCISES

*31. Let Y be an F distributed random variable with $(5,10)$ degrees of freedom. Find the value of k such that (a) $P(Y \geq k) = .05$, and (b) $P(Y \geq k) = .01$.

32. Let Y be an F distributed random variable with $(10,5)$ degrees of freedom. Find the value of k such that (a) $P(Y \geq k) = .05$, and (b) $P(Y \geq k) = .01$.

*33. Let Y be an F distributed random variable with $(5,10)$ degrees of freedom. Find the value of k such that (a) $P(Y \geq k) = .95$, and (b) $P(Y \geq k) = .99$.

34. Let Y be an F distributed random variable with $(10,5)$ degrees of freedom. Find the value of k such that (a) $P(Y \geq k) = .95$, and (b) $P(Y \geq k) = .99$.

*35. Let Y be an F distributed random variable with $(1,10)$ degrees of freedom. Using Table B.3 in Appendix B, find the value of k such that (a) $P(Y \geq k) = .05$, and (b) $P(Y \geq k) = .01$.

36. Let Y be an F distributed random variable with $(1,20)$ degrees of freedom. Using Table B.3 in Appendix B, find the value of k such that (a) $P(Y \geq k) = .05$, and (b) $P(Y \geq k) = .01$.

*37. Let Y be a student-t distributed random variable with 15 degrees of freedom. Using Table B.4 in Appendix B, find the value of k such that (a) $P(-k \leq Y \leq k) = .95$, and (b) $P(-k \leq Y \leq k) = .99$.

38. Let Y be a student-t distributed random variable with 20 degrees of freedom. Using Table B.4 in Appendix B, find the value of k such that (a) $P(-k \geq Y \geq k) = .05$, and (b) $P(-k \geq Y \geq k) = .01$.

15.6 F Distribution and Statistical Inference Pertaining to Two Variances

In Section 15.2 we applied a chi-square distribution to statistical inferences pertaining to the variance of a probability function. We will now show that the F distribution can be applied to statistical inference pertaining to variances of two probability functions. To do so, we first propose:

Theorem 15.6.1

Let s_1^2 and s_2^2 be the sample variances from two normally distributed probability functions with variances σ_1^2 and σ_2^2, respectively. Then,

$$Y = \frac{s_1^2/\sigma_1^2}{s_2^2/\sigma_2^2}$$

is the F distributed random variable with $(m-1, n-1)$ degrees of freedom, where m and n are the sample sizes which have resulted in the values of s_1^2 and s_2^2, respectively.

Since $(m-1)s_1^2/\sigma_1^2 = \chi^2_{m-1}$ and $(n-1)s_2^2/\sigma_2^2 = \chi^2_{n-1}$, the preceding equation may be expressed as

$$Y = \frac{\chi^2_{m-1}/m-1}{\chi^2_{n-1}/n-1},$$

which is the F distributed random variable with $(m-1, n-1)$ degrees of freedom.

Using Theorem 15.6.1, suppose we assume that $\sigma_1^2 = \sigma_2^2$. Then,

$$Y = \frac{s_1^2/\sigma_1^2}{s_2^2/\sigma_2^2} = \frac{s_1^2}{s_2^2}.$$

Thus, we propose:

Theorem 15.6.2

Let s_1^2 and s_2^2 be the sample variances from two normally distributed probability function with the same variances, where m and n are the sample sizes which yield the values of s_1^2 and s_2^2, respectively. Then,

$$Y = \frac{s_1^2}{s_2^2}$$

is the F distributed random variable with $(m-1, n-1)$ degrees of freedom.

EXAMPLE 15.6.1

Assume that two processes produce a certain type of ball bearing. The process which yields a smaller variance in the diameters of the ball bearings is considered to be the superior of the two.

Let σ_1^2 denote the variance of the diameters of the ball bearings produced by Process 1, and let σ_2^2 denote the variance for Process 2. Assume that Process 2 is more costly than Process 1. Thus, if the variances of the two processes are the same, then we will choose Process 1, whereas if $\sigma_1^2 > \sigma_2^2$, we will choose Process 2. Our decision problem, then, may be formulated in terms of the following set of hypotheses:

$$\text{Null Hypothesis:} \quad \sigma_1^2 = \sigma_2^2$$
$$\text{Alternate Hypothesis:} \quad \sigma_1^2 > \sigma_2^2.$$

Assume now that a simple random sample of size m from Process 1 and of size n from Process 2 are selected to test the hypotheses. Let s_1^2 be the variance obtained from the sample from Process 1, and let s_2^2 be the variance obtained from the sample from Process 2. If the null hypothesis is true, then according to Theorem 15.6.2,

$$Y = \frac{s_1^2}{s_2^2}$$

is the F distributed random variable with $(m-1, n-1)$ degrees of freedom. Suppose now that we devise a decision function which specifies that we

Accept the null hypothesis if $Y \leq F_{(m-1,n-1),.05}$;
Reject the null hypothesis, otherwise.

Then, the α risk associated with the decision function is .05. For example, assume that $m = 10$ and $n = 15$. Then, Y is the F distributed random variable with $(9,14)$ degrees of freedom. From the table of F distributions in Appendix B (see Table B.4), we find that

$$F_{(9,14),.05} = 2.65.$$

Thus, if we wish to formulate a decision function yielding $\alpha = .05$, then one such formulation is to:

Accept the null hypothesis if $Y \leq 2.65$;
reject it, otherwise.

Suppose now our samples yield

$$s_1^2 = .0008$$
and
$$s_2^2 = .0005.$$

Then, the actual value of Y is

15.6 F Distribution and Statistical Inference

$$Y = \frac{.0008}{.0005} = 1.6.$$

Thus, we would accept the null hypothesis and select Process 1.

Is the form of the decision function proposed for this example optimal in some sense? Since the alternate hypothesis is that $\sigma_1^2 < \sigma_2^2$, it must be apparent that we should reject the null hypothesis if

$$Y = \frac{s_1^2}{s_2^2} > k,$$

where k is an arbitrary number.

EXERCISES

*39. Let X_1 and X_2 be two normally distributed random variables whose variances are given as σ_1^2 and σ_2^2. Two hypotheses are proposed pertaining to the variances of the random variables:

$$\text{Null Hypothesis:} \quad \sigma_1^2 = \sigma_2^2$$
$$\text{Alternate Hypothesis:} \quad \sigma_1^2 > \sigma_2^2.$$

You are to take a simple random sample of 11 from the probability function of X_1 and a simple random sample of 16 from the probability function of X_2. Formulate a decision function which you believe to be optimal, given that you are willing to accept a 5 percent α risk.

*40. Refer to Exercise 39 and suppose your samples yield

$$s_1^2 = .0003$$
and $\quad s_2^2 = .0002.$

Should you accept or reject the null hypothesis?

41. Let X_1 and X_2 be two normally distributed random variables whose variances are given as σ_1^2 and σ_2^2. The following two hypotheses are proposed pertaining to these variances:

$$\text{Null Hypothesis:} \quad \sigma_1^2 = \sigma_2^2$$
$$\text{Alternate Hypothesis:} \quad \sigma_1^2 < \sigma_2^2.$$

You are to select a simple random sample of 25 from the probability function of X_1 and a simple random sample of 16 from the probability function of X_2. Formulate a decision function which you believe to be optimal, given that you are willing to accept a 1 percent α risk.

42. Refer to Exercise 41 and suppose your sample yields

$$s_1^2 = .001$$
and $\quad s_1^2 = .003.$

Should you accept or reject the null hypothesis?

*43. Two canning processes are currently being used by a cannery. One of the two processes is cheaper to operate than the other one. According to a government regulation, the variance in the net weight of the cans must be within a specified limit; therefore, the cannery wishes to know whether the variance in the net weight of the cans produced by the cheaper process is as small as the variance for the more expensive process. Assume that the net weight of the cans from the two processes is normally distributed. How would you establish a null and an alternate hypothesis for the problem? Assume that the decision with regard to these hypotheses is to be made on the basis of simple random samples of 25 cans from each of the processes. Devise a decision function which you believe to be optimal, given that you are willing to accept a .01 probability of erroneously concluding that the variance of the cheaper process is larger than that of the more expensive process, even though it is not, in fact, true.

*44. Refer to Exercise 43 and suppose that the actual samples from the processes yield the following results:

>   Sample variance from cheaper process: .015 ounces
>   Sample variance from expensive process: .010 ounces.

What conclusion do you reach with regard to the hypotheses given in Exercise 43?

15.7 Chi-Square and Theory of Large Sampling

We have shown that the normal probability function plays an important role in the theory of large sampling because of the central limit theorem. We will now show that the chi-square probability function plays an important role in testing hypotheses with a large sample because of the following theorem.

Theorem 15.7.1

Let $X_1, \ldots, X_n$ be a random sample of size n from a probability function which satisfies so-called *regularity conditions*. Let the parameters of the probability function be denoted by $\theta_1, \theta_2, \ldots, \theta_k$, and let the null hypothesis pertaining to the probability function be

$$\theta_1 = \theta_1^0, \ldots, \theta_k = \theta_k^0.$$

Then, $-2 \log \lambda$ approaches the chi-square distributed random variable with k degrees of freedom as n becomes larger, where λ is the generalized likelihood-ratio.

The significance of this theorem may be stated as follows. We know from our discussion in Chapter 13 that $0 \leq \lambda \leq 1$. Suppose $\lambda = 1$. Then, $-2 \log \lambda = 0$. On the other hand, as λ approaches 0, $-2 \log \lambda$ approaches an infinitely large positive number. Suppose now that we wish to establish a decision based on the generalized likelihood-ratio test, which will yield an α risk of a specified value. This can be done by formulating the decision function so that we:

15.7 Chi-Square and Theory of Large Sampling   403

Accept the null hypothesis if $-2 \log \lambda \leq \chi^2_{k,\alpha}$;
Reject the null hypothesis otherwise.

According to Theorem 15.7.1, if we resort to formulating this decision function, the probability of erroneously rejecting the null hypothesis is approximately equal to α when the sample size n is large.

EXAMPLE 15.7.1

Let us now return to Example 13.6.5 and re-evaluate the problem of testing the null hypothesis, $\pi = \pi_0$. Suppose we denote the likelihood function of P as

$$L(P) = C(n, nP) \pi^{nP} (1 - \pi)^{n-nP}.$$

Then, given the sample yields P, the value of π which maximizes $L(P)$ is equal to P. Thus,

$$\lambda = \frac{\pi_0^{nP}(1 - \pi_0)^{n-nP}}{P^{nP}(1 - P)^{n-nP}}.$$

We illustrated in Chapter 13 how to obtain the probability function of λ for a small n and, in turn, how to derive a suitable decision function. At the same time, we pointed out that finding the exact probability function of λ would be very difficult when n is large.

Theorem 15.7.1 now tells us that $-2 \log \lambda$ will be approximately chi-square distributed with 1 degree of freedom when n is large. Thus, we are spared from having to ascertain the exact probability function of λ for a large n.

Assume, for example, that we wish to formulate the decision function with a specified α risk. Then, we first transform

$$-2 \log \lambda = 2 \left[nP \log \left(\frac{P}{\pi_0} \right) + (n - nP) \log \left(\frac{1 - P}{1 - \pi_0} \right) \right],$$

and, in turn, obtain the decision function that we:

Accept the null hypothesis if $2\{nP \log (P/\pi_0) + (n - nP) \log [(1 - P)/(1 - \pi_0)]\} \leq \chi^2_{1,.05}$; reject it otherwise.

To be more specific, assume that $\pi_0 = .5$, $n = 100$, $\alpha = .05$, and $P = .4$. Then,

$$\chi^2_{1,.05} = 3.84$$

and

$$2 \left[nP \log \left(\frac{P}{\pi_0} \right) + (n - nP) \log \left(\frac{1 - P}{1 - \pi_0} \right) \right] = 2(40 \log .8 + 60 \log 1.2)$$
$$= 2[(40)(-.2231) + (60)(.1823)] = 4.024.$$

Thus, $-2 \log \lambda = 4.024$ and is larger than $\chi^2_{1,.05} = 3.84$. Consequently, we should reject the null hypothesis.

There is another approach to testing the given hypothesis, which perhaps is more intuitively appealing than the approach discussed here. We will evaluate this other approach in the following section.

EXERCISES

*45. Let X be a Bernoulli-distributed random variable. The following hypotheses are established pertaining to X:

$$\text{Null Hypothesis: } \pi = .4$$
$$\text{Alternate Hypothesis: } \pi \neq .4.$$

Suppose you wish to establish a decision function based on the generalized likelihood-ratio test. Assume that you are willing to accept a 5 percent α risk. What is then the corresponding decision function, if the sample size is to be 100?

*46. Using the situation given in Exercise 45, assume that the actual sample from the probability function yields $P = .5$. Should you accept or reject the null hypothesis given in Exercise 45?

*47. Using the situation given in Exercise 45, suppose that the actual sample from the probability function yields $P = .3$. Should you accept or reject the null hypothesis given in Exercise 45?

48. Let X be a Bernoulli-distributed random variable. The following hypotheses are established pertaining to X:

$$\text{Null Hypothesis: } \pi = .2$$
$$\text{Alternate Hypothesis: } \pi \neq .2.$$

Suppose you wish to establish a decision function based on the generalized likelihood-ratio test. Assume that you are willing to accept a 1 percent α risk. What is then the corresponding decision function, if the sample size is to be 100?

49. Using the situation given in Exercise 48, assume that the actual sample from the probability function yields $P = .3$. Should you accept or reject the null hypothesis given in Exercise 48?

50. Using the situation given in Exercise 48, suppose that the actual sample from the probability function yields $P = .2$. Should you accept or reject the null hypothesis given in Exercise 48?

15.8 Testing "Goodness-of-Fit"

A problem that frequently arises is that we must discover whether a certain observed frequency distribution can be assumed to have resulted from a specified probability function. For example, we might wish to evaluate whether the daily sales records of an item in a retail store can be assumed to be normally distributed; or we might wish to evaluate whether the failure record of a given type of bulb can be assumed to be exponentially distributed; or we might wish to evaluate whether the incoming telephone calls at a switchboard are Poisson-distributed. Comparison of an observed frequency distribution with a specific probability function is called the test of "goodness-of-fit." In theory, the test can be carried out for any sample size. However, the theory underlying the test for small samples is very involved, therefore, we will discuss the procedure only with regard to large samples.

Perhaps the most simple situation to which we can apply the underlying statistical concept is in the testing of hypotheses such as:

Null Hypothesis: The probability function sampled is Bernoulli-distributed with $\pi = \pi_1$.
Alternate Hypothesis: The probability function sampled is Bernoulli-distributed with $\pi \neq \pi_1$.

Let $X_1, \ldots, X_n$ be the set of Bernoulli-distributed random variables which constitute a simple random sample of size n, from a probability function. For our subsequent discussion, we define

$$S = X_1 + \cdots + X_n.$$

We now propose:

Theorem 15.8.1

Let

$$Y = \frac{(S - n\pi)^2}{n\pi} + \frac{[(n - S) - n(1 - \pi)]^2}{n(1 - \pi)}.$$

Then, Y approaches the chi-square distributed random variable with 1 degree of freedom as n becomes larger and larger.

The validity of Theorem 15.8.1 can be shown in the context of Theorem 15.7.1, however, we will not attempt to do so, since we also did not show the validity of Theorem 15.7.1. On the other hand, Theorem 15.8.1 can be justified on an intuitive ground. To do so, we first express

$$Y = (S - n\pi)^2 \left[\frac{(1 - \pi) + \pi}{n\pi(1 - \pi)} \right]$$
$$= \frac{(S - n\pi)^2}{n\pi(1 - \pi)} = \left(\frac{S - n\pi}{\sqrt{n\pi(1 - \pi)}} \right)^2.$$

Since $E(S) = n\pi$ and $V(S) = n\pi(1 - \pi)$, if n is large, then

$$\frac{S - n\pi}{\sqrt{n\pi(1 - \pi)}}$$

approaches the standard normal variable Z and, in turn,

$$\left[\frac{S - n\pi}{\sqrt{n\pi(1 - \pi)}}\right]^2$$

approaches the chi-square distributed random variable with 1 degree of freedom.

The significance of Theorem 15.8.1 for the given hypothesis-testing problem can be stated as follows. Suppose the null hypothesis is, in fact, true; that is, $\pi = \pi_1$. Then,

$$W = \frac{(S - n\pi_1)^2}{n\pi_1} + \frac{[(n - S) - n(1 - \pi_1)]^2}{n\pi_1(1 - \pi_1)}$$

will approach the chi-square distributed random variable with 1 degree of freedom as n becomes larger and larger. Further, $E(W) = 1$, since the expected value of the chi-square distributed random variable is equal to its degree of freedom.

On the other hand, suppose the null hypothesis is not true; that is, $\pi \neq \pi_1$. Assume, for example, that $\pi = \pi_2$, where $\pi_1 \neq \pi_2$. Then,

$$U = \frac{(S - n\pi_2)^2}{n\pi_2} + \frac{[(n - S) - n(1 - \pi_2)]^2}{n\pi_2(1 - \pi_2)}$$

will approach the chi-square distributed random variable with 1 degree of freedom as n becomes larger, whereas W, which we have just defined, will not approach it. Furthermore, we can show that both $E(W) > 1$ and $E(W)$ will increase without limits as n becomes larger and larger, which means that, given α, the rejection region for the decision function can be established in such a way that

$$P(W \geq \chi^2_{1,\alpha} | \pi = \pi_1) = \alpha.$$

It can also be shown that the decision function just formulated is equivalent to the decision function based on the concept of the likelihood-ratio test.

EXAMPLE 15.8.1

Suppose that we wish to test a null hypothesis that a given coin is fair. The null hypothesis and the alternate hypothesis are expressed as

Null Hypothesis: $\pi = .5$
Alternate Hypothesis: $\pi \neq .5$,

where π depicts the probability that the coin will land on "heads."

Assume now that we wish to test the hypothesis in terms of goodness-of-fit. Assume also that we are willing to accept approximately a .05 probability of erroneously rejecting the null hypothesis when it is, in fact, true. The critical value for the rejection region may be established in the following manner. Let

$$W = \frac{(S - .5n)^2}{.5n} + \frac{[(n - S) - .5n]^2}{.5n}.$$

Then, the null hypothesis should be rejected if $W > 3.84$ and accepted otherwise.

Suppose now we have tossed the coin 100 times and have obtained the following results.

	Heads	Tails
Observed	40	60
Expected	50	50

The actual value of W can be calculated as

$$W = \frac{(40-50)^2}{50} + \frac{(60-50)^2}{50} = 4.$$

Thus, we would reject the hypothesis that the coin is fair at $\alpha \cong .05$.

At this point, the perceptive reader may note that we could have tested the hypothesis that the coin is fair without formulating it as a goodness-of-fit problem. We could just as well have tested the hypothesis by using the procedure discussed in the preceding section, which is, in fact, equivalent to the procedure discussed in this section. Nevertheless, we have described the application of the test for goodness-of-fit for the given hypothesis-testing problem, since we believe that our discussion will make it easier for the reader to accept the following theorem, which we present without a proof.

Theorem 15.8.2

Let $X_1, \ldots, X_n$ be the simple random sample of size n from a probability function, and let $E_1, \ldots, E_k$ be the classes into which X_j can fall. Let $\pi_1, \ldots, \pi_k$ denote the probability that X_j will fall into $E_1, \ldots, E_k$, and let $S_1, \ldots, S_k$ be the random variables associated with the number of X_j's falling into $E_1, \ldots, E_k$. Then,

$$Y = \frac{(S_1 - n\pi_1)^2}{n\pi_1} + \cdots + \frac{(S_k - n\pi_k)^2}{n\pi_k}$$

approaches the chi-square distributed random variable with $k-1$ degrees of freedom as n becomes larger.

Instead of trying to explain the validity of this theorem, we will illustrate that it is equivalent to Theorem 15.8.1 when $k = 2$. When $k = 2$, Y becomes

$$Y = \frac{(S_1 - n\pi_1)^2}{n\pi_1} + \frac{(S_2 - n\pi_2)^2}{n\pi_2},$$

according to Theorem 15.8.2. Now let $S_1 = S$ and $\pi_1 = \pi$. Then, $S_2 = n - S$ and $\pi_2 = 1 - \pi$. Then, Y can be expressed as

$$Y = \frac{(S - n\pi)^2}{n\pi} + \frac{[(n - S) - n(1 - \pi)]^2}{n(1 - \pi)},$$

which is the same as we have expressed in Theorem 15.8.1.

We may point out, however, one limitation in using the preceding theorem. The χ^2 approximation proposed by the theorem works well if $n\pi_i$ is large for all i's. However, it has been found through experience and theory that such approximation does not work well if $n\pi_i$ is very small for some i's. It has been, therefore, a well-established practice, if any $n\pi_i$ is less than 5, to combine several classes of E_i into one single class, for example, E_t so that the resulting $n\pi_t$ is at least equal to 5. We will give one such illustration in Example 15.8.3.

EXAMPLE 15.8.2

Suppose that we wish to test a null hypothesis that a given die is fair. Let $\pi_1, \pi_2, \ldots, \pi_6$ depict the probability that the die will land on 1, 2, ..., 6. Then, the null hypothesis may be expressed as

$$\text{Null Hypothesis: } \pi_1 = \pi_2 = \pi_3 = \pi_4 = \pi_5 = \pi_6 = \frac{1}{6}.$$

There are many different ways to express the alternate hypothesis, for example,

$$\text{Alternate Hypothesis: } \pi_1 = \pi_2 \neq \pi_3 = \pi_4 = \pi_5 = \pi_6$$

or

$$\text{Alternate Hypothesis: } \pi_1 > \pi_2 + \pi_3 + \pi_4 = \pi_5 = \pi_6,$$

both of which are composite hypotheses.

The only reasonable course of action for us then is to design a decision function so that we can control the α risk in a suitable way. Let us assume, therefore, that we are willing to accept an α risk of approximately .05. Then, the critical value for the rejection region may be established in the following manner. Let

$$W = \frac{(S_1 - n/6)^2 + (S_2 - n/6)^2 + \cdots + (S_6 - n/6)^2}{n/6},$$

where $S_1, S_2, \ldots, S_6$ denote the number of times that the die lands on 1, 2, ..., 6 out of n tosses. W is, then, the chi-square distributed random variable with 5 degrees of freedom, if the null hypothesis is, in fact, true. The critical value in question is 11.1, and the null hypothesis should be rejected if $W > 11.1$.

Suppose now that we have tossed the die, 600 times and have obtained the following experimental results:

	1	2	3	4	5	6
Observed Frequency	120	115	80	110	90	85
Expected Frequency	100	100	100	100	100	100

15.8 Testing "Goodness-of-Fit"

Then,

$$W = \frac{(120-100)^2 + (115-100)^2 + (80-100)^2}{100}$$
$$+ \frac{(110-100)^2 + (90-100)^2 + (85-100)^2}{100} = 14.5.$$

Since the actual value of W is greater than 11.1, we should reject the null hypothesis that the die is fair at $\alpha \cong .05$. On the other hand, suppose we are willing to accept an α risk of only .01. Then, the critical value would be 15.1, and therefore, the null hypothesis should not be rejected.

EXAMPLE 15.8.3

Let X be a random variable, and assume that the null hypothesis pertaining to X is that it is a Poisson-distributed random variable with a probability function $f(x) = e^{-.5}.5^x/x!$. If the null hypothesis is, in fact, true, then the probability function of X is that given in Table 15.8.1.

TABLE 15.8.1

x	$f(x)$
0	.6065
1	.3033
2	.0758
3	.0072
4	.0016
5	.0002
	1.0000

Assume that we wish to test the given hypothesis by selecting a simple random sample of 100 from the probability function. One way to test the hypothesis is to formulate it as a goodness-of-fit problem.

Let E_1 be the event which occurs if X assumes 0; let E_2 be the event which occurs if X assumes 1, and so on, where E_6 is the event which occurs if X assumes 5. Then, we have $f(\pi_i)$ and $n\pi_i$, as shown in Table 15.8.2.

TABLE 15.8.2

i	$f(\pi_i)$	$n\pi_i$
1	.6065	60.65
2	.3033	30.33
3	.0758	7.58
4	.0072	.72
5	.0016	.16
6	.0002	.02

Now let S_1 be the random variable associated with the number of times 0 is obtained; let S_2 be the random variable associated with the number of times 1 is obtained, and so on. Then, according to Theorem 15.8.2,

$$Y = \frac{(S_1 - 60.65)^2}{60.65} + \frac{(S_2 - 30.33)^2}{30.33} + \cdots + \frac{(S_6 - .02)^2}{.02}$$

is approximately a chi-square distributed random variable with 5 degrees of freedom, if the null hypothesis is, in fact, true. We have, however, pointed out that a chi-square approximation is not very good when some $n\pi_i$ are less than 5. We note that all of $n\pi_4$, $n\pi_5$, and $n\pi_6$ are less than 5. Therefore, we modify E_i as shown in Table 15.8.3, and, in turn, we define the following:

$$Y = \frac{(S_1 - 60.65)^2}{60.65} + \frac{(S_2 - 30.33)^2}{30.33} + \frac{(S_3 - 9.02)^2}{9.02}.$$

Then, Y is a chi-square distributed random variable with 2 degrees of freedom, provided, of course, that the null hypothesis is true.

TABLE 15.8.3

x	i	$f(\pi_i)$	$n\pi_i$
0	1	.6065	60.65
1	2	.3033	30.33
2, 3, 4, 5	3	.0902	9.02

Assume now that we wish to formulate a decision function which will yield a 5 percent α risk. We note from the table of chi-square probability functions (Table B.2 in Appendix B) that $\chi^2_{2,.05} = 5.99$. Thus, the null hypothesis should be rejected if $Y > 5.991$; otherwise, it should be accepted.

Let us now assume that the actual sample yields

$$S_1 = 54,$$
$$S_2 = 40,$$
and $$S_3 = 6.$$

Then, the actual value of Y for the sample may be calculated as

$$Y = \frac{(54 - 60.65)^2}{60.65} + \frac{(40 - 30.33)^2}{30.33} + \frac{(6 - 9.02)^2}{9.02} = 4.823.$$

Therefore, the null hypothesis should be accepted.

EXERCISES

*51. Let X be a Bernoulli-distributed random variable. The following hypotheses are established pertaining to X.

Null Hypothesis: $\pi = .4$
Alternate Hypothesis: $\pi \neq .4$.

You wish to test these hypotheses in terms of goodness-of-fit. Assume that the sample size is 100 and the α risk is .05. Formulate the corresponding decision function.

*52. Refer to Exercise 51 and assume that the actual sample from the probability function yields $S = 50$, where

$$S = X_1 + \cdots + X_{100}.$$

Should you accept or reject the null hypothesis given in Exercise 51? Compare your conclusion for this exercise with your conclusion from Exercise 46.

53. Using the situation in Exercise 51, assume that the actual sample from the probability function yields $S = 30$. Should you accept or reject the null hypothesis given in Exercise 51? Compare your answer for Exercise 47.

*54. At 7 A.M. each evening, three local television stations in a city broadcast the regional and local news for the day. It is hypothesized that all three news programs are equally popular among the city's viewers. Select a simple random sample of 1200 viewers to test the hypothesis, and assume that you are willing to accept a 5 percent probability of erroneously rejecting the hypothesis. What should be then the corresponding decision function?

*55. For the situation described in Exercise 54, suppose the actual sample yields the following results:

Station	Number of Viewers
A	380
B	460
C	360

In the light of this evidence, should you accept or reject the hypothesis proposed in Exercise 54?

*56. Suppose that the null hypothesis pertaining to a random variable X is that it is Poisson-distributed with a probability function of $f(x) = e^{-1}/x!$. You are to test the hypothesis by taking a simple random sample of 200 from the probability function. How many classes of E_i would you establish? Assume that you are willing to accept a 5 percent α risk. Formulate the corresponding decision function.

*57. Using the situation described in Exercise 56, assume that your sample yields the following information:

Value Observed	Number of Times Observed
0	80
1	70
2	40
3	8
4	2

In the light of this information, should you accept or reject the null hypothesis given in Exercise 56?

*58. The null hypothesis pertaining to a random variable X is that it is normally distributed with $\mu = 75$ and $\sigma = 10$. A simple random sample of 1000 observations from the probability function yields the following data.

Value of X	Number of Times Observed
between 45 and 55	30
between 55 and 65	160
between 65 and 75	320
between 75 and 85	360
between 85 and 95	120
between 95 and 105	10
	1000

Assume that you are willing to accept a 5 percent α risk. Should you then accept the null hypothesis, in light of the given sample evidence?

*59. It is believed that the life spans of certain tubes produced by a process are exponentially distributed, with a mean life of 500 hours. A simple random sample of 100 tubes from the process has yielded the following results.

Life Span	Number of Tubes
between 0 and 500	62
between 500 and 1000	22
between 1000 and 1500	10
over 1500	6
	100

In the light of this sample evidence, what conclusion do you draw with regard to the belief that the life spans of the tubes are exponentially distributed, with a mean life span of 500 hours? (Assume that you are willing to accept a 5 percent probability of making an erroneous conclusion.)

Chapter 16
Regression
Analysis

16.1 Multivariate Sampling

So far we have assumed that each sample observation consists of a single measure. However, at times we might wish to obtain several different measures from each sample observation. Usually the reason for such multivariate sampling is to evaluate how one of these measures is related to one or more of the remaining measures.

EXAMPLE 16.1.1

Assume that the admissions director of a university wishes to evaluate the wisdom of basing the university's admission policy on the high school grade-point averages and SAT test scores of its applicants. Therefore, the admissions director selects a random sample of n students from the new sophomore students and obtains from each student:

1. freshman grade-point average
2. high school grade-point average
3. SAT verbal test score
4. SAT quantitative test score.

Thus, the admissions director has, in fact, obtained four different types of measures from each sample observation.
Let

Y_i = freshman grade-point average from the ith student
X_i = high school grade-point average from the ith student
U_i = SAT verbal test score for the ith student
W_i = SAT quantitative test score for the ith student.

Then, his sample result may be given by

$$\{(Y_1, X_1, U_1, W_1), \ldots, (Y_n, X_n, U_n, W_n)\}.$$

EXAMPLE 16.1.2

Assume that an agronomist wishes to evaluate the relationship between the yield of a certain crop and the total amount of fertilizer and water received by the crop. He divides his experimental plot for the crop into n different subplots and administers different quantities of fertilizer and water to these subplots. Let

Y_i = yield from the ith subplot
X_i = fertilizer given to the ith subplot
U_i = water given to the ith subplot.

He will then obtain a set of three measures (Y_i, X_i, U_i) from each sample observation.

16.2 Simple Regression

For the sake of simplicity, we will confine our analysis to situations in which we will take only two measures from each sample observation. Let the set of observations thus obtained be denoted by

$$\{(Y_1, X_1), \ldots, (Y_n, X_n)\}.$$

Then, we may assume that Y_i is the ith observation of the dependent variable Y, and X_i is the ith observation of the independent variable X. Now we are interested in ascertaining the function

$$Y = g(X),$$

which depicts the observed relationship between X and Y.

Four different functions which can be fitted to a given set of observations are shown in Figure 16.2.1. However, these are only a few examples of all the possible functions which we can use to explain the relationship between X and Y.

Two problems confront us now:

1. We must decide upon the structure of the function $g(X)$.
2. Once we decide upon the structure of the function, we need to ascertain the values of the parameters, such as α, β, and γ.

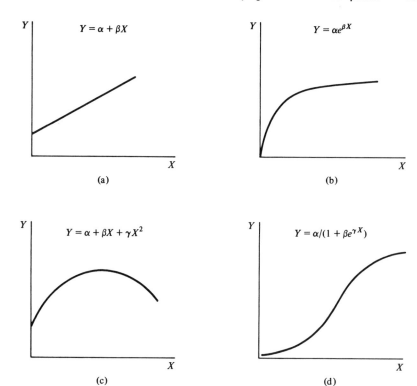

FIGURE 16.2.1 Four possible functions showing relationship between x and y: (a) linear function; (b) exponential function; (c) quadratic function; and (d) logistic function.

We will, however, resolve the first of these problems by assuming that the function to be fitted is linear. Such a function is then said to be a *simple regression function*.

16.3 Underlying Statistical Assumptions

Let us assume now that

$$Y = \alpha + \beta X$$

depicts the true relationship between the dependent variable Y and the independent variable X. Now let Y_i and X_i depict the ith empirically observed values of Y and X, respectively. We *cannot*, however, express that

$$Y_i = \alpha + \beta X_i$$

for the following reason. Even though the value of X_i may be observed accurately, very likely there are errors committed in measuring the value of Y. Consequently, if the true value of Y for the observed value of X_i is denoted by $\tilde{Y}_i$, then,

$$Y_i = \tilde{Y}_i + \varepsilon_i,$$

where ε_i is the error associated with the ith measurement for Y. In turn, the equation

$$Y_i = \alpha + \beta X_i + \varepsilon_i$$

depicts the relationship between the observed values of Y and X.

Since the error in measuring will vary from one sample observation to another, it is reasonable to consider that ε_i is a random variable. The following four assumptions are usually made for ε_i:

1. $E(\varepsilon_i) = 0$ for all i's.
2. $V(\varepsilon_i)$ is the same for all i's.
3. $\mathrm{Cov}\,(\varepsilon_i, \varepsilon_j) = 0$ for all pairs of i and j, except when $i = j$.
4. ε_i is normally distributed for all i's.

We may assume X_i is either a random variable or a *deterministic* (nonrandom) *variable*. For example, suppose that Y_i and X_i depict the yield of the crop and the amount of fertilizer given to the ith subplot, respectively. Then, the value of X_i for the ith sample observation can be controlled by the agronomist, and, therefore, it should be considered to be a deterministic variable.

On the other hand, suppose Y_i and X_i, respectively, depict the freshman grade-point average and high school grade-point average of the ith student selected for the sample. Then, the admissions director really has no control over the value of X_i, and, therefore, we should consider X_i to be a random variable.

Although two different interpretations for X_i are required for the situations just described, two different statistical procedures are not required. Thus, from now on we will simply assume that X_i is a deterministic variable, and we will define:

Definition 16.3.1

Let $Y_1, \ldots, Y_n$ be observable random variables such that

$$Y_i = \alpha + \beta X_i + \varepsilon_i,$$

where α and β are unobservable constants, $X_1, \ldots, X_n$ are observable deterministic variables, and $\varepsilon_1, \ldots, \varepsilon_n$ are unobservable random variables which are uncorrelated with each other and normally distributed with $E(\varepsilon_i) = 0$ and $V(\varepsilon_i) = \sigma^2$ for $i = 1, \ldots, n$. Then, the function

$$Y_i = \alpha + \beta X_i + \varepsilon_i$$

is said to be a *simple linear regression function*.

Clearly, Y_i is a random variable, since it is a function of ε_i, which also is a random variable. We may further deduce:

Theorem 16.3.1
Let $Y_i = \alpha + \beta X_i + \varepsilon_i$. Then, $Y_1, \ldots, Y_n$ are uncorrelated, normally distributed random variables with $E(Y_i) = \alpha + \beta X_i$ and $V(Y_i) = \sigma^2$ for all i's.

The probability functions for Y_i are illustrated in Figure 16.3.1. Each of the three bell-shaped curves depicts the probability function of Y for a given value of X_i.

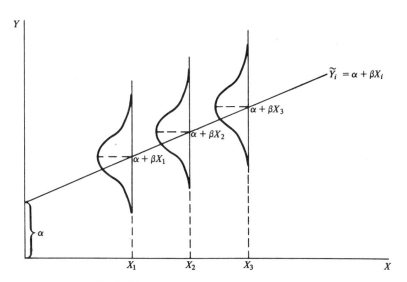

FIGURE 16.3.1 Probability functions for Y_i.

EXERCISES

*1. The data given in Table 16.3.1 were generated by tossing n coins 12 times. Suppose you have defined H_i to be the number of "heads" obtained during the ith toss of n coins and have defined $\varepsilon_i = H_i - E(H_i)$. Then, you can define $Y_i = \alpha + \beta X_i + \varepsilon_i$.
Explain why the function which has generated the data may be considered to be a simple linear regression function.

TABLE 16.3.1

i	X_i	Y_i
1	4	132
2	4	124
3	4	142
4	4	154
5	6	198
6	6	208
7	6	182
8	6	204
9	8	256
10	8	268
11	8	248
12	8	260

16.4 Estimation of Parameters

Given a set of sample observations, our first problem is to estimate the values of α and β. To illustrate the nature of the estimation problem, let us consider the following example.

EXAMPLE 16.4.1

Let H_i be a random variable associated with the number of "heads" obtained by tossing 100 coins on the ith trial. Then, $H_1, \ldots, H_n$ are statistically independent random variables with $E(H_i) = 50$ and $V(H_i) = n\pi(1 - \pi) = 25$, where $n = 100$ and $\pi = .5$. Furthermore, H_i is approximately normally distributed according to the central limit theorem.

Let

$$\varepsilon_i = H_i - 50.$$

Then, $\varepsilon_1, \ldots, \varepsilon_n$ are also statistically independent random variables with $E(\varepsilon_i) = 0$ and $V(\varepsilon_i) = 25$. Furthermore, ε_i is also a normally distributed random variable.

Suppose now that we let

$$Y_i = 10 + 10X_i + \varepsilon_i,$$

where X_i is a deterministic variable, and let $\varepsilon_i = H_i - 50$, where H_i is the random variable generated by tossing 100 coins. Then, the function satisfies the definition of a simple regression function.

By tossing 100 coins 12 times, the set of empirical observations given in Table 16.4.1 are generated for the regression function. For example, the first

16.4 Estimation of Parameters

TABLE 16.4.1

Observation	X_i	H_i	ε_i	Y_i
1	1	41	−9	11
2	1	56	6	26
3	1	48	−2	18
4	1	52	2	22
5	2	47	−3	27
6	2	56	6	36
7	2	46	−4	26
8	2	42	−8	22
9	3	49	−1	39
10	3	54	4	44
11	3	45	−5	35
12	3	55	5	45

set of observations is generated as follows. After tossing 100 coins, we have obtained 41 "heads." Thus, $H_1 = 41$ and $\varepsilon_1 = -9$. We arbitrarily let $X_1 = 1$. Then,

$$Y_1 = 10 + 10X_1 + \varepsilon_1$$
$$= 10 + 10 - 9 = 11.$$

The remaining observations are generated in a similar manner.

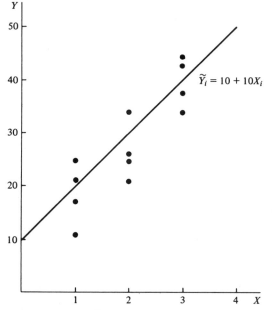

FIGURE 16.4.1 Scatter diagram of sample observations.

A scatter diagram of the sample observations is shown in Figure 16.4.1. The line drawn in the figure portrays the relationship between X and Y in the absence of errors. We will call the equation depicting the line a *prediction equation*.

Suppose now that we are not told that $\alpha = 10$ and $\beta = 10$, but we are asked to estimate the values of α and β from the sample observations. Let us redraw the scatter diagram given above without the line $\hat{Y}_i = 10 + 10X_i$ (see Figure 16.4.2).

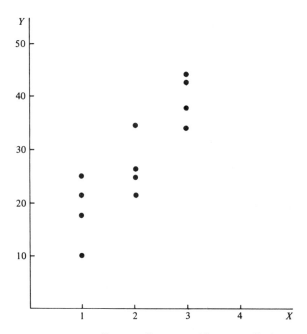

FIGURE 16.4.2 Scatter diagram without prediction equation.

Let **a** and **b** denote the estimators for α and β, respectively. Since, by assumption, we do not know the values of α and β, the values of **a** and **b** can be determined only through examining the dots in the diagram. Perhaps we can draw a line for a prediction equation which appears to fit the given set of dots and then obtain from such a line the values of **a** and **b**. This, of course, means that we could probably draw another line or several different lines which appear to fit the dots equally well and then obtain the corresponding sets of values for **a** and **b**. Are there preferable methods we can use to obtain the values of **a** and **b** from the given set of observations? The answer is "yes," and one such method is known as the *least-square method*.

We will now explain the concept of the *least-square*. To do so, we first propose:

Definition 16.4.1

Let $\{(Y_1,X_1), \ldots , (Y_n,X_n)\}$ be a random sample of n observations, and let **a** and **b** be the estimated values of α and β. Then,

$$S = \sum_{i=1}^{n} [Y_i - (\mathbf{a} + \mathbf{b}X_i)]^2$$

is called the *sum of the squares*.

EXAMPLE 16.4.2

Let us now return to Example 16.4.1 and arbitrarily let

$$\mathbf{a} = 15, \quad \text{and } \mathbf{b} = 8.$$

Then, the sum of the squares may be calculated as shown in Table 16.4.2. Thus, the sum of the squares is 403.

TABLE 16.4.2 Calculations for Sum of the Squares

i	X_i	Y_i	$15 + 8X_i$	$[Y_i - (15 + 8X_i)]^2$
1	1	11	23	144
2	1	26	23	9
3	1	18	23	25
4	1	22	23	1
5	2	27	31	16
6	2	36	31	25
7	2	26	31	25
8	2	22	31	81
9	3	39	39	0
10	3	44	39	25
11	3	35	39	16
12	3	45	39	36
				403

On the other hand, suppose we let $\mathbf{a} = 5$ and $\mathbf{b} = 12$. Then, the sum of the squares may be calculated as shown in Table 16.4.3. Thus, the sum of the squares is 319.

We have illustrated that, given a set of sample observations, the value of the sum of the squares will depend on the values that we assign to **a** and **b**. In other words, S is a function of **a** and **b**, and we may, therefore, express that

$$S = f(\mathbf{a},\mathbf{b})$$

for the given sample. The *least-square method* is that which will yield the

TABLE 16.4.3 Calculations for Sum of the Squares

i	X_i	Y_i	$5 + 12X_i$	$[Y_i - (5 + 12X_i)]^2$
1	1	11	17	36
2	1	26	17	81
3	1	18	17	1
4	1	22	17	25
5	2	27	29	4
6	2	36	29	49
7	2	26	29	9
8	2	22	29	49
9	3	39	41	4
10	3	44	41	9
11	3	35	41	36
12	3	45	41	16
				319

smallest possible S for the given set of sample observations. Stated formally, we have:

Definition 16.4.2
Let $\{(Y_1,X_1), \ldots, (Y_n,X_n)\}$ be a random sample of n observations. The *least-square method* is that which will yield values of **a** and **b** such that the sum of the squares (S) is as small as possible.

The next obvious question is then, "How do we find the values of **a** and **b** which will minimize S?" We now propose:

Theorem 16.4.1
Given the sample observations $\{(Y_1,X_1), \ldots, (Y_n,X_n)\}$, the values of **a** and **b** which will minimize S are found by solving the following set of normal equations:

$$\sum_{i=1}^{n} Y_i = n\mathbf{a} + \left(\sum_{i=1}^{n} X_i\right) \mathbf{b}$$

$$\sum_{i=1}^{n} X_i Y_i = \left(\sum_{i=1}^{n} X_i\right) \mathbf{a} + \left(\sum_{i=1}^{n} X_i^2\right) \mathbf{b}.$$

Before we present a proof of the theorem, we will describe first how we can utilize the theorem to find the values of **a** and **b** which will minimize S.

EXAMPLE 16.4.3

Let us return to the set of sample observations given in Example 16.4.1. Table 16.4.4 indicates our calculations for $\sum_{i=1}^{n} X_i$, $\sum_{i=1}^{n} Y_i$, $\sum_{i=1}^{n} X_i Y_i$, and $\sum_{i=1}^{n} X_i^2$. Thus, $\sum_{i=1}^{n} X_i = 24$, $\sum_{i=1}^{n} Y_i = 351$, $\sum_{i=1}^{n} X_i Y_i = 788$, and $\sum_{i=1}^{n} X_i^2 = 56$.

TABLE 16.4.4

i	X_i	Y_i	$X_i Y_i$	X_i^2
1	1	11	11	1
2	1	26	26	1
3	1	18	18	1
4	1	22	22	1
5	2	27	54	4
6	2	36	72	4
7	2	26	52	4
8	2	22	44	4
9	3	39	117	9
10	3	44	132	9
11	3	35	105	9
12	3	45	135	9
	24	351	788	56

The set of normal equations is, then,

$$351 = 12\mathbf{a} + 24\mathbf{b}$$
$$788 = 24\mathbf{a} + 56\mathbf{b}.$$

When we solve the equations for **a** and **b**, we find that

$$\mathbf{a} = 7.75$$
and
$$\mathbf{b} = 10.75.$$

Since the sum of the squares, based on **a** = 7.75 and **b** = 10.75, should be the smallest possible value that S can assume, obviously it should be smaller than that based on **a** = 5 and **b** = 12 or that based on **a** = 15 and **b** = 8.

The sum of the squares (S) based on **a** = 7.75 and **b** = 10.75 is calculated as shown in Table 16.4.5. As we can see, the sum of the squares (S) is 305.75, which is smaller than the two sum of the squares which we have calculated in Example 16.4.2.

The reader might now ask, "How do we know that the sum of the squares (S), calculated according to Theorem 16.4.1, gives the smallest possible value of S?" The validity of the theorem, however, does not rely on a simple intui-

TABLE 16.4.5

i	X_i	Y_i	$7.75 + 10.75X_i$	$[Y_i - (7.75 + 10.75X_i)]^2$
1	1	11	18.50	56.2500
2	1	26	18.50	56.2500
3	1	18	18.50	.2500
4	1	22	18.50	12.2500
5	2	27	29.25	5.0625
6	2	36	29.25	45.5625
7	2	26	29.25	10.5625
8	2	22	29.25	52.5625
9	3	39	40.00	1.0000
10	3	44	40.00	16.0000
11	3	35	40.00	25.0000
12	3	45	40.00	25.0000
				305.7500

tive explanation. Therefore, we ask the reader, if he is sufficiently familiar with calculus, to continue with the remaining portion of this section, and if he is not, to accept the validity of the theorem on faith and to go on to the next section.

The validity of the theorem may be proved as follows. Let

$$S = \sum_{i=1}^{n} [Y_i - (\mathbf{a} + \mathbf{b}X_i)]^2.$$

Then, the values of **a** and **b** which minimize S are found by solving the following equations:

$$\frac{\partial S}{\partial \mathbf{a}} = 0$$

and

$$\frac{\partial S}{\partial \mathbf{b}} = 0,$$

where $\partial S/\partial \mathbf{a}$ and $\partial S/\partial \mathbf{b}$ are the partial derivatives of S with respect to **a** and **b**. Thus, we find the desired minimum values of **a** and **b** by solving the equations

$$\frac{\partial S}{\partial \mathbf{a}} = 2 \sum_{i=1}^{n} [Y_i - (\mathbf{a} + \mathbf{b}X_i)] = 0$$

and

$$\frac{\partial S}{\partial \mathbf{b}} = 2 \sum_{i=1}^{n} [Y_i - (\mathbf{a} + \mathbf{b}X_i)]X_i = 0.$$

If we rearrange these two equations, we have

and
$$\sum_{i=1}^{n} Y_i = n\mathbf{a} + \left(\sum_{i=1}^{n} X_i\right)\mathbf{b}$$
$$\sum_{i=1}^{n} X_i Y_i = \left(\sum_{i=1}^{n} X_i\right)\mathbf{a} + \left(\sum_{i=1}^{n} X_i^2\right)\mathbf{b},$$

which are the normal equations given in Theorem 16.4.1.

EXERCISES

*2. Using the data given in Exercise 1, assume that we have decided to let $\mathbf{a} = 15$ and $\mathbf{b} = 35$, where $\mathbf{a}$ and $\mathbf{b}$ depict the estimators for α and β. Calculate the sum of the square S corresponding to the given values of $\mathbf{a}$ and $\mathbf{b}$.

3. Using the data given in Exercise 1, assume that we have decided to let $\mathbf{a} = 25$ and $\mathbf{b} = 25$, where, again, $\mathbf{a}$ and $\mathbf{b}$ depict the estimators for α and β. Calculate the sum of the square S corresponding to the given values of $\mathbf{a}$ and $\mathbf{b}$.

*4. Using the data given in Exercise 1, find the values of $\mathbf{a}$ and $\mathbf{b}$ which will minimize the sum of the square S. Then, calculate the sum of the square. Compare the sum of the square obtained for this question with those obtained for Exercises 2 and 3.

16.5 Alternate Method of Estimating Parameters

In our least-square calculations of $\mathbf{a}$ and $\mathbf{b}$, we have used the observed values of Y_i and X_i. We can, however, still make the least-square calculations of $\mathbf{a}$ and $\mathbf{b}$ even if we use a specified function of Y_i and X_i. For example, let

$$x_i = X_i - \bar{X},$$

where $\bar{X} = (X_1 + \cdots + X_n)/n$. Then, we might consider that the sample consists of $\{(Y_1,x_1), \ldots, (Y_n,x_n)\}$, instead of $\{(Y_1,X_1), \ldots, (Y_n,X_n)\}$. Then, we may let the regression function be of the form

$$Y_i = \alpha^* + \beta x_i + \varepsilon_i.$$

The following theorem may then be proposed.

Theorem 16.5.1

Let $Y_i = \alpha^* + \beta x_i + \varepsilon_i$. Then, $Y_1, \ldots, Y_n$ are uncorrelated, normally distributed random variables with $E(Y_i) = \alpha^* + \beta x_i$ and $V(Y_i) = \sigma^2$ for all i's.

The assumption underlying the theorem is, of course, that $\varepsilon_1, \ldots, \varepsilon_n$ are uncorrelated, normally distributed random variables with $E(\varepsilon_i) = 0$ and

$V(\varepsilon_i) = \sigma^2$ for all i's. The reader can compare the similarity between Theorems 16.5.1 and 16.3.1.

Now let $\mathbf{a}^*$ and $\mathbf{b}$ be the estimators for α^* and β, respectively. Then, the sum of the squares may be defined as

$$S = \sum_{i=1}^{n} [Y_i - (\mathbf{a}^* + \mathbf{b}x_i)]^2.$$

We now propose:

Theorem 16.5.2

Given the sample observations, $\{(Y_1, x_1), \ldots, (Y_n, x_n)\}$, the values of $\mathbf{a}^*$ and $\mathbf{b}$ which minimize S are given as

$$\mathbf{a}^* = \frac{\sum_{i=1}^{n} Y_i}{n}$$

and

$$\mathbf{b} = \frac{\sum_{i=1}^{n} x_i Y_i}{\sum_{i=1}^{n} x_i^2}.$$

The proof of this theorem is very similar to the proof for Theorem 16.4.1. We ask the reader, if he is familiar with calculus, to prove the theorem. We will, however, illustrate how we can utilize the theorem to find $\mathbf{a}^*$ and $\mathbf{b}$.

EXAMPLE 16.5.1

Let us return to Example 16.4.3. From the data given in the example, we obtain the values shown in Table 16.5.1.

Thus,

$$\mathbf{a}^* = \frac{\sum_{i=1}^{n} Y_i}{n} = \frac{351}{12} = 29.25$$

and

$$\mathbf{b} = \frac{\sum_{i=1}^{n} x_i Y_i}{\sum_{i=1}^{n} x_i^2} = \frac{86}{8} = 10.75.$$

We now observe that the value of $\mathbf{b}$ is the same in this example as it was in Example 16.4.3, whereas the value of $\mathbf{a}^*$ is 29.25 in this example and the value of $\mathbf{a}$ is 7.75 in Example 16.4.3. This difference is a result of the fact

TABLE 16.5.1

i	x_i	Y_i	$x_i Y_i$	x_i^2
1	−1	11	−11	1
2	−1	26	−26	1
3	−1	18	−18	1
4	−1	22	−22	1
5	0	27	0	0
6	0	36	0	0
7	0	26	0	0
8	0	22	0	0
9	1	39	39	1
10	1	44	44	1
11	1	35	35	1
12	1	45	45	1
	0	351	86	8

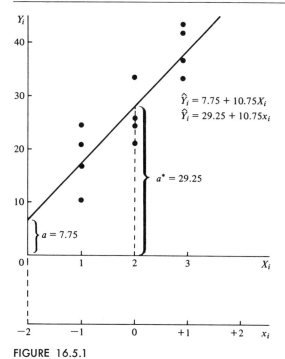

FIGURE 16.5.1

that the sample observations in the two examples are expressed in terms of two different horizontal scales. The apparent discrepancy is explained graphically in Figure 16.5.1. Can we then calculate the value of **a** on the basis of **a*** and **b**? The answer is "yes." First, we observe that $\mathbf{a}^* = \bar{Y}$, since

$$\mathbf{a}^* = \frac{\sum_{i=1}^{n} Y_i}{n} = \bar{Y}.$$

Now, let us take one of the normal equations given in Theorem 16.4.1,

$$\sum_{i=1}^{n} Y_i = n\mathbf{a} + \left(\sum_{i=1}^{n} X_i\right)\mathbf{b},$$

and divide both sides of the equation by n. Then, we obtain

$$\frac{\sum_{i=1}^{n} Y_i}{n} = \mathbf{a} + \mathbf{b}\frac{\left(\sum_{i=1}^{n} X_i\right)}{n}.$$

Thus,
$$\bar{Y} = \mathbf{a} + \mathbf{b}\bar{X}$$

and, in turn,
$$\mathbf{a} = \bar{Y} - \mathbf{b}\bar{X}.$$

Thus, we propose:

Theorem 16.5.3

Let $\mathbf{a}^*$ be the least-square estimate of α^* for $Y_i = \alpha^* + \beta x_i + \varepsilon_i$, and let $\mathbf{a}$ be the least-square estimate of α for $Y_i = \alpha + \beta X_i + \varepsilon_i$. Then,

$$\mathbf{a} = \bar{Y} - \mathbf{b}\bar{X},$$

where $\bar{Y} = \mathbf{a}^*$.

EXAMPLE 16.5.2

Let us return to Example 16.5.1. Since $\bar{Y} = 29.25$, $\bar{X} = 2$, and $\mathbf{b} = 10.75$, we have

$$\mathbf{a} = 29.25 - 2(10.75) = 7.75,$$

which is the same as what we obtained in Example 16.4.3.

EXERCISES

*5. Redefine the regression function given in Exercise 1 as

$$Y_i = \alpha^* + \beta x_i + \varepsilon_i,$$

where $x_i = X_i - \bar{X}$ and $X = (X_1 + \cdots + X_n)/n$. Let $\mathbf{a}^*$ and $\mathbf{b}$ be the estimators for α^* and β. Find the values of $\mathbf{a}^*$ and $\mathbf{b}$ which will minimize the sum of the square S.

*6. Utilizing the value of $\mathbf{a}^*$ obtained in Exercise 5, calculate the value of $\mathbf{a}$ which you have obtained in Exercise 4.

16.6 Mean and Variance of an Estimator

In the preceding section we have defined the least-square values of **a** and **b** as

$$\mathbf{a} = \bar{Y} - \mathbf{b}\bar{X}$$

and

$$\mathbf{b} = \frac{\sum_{i=1}^{n} x_i Y_i}{\sum_{i=1}^{n} x_i^2}.$$

Let us first examine the estimator **b**. Since Y_i is a random variable, **b** must be a random variable. In turn, **a** must be a random variable, since it is a function of two other random variables, $\bar{Y}$ and **b**. This means that we should be able to specify the expected value and the variance of both **a** and **b**. We will discuss **b** first.

Theorem 16.6.1

Let **b** be the least-square estimator of β. Then, the expected value and the variance of **b**, denoted μ_b and σ_b^2, respectively, are

$$\mu_b = \beta$$

and

$$\sigma_b^2 = \frac{\sigma^2}{\sum_{i=1}^{n} x_i^2}.$$

The validity of this theorem may be stated as follows. Let

$$k_i = \frac{x_i}{\sum_{i=1}^{n} x_i^2}.$$

Then,

$$\mathbf{b} = \frac{\sum_{i=1}^{n} x_i Y_i}{\sum_{i=1}^{n} x_i^2} = \sum_{i=1}^{n} k_i Y_i$$

$$= \sum_{i=1}^{n} k_i (\alpha + \beta X_i + \varepsilon_i)$$

$$= \alpha \left(\sum_{i=1}^{n} k_i \right) + \beta \left(\sum_{i=1}^{n} k_i X_i \right) + \sum_{i=1}^{n} k_i \varepsilon_i.$$

However, it can be shown that $\sum_{i=1}^{n} k_i = 0$ and $\sum_{i=1}^{n} k_i X_i = 1$. Thus,

$$\mathbf{b} = \beta + \sum_{i=1}^{n} k_i \varepsilon_i.$$

This means that

$$\mu_b = E\left(\beta + \sum_{i=1}^{n} k_i \varepsilon_i\right)$$

$$= \beta + \sum_{i=1}^{n} (k_i E(\varepsilon_i)) = \beta,$$

since, by assumption, $E(\varepsilon_i) = 0$ for all i's. In turn,

$$\sigma_b^2 = \sum_{i=1}^{n} [k_i^2 V(\varepsilon_i)] = \sum_{i=1}^{n} \left(\frac{x_i}{\sum_{i=1}^{n} x_i^2}\right)^2 \sigma^2$$

$$= \frac{\sigma^2}{\sum_{i=1}^{n} x_i^2},$$

since, by assumption, ε_i is statistically independent and $V(\varepsilon_i) = \sigma^2$ for all i's.

Theorem 16.6.2

Let $\mathbf{a}$ be the least-square estimator for α. Then, the expected value and the variance of $\mathbf{a}$, denoted μ_a and σ_a^2, are

$$\mu_a = \alpha$$

and

$$\sigma_a^2 = \sigma^2 \left(\frac{1}{n} + \frac{\bar{X}}{\sum_{i=1}^{n} x_i^2}\right).$$

We will provide a proof of this theorem for those who wish to follow it through.

We first express that

$$\mathbf{a} = \bar{Y} - \mathbf{b}\bar{X} = \frac{\sum_{i=1}^{n} Y_i}{n} - \bar{X}\left(\sum_{i=1}^{n} k_i Y_i\right).$$

By substituting $(\alpha + \beta X_i + \varepsilon_i)$ for Y_i in equation and by collecting the terms, we obtain

$$\mathbf{a} = \alpha + \sum_{i=1}^{n}\left(\frac{1}{n} - \bar{X}k_i\right)\varepsilon_i.$$

Then

$$\mu_a = E\left[\alpha + \sum_{i=1}^{n}\left(\frac{1}{n} - \bar{X}k_i\right)\varepsilon_i\right]$$

$$= \alpha + \sum_{i=1}^{n}\left[\left(\frac{1}{n} - \bar{X}k_i\right)E(\varepsilon_i)\right] = \alpha,$$

since, by assumption, $E(\varepsilon_i) = 0$ for all i's. Next, we obtain

$$\sigma_a^2 = \frac{\sum_{i=1}^{n} V(\varepsilon_i)}{n^2} + \bar{X}^2\left(\sum_{i=1}^{n} k_i^2 V(\varepsilon_i)\right)$$

$$= \sigma^2\left(\frac{1}{n} + \frac{\bar{X}^2}{\sum_{i=1}^{n} x_i^2}\right),$$

since by assumption, $V(\varepsilon_i) = \sigma^2$ and $\sum_{i=1}^{n} k_i^2 = 1/(\sum_{i=1}^{n} x_i^2)$.

We observe now that, in order to find the values of σ_a^2 and σ_b^2, we also need to know the value of σ^2. For our example problem, by assumption, the value of σ^2 is 25. However, in almost all regression problems, the value of σ^2 is not likely to be known. Therefore, we will have to estimate the value of σ^2 before we can estimate the values of σ_a^2 and σ_b^2.

The reader may recall from our discussion in Chapter 11 that, given the sample of n observations $\{Y_1, \ldots, Y_n\}$, the minimum-variance unbiased estimator of the variance of Y was given as

$$\mathbf{s}^2 = \frac{\sum_{i=1}^{n}(Y_i - \bar{Y})^2}{n - 1}.$$

Similarly, we propose:

Theorem 16.6.3

Given the random sample of n observations $\{(Y_1, X_1), \ldots, (Y_n, X_n)\}$,

$$\mathbf{s}^2 = \frac{\sum_{i=1}^{n}[Y_i - (\mathbf{a} + \mathbf{b}X_i)]^2}{n - 2}$$

is the minimum-variance unbiased estimator of σ^2.

If we let $e_i = Y_i - (a + bX_i)$, then we may express that

$$s^2 = \frac{\sum_{i=1}^{n} e_i^2}{n-2}.$$

EXAMPLE 16.6.1

We will now estimate σ^2 and, in turn, σ_a^2 and σ_b^2. First, from Example 16.4.3, we find that

$$\sum_{i=1}^{12} e_i = 305.75.$$

Therefore,

$$s^2 = \frac{\sum_{i=1}^{12} e_i}{10} = 30.575.$$

How different is the value of s^2 from that of σ^2? Ordinarily, we could not answer such a question, since the value of σ^2 is seldom known. On the other hand, for the given regression problem, we have pointed out that $\sigma^2 = 25$. Thus, the discrepancy between the values of s^2 and σ^2 is 5.575.

Now let s_a^2 and s_b^2 denote the estimators of σ_a^2 and σ_b^2, respectively. Then,

$$s_a^2 = s^2 \left(\frac{1}{12} + \frac{\bar{X}^2}{\sum_{i=1}^{12} x_i^2} \right)$$

$$= 30.575 \left(\frac{1}{12} + \frac{4}{8} \right) = 17.835$$

and

$$s_b^2 = \frac{s^2}{\sum_{i=1}^{12} x_i^2} = \frac{30.575}{8} = 3.822.$$

EXERCISES

*7. For the simple linear regression function given in Exercise 1, find the point-estimate for the variance of ε_i, utilizing the minimum-variance unbiased estimator for $V(\varepsilon_i)$.

*8. On the basis of your calculations in Exercise 7, what is the estimated value of n used to generate the data given in Exercise 1? In order words, what do you guess is the number of coins used to generate the given data?

*9. Assume that, for the data given in Exercise 1, the values of **a** and **b** are to be ascertained by the least-square method. Calculate the estimated values of the variances for **a** and **b**.

16.7 Some Properties of Estimators

In determining the values of **a** and **b**, we have stated that the least-square method possesses some desirable properties. We can now point out some of these properties.

Theorem 16.7.1
Let **a** and **b** be the least-square estimators of α and β, respectively. Then, **a** and **b** are the unbiased estimators of α and β, respectively.

We have already examined the validity of this theorem in the preceding section, and we have pointed out that

$$E(\mathbf{a}) = \alpha, \quad \text{and } E(\mathbf{b}) = \beta.$$

Thus, **a** and **b** are unbiased estimators of α and β.

Theorem 16.7.2
Let **a** and **b** be the least-square estimators of α and β. Then, **a** and **b** are the minimum-variance estimators of α and β among all possible linear, unbiased estimators of α and β.

This theorem offers perhaps one of the most important reasons for favoring the least-square method. It is known as the extension of the Gauss-Markoff theorem on least-squares. We will not, however, prove the theorem, since it would require a good deal of calculus.

Theorem 16.7.3
Let **a** and **b** be the least-square estimators for α and β. Then, **a** and **b** are normally distributed random variables.

The validity of this theorem may be stated as follows. Consider first that

$$\mathbf{b} = \sum_{i=1}^{n} k_i Y_i,$$

where $k_i = x_i/(\Sigma_{i=1}^{n} x_i^2)$. Then, **b** is a linear function of the random variables Y_i. Since Y_i are normally distributed random variables, **b** must also be a normally distributed random variable. Consider next that

$$\mathbf{a} = \bar{Y} - \mathbf{b}\bar{X}.$$

Then, **a** is a linear junction of two other random variables, $\bar{Y}$ and **b**, which are normally distributed. Thus, **a** is a normally distributed random variable. We should, however, point out that **a** and **b** are not statistically independent. In fact, covariance between **a** and **b** may be given as

$$\text{Cov}(\mathbf{a},\mathbf{b}) = \sigma^2 \left(\frac{-\bar{X}}{\sum_{i=1}^{n} x_i^2} \right).$$

On the other hand, it can be shown that **a** and **b** are statistically independent of $\mathbf{s}^2$.

16.8 Inferences for Parameters

The values which we obtain for **a**, **b**, and $\mathbf{s}^2$ are only point estimates for α, β, and σ^2, respectively; therefore, we may either establish confidence intervals for these parameters or test hypotheses pertaining to them.

First, we will consider the statistical inference pertaining to σ^2. To do so, we propose:

Theorem 16.8.1
Let $\mathbf{s}^2 = (\sum_{i=1}^{n} \mathbf{e}_i^2)/(n-2)$. Then, $(n-2)\mathbf{s}^2/\sigma^2$ is chi-square distributed with $n-2$ degrees of freedom.

Although we will not prove this theorem, we ask the reader to compare this theorem with Theorem 15.2.2 for similarity.

EXAMPLE 16.8.1
In Example 16.6.1, we obtained $\mathbf{s}^2 = 30.575$. We will now establish a 95 percent confidence interval for σ^2 around $\mathbf{s}^2$.

On the basis of the preceding theorem, we first derive that

$$P\left(\chi^2_{10,.975} \leq \frac{10\mathbf{s}^2}{\sigma^2} \leq \chi^2_{10,.025}\right) = .95,$$

which may be modified to

$$P\left(\frac{10\mathbf{s}^2}{\chi^2_{10,.025}} \leq \sigma^2 \leq \frac{10\mathbf{s}^2}{\chi^2_{10,.975}}\right) = .95,$$

and, in turn,

$$P\left[\frac{(10)(30.575)}{20.48} \leq \sigma^2 \leq \frac{(10)(30.575)}{3.25}\right] = .95.$$

Thus,

$$P(14.9 \leq \sigma^2 \leq 92.4) = .95,$$

which is the 95 percent confidence interval for σ^2. Recall that, by assumption, $\sigma^2 = 25$. Thus, the confidence interval does in fact include the value of σ^2.

Let us now turn to the problem of making statistical inferences pertaining to α and β. We have already pointed out that **a** and **b** are normally distributed random variables. It appears then that we can evaluate statistical inference pertaining to α and β by using the table of normal probability functions (Table B.1). This would be correct, however, only if we knew the values of σ_a and σ_b. On the other hand, most likely we would be able to determine only the values of $\mathbf{s}_a$ and $\mathbf{s}_b$ in practice, since the value of **s** would have to be used instead of the value of σ.

The reader may recall that we have encountered a similar problem in connection with evaluating statistical inference pertaining to μ of a normal density function. We pointed out that even though the sample mean $\bar{X}$ is normally distributed, we cannot use the table of normal probability functions if the value of $\sigma_{\bar{x}}$ is not known. At the same time, however, we pointed out that the equation

$$\mathbf{t} = \frac{\bar{X} - \mu}{s_{\bar{x}}}$$

is student-t distributed with $n - 1$ degrees of freedom.

We now propose an analogous proposition pertaining to **a** and **b**.

Theorem 16.8.2

Let **a** and **b** be the least-square estimators of α and β. Then,

$$\mathbf{t}_1 = \frac{\mathbf{a} - \alpha}{\mathbf{s}_a}$$

and

$$\mathbf{t}_2 = \frac{\mathbf{b} - \beta}{\mathbf{s}_b}$$

are student-t distributed random variables with $n - 2$ degrees of freedom.

We will illustrate why $\mathbf{t}_2$, for example, is student-t distributed with $n - 2$ degrees of freedom. To do so, we first let

$$\mathbf{t}_2 = \frac{\mathbf{b} - \beta}{\mathbf{s} \Big/ \sqrt{\sum_{i=1}^{n} x_i^2}},$$

which, in turn, may be modified to

$$t_2 = \frac{(b-\beta)/\sigma_b}{s\Big/\sigma_b \sqrt{\sum_{i=1}^{n} x_i^2}} = \frac{(b-\beta)/\sigma_b}{s/\sigma}$$

and, in turn,

$$t_2 = \frac{(b-\beta/\sigma_b)\sqrt{n-2}}{(s/\sigma)\sqrt{n-2}} = \frac{(b-\beta/\sigma_b)\sqrt{n-2}}{\sqrt{(n-2)s^2/\sigma^2}}.$$

Since $(b-\beta)/\sigma_b = Z$ and $(n-2)s^2/\sigma^2 = \chi^2$ with $n-2$ degrees of freedom, we have

$$t_2 = \frac{Z\sqrt{n-2}}{\sqrt{\chi^2_{n-2}}},$$

which satisfies the definition of the student-t distributed random variable with $n-2$ degrees of freedom.

That t_1 in Theorem 16.8.2 is student-t distributed with $n-2$ degrees of freedom may be explained in a similar manner.

EXAMPLE 16.8.2

Let us return to our illustrative regression problem and establish the hypotheses that

Null Hypothesis: $\alpha = 0$
Alternate Hypothesis: $\alpha > 0$

and

Null Hypothesis: $\beta = 0$
Alternate Hypothesis: $\beta > 0$.

The first of these sets of hypotheses may now be tested as follows. On the basis of the preceding theorem, we know that

$$t_1 = \frac{a-0}{s_a}$$

is a student-t distributed random variable with 10 degrees of freedom, if the null hypothesis is true. If we are willing to accept, a .05 α risk, then the null hypothesis $\alpha = 0$ should be rejected if $t_1 > 1.812$. From Examples 16.4.3 and 16.6.1, we find that

$$a = 7.75$$

and

$$s_a = 4.22.$$

Thus,

$$t_1 = \frac{7.75 - 0}{4.22} = 1.836,$$

and, in turn, we should reject the null hypothesis.

Let us next turn to the second set of hypotheses. We know that

$$t_2 = \frac{b - 0}{s_b}$$

is a student-t distributed random variable with 10 degrees of freedom, if the null hypothesis is true. The critical value for the rejection region again is 1.812. From Examples 16.4.3 and 16.6.1, we find that

and
$$b = 10.75$$
$$s_b = 1.95.$$

Thus,

$$t_2 = \frac{10.75 - 0}{1.95} = 5.513,$$

and, in turn, we should reject the null hypothesis.

These decisions have turned out to be the correct decisions, since the true values of the supposedly unknown parameters are $\alpha = 10$ and $\beta = 10$.

EXAMPLE 16.8.3

Instead of testing hypotheses pertaining to α and β, we can establish confidence intervals for them. For example, suppose we wish to establish a 95 percent confidence interval for α and β. The interval in question for α may be established as follows. First, the interval is given by

$$P[a - t_{10,.05}(s_a) \leq \alpha \leq a + t_{10,.05}(s_a)] = .95,$$

where $t_{10,.05} = 2.228$. Thus, if we substitute the values of a, s_a, and $t_{10,.05}$ in this equation, we obtain

$$P(-1.65 \leq \alpha \leq 17.15) = .95,$$

which is the 95 percent confidence interval for α.

The interval for β may be established in a similar way. First, we derive that

$$P(b - t_{10,.05}s_b \leq \beta \leq b + t_{10,.05}s_b) = .95,$$

where $t_{10,.05} = 2.228$. Then, if we substitute the values for b, s_b, and $t_{10,.05}$, we obtain

$$P(6.41 \leq \beta \leq 15.09) = .95,$$

which is the 95 percent confidence interval for β.

We now observe that both of these intervals contain the true values of α and β.

EXERCISES

*10. For the simple linear regression function given in Exercise 1, establish a 95 percent confidence interval for σ^2, where $\sigma^2 = V(\varepsilon_i)$.

*11. For the simple linear regression function given in Exercise 1, the following hypotheses are established:

$$\text{Null Hypothesis: } \alpha = 0$$
$$\text{Alternate Hypothesis: } \alpha > 0.$$

On the basis of the sample observations given in Exercise 1, would you accept the null hypothesis if you are willing to accept a 5 percent α risk?

*12. For the simple linear regression function in Exercise 1, the following hypotheses are established:

$$\text{Null Hypothesis: } \beta = 0$$
$$\text{Alternate Hypothesis: } \beta > 0.$$

Assume that you are willing to accept a 5 percent α risk. Would you then accept the null hypothesis on the basis of the sample evidence given in Exercise 1?

*13. For the simple linear regression function given in Exercise 1, establish a 95 percent confidence interval for α.

*14. For the simple linear regression function given in Exercise 1, establish a 95 percent confidence interval for β.

16.9 Predictions with Regression Equations

In practice we might use a prediction equation for the following prediction problems:

1. to predict the value of $E(Y_i)$, given the value of x_i
2. to predict the value of Y_i, given the value of x_i.

We will evaluate these two problems in the order in which they have been presented.

First, let us discuss the problem of predicting the value of $E(Y_i)$, given the value of x_i. Let $\tilde{Y}_i = \alpha^* + \beta x_i$ and let

$$\hat{Y}_i = \mathbf{a}^* + \mathbf{b} x_i = \bar{Y} + \mathbf{b} x_i.$$

Then, $\hat{Y}_i$ is a point estimator for $\tilde{Y}_i$. We can also show that $E(\hat{Y}_i) = \tilde{Y}_i$, where $E(\mathbf{a}^*) = \alpha^*$ and $E(\mathbf{b}) = \beta$.

Theorem 16.9.1
Let $\hat{Y}_i = \bar{Y} + \mathbf{b}x_i$. Then, the variance of $\hat{Y}_i$, denoted $\sigma_{\hat{y}_i}^2$, is given as

$$\sigma_{\hat{y}_i}^2 = \sigma^2 \left(\frac{1}{n} + \frac{x_i^2}{\sum_{i=1}^{n} x_i^2} \right).$$

The validity of this theorem is not too difficult to prove. We know that
$$V(\hat{Y}_i) = V(\bar{Y}) + x_i^2 V(\mathbf{b}),$$
but that
$$V(\bar{Y}) = \frac{\sigma^2}{n} \quad \text{and} \quad V(\mathbf{b}) = \frac{\sigma^2}{\sum_{i=1}^{n} x_i^2}.$$

Therefore, if we substitute these values into the preceding equation, we have

$$\sigma_{\hat{y}_i}^2 = \sigma^2 \left(\frac{1}{n} + \frac{x_i^2}{\sum_{i=1}^{n} x_i^2} \right),$$

which is the variance of $\hat{Y}_i$.

The formula just obtained shows that, given σ and n, the variance of $\hat{Y}_i$ is also a function of x_i. Since $x_i = X_i - \bar{X}$, the formula indicates that the variance in question will increase as the value of X_i deviates farther and farther from its mean value $\bar{X}$.

We may observe that, in order to find the value of $\sigma_{\hat{y}_i}^2$, we also need to know the value of σ^2. We have, however, pointed out that very likely this is not possible. The alternative then is to obtain an estimate of $\sigma_{\hat{y}}^2$ as

$$s_{\hat{y}_i}^2 = s^2 \left(\frac{1}{n} + \frac{x_i^2}{\sum_{i=1}^{n} x_i^2} \right),$$

where $s_{\hat{y}_i}^2$ denotes the estimated value of $\sigma_{\hat{y}_i}^2$. Then, we can derive:

Theorem 16.9.2
Let
$$U = \frac{\hat{Y}_i - \tilde{Y}_i}{s_{\hat{y}_i}}.$$

Then, U is student-t distributed with $n - 2$ degrees of freedom.

EXAMPLE 16.9.1

Let us now return to our illustrative regression problem on page 423. Assume that we wish to establish a 95 percent confidence interval for $\hat{Y}_i$ for various given values of x_i. Then, we first derive that

$$P\left\{-t_{10,.05} \leq \frac{\hat{Y}_i - \tilde{Y}_i}{s_{\hat{y}_i}} \leq t_{10,.05}\right\} = .95$$

and in turn that

$$P\{\hat{Y}_i - t_{10,.05}s_{\hat{y}_i} \leq Y_i \leq \hat{Y}_i + t_{10,.05}s_{\hat{y}_i}\} = .95.$$

Let $\ell_i^L = \hat{Y}_i - t_{10,.05}s_{\hat{y}_i}$ and $\ell_i^H = \hat{Y}_i + t_{10,.05}s_{\hat{y}_i}$. These values for ℓ_i^L and ℓ_i^H for different x_i are given in Table 16.9.1.

TABLE 16.9.1

x_i	$\hat{Y}_i$	$s_{\hat{y}_i}$	ℓ_i^L	ℓ_i^H
−1.0	18.50	2.52	12.89	24.11
−.5	23.89	1.87	19.72	28.06
.0	29.25	1.60	25.69	32.81
.5	34.61	1.87	30.44	38.78
1.0	40.00	2.52	34.39	45.61

We will now illustrate how a specific interval is established among those given Table 16.9.1. Let $x_i = .5$, for example. Then,

$$\hat{Y}_i = \bar{Y} + bx_i$$
$$= 29.25 + (10.75)(.5) = 34.61.$$

Next, we obtain

$$s_{\hat{y}_i}^2 = s^2 \left(\frac{1}{n} + \frac{x_i^2}{\sum_{i=1}^{n} x_i^2}\right)$$

$$= 30.575 \left(\frac{1}{12} + \frac{.5^2}{8}\right) = 3.500$$

and, in turn, we find that $s_{\hat{y}_i} = \sqrt{3.500} \cong 1.87$. Next, we obtain from Table B.3 in Appendix B that $t_{10,.05} = 2.228$. Finally, we calculate the following:

$$\ell_i^L = \hat{Y}_i - t_{10,.05}s_{\hat{y}_i} = 34.61 - (2.228)(1.87) = 30.44$$
and
$$\ell_i^H = \hat{Y}_i + t_{10,.05}s_{\hat{y}_i} = 34.61 + (2.228)(1.87) = 38.78.$$

The remaining intervals in Table 16.9.1 are established in a similar manner.

16.9 Predictions with Regression Equations 441

In the table we have provided confidence intervals for $\tilde{Y}_i$ for only five arbitrarily selected values of x_i. We can provide such intervals for all different values of x_i, for example, between -1 and 1. Such intervals are shown by the two smooth curves in Figure 16.9.1.

Two observations should be made concerning the confidence intervals depicted by these curves. First, we note that the width of the intervals become larger and larger as $|x|$ becomes larger, because $s_{\hat{y}_i}$ becomes larger as $|x|$ becomes larger. Second, the dotted line in Figure 16.9.1 depicts $\tilde{Y}_i$, the expected value of $\hat{Y}_i$. Our objective is, then, to establish intervals which we would be 95 percent confident would include the value of $\tilde{Y}_i$ for any given value of x_i. We now observe that in this figure the value of $\tilde{Y}_i$ is included in the prediction interval for every x_i between -1 and 1.

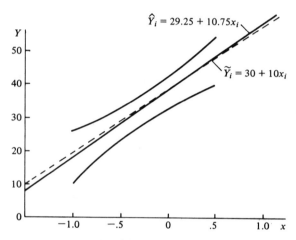

FIGURE 16.9.1 Confidence intervals for $E(Y_i|x_i)$.

We will now discuss the problem of predicting the value of Y_i, given the value of x_i. Assume, for example, that we wish to find the limits of a confidence interval so that

$$P(\ell_i^L \leq Y_i \leq \ell_i^H) = .95.$$

If we knew the value of Y_i, then we could define the limits in question as

and
$$\ell_i^L = \tilde{Y}_i - 1.96\sigma^2$$
$$\ell_i^H = \tilde{Y}_i + 1.96\sigma^2.$$

However, the value of $\tilde{Y}_i$ is not likely to be known. An alternative then is to establish such an interval around the observed value of $\hat{Y}_i$ which in itself is a random variable. To do so, we first establish:

Theorem 16.9.3

Let $Y_d = Y_i - \hat{Y}_i$. Then, $E(Y_d) = 0$ and $V(Y_d)$, denoted $\sigma_{y_d}^2$, is given by

$$\sigma_{y_d}^2 = \sigma^2 \left(1 + \frac{1}{n} + \frac{x_i^2}{\sum_{i=1}^{n} x_i^2} \right).$$

That $E(Y_d) = 0$ is explained as follows. We know that $E(Y_i) = \tilde{Y}_i$ and $E(\hat{Y}_i) = \tilde{Y}_i$. Thus,

$$E(Y_d) = E(Y_i - \hat{Y}_i) = E(Y_i) - E(\hat{Y}_i) = \tilde{Y}_i - \tilde{Y}_i = 0.$$

We also know that
$$V(Y_d) = V(Y_i) + V(\hat{Y}_i).$$

Since
$$V(Y_i) = \sigma^2 \quad \text{and} \quad V(\hat{Y}_i) = \sigma^2 \left(\frac{1}{n} + \frac{x_i^2}{\sum_{i=1}^{n} x_i^2} \right),$$

we can substitute these values and obtain

$$\sigma_{y_d}^2 = \sigma^2 \left(1 + \frac{1}{n} + \frac{x_i^2}{\sum_{i=1}^{n} x_i^2} \right).$$

From our preceding discussion it is apparent that

$$W = \frac{Y_i - \hat{Y}_i}{\sigma_{y_d}}$$

is a normally distributed random variable with a zero mean and unit variance. On the other hand, we are not likely to ascertain the value of σ_{y_d} because of our inability to determine the value of σ. What we will obtain is

$$s_{y_d}^2 = s^2 \left(1 + \frac{1}{n} + \frac{x_i^2}{\sum_{i=1}^{n} x_i^2} \right).$$

which we might use as an estimator for $\sigma_{y_d}^2$.

Now we propose:

Theorem 16.9.4

Let
$$U = \frac{Y_i - \hat{Y}_i}{s_{y_d}}.$$

16.9 Predictions with Regression Equations

Then, U is a student-t distributed random variable with $n - 2$ degrees of freedom.

EXAMPLE 16.9.2

Let us now return to our illustrative regression problem. Assume now that we wish to establish an interval whose limits are ℓ_i^L and ℓ_i^H such that

$$P(\ell_i^L \leq Y_i \leq \ell_i^H) = .95.$$

To do so, we first must derive that

$$P\left\{-t_{10,.05} \leq \frac{Y_i - \hat{Y}_i}{s_{y_d}} \leq t_{10,.05}\right\} = .95$$

and, in turn, that

$$P\{\hat{Y}_i - t_{10,.05}s_{y_d} \leq Y_i \leq \hat{Y}_i + t_{10,.05}s_{y_d}\} = .95.$$

TABLE 16.9.2

x_i	$\hat{Y}_i$	s_{y_d}	ℓ_i^L	ℓ_i^H
-1.0	18.50	6.08	4.96	32.04
$-.5$	23.89	5.84	10.88	36.77
.0	29.25	5.75	16.44	42.06
.5	34.61	5.84	21.60	47.62
1.0	40.00	6.08	26.46	53.54

The intervals for different values of x_i are shown in Table 16.9.2. We will illustrate how a specific interval among those given in the table is established. Let $x_i = .5$. Then,

$$\hat{Y}_i = \bar{Y} + \mathbf{b}x_i = 29.25 + (10.75)(.5) = 34.61.$$

Next, we obtain

$$s_{y_d}^2 = s^2\left(1 + \frac{1}{n} + \frac{x_i^2}{\sum_{i=1}^{n} x_i^2}\right)$$

$$= 30.575\left(1 + \frac{1}{12} + \frac{.5^2}{8}\right) = 34.075,$$

and, in turn, that $s_{y_d} = \sqrt{34.075} = 5.84$. Thus,

$$\ell_i^L = \hat{Y}_i - t_{10,.05}s_{y_d} = 34.61 - (2.228)(5.84) = 21.60$$

and $\quad \ell_i^H = \hat{Y}_i + t_{10,.05}s_{y_d} = 34.61 + (2.228)(5.84) = 47.62.$

The remaining intervals in the table are established in a similar manner. In Figure 16.9.2 we have provided these confidence intervals for all values of x_i between -1 and 1. Two observations should be made pertaining to the intervals shown in the figure: First, we note that the intervals widen as $|x|$ becomes larger, which results from the fact that the value of s_{y_d} becomes larger as $|x|$ becomes larger; Second, if we were to generate additional observations, for example 50 in the same manner that we have generated the first 12 observations, then, about 95 percent of them would fall in the intervals shown in Figure 16.9.2. This means that we should expect about 2 or 3 observations to fall outside of the intervals.

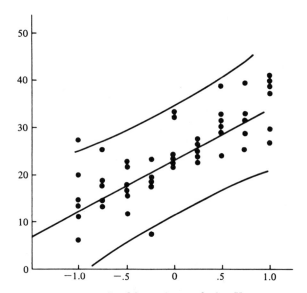

FIGURE 16.9.2 Confidence intervals for Y_i.

The dots shown in Figure 16.9.2 depict the additional 50 observations that we have generated by tossing 100 coins 50 times. We now observe that 48 of these dots fall within the prediction intervals and only 2 of them fall outside of the intervals. Thus, the actual result of our experiment conforms to the theoretical result.

EXERCISES

*15. For the simple linear regression function given in Exercise 1, establish a 95 percent confidence interval for $\tilde{Y}_i = E(Y_i)$ given that (a) $X_i = 4$, (b) $X_i = 5$, (c) $X_i = 6$, (d) $X_i = 7$, and (e) $X_i = 8$.

*16. For the simple linear regression function given in Exercise 1, establish a 95 percent confidence interval for the value of Y_i, given that (a) $X_i = 4$, (b) $X_i = 5$, (c) $X_i = 6$, (d) $X_i = 7$, and (e) $X_i = 8$.

16.10 Re-Examination of Regression Assumptions

We have now completed a description of a simple linear regression analysis. We cannot, however, overemphasize the fact that many of the statistical procedures described in the preceding sections are valid only if the underlying assumptions which we have made pertaining to ε_i are true. We will, therefore, re-examine the significance of these assumptions by evaluating the statistical implications of ignoring them.

Let us now return to the four assumptions pertaining to ε_i we have discussed in Section 16.3. Suppose that all four assumptions are valid. Then, the least-square determined values of **a** and **b** are the minimum-variance estimators for α and β among all possible linear unbiased estimators for α and β. Furthermore, **a** and **b** are, for example, normally distributed random variables, which are statistically independent of s^2, and, in turn,

$$t_1 = \frac{\mathbf{a} - \alpha}{s_a}$$

and

$$t_2 = \frac{\mathbf{b} - \beta}{s_b}$$

are student-t distributed random variables with $n - 2$ degrees of freedom.

Suppose now that we ignore the assumption that ε_i is normally distributed. Then, the least-square values of **a** and **b** are still the minimum-variance estimators among all possible linear unbiased estimators for α and β. On the other hand, **a** and **b** are no longer normally distributed, which means that, in theory, we cannot test any type of statistical inference pertaining to α and β, for example. It has been found, however, in practice that the statistical inferences that we evaluate pertaining to α and β are not very sensitive to the normality assumption for ε_i, particularly if the sample size also happens to be fairly large.

Suppose now that we ignore the assumptions that $V(\varepsilon_i)$ is the same for all i's and that ε_i and ε_j are statistically independent. Then, it can be shown that **a** and **b** are still unbiased estimators for α and β. However, the other properties of the least-square method, which we have elaborated in the preceding sections, are not generally valid; for example, the formulae which we have utilized to find the variances of **a** and **b** can no longer be used.

Finally, suppose that we ignore the assumption that $E(\varepsilon_i) = 0$ for all i's. Then, the least-square values of **a** and **b** are not even unbiased estimators for α and β.

16.11 An Application of Regression Analysis

In the preceding sections we have been primarily concerned with the methodological approach pertaining to regression analysis. We will now describe one practical application of the analysis.

Suppose a construction company builds fine custom houses in California. One problem that the company faces before contracting to build any house is predicting its cost, since the company must specify the total contract price to the customer in advance. The major factor affecting the total construction cost of a house is its size, which is usually measured in terms of the size of living space.

We may assume, however, that, to build any house the construction company must incur an F_i fixed cost and a V_i variable cost per square foot of living space. Then, the total cost Y_i may be expressed as

$$Y_i = F_i + V_i X_i,$$

where X_i denotes the size of living space in terms of the number of square feet. The fixed cost F_i and the unit variable cost V_i are likely to vary from one house to another. Thus, we can assume that F_i and V_i are random variables.

In turn, we may express that

$$F_i = \alpha + \mathbf{u}_i$$
and
$$V_i = \beta + \mathbf{w}_i,$$

where α and β are constants and $\mathbf{u}_i$ and $\mathbf{w}_i$ are random variables. Suppose we assume that

1. $E(\mathbf{u}_i) = 0$ and $E(\mathbf{w}_i) = 0$ for all i's
2. $V(\mathbf{u}_i) = \sigma_u^2$ and $V(\mathbf{w}_i) = \sigma_w^2$ for all i's
3. $\mathbf{u}_i$ and $\mathbf{w}_i$ are statistically independent
4. $\mathbf{u}_i$ and $\mathbf{w}_i$ are normally distributed.

The immediate result of these assumptions is that there exists the average (or expected) amount of fixed cost and unit variable cost, which may be given as

$$E(F_i) = \alpha$$
and
$$E(V_i) = \beta.$$

In turn, we may express that

$$Y_i = \alpha + \mathbf{u}_i + (\beta + \mathbf{w}_i) X_i$$
$$= \alpha + \beta X_i + \mathbf{u}_i + \mathbf{w}_i X_i.$$

16.11 An Application of Regression Analysis

Let
$$\varepsilon_i = \mathbf{u}_i + \mathbf{w}_i X_i.$$
Then,
$$Y_i = \alpha + \beta X_i + \varepsilon_i.$$

We can deduce now that

1. $E(\varepsilon_i) = 0$ for all i's
2. ε_i are statistically independent
3. ε_i are normally distributed

on the basis of the assumptions made pertaining to $\mathbf{u}_i$ and $\mathbf{w}_i$. On the other hand,

4. $V(\varepsilon_i) = \sigma_u^2 + X_i^2 \sigma_w^2,$

which means that $V(\varepsilon_i)$ is a function of X_i. Thus, $V(\varepsilon_i)$ is no longer a constant value but will increase with increasing values of X_i. Even though this is

TABLE 16.11.1

Observation Number	Living Space* X_i	Construction Cost Y_i	x_i
1	2800	$46,700	−300
2	4100	62,600	1000
3	2100	37,000	−1000
4	3200	54,100	100
5	3400	56,400	300
6	3000	48,000	−100
7	2600	43,000	−500
8	3500	57,600	400
9	2100	34,400	−1000
10	4100	66,600	1000
11	3700	57,900	600
12	2500	44,000	−600
13	2800	49,100	−300
14	2900	49,200	−200
15	2900	47,000	−200
16	2500	38,700	−600
17	3800	59,000	700
18	3600	57,100	500
19	2600	46,200	−500
20	3100	57,800	0
21	3400	51,900	300
22	3500	56,600	400
	68,200	$1116,500	0

* Measured in square feet.

likely to be true in reality, for the sake of simplicity we will assume that $V(\varepsilon_i)$ is a constant for all X_i's. This assumption may then be stated as

5. $V(\varepsilon_i) = \sigma^2$ for all i's.

We now observe that assumptions 1, 2, 3, and 5 together constitute the assumptions needed to apply a simple linear regression analysis. Thus, we can now carry out a simple linear regression analysis for the construction company.

Assume that the company has built 22 houses during the last 12 months, and that the sizes of the living spaces vary from 2000 to 4000 square feet. The living spaces and the cost data for these houses are given in the first three columns in Table 16.11.1. The sum of the numbers at the bottom of the second column indicates that the total amount of living space for the 22 houses is 68,200 square feet. The average size of the living space is then

$$\bar{X} = \frac{68{,}200}{22} = 3100.$$

Let $x_i = X_i - 3100$ for each observation. The resulting value of x_i is shown in the last column in the table.

The observed relationship between the cost of building a house and the size of the living space is shown by the dots in Figure 16.11.1.

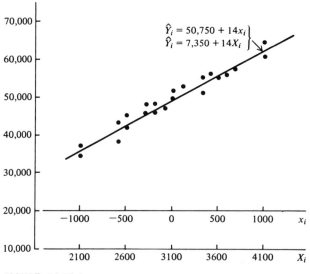

FIGURE 16.11.1

The first step in the application of the regression analysis is to estimate the values of α and β from the data provided in Table 16.11.1 and Figure 16.11.1. To do so, we first obtain the figures shown in Table 16.11.2. Let **a** and **b** be the point-estimates for α and β, respectively. Then,

16.11 An Application of Regression Analysis

$$b = \frac{\Sigma x_i Y_i}{\Sigma x_i^2} = \frac{99{,}390{,}000}{7{,}100{,}000} = 14.00$$

and

$$a = \bar{Y} - b\bar{X} = 50{,}750 - 14(3100) = 7350.$$

TABLE 16.11.2

i	x_i	Y_i	x_i^2	$x_i Y_i$
1	−300	46,700	90,000	−14,010,000
2	1000	62,600	1,000,000	62,600,000
3	−1000	37,000	1,000,000	−37,000,000
4	100	54,100	10,000	5,410,000
5	300	56,400	90,000	16,920,000
6	−100	48,000	10,000	−4,800,000
7	−500	43,000	250,000	−21,500,000
8	400	57,600	160,000	23,040,000
9	−1000	34,400	1,000,000	−34,400,000
10	1000	66,600	1,000,000	66,600,000
11	600	57,900	360,000	34,740,000
12	−600	44,000	360,000	−26,400,000
13	−300	49,100	90,000	−14,730,000
14	−200	49,200	40,000	−9,840,000
15	−200	47,000	40,000	−9,400,000
16	−600	38,700	360,000	−23,220,000
17	700	59,600	490,000	41,720,000
18	500	57,100	250,000	28,550,000
19	−500	46,200	250,000	−23,100,000
20	0	52,800	0	0
21	300	51,900	90,000	15,570,000
22	400	56,600	160,000	22,640,000
	0	1,116,500	7,100,000	99,390,000

Thus, the point-estimate for the fixed cost is \$7350 and the point-estimate for the variable cost per square foot of living space is \$14.00. Now let

$$\hat{Y}_i = 50{,}750 + 14x_i$$

or

$$\hat{Y}_i = 7350 + 14X_i.$$

Then, $\hat{Y}_i$ depicts the regression line generated from the data, which is shown in Figure 16.11.1.

Suppose we plan to evaluate some statistical inferences pertaining to α and β. Then, we need to calculate first the value of s^2 and, in turn, the values of s_a^2 and s_b^2. The calculations required for us to ascertain the value of s^2 are shown in Table 16.11.3. From this table, we obtain

$$s^2 = \frac{\sum_{i=1}^{n}(Y_i - \hat{Y}_i)^2}{n-2} = \frac{70{,}975{,}000}{20} = 3{,}548{,}750$$

TABLE 16.11.3

i	x_i	Y_i	$\hat{Y}_i$	$(Y_i - \hat{Y}_i)^2$
1	−300	46,700	46,550	22,500
2	1000	62,600	64,750	4,622,500
3	−1000	37,000	36,750	62,500
4	100	54,100	52,150	3,802,500
5	300	56,400	54,950	2,102,500
6	−100	48,000	49,350	1,822,500
7	−500	43,000	43,750	562,500
8	400	57,600	56,350	1,562,500
9	−1000	34,400	36,750	5,522,500
10	1000	66,600	64,750	3,422,500
11	600	57,900	59,150	1,562,500
12	−600	44,000	42,350	2,722,500
13	−300	49,100	46,550	6,502,500
14	−200	49,200	47,950	1,562,500
15	−200	47,000	47,950	902,500
16	−600	38,700	42,350	13,322,500
17	700	59,600	60,550	902,500
18	500	57,100	57,750	422,500
19	−500	46,200	43,750	6,002,500
20	0	52,800	50,750	4,202,500
21	300	51,900	54,950	9,302,500
22	400	56,600	56,350	62,500
				70,975,000

and $s = \sqrt{3{,}548{,}750} \cong 1884$. Next we calculate that

$$s_a^2 = s^2 \left(\frac{1}{n} + \frac{\bar{X}^2}{\sum_{i=1}^{n} x_i^2} \right)$$

$$= 3{,}548{,}750 \left(\frac{1}{22} + \frac{3100^2}{7{,}100{,}000} \right) \cong 4{,}964{,}614$$

and $s_a = \sqrt{4{,}964{,}614} \cong 2228$. Then, we calculate that

$$s_b^2 = \frac{s^2}{\sum_{i=1}^{n} x_i^2} = \frac{3{,}548{,}750}{7{,}100{,}000} \cong .500$$

and

$$s_b = \sqrt{.500} \cong .707.$$

16.11 An Application of Regression Analysis

We have pointed out that the values **a** = 7350 and **b** = 14 are the point-estimates for α and β, respectively. We can now establish interval-estimates for α and β as follows. Assume that we wish to establish 95 percent confidence intervals. Then,

$$P\left\{-t_{20,.05} \leq \frac{\mathbf{a} - \alpha}{\mathbf{s}_a} \leq t_{20,.05}\right\} = .95$$

and

$$P\left\{-t_{20,.05} \leq \frac{\mathbf{b} - \beta}{\mathbf{s}_b} \leq t_{20,.05}\right\} = .95.$$

In turn,

$$P\{\mathbf{a} - t_{20,.05}\mathbf{s}_a \leq \alpha \leq \mathbf{a} + t_{20,.05}\mathbf{s}_a\} = .95$$

or $\quad P\{7350 - (2.086)(2228) \leq \alpha \leq 7350 + (2.086)(2228)\} = .95,$

and

$$P\{\mathbf{b} - t_{20,.05}\mathbf{s}_b \leq \beta \leq \mathbf{b} + t_{20,.05}\mathbf{s}_b\} = .95$$

or $\quad P\{14 - (2.086)(.707) \leq \beta \leq 14 + (2.086)(.707)\} = .95.$

Thus, the 95 percent confidence intervals are

$$P(\$2702 \leq \alpha \leq \$11,998) = .95$$

and $\quad P(\$12.53 \leq \beta \leq \$15.47) = .95.$

This means that the construction company can be 95 percent confident that in building any house the fixed cost is likely to be between $2702 and $11,998 and the variable cost per square foot of living space is likely to be between $12.53 and $15.47.

Although the interval-estimates for the fixed and variable costs per square feet should be of interest to the company, the interval-estimates for the total cost of building a house of a given size should also be of interest to the company. Two types of interval-estimates can be made for this purpose:

1. interval-estimate for the average (or expected) value of the total cost of building houses of a given size
2. interval-estimate for the actual value of the total cost of building a single house of a given size.

The interval-estimate for the first objective really amounts to establishing an interval-estimate for $E(Y_i) = \tilde{Y}_i$ for the given value of x_i. The intervals thus established are shown in Table 16.11.4.

We will now illustrate how one of the intervals in this table has been established. Let $x_i = 600$, for example. Then, the point-estimate for $E(Y_i)$ is given as

$$\hat{Y}_i = \bar{Y} + \mathbf{b}x_i$$
$$= 50{,}750 + (14)(600) = 59{,}150.$$

Next, we calculate that

TABLE 16.11.4

x_i	$\hat{Y}_i$	$s_{\hat{y}_i}$	ℓ_i^L	ℓ_i^H
−1000	36,750	813	35,054	38,446
−800	39,550	694	38,103	40,997
−600	42,350	584	41,131	43,569
−400	45,150	491	44,125	46,175
−200	47,950	426	47,062	48,838
0	50,750	402	49,912	50,588
200	53,550	426	52,662	54,438
400	56,350	491	55,325	57,374
600	59,150	584	57,931	60,369
800	61,950	694	60,502	63,397
1000	64,750	813	63,053	66,446

$$s_{\hat{y}_i}^2 = s^2\left(\frac{1}{n} + \frac{x_i^2}{\Sigma x_i^2}\right)$$

$$= 3{,}548{,}750\left(\frac{1}{22} + \frac{600^2}{7{,}100{,}000}\right) = 341{,}243$$

and $s_{\hat{y}_i} \cong \sqrt{341{,}243} = 584$. Since

$$P\left\{-t_{20,.05} \leq \frac{\hat{Y}_i - \tilde{Y}_i}{s_{\hat{y}_i}} \leq t_{20,.05}\right\} = .95$$

and

$$P\{\hat{Y}_i - t_{20,.05}s_{\hat{y}_i} \leq \tilde{Y}_i \leq \hat{Y}_i + t_{20,.05}s_{\hat{y}_i}\} = .95,$$

we have

$$\ell^L = 59{,}150 - (2.086)(584) = 57{,}931$$
and
$$\ell^H = 59{,}150 + (2.086)(584) = 60{,}369.$$

The remaining intervals are established in a similar manner.

The significance of the intervals established for the construction company can now be stated as follows. Suppose the company were to build many houses, each with 3700 square feet of living space (note that $3700 = \bar{X} + x_i$, where $\bar{X} = 3100$ and $x_i = 600$). Then, the company can be 95 percent sure that the average of the total costs of these houses is likely to be between $57,931 and $60,369.

The interval-estimate for the actual, total cost of building a house of a given size is found by establishing an interval-estimate for Y_i for the given x_i. The intervals thus established for a number of different x_i's are shown in Table 16.11.5. We will again illustrate how one of these intervals is established.

16.11 An Application of Regression Analysis

TABLE 16.11.5

x_i	Y_i	s_{y_d}	ℓ_i^L	ℓ_i^H
−1000	36,750	2,052	32,470	41,030
−800	39,550	2,007	35,362	43,738
−600	42,350	1,972	38,236	46,464
−400	45,150	1,947	41,089	49,221
−200	47,950	1,931	43,921	51,979
0	50,750	1,926	46,732	54,768
200	53,550	1,931	49,521	57,579
400	56,350	1,947	52,289	60,411
600	59,150	1,972	55,036	63,264
800	61,950	2,007	57,762	66,138
1000	64,750	2,052	60,470	69,030

Let $x_i = 600$, for example, and let $Y_d = Y_i - \hat{Y}_i$. Then,

$$s_{y_d}^2 = s^2 + s_{\hat{y}_i}^2 = 3{,}548{,}750 + 341{,}243 = 3{,}889{,}993$$

and $\quad s_{y_d} = \sqrt{3{,}889{,}993} = 1{,}972.$

Since

$$P\left\{-t_{20,.05} \leq \frac{Y_i - \hat{Y}_i}{s_{Y_d}} \leq t_{20,.05}\right\} = .95,$$

and, in turn,

$$P\{\hat{Y}_i - t_{20,.05}s_{y_d} \leq Y_i \leq \hat{Y}_i + t_{20,.05}s_{y_d}\} = .95,$$

we have

$$\ell_i^L = 59{,}150 - (2.086)(1972) = 55{,}036$$

and $\quad \ell_i^H = 59{,}150 + (2{,}086)(1972) = 63{,}264.$

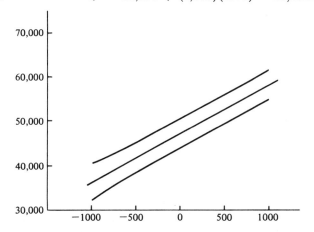

FIGURE 16.11.2

The remaining intervals are established in a similar manner and are shown graphically in Figure 16.11.2. The significance of the intervals given in this figure can be stated as follows. Suppose the construction company is in the process of negotiating the contract for a house with 3700 square feet of living space. Then, it may be 95 percent confident that the total cost of building the house will be between $55,036 and $63,264. Thus, if the company can price the house at more than $63,264, for example, the likelihood of incurring a loss on the house will be very small.

EXERCISES

*17. In the operation of a chemical process, it is known that the yield of the process is influenced by such factors as temperature, pressure, catalysts, and so on. The yield data for a given chemical process have been collected for three different temperature settings, as shown in Table 16.11.6.

TABLE 16.11.6

Experiment Run	Temperature	Yield
1	120	21.94
2	120	21.36
3	120	22.23
4	120	21.92
5	130	23.20
6	130	21.79
7	130	23.40
8	130	23.82
9	140	25.13
10	140	23.34
11	140	24.64
12	140	23.56

a. Utilizing the data provided, ascertain the apparent relationship between the temperature and the yield for the process. On the basis of the relationship that you have observed, what will be the effect on the yield of raising the temperature by 1 degree?

b. Suppose that the process is to be run continuously with a given temperature setting. What is your point-estimate of the variance of the yield for the given temperature setting? What is your 95 percent confidence interval for the variance in question?

c. In (a) you were asked to calculate the effect on the yield of raising the temperature by 1 degree. Establish a 95 percent confidence interval for your answer in (a).

16.11 An Application of Regression Analysis 455

d. Suppose the process is to be run continuously at a temperature setting of 125 degrees. Establish a 95 percent confidence interval for the *average* yield for the given temperature setting.

e. Suppose the process is to be run continuously at a temperature setting of 135 degrees. Establish a 95 percent confidence-interval for the yield for the given temperature setting.

18. A taxicab company owns a large fleet of taxicabs. It is believed that the per-mile operating cost of a taxicab is influenced by its age. A simple random sample of 12 taxicabs has been obtained from the company's fleet, and a careful study has been made to ascertain the operating costs of each taxicab per mile. The results of the study are shown in Table 16.11.7.

TABLE 16.11.7

Taxicab	Age (in months)	Per-mile Operating Cost (in cents)
1	3	9.6
2	17	10.5
3	24	11.9
4	30	14.3
5	19	11.3
6	11	10.8
7	22	11.9
8	20	10.1
9	15	10.3
10	8	10.1
11	35	14.6
12	18	12.8

a. Utilizing the data provided, ascertain the apparent relationship between the age of a taxicab and its per-mile operating costs. On the basis of the relationship which you have observed, how much is per-mile operating cost expected to increase when a taxicab gets one month older?

b. Suppose the company collects the per-mile operating cost data for all of its taxicabs of a given age, for example, 20 months. What is your point-estimate of the variance in the per-mile operating costs for this group of taxicabs? What is your 95 percent confidence interval for the variance in question?

c. In (a) you have ascertained the effect of a taxicab's gaining 1 month on its per-mile operating cost. Establish a 95 percent confidence interval for your answer in (a).

d. Suppose the company collects the operating cost data for all of its taxicabs which are 30 months of age. Establish a 95 percent confidence interval for the mean operating cost for this group.

e. Suppose the company selects a single taxicab which is 20 months old. Establish a 95 percent confidence interval for the operating cost for this particular taxicab.

Appendix to Chapter 16
Computer Program for Simple Regression Analysis

This appendix contains a listing of a computer program for a simple regression analysis and operating instructions for the program.

A.16.1 Description of the Program

The computer program to be presented is completely interactive so that persons with little or no programming experience can use it. It is written in Extended Basic language, as implemented in the WANG 3300 time-sharing system. With minor modifications, if any, the program can be used in any time-sharing system which accepts Extended Basic language.

A listing of the program is now provided in Figure A.16.1.1.

A.16.2 Data-Input Phase

The following three instructions will load and execute the program.

> START
> LOAD XX/YY
> RUN

In the Load instruction, XX and YY, respectively, depict the device number XX and file number YY for the program in the storage.

```
1000 PRINT "**********SIMPLE LINEAR REGRESSION********"
1011 DIM X(25),Y(25),Z(25),K(25),R(25),G(25),P(25),Q(25)
1012 L=14
1015 PRINT
1020 PRINT"ENTER NUMBER OF OBSERVATIONS:";
1030 INPUT N
1040 FOR I=1 TO N
1041 PRINT
1050 PRINT " X(";I;")=";
1060 INPUT X(I)
1070 PRINT " Y(";I;")=";
1071 INPUT Y(I)
1080 NEXT I
1082 PRINT
1084 PRINT " ARE ALL ENTRIES CORRECT?      (YES=1,NO=0)";
1086 INPUT F
1088 IF F=1 THEN 1104
1090 PRINT
1091 PRINT "ENTER INDEX OF INCORRECT OBSERVATION";
1092 INPUT I
1093 PRINT " X(";I;")=";
1094 INPUT X(I)
1095 PRINT "Y(";I;")=";
1096 INPUT Y(I)
1097 PRINT "MORE CHANGES ? (YES=1,NO=0)";
1098 INPUT F
1100 IF F=1 THEN 1091
1104 PRINT
1106 LET S1,S2,S7,V1,V2=0
1110 FOR I= 1 TO N
1120 S1=S1+X(I)
1125 S2=S2+Y(I)
1130 NEXT I
1140 PRINT "SUM (X)=";S1;"      ";"SUM (Y)=";S2
1160 M1=S1/N
1165 M2=S2/N
1170 PRINT
1180 PRINT "MEAN (X) =";M1;"     MEAN (Y) =";M2
1181 PRINT
1195 PRINT " I";TAB(10);"X(I)-XBAR";TAB(10+L);"  Y(I)";TAB(10+2*L);
1196 PRINT "[(X-XBAR)*Y]";TAB(10+3*L);"[(X-XBAR)↑2]"
1200 FOR I = 1 TO N
1210 Z(I)=X(I)-M1
1220 W1 = Z(I)*Y(I)
1230 W2 = (Z(I))↑2
1250 PRINT I;TAB(10);Z(I);TAB(10+L);Y(I);TAB(10+2*L);W1;TAB(10+3*L);W2
1260 V1=V1+W1
1270 V2=V2+W2
1280 NEXT I
1281 PRINT
1290  PRINT "     SUMS OF  CROSS PRODUCTS   "
1295 PRINT "SUM([X-XBAR]*Y)=";V1;"     SUM([X-XBAR]↑2)=";V2
1300 B=V1/V2
1301 A=M2-(B*M1)
1305 PRINT"---------------------------------------"
1310 PRINT "    A=";A;"         B=";B
1320 PRINT"---------------------------------------"
1330 FOR I=1 TO N
1332 S7=S7+(Y(I)-(A+B*X(I)))↑2
1335 NEXT I
1340 S9=S7/(N-2)
```

FIGURE A.16.1.1

```
1350 S8=S9*((1/N)+((M1↑2)/V2))
1355 S3=SQR(S8)
1360 S4=S9/V2
1362 S6=SQR(S4)
1365 T1=A/S3
1366 T2=B/S6
1370 PRINT
1371 PRINT "      SUM OF THE SQUARES=";S7
1372 PRINT
1373 PRINT "            [S↑2]=";S9
1374 PRINT
1375 PRINT "   [SA↑2]=";S8;TAB(25);"[SA]=";S3
1380 PRINT
1385 PRINT "   [SB↑2]=";S4;TAB(25);"[SB]=";S6
1390 PRINT
1395 PRINT "   [T1]  =";T1;TAB(25);"[T2]=";T2
1410 PRINT
1412 PRINT "DO YOU WISH TO COMPUTE CONFIDENCE INTERVALS"
1414 PRINT "FOR THE Y ESTIMATES ?   (YES =1,NO=0) ";
1415 INPUT E
1416 IF E=0 THEN 1650
1418 PRINT
1420 PRINT" ENTER VALUE OF RANDOM VARIABLE 'T':";
1425 INPUT T
1427 PRINT
1430 PRINT"FOR HOW MANY X'S DO WISH TO HAVE"
1431 PRINT "CONFIDENCE INTERVALS COMPUTED ";
1450 INPUT I1
1460 FOR I=1 TO I1
1470 PRINT"    X(";I;")=";
1475 INPUT X(I)
1480 K(I)=M2+B*X(I)
1490 R(I)=S9*((1/N)+(X(I)↑2/V2))
1500 G(I)=SQR(R(I))
1510 P(I)=K(I)-T*G(I)
1515 Q(I)=K(I)+T*G(I)
1520 NEXT I
1561 PRINT
1570 PRINT"   X(I)";TAB(10);"[Y(I)EST]";TAB(10+L);"[SY(I)EST]";TAB(10+2
1571 PRINT " LLI";TAB(10+3*L);" LHI"
1575 FOR I=1TO I1
1580 PRINTX(I);TAB(10);K(I);TAB(10+L);G(I);TAB(10+2*L);
1581 PRINT P(I);TAB(10+3*L);Q(I)
1585 NEXT I
1586 PRINT
1600 PRINT
1610 PRINT "ANOTHER T?       (YES=1,NO=0)";
1620 INPUT.
1630 IF E    THEN 1420
1640 PRIN
1642 PRINT"MORE X'S ? (YES=1,NO=0)";
1643 INPUT E
1644 IF E=1 THEN 1427
16 0 PRINT
1655 PRINT "DO YOU WISH TO COMPUTE CONFIDENCE INTERVALS "
1660 PRINT "FOR   Y ? (YES=1 , NO=0)";
1662 INPUT E
1664 IF E=0 THEN1998
1668 PRINT
1670 PRINT " ENTER VALUE OF RANDOM VARIABLE 'T':";
1675 INPUT T
```

FIGURE A.16.1.1 (continued)

```
1678 PRINT
1680 PRINT"FOR HOW MANY X'S DO YOU WISH TO HAVE"
1685 PRINT "CONFIDENCE INTERVALS COMPUTED ";
1690 INPUT I1
1700 FOR I= 1 TO I1
1710 PRINT"   X(";I;")=";
1800 INPUT X(I)
1820 K(I)=M2+B*X(I)
1830 R(I)=S9*(1+(1/N)+(X(I)↑2/V2))
1840 G(I)=SQR(R(I))
1850 P(I)=K(I)-T*G(I)
1855 Q(I)=K(I)+T*G(I)
1860 NEXT I
1900 PRINT
1910 PRINT " X(I)";TAB(10);" Y(I)";TAB(10+L);" SY(I)";TAB(10+2*L);
1920 PRINT " LLI";TAB(10+3*L);" LHI"
1930 FOR I=1 TO I1
1940 PRINT X(I);TAB(10);K(I);TAB(10+L);G(I);TAB(10+2*L);
1950 PRINT P(I);TAB(10+3*L);Q(I)
1960 NEXT I
1962 PRINT
1964 PRINT " ANOTHER   T ? (YES=1 NO=0)";
1965 INPUT E
1966 IF E=1 THEN 1668
1980 PRINT
1990 PRINT "MORE X'S ? (YES=1,NO=0)";
1992 INPUT E
1994 IF E=1 THEN    1678
1998 PRINT
2000 PRINT "*********************************************"
9999 END
```

FIGURE A.16.1.1 (continued)

After the execution phase begins, the computer will type

ENTER NUMBER OF OBSERVATIONS:?

At this point, the user should type the value of N, where N depicts the number of observations in the sample.

The computer will then type

X(1)?

This is the instruction by the computer to the user to type the value of X_1. When the user types the value of X_1, the computer will next type

Y(1)?

The user should type the value of Y_1 at this point. The values $X_2, Y_2, \ldots, X_n, Y_n$ are read into the computer in similar manner.

After X_n, Y_n have been read into the computer, the computer will type

ARE ALL ENTRIES CORRECT?

At this point, the user should examine whether or not all sample observations have been read correctly into the computer. If so, he should type "1." If some observations have been erroneously read into the computer, he should type "0." If the user types "1," the computer will go to the calculation phase. If the user types "0," then, the computer will go to the input-correction phase and type

ENTER INDEX OF INCORRECT OBSERVATION?

```
**********SIMPLE LINEAR REGRESION********
ENTER NUMBER OF OBSERVATIONS:? 12
     X( 1 )=?1
     Y( 1 )=?122

     X( 2 )=?1
     Y( 2 )=?26

     X( 3 )=?1
     Y( 3 )=?18

     X( 4 )=?1
     Y( 4 )=?22

     X( 5 )=?2
     Y( 5 )=?27

     X( 6 )=?2
     Y( 6 )=?36

     X( 7 )=?2
     Y( 7 )=?26

     X( 8 )=?2
     Y( 8 )=?22

     X( 9 )=?3
     Y( 9 )=?39

     X( 10 )=?3
     Y( 10 )=?44

     X( 11 )=?3
     Y( 11 )=?35

     X( 12 )=?3
     Y( 12 )=?45

     ARE ALL ENTRIES CORRECT?      (YES=1,NO=0)? 0

     ENTER INDEX OF INCORRECT OBSERVATION? 1
       X( 1 )=?1
       Y( 1 )=?11
     MORE CHANGES ? (YES=1,NO=0)? 0
```

FIGURE A.16.2.1

The user then should type the index number for an erroneously read-in observation. For example, suppose the values of X_1 have been erroneously read in. Then, he should type "1." At this point, the computer will type

$$X(1)?$$

The user can now type the correct value of X_1. The computer will then type

$$Y(1)?$$

The user should type the correct value of Y_1 even if the value of Y_1 has been read in correctly the first time.

Next, the computer will type

$$\text{MORE CHANGES?}$$

If there are more changes to be made, the user should type "1"; otherwise, he should type "0." If the user types "1," the computer will ask the user to type the index number of the next erroneously read-in observation. The correction procedure to be followed is the same as the one just described. If the user types "0," the computer will go to the calculation phase.

The manner in which the computer and the user interact during the data-input phase is illustrated by the printout shown in Figure A.16.2.1. The data read into the computer for the printout are those given in Example 16.4.1.

A.16.3 Calculation Phase

During the calculation phase the computer makes a number of routine calculations required in a regression analysis and prints out the results of these calculations.

The notations used in the text and in the computer printout are compared in the following chart.

Text Notations	Computer Notations
ΣX_i	SUM (X)
ΣY_i	SUM (Y)
$\bar{X}$	MEAN (X)
$\bar{Y}$	MEAN (Y)
$(X_i - \bar{X})^2$	(X − XBAR)↑2
a	A
b	B
s^2	[S↑2]
s_a	SA
s_b	SB
t_1	T1
t_2	T2

The printout shown in Figure A.16.3.1 contains the results of various calculations for a simple regression analysis for the data which were read into the computer during the data-input phase. The reader should compare the information contained in the printout with the results of calculations shown in Examples 16.4.3, 16.5.1, 16.6.1, and 16.8.2.

```
SUM (X)= 24        SUM (Y)= 351

MEAN (X) = 2       MEAN (Y) = 29.25

   I       X(I)-XBAR       Y(I)       [(X-XBAR)*Y]    [(X-XBAR)↑2]
   1          -1            11            -11              1
   2          -1            26            -26              1
   3          -1            18            -18              1
   4          -1            22            -22              1
   5           0            27              0              0
   6           0            36              0              0
   7           0            26              0              0
   8           0            22              0              0
   9           1            39             39              1
  10           1            44             44              1
  11           1            35             35              1
  12           1            45             45              1

       SUMS OF  CROSS PRODUCTS
  SUM([X-XBAR]*Y)= 86     SUM([X-XBAR]↑2)= 8
  ----------------------------------------
       A= 7.75          B= 10.75
  ----------------------------------------

       SUM OF THE SQUARES= 305.75

       [S↑2]= 30.575

  [SA↑2]= 17.835417      [SA]= 4.2231999

  [SB↑2]= 3.821875       [SB]= 1.9549616

  [T1]  = 1.8351014      [T2]= 5.4988292
```

FIGURE A.16.3.1

A.16.4 Interval-Prediction Phase

After the computer finishes printing out the outputs shown in the preceding figure, it will type

DO YOU WISH TO COMPUTE CONFIDENCE INTERVAL
FOR THE Y ESTIMATES?

This is an inquiry by the computer as to whether or not the user wishes to calculate confidence intervals for $E(Y_i)$ for various given values of x_i where

$x_i = X_i - \bar{X}$. If the user wishes to calculate such intervals, he should type "1," otherwise, he should type "0." If he types "1," then, the computer will type

ENTER VALUE OF RANDOM VARIABLE 'T':?

Suppose the user wishes to calculate a 95 percent confidence interval. Then, the value of T should be such that $P(-T \leq Y \leq T) = .95$, where Y is a student-t distributed random variable. Thus, if the degree of freedom is 10, for example, then $T = 2.228$. When the user types the value of T, the computer will type

FOR HOW MANY X'S DO WISH TO HAVE
CONFIDENCE INTERVALS COMPUTED?

At this point, the user should type the number of values of x_i for which he wishes to calculate the confidence intervals. When the user types this number, the computer will immediately ask the first value of x_i by typing

X(1)?

The user should then type the first value of x_i where $x_i = X_i - \bar{X}$. The computer will then ask the user to type the value of the second x_i, and so on. After all the values of x_i have been read into the computer, the computer will calculate the confidence intervals for these values and print out the intervals.

The computer will next ask the user whether or not he wishes to calculate the confidence intervals with a different value of T. If the answer is "yes," the user should type "1." In that case, the computer will calculate the corresponding confidence intervals and print them out. If the user types "0," the computer will ask the user if he wishes to calculate the intervals with different values of x_i. If the user types "0," then, the computer will type

DO YOU WISH TO COMPUTE CONFIDENCE INTERVALS FOR Y?

This is an inquiry by the computer as to whether or not the user wishes to calculate confidence intervals for Y_i for various given values of x_i. If the user types "1," then, the computer will ask the user to input the value of T as well as the values of x_i for which the user wishes to calculate the intervals. After these values are read into the computer, the computer will calculate the desired confidence intervals and print out the results.

The computer will then ask the user whether he wishes to calculate additional intervals with a different value of T or a different value of x_i. If the user does not want any more of such intervals to be calculated, he should type "0" to each computer inquiry. Then, the computer will terminate the execution.

The manner by which the computer and the user interact during the interval-prediction phase is illustrated by the printout shown in Figure A.16.4.1. The reader should compare this printout with Tables 16.9.1 and 16.9.2.

```
DO YOU WISH TO COMPUTE CONFIDENCE INTERVALS
FOR THE Y ESTIMATES ?   (YES =1,NO=0):? 1

ENTER VALUE OF RANDOM VARIABLE 'T':? 2.228

FOR HOW MANY X'S DO WISH TO HAVE
CONFIDENCE INTERVALS COMPUTED ? 5
   X( 1 )=?-1
   X( 2 )=?-.5
   X( 3 )=?0
   X( 4 )=?.5
   X( 5 )=?1

   X(I)      [Y(I)EST]     [SY(I)EST]      LLI          LHI
   -1         18.5         2.5238446     12.876874    24.123126
   -.5        23.875       1.8717332     19.704778    28.045222
    0         29.25        1.5962195     25.693623    32.806377
    .5        34.625       1.8717332     30.454778    38.795222
    1         40           2.5238446     34.376874    45.623126

ANOTHER T?       (YES=1,NO=0)?  0

MORE X'S ? (YES=1,NO=0)?  0

DO YOU WISH TO COMPUTE CONFIDENCE INTERVALS
FOR   Y ? (YES=1 , NO=0)? 1

ENTER VALUE OF RANDOM VARIABLE 'T':? 2.228

FOR HOW MANY X'S DO YOU WISH TO HAVE
CONFIDENCE INTERVALS COMPUTED ?5
   X( 1 )=?-1
   X( 2 )=?-.5
   X( 3 )=?0
   X( 4 )=?.5
   X( 5 )=?1

   X(I)       Y(I)         SY(I)          LLI          LHI
   -1         18.5         6.0782227     4.95772      32.04228
   -.5        23.875       5.8376694    10.868673     36.881327
    0         29.25        5.7552512    16.4273       42.0727
    .5        34.625       5.8376694    21.618673     47.631327
    1         40           6.0782227    26.45772      53.54228

ANOTHER  T ? (YES=1 NO=0)?  0

MORE X'S ? (YES=1,NO=0)?  0

*******************************************
```

FIGURE A.16.4.1

Bibliography

PROBABILITY

Introductory

Goldberg, S., *Probability: An Introduction*. Englewood Cliffs, N.J., Prentice-Hall, 1960.

Hodges, J., and Lehman, E., *Basic Concepts of Probability and Statistics*. San Francisco, Holden Day, 1964.

Mosteller, F., Rourke, R., and Thomas, G., Jr., *Probability and Statistics*. Reading, Mass., Addison-Wesley, 1961.

Intermediate

Parzen, E., *Modern Probability Theory and Its Applications*. New York, John Wiley, 1960.

Advanced

Feller, W., *An Introduction to Probability Theory and Its Applications*, Vols. 1 and 2. New York, John Wiley, 1966.

STATISTICS

Introductory

Blackwell, D., *Basic Statistics*. New York, McGraw-Hill, 1969.

Chernoff, H., and Moses, L., *Elementary Decision Theory*. New York, John Wiley, 1959.

Clelland, R., and others, *Basic Statistics with Business Applications*. New York, John Wiley, 1966.

Dyckman, T. R., Smidt, S., and McAdams, A. K., *Management Decision Making under Uncertainty*. New York, Macmillan, 1969.
Ehrenfeld, S., and Littauer, S., *Introduction to Statistical Method*. New York, McGraw-Hill, 1964.
Forester, J., *Statistical Selection of Business Strategies*. Homewood, Ill., Richard D. Irwin, 1968.
Hadley, G., *Probability and Statistical Decision Theory*, San Francisco, Holden Day, 1967.
Hamburg, M., *Statistical Analysis for Decision Making*. New York, Harcourt Brace Javonavich, 1970.
Neter, J., and Wasserman, W., *Fundamental Statistics for Business and Economics*, 3d ed. Boston, Allyn and Bacon, 1966.
Raiffa, H., *Decision Analysis*. Reading, Mass., Addison-Wesley, 1968.
Richmond, S., *Statistical Analysis*, 2d ed. New York, Ronald Press, 1964.
Schlaifer, R., *Probability and Statistics for Business Decisions*. New York, McGraw-Hill Book Co., 1959.
Summers, G., and Peters, W., *Statistical Analysis for Decision Making*. Englewood Cliffs, N.J., Prentice-Hall, 1968.
Wolf, F. L., *Elements of Probability and Statistics*. New York, McGraw-Hill, 1962.
Wonnocott, T. H. and R. J., *Introductory Statistics*. New York, John Wiley, 1969.
Yamane, T., *Statistics, An Introductory Analysis*, 2d ed. New York, Harper & Row, 1967.

Intermediate

Anderson, R. L., and Bancroft, T. A., *Statistical Theory in Research*. New York, McGraw-Hill, 1952.
Hoel, P. G., *Introduction to Mathematical Statistics*, 3d ed. New York, John Wiley, 1962.
Lindgreen, B. W., *Statistical Theory*, 2d ed. New York, Macmillan, 1968.
Mood, A., and Graybill, F., *Introduction to Theory of Statistics*. New York, McGraw-Hill, 1963.
Pratt, J., Raiffa, H., and Schlaifer, R., *Introduction to Statistical Decision Theory*. New York, McGraw-Hill, 1965.
Weiss, L., *Statistical Decision Theory*. New York, McGraw-Hill, 1961.

Advanced

Blackwell, D., and Girshick, M., *Theory of Games and Statistical Decisions*. New York, John Wiley, 1954.
Cramér, H., *Mathematical Methods of Statistics*. Princeton, N.J., Princeton University Press, 1946.
Kendall, M. G., and Stuart, A., *The Advanced Theory of Statistics*, Vol. 1 (2d ed.), 1963; Vol. 2, 1961; Vol. 3, 1966. New York, Hafner Publishing Co.
Lehmann, E., *Testing Statistical Hypotheses*. New York, John Wiley, 1952.
Raiffa, H., and Schlaifer, R., *Applied Statistical Decision Theory*. Cambridge, Mass., Division of Research, Harvard Business School, 1961.
Savage, L. J., *The Foundations of Statistics*. New York, John Wiley, 1954.
Wald, A., *Statistical Decision Functions*. New York, John Wiley, 1950.
Wilks, S. S., *Mathematical Statistics*. New York, John Wiley, 1962.

Appendix A
Sets and Their Operations

In this textbook we have frequently utilized definitions and notations which are more fully developed in *set theory*, a branch of mathematics. We will, therefore, present in this appendix a very elementary treatment of the definitions and notations which we borrow from *set theory*.

A.1 Definitions and Notations

First, we propose:

Definition A.1.1
A *set* is a collection of objects. An object in a *set* is called an *element* of that set.

We will denote a set by an italicized capital letter; for example, A, B, or S. Whether or not we choose A, B, or S to depict a particular set is arbitrary. We will denote an object which can conceivably belong to a set by an italicized, lower case letter; for example, a, b, x, or y. Again, whether or not we choose a, b, x, or y to depict an object is arbitrary.

Let S depict a set and let x depict an object. If x is an element of S, we denote this fact as

$$x \in S,$$

but if x is not an element of S, we denote this fact as

$$x \notin S.$$

There are two different ways to describe a set: we may describe a set by listing all of its elements, or by defining the attribute or attributes which must be possessed by an element of the set.

EXAMPLE A.1.1

Let A be a set whose elements are three individuals whose names are Allen, Bob, and Charles. Then, we have described the set A by listing all of its elements. We might also describe this set in a more simplified manner as

$$A = \{\text{Allen, Bob, Charles}\}.$$

EXAMPLE A.1.2

Let A be a set whose elements are the names of 50 states in the United States. We can describe this set by listing the names of these states; for example,

$$A = \{\text{Alaska}, \ldots \ldots \ldots, \text{Wyoming}\}.$$

Or, we can describe the same set by defining the attribute which must be possessed by the elements of the set; for example,

$$A = \{x | x = \text{name of a state in the United States}\}.$$

The right-hand side of the preceding notation implies that x is an element of the set A if x happens to be a name of a state in the United States.

For the preceding example, whether we describe the set by listing all of its elements or by defining the attribute for its elements is of trivial significance. On the other hand, when the elements of a set happen to be numbers, at times the only way that the set can be described is by defining the attribute for its elements.

EXAMPLE A.1.3

Let S be a set whose elements are all real numbers between 0 and 1. Clearly this set cannot be described by listing all of its elements, since, for example, there are an infinite number of elements in this set. There is, however, another factor, which is beyond the scope of our discussion, that will make it impossible for us to list the elements of this set. Consequently, the only way that we can describe this set is by defining the attribute for the elements of the set. This can be done as

$$S = \{x | 0 \leq x \leq 1\}.$$

The right-hand side of this notation implies that x is an element of S if x is a real number between 0 and 1.

Definition A.1.2

Let A be a set. A is said to be a *finite set* if it contains a finite number of elements; an *infinite set* if it contains an infinite number of elements; and an *empty* or *null set* if it does not contain any element.

Examples of finite and infinite sets have already been presented. For example, the set containing the names of 50 states is a finite set. On the other hand, the set whose elements are real numbers between 0 and 1 is an infinite set. We will now give examples of null sets.

EXAMPLE A.1.4

The household of James Brown consist of James Brown himself, his wife Alice, and his two sons Robert and Charles. Let S be the set consisting of the daughters of James and Alice Brown. Then, such a set must be empty since the Browns do not have any daughter. The set is, therefore, an empty or null set.

EXAMPLE A.1.5

Let S be a set given by the following description:

$$S = \{x | 0 < x < 1 \text{ and } x \text{ is an integer}\}.$$

There is no integer whose value is greater than 0 but less than 1. Therefore, S is a null set.

When a set is empty, we denote it by ϕ, a Greek letter Phi. For example, the equation

$$S = \{\phi\}$$

means that S is a null set.

A.2 Relationships between Sets

Let x and y be two real numbers. Then, we are already familiar with the following notations:

$$x = y; x < y; x > y.$$

These notations are said to depict the relationship between x and y. Now let A and B be two sets. We can also describe certain relationships between A and B.

Definition A.2.1

Let A and B be two sets. A and B are said to be equal if every element of A is also an element of B and, conversely, if every element of B is an element of A.

EXAMPLE A.2.1

Let $A = \{1,2,3\}$ and let $B = \{3,1,2\}$. We observe that every element of A is also an element of B. We also observe that every element of B is an element of A. Thus, according to the preceding definition, A and B are equal.

EXAMPLE A.2.2

Let $A = \{\text{James, Charles, Robert}\}$ and let $B = \{\text{James, Robert}\}$. We observe that every element of B is an element of A, but not every element of A is an element of B. Thus, A and B are not equal.

Definition A.2.2

Let A and B be two sets. A is said to be a subset of B if every element of A is an element of B. B is said to be a subset of A if every element of B is an element of A.

Let us return to Example A.2.1. We have observed that every element of A is an element of B. Thus, A is a subset of B. On the other hand, we also note that B is a subset of A. Let us next return to Example A.2.2. We observe that every element of B is an element of A. Thus, B is a subset of A, but, we note that not every element of A is an element of B. Thus, A is not a subset of B.

Definition A.2.3

Let A and B be two sets. A is said to be a *proper subset* of B if A is a subset of B but B is not a subset of A.

EXAMPLE A.2.3

Let $A = \{1,2\}$ and let $B = \{1,3,2,4\}$. We observe that A is a subset of B but B is not a subset of A. Thus, A is not only a subset of B but also a proper subset of B.

Whenever A is a subset of B, we will denote this fact by the relationship

$$A \subseteq B.$$

In addition, if A also happens to be a proper subset of B, then we will denote this fact by the relationship

$$A \subset B.$$

A.3 Operations with Sets

Let x and y be two real numbers. Again, we are familiar with the following notations:

$$x + y; x - y; x \times y; x \div y.$$

These are called the operations with two real numbers x and y. Now let A and B be two sets. We can also describe certain operations between A and B.

Definition A.3.1
Let A and B be two sets. The intersection of A and B, denoted $A \cap B$, is defined as a set whose elements belong to both A and B.

Figure A.3.1 is called a Venn diagram. We let the rectangle depict A and the circle depict B. Suppose we pick a point in the shaded area of the Venn diagram. The point belongs to A and, at the same time, it also belongs to B. Thus, all the points in the shaded area in question together constitute the intersection of A and B.

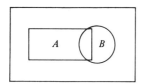

FIGURE A.3.1 Venn diagram.

EXAMPLE A.3.1
Let $A = \{3,5,4,6\}$ and let $B = \{4,7,6,8\}$. We observe, then, that the intersection between A and B is given by the equation

$$A \cap B = \{4,6\}.$$

Note that 4 belongs to both A and B. The same can be said for 6.

EXAMPLE A.3.2
Let $A = \{1,2\}$ and let $B = \{3,4\}$. We observe, then, that there is not any number which belongs to both A and B. Thus, the intersection is empty and we denote this by the equation

$$A \cap B = \{\phi\}.$$

When the intersection of two sets are empty, they are also said to be *disjoint sets*.

Definition A.3.2
Let A and B be two sets. The union of A and B, denoted $A \cup B$, is defined as a set whose elements belong to either A or B.

Let us examine the Venn diagram shown in Figure A.3.2. The rectangle depicts A and the circle depicts B. Suppose we pick any point in the shaded area of the diagram. Regardless of how we pick such a point, we observe that the point must belong to at least one of the two sets A and B. Thus, the entire shaded area constitutes the union of A and B.

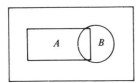

FIGURE A.3.2 Venn diagram.

EXAMPLE A.3.3

Let $A = \{1,2,3,4,5\}$ and let $B = \{4,5,6,7,8\}$. Then,

$$A \cup B = \{1,2,3,4,5,6,7,8\}.$$

Suppose we pick an element from the set $A \cup B$. We observe that the element so chosen must belong to at least one set among A and B.

In any discussion of sets, there arises a need to confine the discussion to a certain framework. We can do this by defining a set which contains all objects which might conceivably be considered in the discussion. Such a set is called a *universal set*. For example, suppose we wish to discuss something about the number system. Then, the universal set for our discussion may be the set of all real numbers. On the other hand, suppose we wish to discuss something about bridge hands. Then, the universal set for our discussion may be all the cards in a bridge deck.

Definition A.3.3

Let U be a universal set and let A be a subset of U. Then, the complement of A in U, denoted A', is a set whose elements belong to U but do not belong to A.

Consider Figure A.3.3. Let the universal set be depicted by the entire Venn diagram. Let the circle depict A. Then, the complement of A is depicted by the unshaded area of the figure.

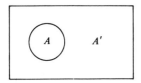

FIGURE A.3.3 Venn diagram.

EXAMPLE A.3.4

Let $U = \{1,2,3,4,5,6,7,8,9,10\}$ and let $A = \{4,5,6\}$. Then,

$$A' = \{1,2,3,7,8,9,10\}.$$

We observe that every element of A' belongs to U but does not belong to A. Thus, A' is the complement of A in U.

EXERCISES

1. Given that $S = \{x|0 \leq x \leq 5$ and x is an integer$\}$, describe the set by listing all of its elements.
2. Given that $S = \{1,2,3,4,5,6,7,8,9,10\}$, describe the set by defining the attributes for the elements of the set.
3. Given that $A = \{$James, Charles$\}$, $B = \{$Charles, James$\}$, and $C = \{$James, Charles, David$\}$,
   a. are A and B equal sets?
   b. is A a subset of B?
   c. is B a proper subset of A?
   d. is A a proper subset of C?
   e. is C a subset of B?
4. Given that $A = \{1,2\}$, $B = \{2,1\}$ and $C = \{3,1,4,2\}$,
   a. are A and B equal sets?
   b. is A a subset of B?
   c. is B a proper subset of A?
   d. is C a subset of A?
5. Given that $A = \{3,4,5\}$ and $B = \{4,3,5,6\}$,
   a. is $A \subseteq B$?
   b. is $A \subset B$?
   c. is $B \subseteq A$?
   d. is $A = B$?
6. Given the following Venn diagram, find (a) $A \cap B$; (b) $A \cap C$; (c) $a \cap b \cap c$; (d) $A \cup B$; (e) $B \cup C$; and (f) $A \cup B \cup C$.

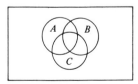

7. Let $A = \{1,2,3\}$, $B = \{3,4,5,6\}$, and $C = \{5,6\}$. Find (a) $A \cap B$; (b) $A \cap C$; (c) $A \cup B$; (d) $B \cup C$; (e) $A \cap B \cap C$; and (f) $A \cup B \cup C$.
8. Let the universal set be $U = \{1,2,3,4,5\}$ and let $A = \{3\}$. Find A'.

Appendix B
List
of Tables

Table B.1 Area under the Normal Curve
Table B.2 Chi-Square
Table B.3 Student-t Distribution
Table B.4 F Distribution
Table B.5 Exponential Functions
Table B.6 Binomial Probabilities
Table B.7 Poisson Probabilities
Table B.8 Unit Normal Loss Function
Table B.9 Random Numbers
Table B.10 Natural Logarithms of Numbers (Base e)
Table B.11 Common Logarithms of Numbers (Base 10)
Table B.12 Powers and Roots

EXAMPLE

Let Z be the standard normal variable. Then,

$$P(0 \leq Z \leq 1.05) = .3531,$$

where $z = 1.05$.

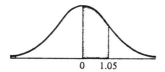

TABLE B.1 Area under the Normal Curve

z	.00	.01	.02	.03	.04	.05	.06	.07	.08	.09
0.0	.0000	.0040	.0080	.0120	.0160	.0199	.0239	.0279	.0319	.0359
0.1	.0398	.0438	.0478	.0517	.0557	.0596	.0636	.0675	.0714	.0753
0.2	.0793	.0832	.0871	.0910	.0948	.0987	.1026	.1064	.1103	.1141
0.3	.1179	.1217	.1255	.1293	.1331	.1368	.1406	.1443	.1480	.1517
0.4	.1554	.1591	.1628	.1664	.1700	.1736	.1772	.1808	.1844	.1879
0.5	.1915	.1950	.1985	.2019	.2054	.2088	.2123	.2157	.2190	.2224
0.6	.2257	.2291	.2324	.2357	.2389	.2422	.2454	.2486	.2518	.2549
0.7	.2580	.2612	.2642	.2673	.2704	.2734	.2764	.2794	.2823	.2852
0.8	.2881	.2910	.2939	.2967	.2995	.3023	.3051	.3078	.3106	.3133
0.9	.3159	.3186	.3212	.3238	.3264	.3289	.3315	.3340	.3365	.3389
1.0	.3413	.3438	.3461	.3485	.3508	.3531	.3554	.3577	.3599	.3621
1.1	.3643	.3665	.3686	.3708	.3729	.3749	.3770	.3790	.3810	.3830
1.2	.3849	.3869	.3888	.3907	.3925	.3944	.3962	.3980	.3997	.4015
1.3	.4032	.4049	.4066	.4082	.4099	.4115	.4131	.4147	.4162	.4177
1.4	.4192	.4207	.4222	.4236	.4251	.4265	.4279	.4292	.4306	.4319
1.5	.4332	.4345	.4357	.4370	.4382	.4394	.4406	.4418	.4429	.4441
1.6	.4452	.4463	.4474	.4484	.4495	.4505	.4515	.4525	.4535	.4545
1.7	.4554	.4564	.4573	.4582	.4591	.4599	.4608	.4616	.4625	.4633
1.8	.4641	.4649	.4656	.4664	.4671	.4678	.4686	.4693	.4699	.4706
1.9	.4713	.4719	.4726	.4732	.4738	.4744	.4750	.4756	.4761	.4767
2.0	.4772	.4778	.4783	.4788	.4793	.4798	.4803	.4808	.4812	.4817
2.1	.4821	.4826	.4830	.4834	.4838	.4842	.4846	.4850	.4854	.4857
2.2	.4861	.4864	.4868	.4871	.4875	.4878	.4881	.4884	.4887	.4890
2.3	.4893	.4896	.4898	.4901	.4904	.4906	.4909	.4911	.4913	.4916
2.4	.4918	.4920	.4922	.4925	.4927	.4929	.4931	.4932	.4934	.4936
2.5	.4938	.4940	.4941	.4943	.4945	.4946	.4948	.4949	.4951	.4952
2.6	.4953	.4955	.4956	.4957	.4959	.4960	.4961	.4962	.4963	.4964
2.7	.4965	.4966	.4967	.4968	.4969	.4970	.4971	.4972	.4973	.4974
2.8	.4974	.4975	.4976	.4977	.4977	.4978	.4979	.4979	.4980	.4981
2.9	.4981	.4982	.4982	.4983	.4984	.4984	.4985	.4985	.4986	.4986
3.0	.49865	.4987	.4987	.4988	.4988	.4989	.4989	.4989	.4990	.4990
4.0	.49997									

EXAMPLE

If Y is chi-square distributed with 10 degrees of freedom, then,

$$P(Y \geq 18.31) = .05,$$

where

$$\chi^2_{10,\ .05} = 18.31.$$

TABLE B.2 Chi-Square

Degrees of freedom	Probability that chi-square value will be exceeded									
	0.995	0.990	0.975	0.950	0.900	0.100	0.050	0.025	0.010	0.005
1	0.0⁴393	0.0³157	0.0⁵982	0.0³393	0.0158	2.71	3.84	5.02	6.63	7.88
2	0.0100	0.0201	0.0506	0.103	0.211	4.61	5.99	7.38	9.21	10.60
3	0.072	0.115	0.216	0.352	0.584	6.25	7.81	9.35	11.34	12.84
4	0.207	0.297	0.484	0.711	1.064	7.78	9.49	11.14	13.28	14.86
5	0.412	0.554	0.831	1.145	1.61	9.24	11.07	12.83	15.09	16.75
6	0.676	0.872	1.24	1.64	2.20	10.64	12.59	14.45	16.81	18.55
7	0.989	1.24	1.69	2.17	2.83	12.02	14.07	16.01	18.48	20.28
8	1.34	1.65	2.18	2.73	3.49	13.36	15.51	17.53	20.09	21.96
9	1.73	2.09	2.70	3.33	4.17	14.68	16.92	19.02	21.67	23.59
10	2.16	2.56	3.25	3.94	4.87	15.99	18.31	20.48	23.21	25.19
11	2.60	3.05	3.82	4.57	5.58	17.28	19.68	21.92	24.72	26.76
12	3.07	3.57	4.40	5.23	6.30	18.55	21.03	23.34	26.22	28.30
13	3.57	4.11	5.01	5.89	7.04	19.81	22.36	24.74	27.69	29.82
14	4.07	4.66	5.63	6.57	7.79	21.06	23.68	26.12	29.14	31.32
15	4.60	5.23	6.26	7.26	8.55	22.31	25.00	27.49	30.58	32.80
16	5.14	5.81	6.91	7.96	9.31	23.54	26.30	28.85	32.00	34.27
17	5.70	6.41	7.56	8.67	10.09	24.77	27.59	30.19	33.41	35.72
18	6.26	7.01	8.23	9.39	10.86	25.99	28.87	31.53	34.81	37.16
19	6.84	7.63	8.91	10.12	11.65	27.20	30.14	32.85	36.19	38.58
20	7.43	8.26	9.59	10.85	12.44	28.41	31.41	34.17	37.57	40.00
21	8.03	8.90	10.28	11.59	13.24	29.62	32.67	35.48	38.93	41.40
22	8.64	9.54	10.98	12.34	14.04	30.81	33.92	36.78	40.29	42.80
23	9.26	10.20	11.69	13.09	14.85	32.01	35.17	38.08	41.64	44.18
24	9.89	10.86	12.40	13.85	15.66	33.20	36.42	39.36	42.98	45.56
25	10.52	11.52	13.12	14.61	16.47	34.38	37.65	40.65	44.31	46.93
26	11.16	12.20	13.84	15.38	17.29	35.56	38.89	41.92	45.64	48.29
27	11.81	12.88	14.57	16.15	18.11	36.74	40.11	43.19	46.96	49.64
28	12.46	13.56	15.31	16.93	18.94	37.92	41.34	44.46	48.28	50.99
29	13.12	14.26	16.05	17.71	19.77	39.09	42.56	45.72	49.59	52.34
30	13.79	14.95	16.79	18.49	20.60	40.26	43.77	46.98	50.89	53.67
40	20.71	22.16	24.43	26.51	29.05	51.80	55.76	59.34	63.69	66.77
50	27.99	29.71	32.36	34.76	37.69	63.17	67.50	71.42	76.15	79.49
60	35.53	37.48	40.48	43.19	46.46	74.40	79.08	83.30	88.38	91.95
70	43.28	45.44	48.76	51.74	55.33	85.53	90.53	95.02	100.4	104.22
80	51.17	53.54	57.15	60.39	64.28	96.58	101.9	106.6	112.3	116.32
90	59.20	61.75	65.65	69.13	73.29	107.6	113.1	118.1	124.1	128.3
100	67.33	70.06	74.22	77.93	82.36	118.5	124.3	129.6	135.8	140.2

SOURCE: By permission of Prof. E. S. Pearson, from Catherine M. Thompson, "Tables of the Percentage Points of the Incomplete Beta Function and of the χ^2 Distribution," *Biometrika*, vol. 32, pp. 168–181, 188–189, 1941.

EXAMPLE

If Y is a student-t distributed random variable with 10 degrees of freedom, then,

$$P(-2.228 \leq Y \leq 2.228) = .95$$

and

$$P(-2.228 \geq Y \geq 2.228) = .05,$$

where

$$t_{10,\ .05} = 2.228.$$

TABLE B.3 Student-t Distribution

Degrees of freedom	Probability											
	0.9	0.8	0.7	0.6	0.5	0.4	0.3	0.2	0.1	0.05	0.02	0.01	0.001
1	0.158	0.325	0.510	0.727	1.000	1.376	1.963	3.078	6.314	12.706	31.821	63.657	636.619
2	0.142	0.289	0.445	0.617	0.816	1.061	1.386	1.886	2.920	4.303	6.965	9.925	31.598
3	0.137	0.277	0.424	0.584	0.765	0.978	1.250	1.638	2.353	3.182	4.541	5.841	12.924
4	0.134	0.271	0.414	0.569	0.741	0.941	1.190	1.533	2.132	2.776	3.747	4.604	8.610
5	0.132	0.267	0.408	0.559	0.727	0.920	1.156	1.476	2.015	2.571	3.365	4.032	6.869
6	0.131	0.265	0.404	0.553	0.718	0.906	1.134	1.440	1.943	2.447	3.143	3.707	5.959
7	0.130	0.263	0.402	0.549	0.711	0.896	1.119	1.415	1.895	2.365	2.998	3.499	5.408
8	0.130	0.262	0.399	0.546	0.706	0.889	1.108	1.397	1.860	2.306	2.896	3.355	5.041
9	0.129	0.261	0.398	0.543	0.703	0.883	1.100	1.383	1.833	2.262	2.821	3.250	4.781
10	0.129	0.260	0.397	0.542	0.700	0.879	1.093	1.372	1.812	2.228	2.764	3.169	4.587
11	0.129	0.260	0.396	0.540	0.697	0.876	1.088	1.363	1.796	2.201	2.718	3.106	4.437
12	0.128	0.259	0.395	0.539	0.695	0.873	1.083	1.356	1.782	2.179	2.681	3.055	4.318
13	0.128	0.259	0.394	0.538	0.694	0.870	1.079	1.350	1.771	2.160	2.650	3.012	4.221
14	0.128	0.258	0.393	0.537	0.692	0.868	1.076	1.345	1.761	2.145	2.624	2.977	4.140
15	0.128	0.258	0.393	0.536	0.691	0.866	1.074	1.341	1.753	2.131	2.602	2.947	4.073
16	0.128	0.258	0.392	0.535	0.690	0.865	1.071	1.337	1.746	2.120	2.583	2.921	4.015
17	0.128	0.257	0.392	0.534	0.689	0.863	1.069	1.333	1.740	2.110	2.567	2.898	3.965
18	0.127	0.257	0.392	0.534	0.688	0.862	1.067	1.330	1.734	2.101	2.552	2.878	3.922
19	0.127	0.257	0.391	0.533	0.688	0.861	1.066	1.328	1.729	2.093	2.539	2.861	3.883
20	0.127	0.257	0.391	0.533	0.687	0.860	1.064	1.325	1.725	2.086	2.528	2.845	3.850
21	0.127	0.257	0.391	0.532	0.686	0.859	1.063	1.323	1.721	2.080	2.518	2.831	3.819
22	0.127	0.256	0.390	0.532	0.686	0.858	1.061	1.321	1.717	2.074	2.508	2.819	3.792
23	0.127	0.256	0.390	0.532	0.685	0.858	1.060	1.319	1.714	2.069	2.500	2.807	3.767
24	0.127	0.256	0.390	0.531	0.685	0.857	1.059	1.318	1.711	2.064	2.492	2.797	3.745
25	0.127	0.256	0.390	0.531	0.684	0.856	1.058	1.316	1.708	2.060	2.485	2.787	3.725
26	0.127	0.256	0.390	0.531	0.684	0.856	1.058	1.315	1.706	2.056	2.479	2.779	3.707
27	0.127	0.256	0.389	0.531	0.684	0.855	1.057	1.314	1.703	2.052	2.473	2.771	3.690
28	0.127	0.256	0.389	0.530	0.683	0.855	1.056	1.313	1.701	2.048	2.467	2.763	3.674
29	0.127	0.256	0.389	0.530	0.683	0.854	1.055	1.311	1.699	2.045	2.462	2.756	3.659
30	0.127	0.256	0.389	0.530	0.683	0.854	1.055	1.310	1.697	2.042	2.457	2.750	3.646
40	0.126	0.255	0.388	0.529	0.681	0.851	1.050	1.303	1.684	2.021	2.423	2.704	3.551
60	0.126	0.254	0.387	0.527	0.679	0.848	1.046	1.296	1.671	2.000	2.390	2.660	3.460
120	0.126	0.254	0.386	0.526	0.677	0.845	1.041	1.289	1.658	1.980	2.358	2.617	3.373
∞	0.126	0.253	0.385	0.524	0.674	0.842	1.036	1.282	1.645	1.960	2.326	2.576	3.291

SOURCE: This table is abridged from Table II of Fisher & Yates: *Statistical Tables for Biological, Agricultural and Medical Research*, published by Oliver & Boyd Ltd., Edinburgh, and by permission of the authors and publishers.

EXAMPLE

Let Y be an F distributed random variable with 5 degrees of freedom for the numerator and 6 degrees of freedom for the denominator. Then,

$$(Y \geq 4.39) = .05,$$

where

$$F_{(5,6),\ .05} = 4.39.$$

TABLE B.4 F Distribution (Upper 5% points)

n \ m	1	2	3	4	5	6	7	8	9	10	12	15	20	24	30	40	60	120	∞
1	161.4	199.5	215.7	224.6	230.2	234.0	236.8	238.9	240.5	241.9	243.9	245.9	248.0	249.1	250.1	251.1	252.2	253.3	254.3
2	18.51	19.00	19.16	19.25	19.30	19.33	19.35	19.37	19.38	19.40	19.41	19.43	19.45	19.45	19.46	19.47	19.48	19.49	19.50
3	10.13	9.55	9.28	9.12	9.01	8.94	8.89	8.85	8.81	8.79	8.74	8.70	8.66	8.64	8.62	8.59	8.57	8.55	8.53
4	7.71	6.94	6.59	6.39	6.26	6.16	6.09	6.04	6.00	5.96	5.91	5.86	5.80	5.77	5.75	5.72	5.69	5.66	5.63
5	6.61	5.79	5.41	5.19	5.05	4.95	4.88	4.82	4.77	4.74	4.68	4.62	4.56	4.53	4.50	4.46	4.43	4.40	4.36
6	5.99	5.14	4.76	4.53	4.39	4.28	4.21	4.15	4.10	4.06	4.00	3.94	3.87	3.84	3.81	3.77	3.74	3.70	3.67
7	5.59	4.74	4.35	4.12	3.97	3.87	3.79	3.73	3.68	3.64	3.57	3.51	3.44	3.41	3.38	3.34	3.30	3.27	3.23
8	5.32	4.46	4.07	3.84	3.69	3.58	3.50	3.44	3.39	3.35	3.28	3.22	3.15	3.12	3.08	3.04	3.01	2.97	2.93
9	5.12	4.26	3.86	3.63	3.48	3.37	3.29	3.23	3.18	3.14	3.07	3.01	2.94	2.90	2.86	2.83	2.79	2.75	2.71
10	4.96	4.10	3.71	3.48	3.33	3.22	3.14	3.07	3.02	2.98	2.91	2.85	2.77	2.74	2.70	2.66	2.62	2.58	2.54
11	4.84	3.98	3.59	3.36	3.20	3.09	3.01	2.95	2.90	2.85	2.79	2.72	2.65	2.61	2.57	2.53	2.49	2.45	2.40
12	4.75	3.89	3.49	3.26	3.11	3.00	2.91	2.85	2.80	2.75	2.69	2.62	2.54	2.51	2.47	2.43	2.38	2.34	2.30
13	4.67	3.81	3.41	3.18	3.03	2.92	2.83	2.77	2.71	2.67	2.60	2.53	2.46	2.42	2.38	2.34	2.30	2.25	2.21
14	4.60	3.74	3.34	3.11	2.96	2.85	2.76	2.70	2.65	2.60	2.53	2.46	2.39	2.35	2.31	2.27	2.22	2.18	2.13
15	4.54	3.68	3.29	3.06	2.90	2.79	2.71	2.64	2.59	2.54	2.48	2.40	2.33	2.29	2.25	2.20	2.16	2.11	2.07
16	4.49	3.63	3.24	3.01	2.85	2.74	2.66	2.59	2.54	2.49	2.42	2.35	2.28	2.24	2.19	2.15	2.11	2.06	2.01
17	4.45	3.59	3.20	2.96	2.81	2.70	2.61	2.55	2.49	2.45	2.38	2.31	2.23	2.19	2.15	2.10	2.06	2.01	1.96
18	4.41	3.55	3.16	2.93	2.77	2.66	2.58	2.51	2.46	2.41	2.34	2.27	2.19	2.15	2.11	2.06	2.02	1.97	1.92
19	4.38	3.52	3.13	2.90	2.74	2.63	2.54	2.48	2.42	2.38	2.31	2.23	2.16	2.11	2.07	2.03	1.98	1.93	1.88
20	4.35	3.49	3.10	2.87	2.71	2.60	2.51	2.45	2.39	2.35	2.28	2.20	2.12	2.08	2.04	1.99	1.95	1.90	1.84
21	4.32	3.47	3.07	2.84	2.68	2.57	2.49	2.42	2.37	2.32	2.25	2.18	2.10	2.05	2.01	1.96	1.92	1.87	1.81
22	4.30	3.44	3.05	2.82	2.66	2.55	2.46	2.40	2.34	2.30	2.23	2.15	2.07	2.03	1.98	1.94	1.89	1.84	1.78
23	4.28	3.42	3.03	2.80	2.64	2.53	2.44	2.37	2.32	2.27	2.20	2.13	2.05	2.01	1.96	1.91	1.86	1.81	1.76
24	4.26	3.40	3.01	2.78	2.62	2.51	2.42	2.36	2.30	2.25	2.18	2.11	2.03	1.98	1.94	1.89	1.84	1.79	1.73
25	4.24	3.39	2.99	2.76	2.60	2.49	2.40	2.34	2.28	2.24	2.16	2.09	2.01	1.96	1.92	1.87	1.82	1.77	1.71
26	4.23	3.37	2.98	2.74	2.59	2.47	2.39	2.32	2.27	2.22	2.15	2.07	1.99	1.95	1.90	1.85	1.80	1.75	1.69
27	4.21	3.35	2.96	2.73	2.57	2.46	2.37	2.31	2.25	2.20	2.13	2.06	1.97	1.93	1.88	1.84	1.79	1.73	1.67
28	4.20	3.34	2.95	2.71	2.56	2.45	2.36	2.29	2.24	2.19	2.12	2.04	1.96	1.91	1.87	1.82	1.77	1.71	1.65
29	4.18	3.33	2.93	2.70	2.55	2.43	2.35	2.28	2.22	2.18	2.10	2.03	1.94	1.90	1.85	1.81	1.75	1.70	1.64
30	4.17	3.32	2.92	2.69	2.53	2.42	2.33	2.27	2.21	2.16	2.09	2.01	1.93	1.89	1.84	1.79	1.74	1.68	1.62
40	4.08	3.23	2.84	2.61	2.45	2.34	2.25	2.18	2.12	2.08	2.00	1.92	1.84	1.79	1.74	1.69	1.64	1.58	1.51
60	4.00	3.15	2.76	2.53	2.37	2.25	2.17	2.10	2.04	1.99	1.92	1.84	1.75	1.70	1.65	1.59	1.53	1.47	1.39
120	3.92	3.07	2.68	2.45	2.29	2.17	2.09	2.02	1.96	1.91	1.83	1.75	1.66	1.61	1.55	1.50	1.43	1.35	1.25
∞	3.84	3.00	2.60	2.37	2.21	2.10	2.01	1.94	1.88	1.83	1.75	1.67	1.57	1.52	1.46	1.39	1.32	1.22	1.00

m = degrees of freedom for numerator n = degrees of freedom for denominator

EXAMPLE

Let Y be an F distributed random variable with 5 degrees of freedom for the numerator and 6 degrees of freedom for the denominator. Then,

$$P(Y \geq 8.75) = .01,$$

where

$$F_{(5,6),\,.05} = 8.75.$$

TABLE B.4 F Distribution (Cont'd) (Upper 1% points)

n \ m	1	2	3	4	5	6	7	8	9	10	12	15	20	24	30	40	60	120	∞
1	4052	4999·5	5403	5625	5764	5859	5928	5982	6022	6056	6106	6157	6209	6235	6261	6287	6313	6339	6366
2	98·50	99·00	99·17	99·25	99·30	99·33	99·36	99·37	99·39	99·40	99·42	99·43	99·45	99·46	99·47	99·47	99·48	99·49	99·50
3	34·12	30·82	29·46	28·71	28·24	27·91	27·67	27·49	27·35	27·23	27·05	26·87	26·69	26·60	26·50	26·41	26·32	26·22	26·13
4	21·20	18·00	16·69	15·98	15·52	15·21	14·98	14·80	14·66	14·55	14·37	14·20	14·02	13·93	13·84	13·75	13·65	13·56	13·46
5	16·26	13·27	12·06	11·39	10·97	10·67	10·46	10·29	10·16	10·05	9·89	9·72	9·55	9·47	9·38	9·29	9·20	9·11	9·02
6	13·75	10·92	9·78	9·15	8·75	8·47	8·26	8·10	7·98	7·87	7·72	7·56	7·40	7·31	7·23	7·14	7·06	6·97	6·88
7	12·25	9·55	8·45	7·85	7·46	7·19	6·99	6·84	6·72	6·62	6·47	6·31	6·16	6·07	5·99	5·91	5·82	5·74	5·65
8	11·26	8·65	7·59	7·01	6·63	6·37	6·18	6·03	5·91	5·81	5·67	5·52	5·36	5·28	5·20	5·12	5·03	4·95	4·86
9	10·56	8·02	6·99	6·42	6·06	5·80	5·61	5·47	5·35	5·26	5·11	4·96	4·81	4·73	4·65	4·57	4·48	4·40	4·31
10	10·04	7·56	6·55	5·99	5·64	5·39	5·20	5·06	4·94	4·85	4·71	4·56	4·41	4·33	4·25	4·17	4·08	4·00	3·91
11	9·65	7·21	6·22	5·67	5·32	5·07	4·89	4·74	4·63	4·54	4·40	4·25	4·10	4·02	3·94	3·86	3·78	3·69	3·60
12	9·33	6·93	5·95	5·41	5·06	4·82	4·64	4·50	4·39	4·30	4·16	4·01	3·86	3·78	3·70	3·62	3·54	3·45	3·36
13	9·07	6·70	5·74	5·21	4·86	4·62	4·44	4·30	4·19	4·10	3·96	3·82	3·66	3·59	3·51	3·43	3·34	3·25	3·17
14	8·86	6·51	5·56	5·04	4·69	4·46	4·28	4·14	4·03	3·94	3·80	3·66	3·51	3·43	3·35	3·27	3·18	3·09	3·00
15	8·68	6·36	5·42	4·89	4·56	4·32	4·14	4·00	3·89	3·80	3·67	3·52	3·37	3·29	3·21	3·13	3·05	2·96	2·87
16	8·53	6·23	5·29	4·77	4·44	4·20	4·03	3·89	3·78	3·69	3·55	3·41	3·26	3·18	3·10	3·02	2·93	2·84	2·75
17	8·40	6·11	5·18	4·67	4·34	4·10	3·93	3·79	3·68	3·59	3·46	3·31	3·16	3·08	3·00	2·92	2·83	2·75	2·65
18	8·29	6·01	5·09	4·58	4·25	4·01	3·84	3·71	3·60	3·51	3·37	3·23	3·08	3·00	2·92	2·84	2·75	2·66	2·57
19	8·18	5·93	5·01	4·50	4·17	3·94	3·77	3·63	3·52	3·43	3·30	3·15	3·00	2·92	2·84	2·76	2·67	2·58	2·49
20	8·10	5·85	4·94	4·43	4·10	3·87	3·70	3·56	3·46	3·37	3·23	3·09	2·94	2·86	2·78	2·69	2·61	2·52	2·42
21	8·02	5·78	4·87	4·37	4·04	3·81	3·64	3·51	3·40	3·31	3·17	3·03	2·88	2·80	2·72	2·64	2·55	2·46	2·36
22	7·95	5·72	4·82	4·31	3·99	3·76	3·59	3·45	3·35	3·26	3·12	2·98	2·83	2·75	2·67	2·58	2·50	2·40	2·31
23	7·88	5·66	4·76	4·26	3·94	3·71	3·54	3·41	3·30	3·21	3·07	2·93	2·78	2·70	2·62	2·54	2·45	2·35	2·26
24	7·82	5·61	4·72	4·22	3·90	3·67	3·50	3·36	3·26	3·17	3·03	2·89	2·74	2·66	2·58	2·49	2·40	2·31	2·21
25	7·77	5·57	4·68	4·18	3·85	3·63	3·46	3·32	3·22	3·13	2·99	2·85	2·70	2·62	2·54	2·45	2·36	2·27	2·17
26	7·72	5·53	4·64	4·14	3·82	3·59	3·42	3·29	3·18	3·09	2·96	2·81	2·66	2·58	2·50	2·42	2·33	2·23	2·13
27	7·68	5·49	4·60	4·11	3·78	3·56	3·39	3·26	3·15	3·06	2·93	2·78	2·63	2·55	2·47	2·38	2·29	2·20	2·10
28	7·64	5·45	4·57	4·07	3·75	3·53	3·36	3·23	3·12	3·03	2·90	2·75	2·60	2·52	2·44	2·35	2·26	2·17	2·06
29	7·60	5·42	4·54	4·04	3·73	3·50	3·33	3·20	3·09	3·00	2·87	2·73	2·57	2·49	2·41	2·33	2·23	2·14	2·03
30	7·56	5·39	4·51	4·02	3·70	3·47	3·30	3·17	3·07	2·98	2·84	2·70	2·55	2·47	2·39	2·30	2·21	2·11	2·01
40	7·31	5·18	4·31	3·83	3·51	3·29	3·12	2·99	2·89	2·80	2·66	2·52	2·37	2·29	2·20	2·11	2·02	1·92	1·80
60	7·08	4·98	4·13	3·65	3·34	3·12	2·95	2·82	2·72	2·63	2·50	2·35	2·20	2·12	2·03	1·94	1·84	1·73	1·60
120	6·85	4·79	3·95	3·48	3·17	2·96	2·79	2·66	2·56	2·47	2·34	2·19	2·03	1·95	1·86	1·76	1·66	1·53	1·38
∞	6·63	4·61	3·78	3·32	3·02	2·80	2·64	2·51	2·41	2·32	2·18	2·04	1·88	1·79	1·70	1·59	1·47	1·32	1·00

SOURCE: This table is abridged from Table 18 of the *Biometrika Tables for Statisticians*, Vol. 1 (1st ed.), edited by E. S. Pearson and H. O. Hartley. Reproduced with the kind permission of E. S. Pearson and the trustees of *Biometrika*.

TABLE B.5 Exponential Functions

x	e^x	$\text{Log}_{10}(e^x)$	e^{-x}	x	e^x	$\text{Log}_{10}(e^x)$	e^{-x}
0.00	1.0000	0.00000	1.000000	**0.50**	1.6487	0.21715	0.606531
0.01	1.0101	.00434	0.990050	0.51	1.6653	.22149	.600496
0.02	1.0202	.00869	.980199	0.52	1.6820	.22583	.594521
0.03	1.0305	.01303	.970446	0.53	1.6989	.23018	.588605
0.04	1.0408	.01737	.960789	0.54	1.7160	.23452	.582748
0.05	1.0513	0.02171	0.951229	**0.55**	1.7333	0.23886	0.576950
0.06	1.0618	.02606	.941765	0.56	1.7507	.24320	.571209
0.07	1.0725	.03040	.932394	0.57	1.7683	.24755	.565525
0.08	1.0833	.03474	.923116	0.58	1.7860	.25189	.559898
0.09	1.0942	.03909	.913931	0.59	1.8040	.25623	.554327
0.10	1.1052	0.04343	0.904837	**0.60**	1.8221	0.26058	0.548812
0.11	1.1163	.04777	.895834	0.61	1.8404	.26492	.543351
0.12	1.1275	.05212	.886920	0.62	1.8589	.26926	.537944
0.13	1.1388	.05646	.878095	0.63	1.8776	.27361	.532592
0.14	1.1503	.06080	.869358	0.64	1.8965	.27795	.527292
0.15	1.1618	0.06514	0.860708	**0.65**	1.9155	0.28229	0.522046
0.16	1.1735	.06949	.852144	0.66	1.9348	.28663	.516851
0.17	1.1853	.07383	.843665	0.67	1.9542	.29098	.511709
0.18	1.1972	.07817	.835270	0.68	1.9739	.29532	.506617
0.19	1.2092	.08252	.826959	0.69	1.9937	.29966	.501576
0.20	1.2214	0.08686	0.818731	**0.70**	2.0138	0.30401	0.496585
0.21	1.2337	.09120	.810584	0.71	2.0340	.30835	.491644
0.22	1.2461	.09554	.802519	0.72	2.0544	.31269	.486752
0.23	1.2586	.09989	.794534	0.73	2.0751	.31703	.481909
0.24	1.2712	.10423	.786628	0.74	2.0959	.32138	.477114
0.25	1.2840	0.10857	0.778801	**0.75**	2.1170	0.32572	0.472367
0.26	1.2969	.11292	.771052	0.76	2.1383	.33006	.467666
0.27	1.3100	.11726	.763379	0.77	2.1598	.33441	.463013
0.28	1.3231	.12160	.755784	0.78	2.1815	.33875	.458406
0.29	1.3364	.12595	.748264	0.79	2.2034	.34309	.453845
0.30	1.3499	0.13029	0.740818	**0.80**	2.2255	0.34744	0.449329
0.31	1.3634	.13463	.733447	0.81	2.2479	.35178	.444858
0.32	1.3771	.13897	.726149	0.82	2.2705	.35612	.440432
0.33	1.3910	.14332	.718924	0.83	2.2933	.36046	.436049
0.34	1.4049	.14766	.711770	0.84	2.3164	.36481	.431711
0.35	1.4191	0.15200	0.704688	**0.85**	2.3396	0.36915	0.427415
0.36	1.4333	.15635	.697676	0.86	2.3632	.37349	.423162
0.37	1.4477	.16069	.690734	0.87	2.3869	.37784	.418952
0.38	1.4623	.16503	.683861	0.88	2.4109	.38218	.414783
0.39	1.4770	.16937	.677057	0.89	2.4351	.38652	.410656
0.40	1.4918	0.17372	0.670320	**0.90**	2.4596	0.39087	0.406570
0.41	1.5068	.17806	.663650	0.91	2.4843	.39521	.402524
0.42	1.5220	.18240	.657047	0.92	2.5093	.39955	.398519
0.43	1.5373	.18675	.650509	0.93	2.5345	.40389	.394554
0.44	1.5527	.19109	.644036	0.94	2.5600	.40824	.390628
0.45	1.5683	0.19543	0.637628	**0.95**	2.5857	0.41258	0.386741
0.46	1.5841	.19978	.631284	0.96	2.6117	.41692	.382893
0.47	1.6000	.20412	.625002	0.97	2.6379	.42127	.379083
0.48	1.6161	.20846	.618783	0.98	2.6645	.42561	.375311
0.49	1.6323	.21280	.612626	0.99	2.6912	.42995	.371577
0.50	1.6487	0.21715	0.606531	**1.00**	2.7183	0.43429	0.367879

TABLE B.5 Exponential Functions (*Cont'd*)

x	e^x	$\text{Log}_{10}(e^x)$	e^{-x}	x	e^x	$\text{Log}_{10}(e^x)$	e^{-x}
1.00	2.7183	0.43429	0.367879	**1.50**	4.4817	0.65144	0.223130
1.01	2.7456	.43864	.364219	1.51	4.5267	.65578	.220910
1.02	2.7732	.44298	.360595	1.52	4.5722	.66013	.218712
1.03	2.8011	.44732	.357007	1.53	4.6182	.66447	.216536
1.04	2.8292	.45167	.353455	1.54	4.6646	.66881	.214381
1.05	2.8577	0.45601	0.349938	**1.55**	4.7115	0.67316	0.212248
1.06	2.8864	.46035	.346456	1.56	4.7588	.67750	.210136
1.07	2.9154	.46470	.343009	1.57	4.8066	.68184	.208045
1.08	2.9447	.46904	.339596	1.58	4.8550	.68619	.205975
1.09	2.9743	.47338	.336216	1.59	4.9037	.69053	.203926
1.10	3.0042	0.47772	0.332871	**1.60**	4.9530	0.69487	0.201897
1.11	3.0344	.48207	.329559	1.61	5.0028	.69921	.199888
1.12	3.0649	.48641	.326280	1.62	5.0531	.70356	.197899
1.13	3.0957	.49075	.323033	1.63	5.1039	.70790	.195930
1.14	3.1268	.49510	.319819	1.64	5.1552	.71224	.193980
1.15	3.1582	0.49944	0.316637	**1.65**	5.2070	0.71659	0.192050
1.16	3.1899	.50378	.313486	1.66	5.2593	.72093	.190139
1.17	3.2220	.50812	.310367	1.67	5.3122	.72527	.188247
1.18	3.2544	.51247	.307279	1.68	5.3656	.72961	.186374
1.19	3.2871	.51681	.304221	1.69	5.4195	.73396	.184520
1.20	3.3201	0.52115	0.301194	**1.70**	5.4739	0.73830	0.182684
1.21	3.3535	.52550	.298197	1.71	5.5290	.74264	.180866
1.22	3.3872	.52984	.295230	1.72	5.5845	.74699	.179066
1.23	3.4212	.53418	.292293	1.73	5.6407	.75133	.177284
1.24	3.4556	.53853	.289384	1 74	5.6973	75567	175520
1.25	3.4903	0.54287	0.286505	**1.75**	5.7546	0.76002	0.173774
1.26	3.5254	.54721	.283654	1.76	5.8124	.76436	.172045
1.27	3.5609	.55155	.280832	1 77	5.8709	.76870	170333
1.28	3.5966	.55590	.278037	1.78	5.9299	.77304	168638
1.29	3.6328	.56024	.275271	1.79	5.9895	.77739	.166960
1.30	3.6693	0.56458	0.272532	**1.80**	6.0496	0.78173	0.165299
1.31	3.7062	.56893	.269820	1.81	6.1104	.78607	.163654
1.32	3.7434	.57327	.267135	1.82	6.1719	.79042	.162026
1.33	3.7810	.57761	.264477	1.83	6.2339	.79476	.160414
1.34	3.8190	.58195	.261846	1.84	6.2965	79910	.158817
1.35	3.8574	0.58630	0.259240	**1.85**	6.3598	0.80344	0.157237
1.36	3.8962	.59064	.256661	1.86	6.4237	.80779	.155673
1.37	3.9354	.59498	.254107	1.87	6.4383	.81213	.154124
1.38	3.9749	.59933	.251579	1.88	6.5535	.81647	.152590
1.39	4.0149	.60367	.249075	1.89	6.6194	.82082	.151072
1.40	4.0552	0.60801	0.246597	**1.90**	6.6859	0.82516	0.149569
1.41	4.0960	.61236	.244143	1 91	6.7531	.82950	.148080
1.42	4.1371	.61670	.241714	1.92	6.8210	.83385	.146607
1.43	4.1787	.62104	.239309	1.93	6.8895	.83819	.145148
1.44	4.2207	.62538	.236928	1.94	6.9588	.84253	.143704
1.45	4.2631	0.62973	0.234570	**1.95**	7.0287	0.84687	0.142274
1.46	4.3060	.63407	.232236	1.96	7.0993	.85122	.140858
1.47	4.3492	.63841	.229925	1.97	7.1707	.85556	.139457
1.48	4.3929	.64276	.227638	1.98	7.2427	.85990	.138069
1.49	4.4371	.64710	.225373	1.99	7.3155	.86425	.136695
1.50	4.4817	0.65144	0.223130	**2.00**	7.3891	0.86859	0.135335

TABLE B.5 Exponential Functions (*Cont'd*)

x	e^x	$\text{Log}_{10}(e^x)$	e^{-x}	x	e^x	$\text{Log}_{10}(e^x)$	e^{-x}
2.00	7.3891	0.86859	0.135335	**2.50**	12.182	1.08574	0.082085
2.01	7.4633	.87293	.133989	2.51	12.305	1.09008	.081268
2.02	7.5383	.87727	.132655	2.52	12.429	1.09442	.080460
2.03	7.6141	.88162	.131336	2.53	12.554	1.09877	.079659
2.04	7.6906	.88596	.130029	2.54	12.680	1.10311	.078866
2.05	7.7679	0.89030	0.128735	**2.55**	12.807	1.10745	0.078082
2.06	7.8460	.89465	.127454	2.56	12.936	1.11179	.077305
2.07	7.9248	.89899	.126186	2.57	13.066	1.11614	.076536
2.08	8.0045	.90333	.124930	2.58	13.197	1.12048	.075774
2.09	8.0849	.90768	.123687	2.59	13.330	1.12482	.075020
2.10	8.1662	0.91202	0.122456	**2.60**	13.464	1.12917	0.074274
2.11	8.2482	.91636	.121238	2.61	13.599	1.13351	.073535
2.12	8.3311	.92070	.120032	2.62	13.736	1.13785	.072803
2.13	8.4149	.92505	.118837	2.63	13.874	1.14219	.072078
2.14	8.4994	.92939	.117655	2.64	14.013	1.14654	.071361
2.15	8.5849	0.93373	0.116484	**2.65**	14.154	1.15088	0.070651
2.16	8.6711	.93808	.115325	2.66	14.296	1.15522	.069948
2.17	8.7583	.94242	.114178	2.67	14.440	1.15957	.069252
2.18	8.8463	.94676	.113042	2.68	14.585	1.16391	.068563
2.19	8.9352	.95110	.111917	2.69	14.732	1.16825	.067881
2.20	9.0250	0.95545	0.110803	**2.70**	14.880	1.17260	0.067206
2.21	9.1157	.95979	.109701	2.71	15.029	1.17694	.066537
2.22	9.2073	.96413	.108609	2.72	15.180	1.18128	.065875
2.23	9.2999	.96848	.107528	2.73	15.333	1.18562	.065219
2.24	9.3933	.97282	.106459	2.74	15.487	1.18997	.064570
2.25	9.4877	0.97716	0.105399	**2.75**	15.643	1.19431	0.063928
2.26	9.5831	.98151	.104350	2.76	15.800	1.19865	.063292
2.27	9.6794	.98585	.103312	2.77	15.959	1.20300	.062662
2.28	9.7767	.99019	.102284	2.78	16.119	1.20734	.062039
2.29	9.8749	.99453	.101266	2.79	16.281	1.21168	.061421
2.30	9.9742	0.99888	0.100259	**2.80**	16.445	1.21602	0.060810
2.31	10.074	1.00322	.099261	2.81	16.610	1.22037	.060205
2.32	10.176	1.00756	.098274	2.82	16.777	1.22471	.059606
2.33	10.278	1.01191	.097296	2.83	16.945	1.22905	.059013
2.34	10.381	1.01625	.096328	2.84	17.116	1.23340	.058426
2.35	10.486	1.02059	0.095369	**2.85**	17.288	1.23774	0.057844
2.36	10.591	1.02493	.094420	2.86	17.462	1.24208	.057269
2.37	10.697	1.02928	.093481	2.87	17.637	1.24643	.056699
2.38	10.805	1.03362	.092551	2.88	17.814	1.25077	.056135
2.39	10.913	1.03796	.091630	2.89	17.993	1.25511	.055576
2.40	11.023	1.04231	0.090718	**2.90**	18.174	1.25945	0.055023
2.41	11.134	1.04665	.089815	2.91	18.357	1.26380	.054476
2.42	11.246	1.05099	.088922	2.92	18.541	1.26814	.053934
2.43	11.359	1.05534	.088037	2.93	18.728	1.27248	.053397
2.44	11.473	1.05968	.087161	2.94	18.916	1.27683	.052866
2.45	11.588	1.06402	0.086294	**2.95**	19.106	1.28117	0.052340
2.46	11.705	1.06836	.085435	2.96	19.298	1.28551	.051819
2.47	11.822	1.07271	.084585	2.97	19.492	1.28985	.051303
2.48	11.941	1.07705	.083743	2.98	19.688	1.29420	.050793
2.49	12.061	1.08139	.082910	2.99	19.886	1.29854	.050287
2.50	12.182	1.08574	0.082085	**3.00**	20.086	1.30288	0.049787

TABLE B.5 Exponential Functions (*Cont'd*)

x	e^x	$Log_{10}(e^x)$	e^{-x}	x	e^x	$Log_{10}(e^x)$	e^{-x}
3.00	20.086	1.30288	0.049787	**3.50**	33.115	1.52003	0.030197
3.01	20.287	1.30723	.049292	3.51	33.448	1.52437	.029897
3.02	20.491	1.31157	.048801	3.52	33.784	1.52872	.029599
3.03	20.697	1.31591	.048316	3.53	34.124	1.53306	.029305
3.04	20.905	1.32026	.047835	3.54	34.467	1.53740	.029013
3.05	21.115	1.32460	0.047359	**3.55**	34.813	1.54175	0.028725
3.06	21.328	1.32894	.046888	3.56	35.163	1.54609	.028439
3.07	21.542	1.33328	.046421	3.57	35.517	1.55043	.028156
3.08	21.758	1.33763	.045959	3.58	35.874	1.55477	.027876
3.09	21.977	1.34197	.045502	3.59	36.234	1.55912	.027598
3.10	22.198	1.34631	0.045049	**3.60**	36.598	1.56346	0.027324
3.11	22.421	1.35066	.044601	3.61	36.966	1.56780	.027052
3.12	22.646	1.35500	.044157	3.62	37.338	1.57215	.026783
3.13	22.874	1.35934	.043718	3.63	37.713	1.57649	.026516
3.14	23.104	1.36368	.043283	3.64	38.092	1.58083	.026252
3.15	23.336	1.36803	0.042852	**3.65**	38.475	1.58517	0.025991
3.16	23.571	1.37237	.042426	3.66	38.861	1.58952	.025733
3.17	23.807	1.37671	.042004	3.67	39.252	1.59386	.025476
3.18	24.047	1.38106	.041586	3.68	39.646	1.59820	.025223
3.19	24.288	1.38540	.041172	3.69	40.045	1.60255	.024972
3.20	24.533	1.38974	0.040762	**3.70**	40.447	1.60689	0.024724
3.21	24.779	1.39409	.040357	3.71	40.854	1.61123	.024478
3.22	25.028	1.39843	.039955	3.72	41.264	1.61558	.024234
3.23	25.280	1.40277	.039557	3.73	41.679	1.61992	.023993
3.24	25.534	1.40711	.039164	3.74	42.098	1.62426	.023754
3.25	25.790	1.41146	0.038774	**3.75**	42.521	1.62860	0.023518
3.26	26.050	1.41580	.038388	3.76	42.948	1.63295	.023284
3.27	26.311	1.42014	.038006	3.77	43.380	1.63729	.023052
3.28	26.576	1.42449	.037628	3.78	43.816	1.64163	.022823
3.29	26.843	1.42883	.037254	3.79	44.256	1.64598	.022596
3.30	27.113	1.43317	0.036883	**3.80**	44.701	1.65032	0.022371
3.31	27.385	1.43751	.036516	3.81	45.150	1.65466	.022148
3.32	27.660	1.44186	.036153	3.82	45.604	1.65900	.021928
3.33	27.938	1.44620	.035793	3.83	46.063	1.66335	.021710
3.34	28.219	1.45054	.035437	3.84	46.525	1.66769	.021494
3.35	28.503	1.45489	0.035084	**3.85**	46.993	1.67203	0.021280
3.36	28.789	1.45923	.034735	3.86	47.465	1.67638	.021068
3.37	29.079	1.46357	.034390	3.87	47.942	1.68072	.020858
3.38	29.371	1.46792	.034047	3.88	48.424	1.68506	.020651
3.39	29.666	1.47226	.033709	3.89	48.911	1.68941	.020445
3.40	29.964	1.47660	0.033373	**3.90**	49.402	1.69375	0.020242
3.41	30.265	1.48094	.033041	3.91	49.899	1.69809	.020041
3.42	30.569	1.48529	.032712	3.92	50.400	1.70243	.019841
3.43	30.877	1.48963	.032387	3.93	50.907	1.70678	.019644
3.44	31.187	1.49397	.032065	3.94	51.419	1.71112	.019448
3.45	31.500	1.49832	0.031746	**3.95**	51.935	1.71546	0.019255
3.46	31.817	1.50266	.031430	3.96	52.457	1.71981	.019063
3.47	32.137	1.50700	.031117	3.97	52.985	1.72415	.018873
3.48	32.460	1.51134	.030807	3.98	53.517	1.72849	.018686
3.49	32.786	1.51569	.030501	3.99	54.055	1.73283	.018500
3.50	33.115	1.52003	0.030197	**4.00**	54.598	1.73718	0.018316

TABLE B.5 Exponential Functions (Cont'd)

x	e^x	$\text{Log}_{10}(e^x)$	e^{-x}	x	e^x	$\text{Log}_{10}(e^x)$	e^{-x}
4.00	54.598	1.73718	0.018316	4.50	90.017	1.95433	0.011109
4.01	55.147	1.74152	.018133	4.51	90.922	1.95867	.010998
4.02	55.701	1.74586	.017953	4.52	91.836	1.96301	.010889
4.03	56.261	1.75021	.017774	4.53	92.759	1.96735	.010781
4.04	56.826	1.75455	.017597	4.54	93.691	1.97170	.010673
4.05	57.397	1.75889	0.017422	4.55	94.632	1.97604	0.010567
4.06	57.974	1.76324	.017249	4.56	95.583	1.98038	.010462
4.07	58.557	1.76758	.017077	4.57	96.544	1.98473	.010358
4.08	59.145	1.77192	.016907	4.58	97.514	1.98907	.010255
4.09	59.740	1.77626	.016739	4.59	98.494	1.99341	.010153
4.10	60.340	1.78061	0.016573	4.60	99.484	1.99775	0.010052
4.11	60.947	1.78495	.016408	4.61	100.48	2.00210	.009952
4.12	61.559	1.78929	.016245	4.62	101.49	2.00644	.009853
4.13	62.178	1.79364	.016083	4.63	102.51	2.01078	.009755
4.14	62.803	1.79798	.015923	4.64	103.54	2.01513	.009658
4.15	63.434	1.80232	0.015764	4.65	104.58	2.01947	0.009562
4.16	64.072	1.80667	.015608	4.66	105.64	2.02381	.009466
4.17	64.715	1.81101	.015452	4.67	106.70	2.02816	.009372
4.18	65.366	1.81535	.015299	4.68	107.77	2.03250	.009279
4.19	66.023	1.81969	.015146	4.69	108.85	2.03684	.009187
4.20	66.686	1.82404	0.014996	4.70	109.95	2.04118	0.009095
4.21	67.357	1.82838	.014846	4.71	111.05	2.04553	.009005
4.22	68.033	1.83272	.014699	4.72	112.17	2.04987	.008915
4.23	68.717	1.83707	.014552	4.73	113.30	2.05421	.008826
4.24	69.408	1.84141	.014408	4.74	114.43	2.05856	.008739
4.25	70.105	1.84575	0.014264	4.75	115.58	2.06290	0.008652
4.26	70.810	1.85009	.014122	4.76	116.75	2.06724	.008566
4.27	71.522	1.85444	.013982	4.77	117.92	2.07158	.008480
4.28	72.240	1.85878	.013843	4.78	119.10	2.07593	.008396
4.29	72.966	1.86312	.013705	4.79	120.30	2.08027	.008312
4.30	73.700	1.86747	0.013569	4.80	121.51	2.08461	0.008230
4.31	74.440	1.87181	.013434	4.81	122.73	2.08896	.008148
4.32	75.189	1.87615	.013300	4.82	123.97	2.09330	.008067
4.33	75.944	1.88050	.013168	4.83	125.21	2.09764	.007987
4.34	76.708	1.88484	.013037	4.84	126.47	2.10199	.007907
4.35	77.478	1.88918	0.012907	4.85	127.74	2.10633	0.007828
4.36	78.257	1.89352	.012778	4.86	129.02	2.11067	.007750
4.37	79.044	1.89787	.012651	4.87	130.32	2.11501	.007673
4.38	79.838	1.90221	.012525	4.88	131.63	2.11936	.007597
4.39	80.640	1.90655	.012401	4.89	132.95	2.12370	.007521
4.40	81.451	1.91090	0.012277	4.90	134.29	2.12804	0.007447
4.41	82.269	1.91524	.012155	4.91	135.64	2.13239	.007372
4.42	83.096	1.91958	.012034	4.92	137.00	2.13673	.007299
4.43	83.931	1.92392	.011914	4.93	138.38	2.14107	.007227
4.44	84.775	1.92827	.011796	4.94	139.77	2.14541	.007155
4.45	85.627	1.93261	0.011679	4.95	141.17	2.14976	0.007083
4.46	86.488	1.93695	.011562	4.96	142.59	2.15410	.007013
4.47	87.357	1.94130	.011447	4.97	144.03	2.15844	.006943
4.48	88.235	1.94564	.011333	4.98	145.47	2.16279	.006874
4.49	89.121	1.94998	.011221	4.99	146.94	2.16713	.006806
4.50	90.017	1.95433	0.011109	5.00	148.41	2.17147	0.006738

TABLE B.5 Exponential Functions (*Cont'd*)

x	e^x	$\text{Log}_{10}(e^x)$	e^{-x}	x	e^x	$\text{Log}_{10}(e^x)$	e^{-x}
5.00	148.41	2.17147	0.006738	**5.50**	244.69	2.38862	0.0040868
5.01	149.90	2.17582	.006671	5.55	257.24	2.41033	.0038875
5.02	151.41	2.18016	.006605	5.60	270.43	2.43205	.0036979
5.03	152.93	2.18450	.006539	5.65	284.29	2.45376	.0035175
5.04	154.47	2.18884	.006474	5.70	298.87	2.47548	.0033460
5.05	156.02	2.19319	0.006409	**5.75**	314.19	2.49719	0.0031828
5.06	157.59	2.19753	.006346	5.80	330.30	2.51891	.0030276
5.07	159.17	2.20187	.006282	5.85	347.23	2.54062	.0028799
5.08	160.77	2.20622	.006220	5.90	365.04	2.56234	.0027394
5.09	162.39	2.21056	.006158	5.95	383.75	2.58405	.0026058
5.10	164.02	2.21490	0.006097	**6.00**	403.43	2.60577	0.0024788
5.11	165.67	2.21924	.006036	6.05	424.11	2.62748	.0023579
5.12	167.34	2.22359	.005976	6.10	445.86	2.64920	.0022429
5.13	169.02	2.22793	.005917	6.15	468.72	2.67091	.0021335
5.14	170.72	2.23227	.005858	6.20	492.75	2.69263	.0020294
5.15	172.43	2.23662	0.005799	**6.25**	518.01	2.71434	0.0019305
5.16	174.16	2.24096	.005742	6.30	544.57	2.73606	.0018363
5.17	175.91	2.24530	.005685	6.35	572.49	2.75777	.0017467
5.18	177.68	2.24965	.005628	6.40	601.85	2.77948	.0016616
5.19	179.47	2.25399	.005572	6.45	632.70	2.80120	.0015805
5.20	181.27	2.25833	0.005517	**6.50**	665.14	2.82291	0.0015034
5.21	183.09	2.26267	.005462	6.55	699.24	2.84463	.0014301
5.22	184.93	2.26702	.005407	6.60	735.10	2.86634	.0013604
5.23	186.79	2.27136	.005354	6.65	772.78	2.88806	.0012940
5.24	188.67	2.27570	.005300	6.70	812.41	2.90977	.0012309
5.25	190.57	2.28005	0.005248	**6.75**	854.06	2.93149	0.0011709
5.26	192.48	2.28439	.005195	6.80	897.85	2.95320	.0011138
5.27	194.42	2.28873	.005144	6.85	943.88	2.97492	.0010595
5.28	196.37	2.29307	.005092	6.90	992.27	2.99663	.0010078
5.29	198.34	2.29742	.005042	6.95	1043.1	3.01835	.0009586
5.30	200.34	2.30176	0.004992	**7.00**	1096.6	3.04006	0.0009119
5.31	202.35	2.30610	.004942	7.05	1152.9	3.06178	.0008674
5.32	204.38	2.31045	.004893	7.10	1212.0	3.08349	.0008251
5.33	206.44	2.31479	.004844	7.15	1274.1	3.10521	.0007849
5.34	208.51	2.31913	.004796	7.20	1339.4	3.12692	.0007466
5.35	210.61	2.32348	0.004748	**7.25**	1408.1	3.14863	0.0007102
5.36	212.72	2.32782	.004701	7.30	1480.3	3.17035	.0006755
5.37	214.86	2.33216	.004654	7.35	1556.2	3.19206	.0006426
5.38	217.02	2.33650	.004608	7.40	1636.0	3.21378	.0006113
5.39	219.20	2.34085	.004562	7.45	1719.9	3.23549	.0005814
5.40	221.41	2.34519	0.004517	**7.50**	1808.0	3.25721	0.0005531
5.41	223.63	2.34953	.004472	7.55	1900.7	3.27892	.0005261
5.42	225.88	2.35388	.004427	7.60	1998.2	3.30064	.0005005
5.43	228.15	2.35822	.004383	7.65	2100.6	3.32235	.0004760
5.44	230.44	2.36256	.004339	7.70	2208.3	3.34407	.0004528
5.45	232.76	2.36690	0.004296	**7.75**	2321.6	3.36578	0.0004307
5.46	235.10	2.37125	.004254	7.80	2440.6	3.38750	.0004097
5.47	237.46	2.37559	.004211	7.85	2565.7	3.40921	.0003898
5.48	239.85	2.37993	.004169	7.90	2697.3	3.43093	.0003707
5.49	242.26	2.38428	.004128	7.95	2835.6	3.45264	.0003527
5.50	244.69	2.38862	0.004087	**8.00**	2981.0	3.47436	0.0003355

TABLE B.5 Exponential Functions (*Cont'd*)

x	e^x	$\text{Log}_{10}(e^x)$	e^{-x}	x	e^x	$\text{Log}_{10}(e^x)$	e^{-x}
8.00	2981.0	3.47436	0.0003355	9.00	8103.1	3.90865	0.0001234
8.05	3133.8	3.49607	.0003191	9.05	8518.5	3.93037	.0001174
8.10	3294.5	3.51779	.0003035	9.10	8955.3	3.95208	.0001117
8.15	3463.4	3.53950	.0002887	9.15	9414.4	3.97379	.0001062
8.20	3641.0	3.56121	.0002747	9.20	9897.1	3.99551	.0001010
8.25	3827.6	3.58293	0.0002613	9.25	10405	4.01722	0.0000961
8.30	4023.9	3.60464	.0002485	9.30	10938	4.03894	.0000914
8.35	4230.2	3.62636	.0002364	9.35	11499	4.06065	.0000870
8.40	4447.1	3.64807	.0002249	9.40	12088	4.08237	.0000827
8.45	4675.1	3.66979	.0002139	9.45	12708	4.10408	.0000787
8.50	4914.8	3.69150	0.0002035	9.50	13360	4.12580	0.0000749
8.55	5166.8	3.71322	.0001935	9.55	14045	4.14751	.0000712
8.60	5431.7	3.73493	.0001841	9.60	14765	4.16923	.0000677
8.65	5710.1	3.75665	.0001751	9.65	15522	4.19094	.0000644
8.70	6002.9	3.77836	.0001666	9.70	16318	4.21266	.0000613
8.75	6310.7	3.80008	0.0001585	9.75	17154	4.23437	0.0000583
8.80	6634.2	3.82179	.0001507	9.80	18034	4.25609	.0000555
8.85	6974.4	3.84351	.0001434	9.85	18958	4.27780	.0000527
8.90	7332.0	3.86522	.0001364	9.90	19930	4.29952	.0000502
8.95	7707.9	3.88694	.0001297	9.95	20952	4.32123	0.0000477
9.00	8103.1	3.90865	0.0001234	10.00	22026	4.34294	0.0000454

SOURCE: From *Handbook of Probability and Statistics*, 2d ed., 1968, The Chemical Rubber Co., Cleveland, Ohio, with permission from the publisher.

TABLE B.6 Binomial Probabilities $b(k,n,\pi) = C(n,k)\pi^k(1-\pi)^{n-k}$

						π					
n	k	.05	.10	.15	.20	.25	.30	.35	.40	.45	.50
1	0	.9500	.9000	.8500	.8000	.7500	.7000	.6500	.6000	.5500	.5000
	1	.0500	.1000	.1500	.2000	.2500	.3000	.3500	.4000	.4500	.5000
2	0	.9025	.8100	.7225	.6400	.5625	.4900	.4225	.3600	.3025	.2500
	1	.0950	.1800	.2550	.3200	.3750	.4200	.4550	.4800	.4950	.5000
	2	.0025	.0100	.0225	.0400	.0625	.0900	.1225	.1600	.2025	.2500
3	0	.8574	.7290	.6141	.5120	.4219	.3430	.2746	.2160	.1664	.1250
	1	.1354	.2430	.3251	.3840	.4219	.4410	.4436	.4320	.4084	.3750
	2	.0071	.0270	.0574	.0960	.1406	.1890	.2389	.2880	.3341	.3750
	3	.0001	.0010	.0034	.0080	.0156	.0270	.0429	.0640	.0911	.1250
4	0	.8145	.6561	.5220	.4096	.3164	.2401	.1785	.1296	.0915	.0625
	1	.1715	.2916	.3685	.4096	.4219	.4116	.3845	.3456	.2995	.2500
	2	.0135	.0486	.0975	.1536	.2109	.2646	.3105	.3456	.3675	.3750
	3	.0005	.0036	.0115	.0256	.0469	.0756	.1115	.1536	.2005	.2500
	4	.0000	.0001	.0005	.0016	.0039	.0081	.0150	.0256	.0410	.0625
5	0	.7738	.5905	.4437	.3277	.2373	.1681	.1160	.0778	.0503	.0312
	1	.2036	.3280	.3915	.4096	.3955	.3602	.3124	.2592	.2059	.1562
	2	.0214	.0729	.1382	.2048	.2637	.3087	.3364	.3456	.3369	.3125
	3	.0011	.0081	.0244	.0512	.0879	.1323	.1811	.2304	.2757	.3125
	4	.0000	.0004	.0022	.0064	.0146	.0284	.0488	.0768	.1128	.1562
	5	.0000	.0000	.0001	.0003	.0010	.0024	.0053	.0102	.0185	.0312
6	0	.7351	.5314	.3771	.2621	.1780	.1176	.0754	.0467	.0277	.0156
	1	.2321	.3543	.3993	.3932	.3560	.3025	.2437	.1866	.1359	.0938
	2	.0305	.0984	.1762	.2458	.2966	.3241	.3280	.3110	.2780	.2344
	3	.0021	.0146	.0415	.0819	.1318	.1852	.2355	.2765	.3032	.3125
	4	.0001	.0012	.0055	.0154	.0330	.0595	.0951	.1382	.1861	.2344
	5	.0000	.0001	.0004	.0015	.0044	.0102	.0205	.0369	.0609	.0938
	6	.0000	.0000	.0000	.0001	.0002	.0007	.0018	.0041	.0083	.0156
7	0	.6983	.4783	.3206	.2097	.1335	.0824	.0490	.0280	.0152	.0078
	1	.2573	.3720	.3960	.3670	.3115	.2471	.1848	.1306	.0872	.0547
	2	.0406	.1240	.2097	.2753	.3115	.3177	.2985	.2613	.2140	.1641
	3	.0036	.0230	.0617	.1147	.1730	.2269	.2679	.2903	.2918	.2734
	4	.0002	.0026	.0109	.0287	.0577	.0972	.1442	.1935	.2388	.2734
	5	.0000	.0002	.0012	.0043	.0115	.0250	.0466	.0774	.1172	.1641
	6	.0000	.0000	.0001	.0004	.0013	.0036	.0084	.0172	.0320	.0547
	7	.0000	.0000	.0000	.0000	.0001	.0002	.0006	.0016	.0037	.0078
8	0	.6634	.4305	.2725	.1678	.1001	.0576	.0319	.0168	.0084	.0039
	1	.2793	.3826	.3847	.3355	.2670	.1977	.1373	.0896	.0548	.0312
	2	.0515	.1488	.2376	.2936	.3115	.2965	.2587	.2090	.1569	.1094
	3	.0054	.0331	.0839	.1468	.2076	.2541	.2786	.2787	.2568	.2188
	4	.0004	.0046	.0185	.0459	.0865	.1361	.1875	.2322	.2627	.2734
	5	.0000	.0004	.0026	.0092	.0231	.0467	.0808	.1239	.1719	.2188
	6	.0000	.0000	.0002	.0011	.0038	.0100	.0217	.0413	.0703	.1094
	7	.0000	.0000	.0000	.0001	.0004	.0012	.0033	.0079	.0164	.0312
	8	.0000	.0000	.0000	.0000	.0000	.0000	.0001	.0002	.0017	.0039

Linear interpolations with respect to π will in general be accurate at most to two decimal places.

TABLE B.6 Binomial Probabilities (*Cont'd*)

						π					
n	k	.05	.10	.15	.20	.25	.30	.35	.40	.45	.50
9	0	.6302	.3874	.2316	.1342	.0751	.0404	.0207	.0101	.0046	.0020
	1	.2985	.3874	.3679	.3020	.2253	.1556	.1004	.0605	.0339	.0176
	2	.0629	.1722	.2597	.3020	.3003	.2668	.2162	.1612	.1110	.0703
	3	.0077	.0446	.1069	.1762	.2336	.2668	.2716	.2508	.2119	.1641
	4	.0006	.0074	.0283	.0661	.1168	.1715	.2194	.2508	.2600	.2461
	5	.0000	.0008	.0050	.0165	.0389	.0735	.1181	.1672	.2128	.2461
	6	.0000	.0001	.0006	.0028	.0087	.0210	.0424	.0743	.1160	.1641
	7	.0000	.0000	.0000	.0003	.0012	.0039	.0098	.0212	.0407	.0703
	8	.0000	.0000	.0000	.0000	.0001	.0004	.0013	.0035	.0083	.0176
	9	.0000	.0000	.0000	.0000	.0000	.0000	.0001	.0003	.0008	.0020
10	0	.5987	.3487	.1969	.1074	.0563	.0282	.0135	.0060	.0025	.0010
	1	.3151	.3874	.3474	.2684	.1877	.1211	.0725	.0403	.0207	.0098
	2	.0746	.1937	.2759	.3020	.2816	.2335	.1757	.1209	.0763	.0439
	3	.0105	.0574	.1298	.2013	.2503	.2668	.2522	.2150	.1665	.1172
	4	.0010	.0112	.0401	.0881	.1460	.2001	.2377	.2508	.2384	.2051
	5	.0001	.0015	.0085	.0264	.0584	.1029	.1536	.2007	.2340	.2461
	6	.0000	.0001	.0012	.0055	.0162	.0368	.0689	.1115	.1596	.2051
	7	.0000	.0000	.0001	.0008	.0031	.0090	.0212	.0425	.0746	.1172
	8	.0000	.0000	.0000	.0001	.0004	.0014	.0043	.0106	.0229	.0439
	9	.0000	.0000	.0000	.0000	.0000	.0001	.0005	.0016	.0042	.0098
	10	.0000	.0000	.0000	.0000	.0000	.0000	.0000	.0001	.0003	.0010
11	0	.5688	.3138	.1673	.0859	.0422	.0198	.0088	.0036	.0014	.0004
	1	.3293	.3835	.3248	.2362	.1549	.0932	.0518	.0266	.0125	.0055
	2	.0867	.2131	.2866	.2953	.2581	.1998	.1395	.0887	.0513	.0269
	3	.0137	.0710	.1517	.2215	.2581	.2568	.2254	.1774	.1259	.0806
	4	.0014	.0158	.0536	.1107	.1721	.2201	.2428	.2365	.2060	.1611
	5	.0001	.0025	.0132	.0388	.0803	.1321	.1830	.2207	.2360	.2256
	6	.0000	.0003	.0023	.0097	.0268	.0566	.0985	.1471	.1931	.2256
	7	.0000	.0000	.0003	.0017	.0064	.0173	.0379	.0701	.1128	.1611
	8	.0000	.0000	.0000	.0002	.0011	.0037	.0102	.0234	.0462	.0806
	9	.0000	.0000	.0000	.0000	.0001	.0005	.0018	.0052	.0126	.0269
	10	.0000	.0000	.0000	.0000	.0000	.0000	.0002	.0007	.0021	.0054
	11	.0000	.0000	.0000	.0000	.0000	.0000	.0000	.0000	.0002	.0005
12	0	.5404	.2824	.1422	.0687	.0317	.0138	.0057	.0022	.0008	.0002
	1	.3413	.3766	.3012	.2062	.1267	.0712	.0368	.0174	.0075	.0029
	2	.0988	.2301	.2924	.2835	.2323	.1678	.1088	.0639	.0339	.0161
	3	.0173	.0852	.1720	.2362	.2581	.2397	.1954	.1419	.0923	.0537
	4	.0021	.0213	.0683	.1329	.1936	.2311	.2367	.2128	.1700	.1208
	5	.0002	.0038	.0193	.0532	.1032	.1585	.2039	.2270	.2225	.1934
	6	.0000	.0005	.0040	.0155	.0401	.0792	.1281	.1766	.2124	.2256
	7	.0000	.0000	.0006	.0033	.0115	.0291	.0591	.1009	.1489	.1934
	8	.0000	.0000	.0001	.0005	.0024	.0078	.0199	.0420	.0762	.1208
	9	.0000	.0000	.0000	.0001	.0004	.0015	.0048	.0125	.0277	.0537
	10	.0000	.0000	.0000	.0000	.0000	.0002	.0008	.0025	.0068	.0161
	11	.0000	.0000	.0000	.0000	.0000	.0000	.0001	.0003	.0010	.0029
	12	.0000	.0000	.0000	.0000	.0000	.0000	.0000	.0000	.0001	.0002

TABLE B.6 Binomial Probabilities (*Cont'd*)

n	k	.05	.10	.15	.20	π .25	.30	.35	.40	.45	.50
13	0	.5133	.2542	.1209	.0550	.0238	.0097	.0037	.0013	.0004	.0001
	1	.3512	.3672	.2774	.1787	.1029	.0540	.0259	.0113	.0045	.0016
	2	.1109	.2448	.2937	.2680	.2059	.1388	.0836	.0453	.0220	.0095
	3	.0214	.0997	.1900	.2457	.2517	.2181	.1651	.1107	.0660	.0349
	4	.0028	.0277	.0838	.1535	.2097	.2337	.2222	.1845	.1350	.0873
	5	.0003	.0055	.0266	.0691	.1258	.1803	.2154	.2214	.1989	.1571
	6	.0000	.0008	.0063	.0230	.0559	.1030	.1546	.1968	.2169	.2095
	7	.0000	.0001	.0011	.0058	.0186	.0442	.0833	.1312	.1775	.2095
	8	.0000	.0000	.0001	.0011	.0047	.0142	.0336	.0656	.1089	.1571
	9	.0000	.0000	.0000	.0001	.0009	.0034	.0101	.0243	.0495	.0873
	10	.0000	.0000	.0000	.0000	.0001	.0006	.0022	.0065	.0162	.0349
	11	.0000	.0000	.0000	.0000	.0000	.0001	.0003	.0012	.0036	.0095
	12	.0000	.0000	.0000	.0000	.0000	.0000	.0000	.0001	.0005	.0016
	13	.0000	.0000	.0000	.0000	.0000	.0000	.0000	.0000	.0000	.0001
14	0	.4877	.2288	.1028	.0440	.0178	.0068	.0024	.0008	.0002	.0001
	1	.3593	.3559	.2539	.1539	.0832	.0407	.0181	.0073	.0027	.0009
	2	.1229	.2570	.2912	.2501	.1802	.1134	.0634	.0317	.0141	.0056
	3	.0259	.1142	.2056	.2501	.2402	.1943	.1366	.0845	.0462	.0222
	4	.0037	.0349	.0998	.1720	.2202	.2290	.2022	.1549	.1040	.0611
	5	.0004	.0078	.0352	.0860	.1468	.1963	.2178	.2066	.1701	.1222
	6	.0000	.0013	.0093	.0322	.0734	.1262	.1759	.2066	.2088	.1833
	7	.0000	.0002	.0019	.0092	.0280	.0618	.1082	.1574	.1952	.2095
	8	.0000	.0000	.0003	.0020	.0082	.0232	.0510	.0918	.1398	.1833
	9	.0000	.0000	.0000	.0003	.0018	.0066	.0183	.0408	.0762	.1222
	10	.0000	.0000	.0000	.0000	.0003	.0014	.0049	.0136	.0312	.0611
	11	.0000	.0000	.0000	.0000	.0000	.0002	.0010	.0033	.0093	.0222
	12	.0000	.0000	.0000	.0000	.0000	.0000	.0001	.0005	.0019	.0056
	13	.0000	.0000	.0000	.0000	.0000	.0000	.0000	.0001	.0002	.0009
	14	.0000	.0000	.0000	.0000	.0000	.0000	.0000	.0000	.0000	.0001
15	0	.4633	.2059	.0874	.0352	.0134	.0047	.0016	.0005	.0001	.0000
	1	.3658	.3432	.2312	.1319	.0668	.0305	.0126	.0047	.0016	.0005
	2	.1348	.2669	.2856	.2309	.1559	.0916	.0476	.0219	.0090	.0032
	3	.0307	.1285	.2184	.2501	.2252	.1700	.1110	.0634	.0318	.0139
	4	.0049	.0428	.1156	.1876	.2252	.2186	.1792	.1268	.0780	.0417
	5	.0006	.0105	.0449	.1032	.1651	.2061	.2123	.1859	.1404	.0916
	6	.0000	.0019	.0132	.0430	.0917	.1472	.1906	.2066	.1914	.1527
	7	.0000	.0003	.0030	.0138	.0393	.0811	.1319	.1771	.2013	.1964
	8	.0000	.0000	.0005	.0035	.0131	.0348	.0710	.1181	.1647	.1964
	9	.0000	.0000	.0001	.0007	.0034	.0116	.0298	.0612	.1048	.1527
	10	.0000	.0000	.0000	.0001	.0007	.0030	.0096	.0245	.0515	.0916
	11	.0000	.0000	.0000	.0000	.0001	.0006	.0024	.0074	.0191	.0417
	12	.0000	.0000	.0000	.0000	.0000	.0001	.0004	.0016	.0052	.0139
	13	.0000	.0000	.0000	.0000	.0000	.0000	.0001	.0003	.0010	.0032
	14	.0000	.0000	.0000	.0000	.0000	.0000	.0000	.0000	.0001	.0005
	15	.0000	.0000	.0000	.0000	.0000	.0000	.0000	.0000	.0000	.0000

TABLE B.6 Binomial Probabilities (Cont'd)

n	k	.05	.10	.15	.20	.25	.30	.35	.40	.45	.50
16	0	.4401	.1853	.0743	.0281	.0100	.0033	.0010	.0003	.0001	.0000
	1	.3706	.3294	.2097	.1126	.0535	.0228	.0087	.0030	.0009	.0002
	2	.1463	.2745	.2775	.2111	.1336	.0732	.0353	.0150	.0056	.0018
	3	.0359	.1423	.2285	.2463	.2079	.1465	.0888	.0468	.0215	.0085
	4	.0061	.0514	.1311	.2001	.2252	.2040	.1553	.1014	.0572	.0278
	5	.0008	.0137	.0555	.1201	.1802	.2099	.2008	.1623	.1123	.0667
	6	.0001	.0028	.0180	.0550	.1101	.1649	.1982	.1983	.1684	.1222
	7	.0000	.0004	.0045	.0197	.0524	.1010	.1524	.1889	.1969	.1746
	8	.0000	.0001	.0009	.0055	.0197	.0487	.0923	.1417	.1812	.1964
	9	.0000	.0000	.0001	.0012	.0058	.0185	.0442	.0840	.1318	.1746
	10	.0000	.0000	.0000	.0002	.0014	.0056	.0167	.0392	.0755	.1222
	11	.0000	.0000	.0000	.0000	.0002	.0013	.0049	.0142	.0337	.0667
	12	.0000	.0000	.0000	.0000	.0000	.0002	.0011	.0040	.0115	.0278
	13	.0000	.0000	.0000	.0000	.0000	.0000	.0002	.0008	.0029	.0085
	14	.0000	.0000	.0000	.0000	.0000	.0000	.0000	.0001	.0005	.0018
	15	.0000	.0000	.0000	.0000	.0000	.0000	.0000	.0000	.0001	.0002
	16	.0000	.0000	.0000	.0000	.0000	.0000	.0000	.0000	.0000	.0000
17	0	.4181	.1668	.0631	.0225	.0075	.0023	.0007	.0002	.0000	.0000
	1	.3741	.3150	.1893	.0957	.0426	.0169	.0060	.0019	.0005	.0001
	2	.1575	.2800	.2673	.1914	.1136	.0581	.0260	.0102	.0035	.0010
	3	.0415	.1556	.2359	.2393	.1893	.1245	.0701	.0341	.0144	.0052
	4	.9076	.0605	.1457	.2093	.2209	.1868	.1320	.0796	.0411	.0182
	5	.0010	.0175	.0668	.1361	.1914	.2081	.1849	.1379	.0875	.0472
	6	.0001	.0039	.0236	.0680	.1276	.1784	.1991	.1839	.1432	.0944
	7	.0000	.0007	.0065	.0267	.0668	.1201	.1685	.1927	.1841	.1484
	8	.0000	.0001	.0014	.0084	.0279	.0644	.1134	.1606	.1883	.1855
	9	.0000	.0000	.0003	.0021	.0093	.0276	.0611	.1070	.1540	.1855
	10	.0000	.0000	.0000	.0004	.0025	.0095	.0263	.0571	.1008	.1484
	11	.0000	.0000	.0000	.0001	.0005	.0026	.0090	.0242	.0525	.0944
	12	.0000	.0000	.0000	.0000	.0001	.0006	.0024	.0081	.0215	.0472
	13	.0000	.0000	.0000	.0000	.0000	.0001	.0005	.0021	.0068	.0182
	14	.0000	.0000	.0000	.0000	.0000	.0000	.0001	.0004	.0016	.0052
	15	.0000	.0000	.0000	.0000	.0000	.0000	.0000	.0001	.0003	.0010
	16	.0000	.0000	.0000	.0000	.0000	.0000	.0000	.0000	.0000	.0001
	17	.0000	.0000	.0000	.0000	.0000	.0000	.0000	.0000	.0000	.0000
18	0	.3972	.1501	.0536	.0180	.0056	.0016	.0004	.0001	.0000	.0000
	1	.3763	.3002	.1704	.0811	.0338	.0126	.0042	.0012	.0003	.0001
	2	.1683	.2835	.2556	.1723	.0958	.0458	.0190	.0069	.0022	.0006
	3	.0473	.1680	.2406	.2297	.1704	.1046	.0547	.0246	.0095	.0031
	4	.0093	.0700	.1592	.2153	.2130	.1681	.1104	.0614	.0291	.0117
	5	.0014	.0218	.0787	.1507	.1988	.2017	.1664	.1146	.0666	.0327
	6	.0002	.0052	.0301	.0816	.1436	.1873	.1941	.1655	.1181	.0708
	7	.0000	.0010	.0091	.0350	.0820	.1376	.1792	.1892	.1657	.1214
	8	.0000	.0002	.0022	.0120	.0376	.0811	.1327	.1734	.1864	.1669
	9	.0000	.0000	.0004	.0033	.0139	.0386	.0794	.1284	.1694	.1855
	10	.0000	.0000	.0001	.0008	.0042	.0149	.0385	.0771	.1248	.1669
	11	.0000	.0000	.0000	.0001	.0010	.0046	.0151	.0374	.0742	.1214

TABLE B.6 Binomial Probabilities (*Cont'd*)

n	k	.05	.10	.15	.20	π .25	.30	.35	.40	.45	.50
18	12	.0000	.0000	.0000	.0000	.0002	.0012	.0047	.0145	.0354	.0708
	13	.0000	.0000	.0000	.0000	.0000	.0002	.0012	.0045	.0134	.0327
	14	.0000	.0000	.0000	.0000	.0000	.0000	.0002	.0011	.0039	.0117
	15	.0000	.0000	.0000	.0000	.0000	.0000	.0000	.0002	.0009	.0031
	16	.0000	.0000	.0000	.0000	.0000	.0000	.0000	.0000	.0001	.0006
	17	.0000	.0000	.0000	.0000	.0000	.0000	.0000	.0000	.0000	.0001
	18	.0000	.0000	.0000	.0000	.0000	.0000	.0000	.0000	.0000	.0000
19	0	.3774	.1351	.0456	.0144	.0042	.0011	.0003	.0001	.0000	.0000
	1	.3774	.2852	.1529	.0685	.0268	.0093	.0029	.0008	.0002	.0000
	2	.1787	.2852	.2428	.1540	.0803	.0358	.0138	.0046	.0013	.0003
	3	.0533	.1796	.2428	.2182	.1517	.0869	.0422	.0175	.0062	.0018
	4	.0112	.0798	.1714	.2182	.2023	.1491	.0909	.0467	.0203	.0074
	5	.0018	.0266	.0907	.1636	.2023	.1916	.1468	.0933	.0497	.0222
	6	.0002	.0069	.0374	.0955	.1574	.1916	.1844	.1451	.0949	.0518
	7	.0000	.0014	.0122	.0443	.0974	.1525	.1844	.1797	.1443	.0961
	8	.0000	.0002	.0032	.0166	.0487	.0981	.1489	.1797	.1771	.1442
	9	.0000	.0000	.0007	.0051	.0198	.0514	.0980	.1464	.1771	.1762
	10	.0000	.0000	.0001	.0013	.0066	.0220	.0528	.0976	.1449	.1762
	11	.0000	.0000	.0000	.0003	.0018	.0077	.0233	.0532	.0970	.1442
	12	.0000	.0000	.0000	.0000	.0004	.0022	.0083	.0237	.0529	.0961
	13	.0000	.0000	.0000	.0000	.0001	.0005	.0024	.0085	.0233	.0518
	14	.0000	.0000	.0000	.0000	.0000	.0001	.0006	.0024	.0082	.0222
	15	.0000	.0000	.0000	.0000	.0000	.0000	.0001	.0005	.0022	.0074
	16	.0000	.0000	.0000	.0000	.0000	.0000	.0000	.0001	.0005	.0018
	17	.0000	.0000	.0000	.0000	.0000	.0000	.0000	.0000	.0001	.0003
	18	.0000	.0000	.0000	.0000	.0000	.0000	.0000	.0000	.0000	.0000
	19	.0000	.0000	.0000	.0000	.0000	.0000	.0000	.0000	.0000	.0000
20	0	.3585	.1216	.0388	.0115	.0032	.0008	.0002	.0000	.0000	.0000
	1	.3774	.2702	.1368	.0576	.0211	.0068	.0020	.0005	.0001	.0000
	2	.1887	.2852	.2293	.1369	.0669	.0278	.0100	.0031	.0008	.0002
	3	.0596	.1901	.2428	.2054	.1339	.0716	.0323	.0123	.0040	.0011
	4	.0133	.0898	.1821	.2182	.1897	.1304	.0738	.0350	.0139	.0046
	5	.0022	.0319	.1028	.1746	.2023	.1789	.1272	.0746	.0365	.0148
	6	.0003	.0089	.0454	.1091	.1686	.1916	.1712	.1244	.0746	.0370
	7	.0000	.0020	.0160	.0545	.1124	.1643	.1844	.1659	.1221	.0739
	8	.0000	.0004	.0046	.0222	.0609	.1144	.1614	.1797	.1623	.1201
	9	.0000	.0001	.0011	.0074	.0271	.0654	.1158	.1597	.1771	.1602
	10	.0000	.0000	.0002	.0020	.0099	.0308	.0686	.1171	.1593	.1762
	11	.0000	.0000	.0000	.0005	.0030	.0120	.0336	.0710	.1185	.1602
	12	.0000	.0000	.0000	.0001	.0008	.0039	.0136	.0355	.0727	.1201
	13	.0000	.0000	.0000	.0000	.0002	.0010	.0045	.0146	.0366	.0739
	14	.0000	.0000	.0000	.0000	.0000	.0002	.0012	.0049	.0150	.0370
	15	.0000	.0000	.0000	.0000	.0000	.0000	.0003	.0013	.0049	.0148
	16	.0000	.0000	.0000	.0000	.0000	.0000	.0000	.0003	.0013	.0046
	17	.0000	.0000	.0000	.0000	.0000	.0000	.0000	.0000	.0002	.0011
	18	.0000	.0000	.0000	.0000	.0000	.0000	.0000	.0000	.0000	.0002
	19	.0000	.0000	.0000	.0000	.0000	.0000	.0000	.0000	.0000	.0000
	20	.0000	.0000	.0000	.0000	.0000	.0000	.0000	.0000	.0000	.0000

SOURCE: From *Handbook of Probability and Statistics*, 2d ed., 1968, The Chemical Rubber Co., Cleveland, Ohio, with permission from the publisher.

TABLE B.7 Poisson Probabilities $P(X = k|\lambda) = \dfrac{\lambda^k e^{-\lambda}}{k!}$

k	0.1	0.2	0.3	0.4	0.5	λ 0.6	0.7	0.8	0.9	1.0
0	.9048	.8187	.7408	.6703	.6065	.5488	.4966	.4493	.4066	.3679
1	.0905	.1637	.2222	.2681	.3033	.3293	.3476	.3595	.3659	.3679
2	.0045	.0164	.0333	.0536	.0758	.0988	.1217	.1438	.1647	.1839
3	.0002	.0011	.0033	.0072	.0126	.0198	.0284	.0383	.0494	.0613
4	.0000	.0001	.0002	.0007	.0016	.0030	.0050	.0077	.0111	.0153
5	.0000	.0000	.0000	.0001	.0002	.0004	.0007	.0012	.0020	.0031
6	.0000	.0000	.0000	.0000	.0000	.0000	.0001	.0002	.0003	.0005
7	.0000	.0000	.0000	.0000	.0000	.0000	.0000	.0000	.0000	.0001

k	1.1	1.2	1.3	1.4	1.5	λ 1.6	1.7	1.8	1.9	2.0
0	.3329	.3012	.2725	.2466	.2231	.2019	.1827	.1653	.1496	.1353
1	.3662	.3614	.3543	.3452	.3347	.3230	.3106	.2975	.2842	.2707
2	.2014	.2169	.2303	.2417	.2510	.2584	.2640	.2678	.2700	.2707
3	.0738	.0867	.0998	.1128	.1255	.1378	.1496	.1607	.1710	.1804
4	.0203	.0260	.0324	.0395	.0471	.0551	.0636	.0723	.0812	.0902
5	.0045	.0062	.0084	.0111	.0141	.0176	.0216	.0260	.0309	.0361
6	.0008	.0012	.0018	.0026	.0035	.0047	.0061	.0078	.0098	.0120
7	.0001	.0002	.0003	.0005	.0008	.0011	.0015	.0020	.0027	.0034
8	.0000	.0000	.0000	.0001	.0001	.0002	.0003	.0005	.0006	.0009
9	.0000	.0000	.0000	.0000	.0000	.0000	.0001	.0001	.0001	.0002

k	2.1	2.2	2.3	2.4	2.5	λ 2.6	2.7	2.8	2.9	3.0
0	.1225	.1108	.1003	.0907	.0821	.0743	.0672	.0608	.0550	.0498
1	.2572	.2438	.2306	.2177	.2052	.1931	.1815	.1703	.1596	.1494
2	.2700	.2681	.2652	.2613	.2565	.2510	.2450	.2384	.2314	.2240
3	.1890	.1966	.2033	.2090	.2138	.2176	.2205	.2225	.2237	.2240
4	.0992	.1082	.1169	.1254	.1336	.1414	.1488	.1557	.1622	.1680
5	.0417	.0476	.0538	.0602	.0668	.0735	.0804	.0872	.0940	.1008
6	.0146	.0174	.0206	.0241	.0278	.0319	.0362	.0407	.0455	.0504
7	.0044	.0055	.0068	.0083	.0099	.0118	.0139	.0163	.0188	.0216
8	.0011	.0015	.0019	.0025	.0031	.0038	.0047	.0057	.0068	.0081
9	.0003	.0004	.0005	.0007	.0009	.0011	.0014	.0018	.0022	.0027
10	.0001	.0001	.0001	.0002	.0002	.0003	.0004	.0005	.0006	.0008
11	.0000	.0000	.0000	.0000	.0000	.0001	.0001	.0001	.0002	.0002
12	.0000	.0000	.0000	.0000	.0000	.0000	.0000	.0000	.0000	.0001

k	3.1	3.2	3.3	3.4	3.5	λ 3.6	3.7	3.8	3.9	4.0
0	.0450	.0408	.0369	.0334	.0302	.0273	.0247	.0224	.0202	.0183
1	.1397	.1304	.1217	.1135	.1057	.0984	.0915	.0850	.0789	.0733
2	.2165	.2087	.2008	.1929	.1850	.1771	.1692	.1615	.1539	.1465
3	.2237	.2226	.2209	.2186	.2158	.2125	.2087	.2046	.2001	.1954
4	.1734	.1781	.1823	.1858	.1888	.1912	.1931	.1944	.1951	.1954
5	.1075	.1140	.1203	.1264	.1322	.1377	.1429	.1477	.1522	.1563
6	.0555	.0608	.0662	.0716	.0771	.0826	.0881	.0936	.0989	.1042
7	.0246	.0278	.0312	.0348	.0385	.0425	.0466	.0508	.0551	.0595
8	.0095	.0111	.0129	.0148	.0169	.0191	.0215	.0241	.0269	.0298
9	.0033	.0040	.0047	.0056	.0066	.0076	.0089	.0102	.0116	.0132

TABLE B.7 Poisson Probabilities (Cont'd)

k	3.1	3.2	3.3	3.4	λ 3.5	3.6	3.7	3.8	3.9	4.0
10	.0010	.0013	.0016	.0019	.0023	.0028	.0033	.0039	.0045	.0053
11	.0003	.0004	.0005	.0006	.0007	.0009	.0011	.0013	.0016	.0019
12	.0001	.0001	.0001	.0002	.0002	.0003	.0003	.0004	.0005	.0006
13	.0000	.0000	.0000	.0000	.0001	.0001	.0001	.0001	.0002	.0002
14	.0000	.0000	.0000	.0000	.0000	.0000	.0000	.0000	.0000	.0001

k	4.1	4.2	4.3	4.4	λ 4.5	4.6	4.7	4.8	4.9	5.0
0	.0166	.0150	.0136	.0123	.0111	.0101	.0091	.0082	.0074	.0067
1	.0679	.0630	.0583	.0540	.0500	.0462	.0427	.0395	.0365	.0337
2	.1393	.1323	.1254	.1188	.1125	.1063	.1005	.0948	.0894	.0842
3	.1904	.1852	.1798	.1743	.1687	.1631	.1574	.1517	.1460	.1404
4	.1951	.1944	.1933	.1917	.1898	.1875	.1849	.1820	.1789	.1755
5	.1600	.1633	.1662	.1687	.1708	.1725	.1738	.1747	.1753	.1755
6	.1093	.1143	.1191	.1237	.1281	.1323	.1362	.1398	.1432	.1462
7	.0640	.0686	.0732	.0778	.0824	.0869	.0914	.0959	.1002	.1044
8	.0328	.0360	.0393	.0428	.0463	.0500	.0537	.0575	.0614	.0653
9	.0150	.0168	.0188	.0209	.0232	.0255	.0280	.0307	.0334	.0363
10	.0061	.0071	.0081	.0092	.0104	.0118	.0132	.0147	.0164	.0181
11	.0023	.0027	.0032	.0037	.0043	.0049	.0056	.0064	.0073	.0082
12	.0008	.0009	.0011	.0014	.0016	.0019	.0022	.0026	.0030	.0034
13	.0002	.0003	.0004	.0005	.0006	.0007	.0008	.0009	.0011	.0013
14	.0001	.0001	.0001	.0001	.0002	.0002	.0003	.0003	.0004	.0005
15	.0000	.0000	.0000	.0000	.0001	.0001	.0001	.0001	.0001	.0002

k	5.1	5.2	5.3	5.4	λ 5.5	5.6	5.7	5.8	5.9	6.0
0	.0061	.0055	.0050	.0045	.0041	.0037	.0033	.0030	.0027	.0025
1	.0311	.0287	.0265	.0244	.0225	.0207	.0191	.0176	.0162	.0149
2	.0793	.0746	.0701	.0659	.0618	.0580	.0544	.0509	.0477	.0446
3	.1348	.1293	.1239	.1185	.1133	.1082	.1033	.0985	.0938	.0892
4	.1719	.1681	.1641	.1600	.1558	.1515	.1472	.1428	.1383	.1339
5	.1753	.1748	.1740	.1728	.1714	.1697	.1678	.1656	.1632	.1606
6	.1490	.1515	.1537	.1555	.1571	.1584	.1594	.1601	.1605	.1606
7	.1086	.1125	.1163	.1200	.1234	.1267	.1298	.1326	.1353	.1377
8	.0692	.0731	.0771	.0810	.0849	.0887	.0925	.0962	.0998	.1033
9	.0392	.0423	.0454	.0486	.0519	.0552	.0586	.0620	.0654	.0688
10	.0200	.0220	.0241	.0262	.0285	.0309	.0334	.0359	.0386	.0413
11	.0093	.0104	.0116	.0129	.0143	.0157	.0173	.0190	.0207	.0225
12	.0039	.0045	.0051	.0058	.0065	.0073	.0082	.0092	.0102	.0113
13	.0015	.0018	.0021	.0024	.0028	.0032	.0036	.0041	.0046	.0052
14	.0006	.0007	.0008	.0009	.0011	.0013	.0015	.0017	.0019	.0022
15	.0002	.0002	.0003	.0003	.0004	.0005	.0006	.0007	.0008	.0009
16	.0001	.0001	.0001	.0001	.0001	.0002	.0002	.0002	.0003	.0003
17	.0000	.0000	.0000	.0000	.0000	.0000	.0001	.0001	.0001	.0001

TABLE B.7 Poisson Probabilities (Cont'd)

k	8.1	8.2	8.3	8.4	8.5	8.6	8.7	8.8	8.9	9.0
0	.0003	.0003	.0002	.0002	.0002	.0002	.0002	.0002	.0001	.0001
1	.0025	.0023	.0021	.0019	.0017	.0016	.0014	.0013	.0012	.0011
2	.0100	.0092	.0086	.0079	.0074	.0068	.0063	.0058	.0054	.0050
3	.0269	.0252	.0237	.0222	.0208	.0195	.0183	.0171	.0160	.0150
4	.0544	.0517	.0491	.0466	.0443	.0420	.0398	.0377	.0357	.0337
5	.0882	.0849	.0816	.0784	.0752	.0722	.0692	.0663	.0635	.0607
6	.1191	.1160	.1128	.1097	.1066	.1034	.1003	.0972	.0941	.0911
7	.1378	.1358	.1338	.1317	.1294	.1271	.1247	.1222	.1197	.1171
8	.1395	.1392	.1388	.1382	.1375	.1366	.1356	.1344	.1332	.1318
9	.1256	.1269	.1280	.1290	.1299	.1306	.1311	.1315	.1317	.1318
10	.1017	.1040	.1063	.1084	.1104	.1123	.1140	.1157	.1172	.1186
11	.0749	.0776	.0802	.0828	.0853	.0878	.0902	.0925	.0948	.0970
12	.0505	.0530	.0555	.0579	.0604	.0629	.0654	.0679	.0703	.0728
13	.0315	.0334	.0354	.0374	.0395	.0416	.0438	.0459	.0481	.0504
14	.0182	.0196	.0210	.0225	.0240	.0256	.0272	.0289	.0306	.0324
15	.0098	.0107	.0116	.0126	.0136	.0147	.0158	.0169	.0182	.0194
16	.0050	.0055	.0060	.0066	.0072	.0079	.0086	.0093	.0101	.0109
17	.0024	.0026	.0029	.0033	.0036	.0040	.0044	.0048	.0053	.0058
18	.0011	.0012	.0014	.0015	.0017	.0019	.0021	.0024	.0026	.0029
19	.0005	.0005	.0006	.0007	.0008	.0009	.0010	.0011	.0012	.0014
20	.0002	.0002	.0002	.0003	.0003	.0004	.0004	.0005	.0005	.0006
21	.0001	.0001	.0001	.0001	.0001	.0002	.0002	.0002	.0002	.0003
22	.0000	.0000	.0000	.0000	.0001	.0001	.0001	.0001	.0001	.0001

k	9.1	9.2	9.3	9.4	9.5	9.6	9.7	9.8	9.9	10
0	.0001	.0001	.0001	.0001	.0001	.0001	.0001	.0001	.0001	.0000
1	.0010	.0009	.0009	.0008	.0007	.0007	.0006	.0005	.0005	.0005
2	.0046	.0043	.0040	.0037	.0034	.0031	.0029	.0027	.0025	.0023
3	.0140	.0131	.0123	.0115	.0107	.0100	.0093	.0087	.0081	.0076
4	.0319	.0302	.0285	.0269	.0254	.0240	.0226	.0213	.0201	.0189
5	.0581	.0555	.0530	.0506	.0483	.0460	.0439	.0418	.0398	.0378
6	.0881	.0851	.0822	.0793	.0764	.0736	.0709	.0682	.0656	.0631
7	.1145	.1118	.1091	.1064	.1037	.1010	.0982	.0955	.0928	.0901
8	.1302	.1286	.1269	.1251	.1232	.1212	.1191	.1170	.1148	.1126
9	.1317	.1315	.1311	.1306	.1300	.1293	.1284	.1274	.1263	.1251
10	.1198	.1210	.1219	.1228	.1235	.1241	.1245	.1249	.1250	.1251
11	.0991	.1012	.1031	.1049	.1067	.1083	.1098	.1112	.1125	.1137
12	.0752	.0776	.0799	.0822	.0844	.0866	.0888	.0908	.0928	.0948
13	.0526	.0549	.0572	.0594	.0617	.0640	.0662	.0685	.0707	.0729
14	.0342	.0361	.0380	.0399	.0419	.0439	.0459	.0479	.0500	.0521
15	.0208	.0221	.0235	.0250	.0265	.0281	.0297	.0313	.0330	.0347
16	.0118	.0127	.0137	.0147	.0157	.0168	.0180	.0192	.0204	.0217
17	.0063	.0069	.0075	.0081	.0088	.0095	.0103	.0111	.0119	.0128
18	.0032	.0035	.0039	.0042	.0046	.0051	.0055	.0060	.0065	.0071
19	.0015	.0017	.0019	.0021	.0023	.0026	.0028	.0031	.0034	.0037

TABLE B.7 Poisson Probabilities (*Cont'd*)

k	6.1	6.2	6.3	6.4	6.5	λ 6.6	6.7	6.8	6.9	7.0
0	.0022	.0020	.0018	.0017	.0015	.0014	.0012	.0011	.0010	.0009
1	.0137	.0126	.0116	.0106	.0098	.0090	.0082	.0076	.0070	.0064
2	.0417	.0390	.0364	.0340	.0318	.0296	.0276	.0258	.0240	.0223
3	.0848	.0806	.0765	.0726	.0688	.0652	.0617	.0584	.0552	.0521
4	.1294	.1249	.1205	.1162	.1118	.1076	.1034	.0992	.0952	.0912
5	.1579	.1549	.1519	.1487	.1454	.1420	.1385	.1349	.1314	.1277
6	.1605	.1601	.1595	.1586	.1575	.1562	.1546	.1529	.1511	.1490
7	.1399	.1418	.1435	.1450	.1462	.1472	.1480	.1486	.1489	.1490
8	.1066	.1099	.1130	.1160	.1188	.1215	.1240	.1263	.1284	.1304
9	.0723	.0757	.0791	.0825	.0858	.0891	.0923	.0954	.0985	.1014
10	.0441	.0469	.0498	.0528	.0558	.0588	.0618	.0649	.0679	.0710
11	.0245	.0265	.0285	.0307	.0330	.0353	.0377	.0401	.0426	.0452
12	.0124	.0137	.0150	.0164	.0179	.0194	.0210	.0227	.0245	.0264
13	.0058	.0065	.0073	.0081	.0089	.0098	.0108	.0119	.0130	.0142
14	.0025	.0029	.0033	.0037	.0041	.0046	.0052	.0058	.0064	.0071
15	.0010	.0012	.0014	.0016	.0018	.0020	.0023	.0026	.0029	.0033
16	.0004	.0005	.0005	.0006	.0007	.0008	.0010	.0011	.0013	.0014
17	.0001	.0002	.0002	.0002	.0003	.0003	.0004	.0004	.0005	.0006
18	.0000	.0001	.0001	.0001	.0001	.0001	.0001	.0002	.0002	.0002
19	.0000	.0000	.0000	.0000	.0000	.0000	.0000	.0001	.0001	.0001

k	7.1	7.2	7.3	7.4	7.5	λ 7.6	7.7	7.8	7.9	8.0
0	.0008	.0007	.0007	.0006	.0006	.0005	.0005	.0004	.0004	.0003
1	.0059	.0054	.0049	.0045	.0041	.0038	.0035	.0032	.0029	.0027
2	.0208	.0194	.0180	.0167	.0156	.0145	.0134	.0125	.0116	.0107
3	.0492	.0464	.0438	.0413	.0389	.0366	.0345	.0324	.0305	.0286
4	.0874	.0836	.0799	.0764	.0729	.0696	.0663	.0632	.0602	.0573
5	.1241	.1204	.1167	.1130	.1094	.1057	.1021	.0986	.0951	.0916
6	.1468	.1445	.1420	.1394	.1367	.1339	.1311	.1282	.1252	.1221
7	.1489	.1486	.1481	.1474	.1465	.1454	.1442	.1428	.1413	.1396
8	.1321	.1337	.1351	.1363	.1373	.1382	.1388	.1392	.1395	.1396
9	.1042	.1070	.1096	.1121	.1144	.1167	.1187	.1207	.1224	.1241
10	.0740	.0770	.0800	.0829	.0858	.0887	.0914	.0941	.0967	.0993
11	.0478	.0504	.0531	.0558	.0585	.0613	.0640	.0667	.0695	.0722
12	.0283	.0303	.0323	.0344	.0366	.0388	.0411	.0434	.0457	.0481
13	.0154	.0168	.0181	.0196	.0211	.0227	.0243	.0260	.0278	.0296
14	.0078	.0086	.0095	.0104	.0113	.0123	.0134	.0145	.0157	.0169
15	.0037	.0041	.0046	.0051	.0057	.0062	.0069	.0075	.0083	.0090
16	.0016	.0019	.0021	.0024	.0026	.0030	.0033	.0037	.0041	.0045
17	.0007	.0008	.0009	.0010	.0012	.0013	.0015	.0017	.0019	.0021
18	.0003	.0003	.0004	.0004	.0005	.0006	.0006	.0007	.0008	.0009
19	.0001	.0001	.0001	.0002	.0002	.0002	.0003	.0003	.0003	.0004
20	.0000	.0000	.0001	.0001	.0001	.0001	.0001	.0001	.0001	.0002
21	.0000	.0000	.0000	.0000	.0000	.0000	.0000	.0000	.0001	.0001

TABLE B.7 Poisson Probabilities (*Cont'd*)

k	9.1	9.2	9.3	9.4	9.5	λ 9.6	9.7	9.8	9.9	10
20	.0007	.0008	.0009	.0010	.0011	.0012	.0014	.0015	.0017	.0019
21	.0003	.0003	.0004	.0004	.0005	.0006	.0006	.0007	.0008	.0009
22	.0001	.0001	.0002	.0002	.0002	.0002	.0003	.0003	.0004	.0004
23	.0000	.0001	.0001	.0001	.0001	.0001	.0001	.0001	.0002	.0002
24	.0000	.0000	.0000	.0000	.0000	.0000	.0000	.0001	.0001	.0001

k	11	12	13	14	15	λ 16	17	18	19	20
0	.0000	.0000	.0000	.0000	.0000	.0000	.0000	.0000	.0000	.0000
1	.0002	.0001	.0000	.0000	.0000	.0000	.0000	.0000	.0000	.0000
2	.0010	.0004	.0002	.0001	.0000	.0000	.0000	.0000	.0000	.0000
3	.0037	.0018	.0008	.0004	.0002	.0001	.0000	.0000	.0000	.0000
4	.0102	.0053	.0027	.0013	.0006	.0003	.0001	.0001	.0000	.0000
5	.0224	.0127	.0070	.0037	.0019	.0010	.0005	.0002	.0001	.0001
6	.0411	.0255	.0152	.0087	.0048	.0026	.0014	.0007	.0004	.0002
7	.0646	.0437	.0281	.0174	.0104	.0060	.0034	.0018	.0010	.0005
8	.0888	.0655	.0457	.0304	.0194	.0120	.0072	.0042	.0024	.0013
9	.1085	.0874	.0661	.0473	.0324	.0213	.0135	.0083	.0050	.0029
10	.1194	.1048	.0859	.0663	.0486	.0341	.0230	.0150	.0095	.0058
11	.1194	.1144	.1015	.0844	.0663	.0496	.0355	.0245	.0164	.0106
12	.1094	.1144	.1099	.0984	.0829	.0661	.0504	.0368	.0259	.0176
13	.0926	.1056	.1099	.1060	.0956	.0814	.0658	.0509	.0378	.0271
14	.0728	.0905	.1021	.1060	.1024	.0930	.0800	.0655	.0514	.0387
15	.0534	.0724	.0885	.0989	.1024	.0992	.0906	.0786	.0650	.0516
16	.0367	.0543	.0719	.0866	.0960	.0992	.0963	.0884	.0772	.0646
17	.0237	.0383	.0550	.0713	.0847	.0934	.0963	.0936	.0863	.0760
18	.0145	.0256	.0397	.0554	.0706	.0830	.0909	.0936	.0911	.0844
19	.0084	.0161	.0272	.0409	.0557	.0699	.0814	.0887	.0911	.0888
20	.0046	.0097	.0177	.0286	.0418	.0559	.0692	.0798	.0866	.0888
21	.0024	.0055	.0109	.0191	.0299	.0426	.0560	.0684	.0783	.0846
22	.0012	.0030	.0065	.0121	.0204	.0310	.0433	.0560	.0676	.0769
23	.0006	.0016	.0037	.0074	.0133	.0216	.0320	.0438	.0559	.0669
24	.0003	.0008	.0020	.0043	.0083	.0144	.0226	.0328	.0442	.0557
25	.0001	.0004	.0010	.0024	.0050	.0092	.0154	.0237	.0336	.0446
26	.0000	.0002	.0005	.0013	.0029	.0057	.0101	.0164	.0246	.0343
27	.0000	.0001	.0002	.0007	.0016	.0034	.0063	.0109	.0173	.0254
28	.0000	.0000	.0001	.0003	.0009	.0019	.0038	.0070	.0117	.0181
29	.0000	.0000	.0001	.0002	.0004	.0011	.0023	.0044	.0077	.0125
30	.0000	.0000	.0000	.0001	.0002	.0006	.0013	.0026	.0049	.0083
31	.0000	.0000	.0000	.0000	.0001	.0003	.0007	.0015	.0030	.0054
32	.0000	.0000	.0000	.0000	.0001	.0001	.0004	.0009	.0018	.0034
33	.0000	.0000	.0000	.0000	.0000	.0001	.0002	.0005	.0010	.0020
34	.0000	.0000	.0000	.0000	.0000	.0000	.0001	.0002	.0006	.0012
35	.0000	.0000	.0000	.0000	.0000	.0000	.0000	.0001	.0003	.0007
36	.0000	.0000	.0000	.0000	.0000	.0000	.0000	.0001	.0002	.0004
37	.0000	.0000	.0000	.0000	.0000	.0000	.0000	.0000	.0001	.0002
38	.0000	.0000	.0000	.0000	.0000	.0000	.0000	.0000	.0000	.0001
39	.0000	.0000	.0000	.0000	.0000	.0000	.0000	.0000	.0000	.0001

SOURCE: From *Handbook of Probability and Statistics*, 2d ed., 1968, The Chemical Rubber Co., Cleveland, Ohio, with permission from the publisher.

TABLE B.8 Unit Normal Loss Function $G(|h|)$

h	0.00	0.01	0.02	0.03	0.04	0.05	0.06	0.07	0.08	0.09
0.0	0.3989	0.3940	0.3890	0.3841	0.3793	0.3744	0.3697	0.3649	0.3602	0.3556
0.1	0.3509	0.3464	0.3418	0.3373	0.3328	0.3284	0.3240	0.3197	0.3154	0.3111
0.2	0.3069	0.3027	0.2986	0.2944	0.2904	0.2863	0.2824	0.2784	0.2745	0.2706
0.3	0.2668	0.2630	0.2592	0.2555	0.2518	0.2481	0.2445	0.2409	0.2374	0.2339
0.4	0.2304	0.2270	0.2236	0.2203	0.2169	0.2137	0.2104	0.2072	0.2040	0.2009
0.5	0.1978	0.1947	0.1917	0.1887	0.1857	0.1828	0.1799	0.1771	0.1742	0.1714
0.6	0.1687	0.1659	0.1633	0.1606	0.1580	0.1554	0.1528	0.1503	0.1478	0.1453
0.7	0.1429	0.1405	0.1381	0.1358	0.1334	0.1312	0.1289	0.1267	0.1245	0.1223
0.8	0.1202	0.1181	0.1160	0.1140	0.1120	0.1100	0.1080	0.1061	0.1042	0.1023
0.9	0.1004	0.09860	0.09680	0.09503	0.09328	0.09156	0.08986	0.08819	0.08654	0.08491
1.0	0.08332	0.08174	0.08019	0.07866	0.07716	0.07568	0.07422	0.07279	0.07138	0.06999
1.1	0.06862	0.06727	0.06595	0.06465	0.06336	0.06210	0.06086	0.05964	0.05844	0.05726
1.2	0.05610	0.05496	0.05384	0.05274	0.05165	0.05059	0.04954	0.04851	0.04750	0.04650
1.3	0.04553	0.04457	0.04363	0.04270	0.04179	0.04090	0.04002	0.03916	0.03831	0.03748
1.4	0.03667	0.03587	0.03508	0.03431	0.03356	0.03281	0.03208	0.03137	0.03067	0.02998
1.5	0.02931	0.02865	0.02800	0.02736	0.02674	0.02612	0.02552	0.02494	0.02436	0.02380
1.6	0.02324	0.02270	0.02217	0.02165	0.02114	0.02064	0.02015	0.01967	0.01920	0.01874
1.7	0.01829	0.01785	0.01742	0.01699	0.01658	0.01617	0.01578	0.01539	0.01501	0.01464
1.8	0.01428	0.01392	0.01357	0.01323	0.01290	0.01257	0.01226	0.01195	0.01164	0.01134
1.9	0.01105	0.01077	0.01049	0.01022	0.0²9957	0.0²9698	0.0²9445	0.0²9198	0.0²8957	0.0²8721
2.0	0.0²8491	0.0²8266	0.0²8046	0.0²7832	0.0²7623	0.0²7418	0.0²7219	0.0²7024	0.0²6835	0.0²6649
2.1	0.0²6468	0.0²6292	0.0²6120	0.0²5952	0.0²5788	0.0²5628	0.0²5472	0.0²5320	0.0²5172	0.0²5028
2.2	0.0²4887	0.0²4750	0.0²4616	0.0²4486	0.0²4358	0.0²4235	0.0²4114	0.0²3996	0.0²3882	0.0²3770
2.3	0.0²3662	0.0²3556	0.0²3453	0.0²3352	0.0²3255	0.0²3159	0.0²3067	0.0²2977	0.0²2889	0.0²2804
2.4	0.0²2720	0.0²2640	0.0²2561	0.0²2484	0.0²2410	0.0²2337	0.0²2267	0.0²2199	0.0²2132	0.0²2067

TABLE B.8 Unit Normal Loss Function *(Cont'd)*

	.00	.01	.02	.03	.04	.05	.06	.07	.08	.09
2.5	$0.0^2 2004$	$0.0^2 1943$	$0.0^2 1883$	$0.0^2 1826$	$0.0^2 1769$	$0.0^2 1715$	$0.0^2 1662$	$0.0^2 1610$	$0.0^2 1560$	$0.0^2 1511$
2.6	$0.0^2 1464$	$0.0^2 1418$	$0.0^2 1373$	$0.0^2 1330$	$0.0^2 1288$	$0.0^2 1247$	$0.0^2 1207$	$0.0^2 1169$	$0.0^2 1132$	$0.0^2 1095$
2.7	$0.0^2 1060$	$0.0^2 1026$	$0.0^3 9928$	$0.0^3 9607$	$0.0^3 9295$	$0.0^3 8992$	$0.0^3 8699$	$0.0^3 8414$	$0.0^3 8138$	$0.0^3 7870$
2.8	$0.0^3 7611$	$0.0^3 7359$	$0.0^3 7115$	$0.0^3 6879$	$0.0^3 6650$	$0.0^3 6428$	$0.0^3 6213$	$0.0^3 6004$	$0.0^3 5802$	$0.0^3 5606$
2.9	$0.0^3 5417$	$0.0^3 5233$	$0.0^3 5055$	$0.0^3 4883$	$0.0^3 4716$	$0.0^3 4555$	$0.0^3 4398$	$0.0^3 4247$	$0.0^3 4101$	$0.0^3 3959$
3.0	$0.0^3 3822$	$0.0^3 3689$	$0.0^3 3560$	$0.0^3 3436$	$0.0^3 3316$	$0.0^3 3199$	$0.0^3 3087$	$0.0^3 2978$	$0.0^3 2873$	$0.0^3 2771$
3.1	$0.0^3 2673$	$0.0^3 2577$	$0.0^3 2485$	$0.0^3 2396$	$0.0^3 2311$	$0.0^3 2227$	$0.0^3 2147$	$0.0^3 2070$	$0.0^3 1995$	$0.0^3 1922$
3.2	$0.0^3 1852$	$0.0^3 1785$	$0.0^3 1720$	$0.0^3 1657$	$0.0^3 1596$	$0.0^3 1537$	$0.0^3 1480$	$0.0^3 1426$	$0.0^3 1373$	$0.0^3 1322$
3.3	$0.0^3 1273$	$0.0^3 1225$	$0.0^3 1179$	$0.0^3 1135$	$0.0^3 1093$	$0.0^3 1051$	$0.0^3 1012$	$0.0^4 9734$	$0.0^4 9365$	$0.0^4 9009$
3.4	$0.0^4 8666$	$0.0^4 8335$	$0.0^4 8016$	$0.0^4 7709$	$0.0^4 7413$	$0.0^4 7127$	$0.0^4 6852$	$0.0^4 6587$	$0.0^4 6331$	$0.0^4 6085$
3.5	$0.0^4 5848$	$0.0^4 5620$	$0.0^4 5400$	$0.0^4 5188$	$0.0^4 4984$	$0.0^4 4788$	$0.0^4 4599$	$0.0^4 4417$	$0.0^4 4242$	$0.0^4 4073$
3.6	$0.0^4 3911$	$0.0^4 3755$	$0.0^4 3605$	$0.0^4 3460$	$0.0^4 3321$	$0.0^4 3188$	$0.0^4 3059$	$0.0^4 2935$	$0.0^4 2816$	$0.0^4 2702$
3.7	$0.0^4 2592$	$0.0^4 2486$	$0.0^4 2385$	$0.0^4 2287$	$0.0^4 2193$	$0.0^4 2103$	$0.0^4 2016$	$0.0^4 1933$	$0.0^4 1853$	$0.0^4 1776$
3.8	$0.0^4 1702$	$0.0^4 1632$	$0.0^4 1563$	$0.0^4 1498$	$0.0^4 1435$	$0.0^4 1375$	$0.0^4 1317$	$0.0^4 1262$	$0.0^4 1208$	$0.0^4 1157$
3.9	$0.0^4 1108$	$0.0^4 1061$	$0.0^4 1016$	$0.0^5 9723$	$0.0^5 9307$	$0.0^5 8908$	$0.0^5 8525$	$0.0^5 8158$	$0.0^5 7806$	$0.0^5 7469$
4.0	$0.0^5 7145$	$0.0^5 6835$	$0.0^5 6538$	$0.0^5 6253$	$0.0^5 5980$	$0.0^5 5718$	$0.0^5 5468$	$0.0^5 5227$	$0.0^5 4997$	$0.0^5 4777$
4.1	$0.0^5 4566$	$0.0^5 4364$	$0.0^5 4170$	$0.0^5 3985$	$0.0^5 3807$	$0.0^5 3637$	$0.0^5 3475$	$0.0^5 3319$	$0.0^5 3170$	$0.0^5 3027$
4.2	$0.0^5 2891$	$0.0^5 2760$	$0.0^5 2635$	$0.0^5 2516$	$0.0^5 2402$	$0.0^5 2292$	$0.0^5 2188$	$0.0^5 2088$	$0.0^5 1992$	$0.0^5 1901$
4.3	$0.0^5 1814$	$0.0^5 1730$	$0.0^5 1650$	$0.0^5 1574$	$0.0^5 1501$	$0.0^5 1431$	$0.0^5 1365$	$0.0^5 1301$	$0.0^5 1241$	$0.0^5 1183$
4.4	$0.0^5 1127$	$0.0^5 1074$	$0.0^5 1024$	$0.0^6 9756$	$0.0^6 9296$	$0.0^6 8857$	$0.0^6 8437$	$0.0^6 8037$	$0.0^6 7655$	$0.0^6 7290$
4.5	$0.0^6 6942$	$0.0^6 6610$	$0.0^6 6294$	$0.0^6 5992$	$0.0^6 5704$	$0.0^6 5429$	$0.0^6 5167$	$0.0^6 4917$	$0.0^6 4679$	$0.0^6 4452$
4.6	$0.0^6 4236$	$0.0^6 4029$	$0.0^6 3833$	$0.0^6 3645$	$0.0^6 3467$	$0.0^6 3297$	$0.0^6 3135$	$0.0^6 2981$	$0.0^6 2834$	$0.0^6 2694$
4.7	$0.0^6 2560$	$0.0^6 2433$	$0.0^6 2313$	$0.0^6 2197$	$0.0^6 2088$	$0.0^6 1984$	$0.0^6 1884$	$0.0^6 1790$	$0.0^6 1700$	$0.0^6 1615$
4.8	$0.0^6 1533$	$0.0^6 1456$	$0.0^6 1382$	$0.0^6 1312$	$0.0^6 1246$	$0.0^6 1182$	$0.0^6 1122$	$0.0^6 1065$	$0.0^6 1011$	$0.0^7 9588$
4.9	$0.0^7 9096$	$0.0^7 8629$	$0.0^7 8185$	$0.0^7 7763$	$0.0^7 7362$	$0.0^7 6982$	$0.0^7 6620$	$0.0^7 6276$	$0.0^7 5950$	$0.0^7 5640$

$$G(-h) = h + G(h)$$

Examples: $G(3.57) = 0.0^4 4417 = 0.00004417$
$G(-3.57) = 3.57004417$

SOURCE: Reproduced from R. Schlaifer, *Probability and Statistics for Business Decisions*, McGraw-Hill Book Co., New York, 1959, by permission of the president and fellows of Harvard College.

TABLE B.9 Random Numbers

```
09 18 82 00 97   32 82 53 95 27   04 22 08 63 04   83 38 98 73 74   64 27 85 80 44
90 04 58 54 97   51 98 15 06 54   94 93 88 19 97   91 87 07 61 50   68 47 66 46 59
73 18 95 02 07   47 67 72 62 69   62 29 06 44 64   27 12 46 70 18   41 36 18 27 60
75 76 87 64 90   20 97 18 17 49   90 42 91 22 72   95 37 50 58 71   93 82 34 31 78
54 01 64 40 56   66 28 13 10 03   00 68 22 73 98   20 71 45 32 95   07 70 61 78 13

08 35 86 99 10   78 54 24 27 85   13 66 15 88 73   04 61 89 75 53   31 22 30 84 20
28 30 60 32 64   81 33 31 05 91   40 51 00 78 93   32 60 46 04 75   94 11 90 18 40
53 84 08 62 33   81 59 41 36 28   51 21 59 02 90   28 46 66 87 95   77 76 22 07 91
91 75 75 37 41   61 61 36 22 69   50 26 39 02 12   55 78 17 65 14   83 48 34 70 55
89 41 59 26 94   00 39 75 83 91   12 60 71 76 46   48 94 97 23 06   94 54 13 74 08

77 51 30 38 20   86 83 42 99 01   68 41 48 27 74   51 90 81 39 80   72 89 35 55 07
19 50 23 71 74   69 97 92 02 88   55 21 02 97 73   74 28 77 52 51   65 34 46 74 15
21 81 85 93 13   93 27 88 17 57   05 68 67 31 56   07 08 28 50 46   31 85 33 84 52
51 47 46 64 99   68 10 72 36 21   94 04 99 13 45   42 83 60 91 91   08 00 74 54 49
99 55 96 83 31   62 53 52 41 70   69 77 71 28 30   74 81 97 81 42   43 86 07 28 34

33 71 34 80 07   93 58 47 28 69   51 92 66 47 21   58 30 32 98 22   93 17 49 39 72
85 27 48 68 93   11 30 32 92 70   28 83 43 41 37   73 51 59 04 00   71 14 84 36 43
84 13 38 96 40   44 03 55 21 66   73 85 27 00 91   61 22 26 05 61   62 32 71 84 23
56 73 21 62 34   17 39 59 61 31   10 12 39 16 22   85 49 65 75 60   81 60 41 88 80
65 13 85 68 06   87 64 88 52 61   34 31 36 58 61   45 87 52 10 69   85 64 44 72 77

38 00 10 21 76   81 71 91 17 11   71 60 29 29 37   74 21 96 40 49   65 58 44 96 98
37 40 29 63 97   01 30 47 75 86   56 27 11 00 86   47 32 46 26 05   40 03 03 74 38
97 12 54 03 48   87 08 33 14 17   21 81 53 92 50   75 23 76 20 47   15 50 12 95 78
21 82 64 11 34   47 14 33 40 72   64 63 88 59 02   49 13 90 64 41   03 85 65 45 52
73 13 54 27 42   95 71 90 90 35   85 79 47 42 96   08 78 98 81 56   64 69 11 92 02

07 63 87 79 29   03 06 11 80 72   96 20 74 41 56   23 82 19 95 38   04 71 36 69 94
60 52 88 34 41   07 95 41 98 14   59 17 52 06 95   05 53 35 21 39   61 21 20 64 55
83 59 63 56 55   06 95 89 29 83   05 12 80 97 19   77 43 35 37 83   92 30 15 04 98
10 85 06 27 46   99 59 91 05 07   13 49 90 63 19   53 07 57 18 39   06 41 01 93 62
39 82 09 89 52   43 62 26 31 47   64 42 18 08 14   43 80 00 93 51   31 02 47 31 67

59 58 00 64 78   75 56 97 88 00   88 83 55 44 86   23 76 80 61 56   04 11 10 84 08
38 50 80 73 41   23 79 34 87 63   90 82 29 70 22   17 71 90 42 07   95 95 44 99 53
30 69 27 06 68   94 68 81 61 27   56 19 68 00 91   82 06 76 34 00   05 46 26 92 00
65 44 39 56 59   18 28 82 74 37   49 63 22 40 41   08 33 76 56 76   96 29 99 08 36
27 26 75 02 64   13 19 27 22 94   07 47 74 46 06   17 98 54 89 11   97 34 13 03 58

91 30 70 69 91   19 07 22 42 10   36 69 95 37 28   28 82 53 57 93   28 97 66 62 52
68 43 49 46 88   84 47 31 36 22   62 12 69 84 08   12 84 38 25 90   09 81 59 31 46
48 90 81 58 77   54 74 52 45 91   35 70 00 47 54   83 82 45 26 92   54 13 05 51 60
06 91 34 51 97   42 67 27 86 01   11 88 30 95 28   63 01 19 89 01   14 97 44 03 44
10 45 51 60 19   14 21 03 37 12   91 34 23 78 21   88 32 58 08 51   43 66 77 08 83

12 88 39 73 43   65 02 76 11 84   04 28 50 13 92   17 97 41 50 77   90 71 22 67 69
21 77 83 09 76   38 80 73 69 61   31 64 94 20 96   63 28 10 20 23   08 81 64 74 49
19 52 35 95 15   65 12 25 96 59   86 28 36 82 58   69 57 21 37 98   16 43 59 15 29
67 24 55 26 70   35 58 31 65 63   79 24 68 66 86   76 46 33 42 22   26 65 59 08 02
60 58 44 73 77   07 50 03 79 92   45 13 42 65 29   26 76 08 36 37   41 32 64 43 44

53 85 34 13 77   36 06 69 48 50   58 83 87 38 59   49 36 47 33 31   96 24 04 36 42
24 63 73 87 36   74 38 48 93 42   52 62 30 79 92   12 36 91 86 01   03 74 28 38 73
83 08 01 24 51   38 99 22 28 15   07 75 95 17 77   97 37 72 75 85   51 97 23 78 67
16 44 42 43 34   36 15 19 90 73   27 49 37 09 39   85 13 03 25 52   54 84 65 47 59
60 79 01 81 57   57 17 86 57 62   11 16 17 85 76   45 81 95 29 79   65 13 00 48 60
```

TABLE B.9 Random Numbers (Cont'd)

```
10 09 73 25 33   76 52 01 35 86   34 67 35 48 76   80 95 90 91 17   39 29 27 49 45
37 54 20 48 05   64 89 47 42 96   24 80 52 40 37   20 63 61 04 02   00 82 29 16 65
08 42 26 89 53   19 64 50 93 03   23 20 90 25 60   15 95 33 47 64   35 08 03 36 06
99 01 90 25 29   09 37 67 07 15   38 31 13 11 65   88 67 67 43 97   04 43 62 76 59
12 80 79 99 70   80 15 73 61 47   64 03 23 66 53   98 95 11 68 77   12 17 17 68 33

66 06 57 47 17   34 07 27 68 50   36 69 73 61 70   65 81 33 98 85   11 19 92 91 70
31 06 01 08 05   45 57 18 24 06   35 30 34 26 14   86 79 90 74 39   23 40 30 97 32
85 26 97 76 02   02 05 16 56 92   68 66 57 48 18   73 05 38 52 47   18 62 38 85 79
63 57 33 21 35   05 32 54 70 48   90 55 35 75 48   28 46 82 87 09   83 49 12 56 24
73 79 64 57 53   03 52 96 47 78   35 80 83 42 82   60 93 52 03 44   35 27 38 84 35

98 52 01 77 67   14 90 56 86 07   22 10 94 05 58   60 97 09 34 33   50 50 07 39 98
11 80 50 54 31   39 80 82 77 32   50 72 56 82 48   29 40 52 42 01   52 77 56 78 51
83 45 29 96 34   06 28 89 80 83   13 74 67 00 78   18 47 54 06 10   68 71 17 78 17
88 68 54 02 00   86 50 75 84 01   36 76 66 79 51   90 36 47 64 93   29 60 91 10 62
99 59 46 73 48   87 51 76 49 69   91 82 60 89 28   93 78 56 13 68   23 47 83 41 13

65 48 11 76 74   17 46 85 09 50   58 04 77 69 74   73 03 95 71 86   40 21 81 65 44
80 12 43 56 35   17 72 70 80 15   45 31 82 23 74   21 11 57 82 53   14 38 55 37 63
74 35 09 98 17   77 40 27 72 14   43 23 60 02 10   45 52 16 42 37   96 28 60 26 55
69 91 62 68 03   66 25 22 91 48   36 93 68 72 03   76 62 11 39 90   94 40 05 64 18
09 89 32 05 05   14 22 56 85 14   46 42 75 67 88   96 29 77 88 22   54 38 21 45 98

91 49 91 45 23   68 47 92 76 86   46 16 28 35 54   94 75 08 99 23   37 08 92 00 48
80 33 69 45 98   26 94 03 68 58   70 29 73 41 35   53 14 03 33 40   42 05 08 23 41
44 10 48 19 49   85 15 74 79 54   32 97 92 65 75   57 60 04 08 81   22 22 20 64 13
12 55 07 37 42   11 10 00 20 40   12 86 07 46 97   96 64 48 94 39   28 70 72 58 15
63 60 64 93 29   16 50 53 44 84   40 21 95 25 63   43 65 17 70 82   07 20 73 17 90

61 19 69 04 46   26 45 74 77 74   51 92 43 37 29   65 39 45 95 93   42 58 26 05 27
15 47 44 52 66   95 27 07 99 53   59 36 78 38 48   82 39 61 01 18   33 21 15 94 66
94 55 72 85 73   67 89 75 43 87   54 62 24 44 31   91 19 04 25 92   92 92 74 59 73
42 48 11 62 13   97 34 40 87 21   16 86 84 87 67   03 07 11 20 59   25 70 14 66 70
23 52 37 83 17   73 20 88 98 37   68 93 59 14 16   26 25 22 96 63   05 52 28 25 62

04 49 35 24 94   75 24 63 38 24   45 86 25 10 25   61 96 27 93 35   65 33 71 24 72
00 54 99 76 54   64 05 18 81 59   96 11 96 38 96   54 69 28 23 91   23 28 72 95 29
35 96 31 53 07   26 89 80 93 54   33 35 13 54 62   77 97 45 00 24   90 10 33 93 33
59 80 80 83 91   45 42 72 68 42   83 60 94 97 00   13 02 12 48 92   78 56 52 01 06
46 05 88 52 36   01 39 09 22 86   77 28 14 40 77   93 91 08 36 47   70 61 74 29 41

32 17 90 05 97   87 37 92 52 41   05 56 70 70 07   86 74 31 71 57   85 39 41 18 38
69 23 46 14 06   20 11 74 52 04   15 95 66 00 00   18 74 39 24 23   97 11 89 63 38
19 56 54 14 30   01 75 87 53 79   40 41 92 15 85   66 67 43 68 06   84 96 28 52 07
45 15 51 49 38   19 47 60 72 46   43 66 79 45 43   59 04 79 00 33   20 82 66 95 41
94 86 43 19 94   36 16 81 08 51   34 88 88 15 53   01 54 03 54 56   05 01 45 11 76

98 08 62 48 26   45 24 02 84 04   44 99 90 88 96   39 09 47 34 07   35 44 13 18 80
33 18 51 62 32   41 94 15 09 49   89 43 54 85 81   88 69 54 19 94   37 54 87 30 43
80 95 10 04 06   96 38 27 07 74   20 15 12 33 87   25 01 62 52 98   94 62 46 11 71
79 75 24 91 40   71 96 12 82 96   69 86 10 25 91   74 85 22 05 39   00 38 75 95 79
18 63 33 25 37   98 14 50 65 71   31 01 02 46 74   05 45 56 14 27   77 93 89 19 36

74 02 94 39 02   77 55 73 22 70   97 79 01 71 19   52 52 75 80 21   80 81 45 17 48
54 17 84 56 11   80 99 33 71 43   05 33 51 29 69   56 12 71 92 55   36 04 09 03 24
11 66 44 98 83   52 07 98 48 27   59 38 17 15 39   09 97 33 34 40   88 46 12 33 56
48 32 47 79 28   31 24 96 47 10   02 29 53 68 70   32 30 75 75 46   15 02 00 99 94
69 07 49 41 38   87 63 79 19 76   35 58 40 44 01   10 51 82 16 15   01 84 87 69 38
```

SOURCE: This table is reproduced with permission from tables of the RAND Corporation from *A Million Random Digits with 100,000 Normal Deviates*, New York, The Free Press, 1955.

TABLE B.10 Natural Logarithms of Numbers (Base e = 2.7183)

0.000-0.499

N	0	1	2	3	4	5	6	7	8	9
0.00	$-\infty$	−6‡ .90776	−6 .21461	−5 .80914	−5 .52146	−5 .29832	−5 .11600	−4 .96185	−4 .82831	−4 .71053
.01	−4.60517	.50986	.42285	.34281	.26870	.19971	.13517	.07454	.01738	*.96332
.02	−3.91202	.86323	.81671	.77226	.72970	.68888	.64966	.61192	.57555	.54046
.03	.50656	.47377	.44202	.41125	.38139	.35241	.32424	.29684	.27017	.24419
.04	.21888	.19418	.17009	.14656	.12357	.10109	.07911	.05761	.03655	.01593
.05	−2.99573	.97593	.95651	.93746	.91877	.90042	.88240	.86470	.84731	.83022
.06	.81341	.79688	.78062	.76462	.74887	.73337	.71810	.70306	.68825	.67365
.07	.65926	.64508	.63109	.61730	.60369	.59027	.57702	.56395	.55105	.53831
.08	.52573	.51331	.50104	.48891	.47694	.46510	.45341	.44185	.43042	.41912
.09	.40795	.39690	.38597	.37516	.36446	.35388	.34341	.33304	.32279	.31264
0.10	−2.30259	.29263	.28278	.27303	.26336	.25379	.24432	.23493	.22562	.21641
.11	.20727	.19823	.18926	.18037	.17156	.16282	.15417	.14558	.13707	.12863
.12	.12026	.11196	.10373	.09557	.08747	.07944	.07147	.06357	.05573	.04794
.13	.04022	.03256	.02495	.01741	.00992	.00248	*.99510	*.98777	*.98050	*.97328
.14	−1.96611	.95900	.95193	.94491	.93794	.93102	.92415	.91732	.91054	.90381
.15	.89712	.89048	.88387	.87732	.87080	.86433	.85790	.85151	.84516	.83885
.16	.83258	.82635	.82016	.81401	.80789	.80181	.79577	.78976	.78379	.77786
.17	.77196	.76609	.76026	.75446	.74870	.74297	.73727	.73161	.72597	.72037
.18	.71480	.70926	.70375	.69827	.69282	.68740	.68201	.67665	.67131	.66601
.19	.66073	.65548	.65026	.64507	.63990	.63476	.62964	.62455	.61949	.61445
0.20	−1.60944	.60445	.59949	.59455	.58964	.58475	.57988	.57504	.57022	.56542
.21	.56065	.55590	.55117	.54646	.54178	.53712	.53248	.52786	.52326	.51868
.22	.51413	.50959	.50508	.50058	.49611	.49165	.48722	.48281	.47841	.47403
.23	.46968	.46534	.46102	.45672	.45243	.44817	.44392	.43970	.43548	.43129
.24	.42712	.42296	.41882	.41469	.41059	.40650	.40242	.39837	.39433	.39030
.25	.38629	.38230	.37833	.37437	.37042	.36649	.36258	.35868	.35480	.35093
.26	.34707	.34323	.33941	.33560	.33181	.32803	.32426	.32051	.31677	.31304
.27	.30933	.30564	.30195	.29828	.29463	.29098	.28735	.28374	.28013	.27654
.28	.27297	.26940	.26585	.26231	.25878	.25527	.25176	.24827	.24479	.24133
.29	.23787	.23443	.23100	.22758	.22418	.22078	.21740	.21402	.21066	.20731
0.30	−1.20397	.20065	.19733	.19402	.19073	.18744	.18417	.18091	.17766	.17441
.31	.17118	.16796	.16475	.16155	.15836	.15518	.15201	.14885	.14570	.14256
.32	.13943	.13631	.13320	.13010	.12701	.12393	.12086	.11780	.11474	.11170
.33	.10866	.10564	.10262	.09961	.09661	.09362	.09064	.08767	.08471	.08176
.34	.07881	.07587	.07294	.07002	.06711	.06421	.06132	.05843	.05555	.05268
.35	−1.04982	.04697	.04412	.04129	.03846	.03564	.03282	.03002	.02722	.02443
.36	.02165	.01888	.01611	.01335	.01060	.00786	.00512	.00239	*.99967	*.99696
.37	−0.99425	.99155	.98886	.98618	.98350	.98083	.97817	.97551	.97286	.97022
.38	.96758	.96496	.96233	.95972	.95711	.95451	.95192	.94933	.94675	.94418
.39	.94161	.93905	.93649	.93395	.93140	.92887	.92634	.92382	.92130	.91879
0.40	−0.91629	.91379	.91130	.90882	.90634	.90387	.90140	.89894	.89649	.89404
.41	.89160	.88916	.88673	.88431	.88189	.87948	.87707	.87467	.87227	.86988
.42	.86750	.86512	.86275	.86038	.85802	.85567	.85332	.85097	.84863	.84630
.43	.84397	.84165	.83933	.83702	.83471	.83241	.83011	.82782	.82554	.82326
.44	.82098	.81871	.81645	.81419	.81193	.80968	.80744	.80520	.80296	.80073
.45	.79851	.79629	.79407	.79186	.78966	.78746	.78526	.78307	.78089	.77871
.46	.77653	.77436	.77219	.77003	.76787	.76572	.76357	.76143	.75929	.75715
.47	.75502	.75290	.75078	.74866	.74655	.74444	.74234	.74024	.73814	.73605
.48	.73397	.73189	.72981	.72774	.72567	.72361	.72155	.71949	.71744	.71539
.49	.71335	.71131	.70928	.70725	.70522	.70320	.70118	.69917	.69716	.69515

‡ Note that the whole number values are given above the decimal values for the first line. In the second and following lines they are given at the left. All decimal values are negative on this page.

TABLE B.10 Natural Logarithms of Numbers (*Cont'd*)

0.500–0.999

N	0	1	2	3	4	5	6	7	8	9
0.50	−0.69315	.69115	.68916	.68717	.68518	.68320	.68122	.67924	.67727	.67531
.51	.67334	.67139	.66943	.66748	.66553	.66359	.66165	.65971	.65778	.65585
.52	.65393	.65201	.65009	.64817	.64626	.64436	.64245	.64055	.63866	.63677
.53	.63488	.63299	.63111	.62923	.62736	.62549	.62362	.62176	.61990	.61804
.54	.61619	.61434	.61249	.61065	.60881	.60697	.60514	.60331	.60148	.59966
.55	.59784	.59602	.59421	.59240	.59059	.58879	.58699	.58519	.58340	.58161
.56	.57982	.57803	.57625	.57448	.57270	.57093	.56916	.56740	.56563	.56387
.57	.56212	.56037	.55862	.55687	.55513	.55339	.55165	.54991	.54818	.54645
.58	.54473	.54300	.54128	.53957	.53785	.53614	.53444	.53273	.53103	.52933
.59	.52763	.52594	.52425	.52256	.52088	.51919	.51751	.51584	.51416	.51249
0.60	−0.51083	.50916	.50750	.50584	.50418	.50253	.50088	.49923	.49758	.49594
.61	.49430	.49266	.49102	.48939	.48776	.48613	.48451	.48289	.48127	.47965
.62	.47804	.47642	.47482	.47321	.47160	.47000	.46840	.46681	.46522	.46362
.63	.46204	.46045	.45887	.45728	.45571	.45413	.45256	.45099	.44942	.44785
.64	.44629	.44473	.44317	.44161	.44006	.43850	.43696	.43541	.43386	.43232
.65	.43078	.42925	.42771	.42618	.42465	.42312	.42159	.42007	.41855	.41703
.66	.41552	.41400	.41249	.41098	.40947	.40797	.40647	.40497	.40347	.40197
.67	.40048	.39899	.39750	.39601	.39453	.39304	.39156	.39008	.38861	.38713
.68	.38566	.38419	.38273	.38126	.37980	.37834	.37688	.37542	.37397	.37251
.69	.37106	.36962	.36817	.36673	.36528	.36384	.36241	.36097	.35954	.35810
0.70	−0.35667	.35525	.35382	.35240	.35098	.34956	.34814	.34672	.34531	.34390
.71	.34249	.34108	.33968	.33827	.33687	.33547	.33408	.33268	.33129	.32989
.72	.32850	.32712	.32573	.32435	.32296	.32158	.32021	.31883	.31745	.31608
.73	.31471	.31334	.31197	.31061	.30925	.30788	.30653	.30517	.30381	.30246
.74	.30111	.29975	.29841	.29706	.29571	.29437	.29303	.29169	.29035	.28902
.75	.28768	.28635	.28502	.28369	.28236	.28104	.27971	.27839	.27707	.27575
.76	.27444	.27312	.27181	.27050	.26919	.26788	.26657	.26527	.26397	.26266
.77	.26136	.26007	.25877	.25748	.25618	.25489	.25360	.25231	.25103	.24974
.78	.24846	.24718	.24590	.24462	.24335	.24207	.24080	.23953	.23826	.23699
.79	.23572	.23446	.23319	.23193	.23067	.22941	.22816	.22690	.22565	.22439
0.80	−0.22314	.22189	.22065	.21940	.21816	.21691	.21567	.21443	.21319	.21196
.81	.21072	.20949	.20825	.20702	.20579	.20457	.20334	.20212	.20089	.19967
.82	.19845	.19723	.19601	.19480	.19358	.19237	.19116	.18995	.18874	.18754
.83	.18633	.18513	.18392	.18272	.18152	.18032	.17913	.17793	.17674	.17554
.84	.17435	.17316	.17198	.17079	.16960	.16842	.16724	.16605	.16487	.16370
.85	−0.16252	.16134	.16017	.15900	.15782	.15665	.15548	.15432	.15315	.15199
.86	.15082	.14966	.14850	.14734	.14618	.14503	.14387	.14272	.14156	.14041
.87	.13926	.13811	.13697	.13582	.13467	.13353	.13239	.13125	.13011	.12897
.88	.12783	.12670	.12556	.12443	.12330	.12217	.12104	.11991	.11878	.11766
.89	.11653	.11541	.11429	.11317	.11205	.11093	.10981	.10870	.10759	.10647
0.90	−0.10536	.10425	.10314	.10203	.10093	.09982	.09872	.09761	.09651	.09541
.91	.09431	.09321	.09212	.09102	.08992	.08883	.08774	.08665	.08556	.08447
.92	.08338	.08230	.08121	.08013	.07904	.07796	.07688	.07580	.07472	.07365
.93	.07257	.07150	.07042	.06935	.06828	.06721	.06614	.06507	.06401	.06294
.94	.06188	.06081	.05975	.05869	.05763	.05657	.05551	.05446	.05340	.05235
.95	.05129	.05024	.04919	.04814	.04709	.04604	.04500	.04395	.04291	.04186
.96	.04082	.03978	.03874	.03770	.03666	.03563	.03459	.03356	.03252	.03149
.97	.03046	.02943	.02840	.02737	.02634	.02532	.02429	.02327	.02225	.02122
.98	.02020	.01918	.01816	.01715	.01613	.01511	.01410	.01309	.01207	.01106
.99	.01005	.00904	.00803	.00702	.00602	.00501	.00401	.00300	.00200	.00100

TABLE B.10 Natural Logarithms of Numbers (Cont'd)

To find the natural logarithm of a number which is 1/10, 1/100, 1/1000, etc. of a number whose logarithm is given, subtract from the given logarithm log. 10, 2 log. 10, 3 log. 10, etc.
To find the natural logarithm of a number which is 10, 100, 1000, etc. times a number whose logarithm is given, add to the given logarithm log. 10, 2 log. 10, 3 log. 10, etc.

```
log. 10  =  2.30258 50930          6 log. 10 = 13.81551 05580
2 log. 10 =  4.60517 01860         7 log. 10 = 16.11809 56510
3 log. 10 =  6.90775 52790         8 log. 10 = 18.42068 07440
4 log. 10 =  9.21034 03720         9 log. 10 = 20.72326 58369
5 log. 10 = 11.51292 54650        10 log. 10 = 23.02585 09299
```

See preceding table for logarithms for numbers between 0.000 and 0.999.

1.00–4.99

N	0	1	2	3	4	5	6	7	8	9
1.0	0.00000	.00995	.01980	.02956	.03922	.04879	.05827	.06766	.07696	.08618
.1	.09531	.10436	.11333	.12222	.13103	.13976	.14842	.15700	.16551	.17395
.2	.18232	.19062	.19885	.20701	.21511	.22314	.23111	.23902	.24686	.25464
.3	.26236	.27003	.27763	.28518	.29267	.30010	.30748	.31481	.32208	.32930
.4	.33647	.34359	.35066	.35767	.36464	.37156	.37844	.38526	.39204	.39878
.5	.40547	.41211	.41871	.42527	.43178	.43825	.44469	.45108	.45742	.46373
.6	.47000	.47623	.48243	.48858	.49470	.50078	.50682	.51282	.51879	.52473
.7	.53063	.53649	.54232	.54812	.55389	.55962	.56531	.57098	.57661	.58222
.8	.58779	.59333	.59884	.60432	.60977	.61519	.62058	.62594	.63127	.63658
.9	.64185	.64710	.65233	.65752	.66269	.66783	.67294	.67803	.68310	.68813
2.0	0.69315	.69813	.70310	.70804	.71295	.71784	.72271	.72755	.73237	.73716
.1	.74194	.74669	.75142	.75612	.76081	.76547	.77011	.77473	.77932	.78390
.2	.78846	.79299	.79751	.80200	.80648	.81093	.81536	.81978	.82418	.82855
.3	.83291	.83725	.84157	.84587	.85015	.85442	.85866	.86289	.86710	.87129
.4	.87547	.87963	.88377	.88789	.89200	.89609	.90016	.90422	.90826	.91228
.5	.91629	.92028	.92426	.92822	.93216	.93609	.94001	.94391	.94779	.95166
.6	.95551	.95935	.96317	.96698	.97078	.97456	.97833	.98208	.98582	.98954
.7	.99325	.99695	*.00063	*.00430	*.00796	*.01160	*.01523	*.01885	*.02245	*.02604
.8	1.02962	.03318	.03674	.04028	.04380	.04732	.05082	.05431	.05779	.06126
.9	.06471	.06815	.07158	.07500	.07841	.08181	.08519	.08856	.09192	.09527
3.0	1.09861	.10194	.10526	.10856	.11186	.11514	.11841	.12168	.12493	.12817
.1	.13140	.13462	.13783	.14103	.14422	.14740	.15057	.15373	.15688	.16002
.2	.16315	.16627	.16938	.17248	.17557	.17865	.18173	.18479	.18784	.19089
.3	.19392	.19695	.19996	.20297	.20597	.20896	.21194	.21491	.21788	.22083
.4	.22378	.22671	.22964	.23256	.23547	.23837	.24127	.24415	.24703	.24990
.5	.25276	.25562	.25846	.26130	.26413	.26695	.26976	.27257	.27536	.27815
.6	.28093	.28371	.28647	.28923	.29198	.29473	.29746	.30019	.30291	.30563
.7	.30833	.31103	.31372	.31641	.31909	.32176	.32442	.32708	.32972	.33237
.8	.33500	.33763	.34025	.34286	.34547	.34807	.35067	.35325	.35584	.35841
.9	.36098	.36354	.36609	.36864	.37118	.37372	.37624	.37877	.38128	.38379
4.0	1.38629	.38879	.39128	.39377	.39624	.39872	.40118	.40364	.40610	.40854
.1	.41099	.41342	.41585	.41828	.42070	.42311	.42552	.42792	.43031	.43270
.2	.43508	.43746	.43984	.44220	.44456	.44692	.44927	.45161	.45395	.45629
.3	.45862	.46094	.46326	.46557	.46787	.47018	.47247	.47476	.47705	.47933
.4	.48160	.48387	.48614	.48840	.49065	.49290	.49515	.49739	.49962	.50185
.5	.50408	.50630	.50851	.51072	.51293	.51513	.51732	.51951	.52170	.52388
.6	.52606	.52823	.53039	.53256	.53471	.53687	.53902	.54116	.54330	.54543
.7	.54756	.54969	.55181	.55393	.55604	.55814	.56025	.56235	.56444	.56653
.8	.56862	.57070	.57277	.57485	.57691	.57898	.58104	.58309	.58515	.58719
.9	.58924	.59127	.59331	.59534	.59737	.59939	.60141	.60342	.60543	.60744

TABLE B.10 Natural Logarithms of Numbers (*Cont'd*)

5.00–9.99

N	0	1	2	3	4	5	6	7	8	9
5.0	1.60944	.61144	.61343	.61542	.61741	.61939	.62137	.62334	.62531	.62728
.1	.62924	.63120	.63315	.63511	.63705	.63900	.64094	.64287	.64481	.64673
.2	.64866	.65058	.65250	.65441	.65632	.65823	.66013	.66203	.66393	.66582
.3	.66771	.66959	.67147	.67335	.67523	.67710	.67896	.68083	.68269	.68455
.4	.68640	.68825	.69010	.69194	.69378	.69562	.69745	.69928	.70111	.70293
.5	.70475	.70656	.70838	.71019	.71199	.71380	.71560	.71740	.71919	.72098
.6	.72277	.72455	.72633	.72811	.72988	.73166	.73342	.73519	.73695	.73871
.7	.74047	.74222	.74397	.74572	.74746	.74920	.75094	.75267	.75440	.75613
.8	.75786	.75958	.76130	.76302	.76473	.76644	.76815	.76985	.77156	.77326
.9	.77495	.77665	.77834	.78002	.78171	.78339	.78507	.78675	.78842	.79009
6.0	1.79176	.79342	.79509	.79675	.79840	.80006	.80171	.80336	.80500	.80665
.1	.80829	.80993	.81156	.81319	.81482	.81645	.81808	.81970	.82132	.82294
.2	.82455	.82616	.82777	.82938	.83098	.83258	.83418	.83578	.83737	.83896
.3	.84055	.84214	.84372	.84530	.84688	.84845	.85003	.85160	.85317	.85473
.4	.85630	.85786	.85942	.86097	.86253	.86408	.86563	.86718	.86872	.87026
.5	.87180	.87334	.87487	.87641	.87794	.87947	.88099	.88251	.88403	.88555
.6	.88707	.88858	.89010	.89160	.89311	.89462	.89612	.89762	.89912	.90061
.7	.90211	.90360	.90509	.90658	.90806	.90954	.91102	.91250	.91398	.91545
.8	.91692	.91839	.91986	.92132	.92279	.92425	.92571	.92716	.92862	.93007
.9	.93152	.93297	.93442	.93586	.93730	.93874	.94018	.94162	.94305	.94448
7.0	1.94591	.94734	.94876	.95019	.95161	.95303	.95445	.95586	.95727	.95869
.1	.96009	.96150	.96291	.96431	.96571	.96711	.96851	.96991	.97130	.97269
.2	.97408	.97547	.97685	.97824	.97962	.98100	.98238	.98376	.98513	.98650
.3	.98787	.98924	.99061	.99198	.99334	.99470	.99606	.99742	.99877	*.00013
.4	2.00148	.00283	.00418	.00553	.00687	.00821	.00956	.01089	.01223	.01357
.5	.01490	.01624	.01757	.01890	.02022	.02155	.02287	.02419	.02551	.02683
.6	.02815	.02946	.03078	.03209	.03340	.03471	.03601	.03732	.03862	.03992
.7	.04122	.04252	.04381	.04511	.04640	.04769	.04898	.05027	.05156	.05284
.8	.05412	.05540	.05668	.05796	.05924	.06051	.06179	.06306	.06433	.06560
.9	.06686	.06813	.06939	.07065	.07191	.07317	.07443	.07568	.07694	.07819
8.0	2.07944	.08069	.08194	.08318	.08443	.08567	.08691	.08815	.08939	.09063
.1	.09186	.09310	.09433	.09556	.09679	.09802	.09924	.10047	.10169	.10291
.2	.10413	.10535	.10657	.10779	.10900	.11021	.11142	.11263	.11384	.11505
.3	.11626	.11746	.11866	.11986	.12106	.12226	.12346	.12465	.12585	.12704
.4	.12823	.12942	.13061	.13180	.13298	.13417	.13535	.13653	.13771	.13889
.5	.14007	.14124	.14242	.14359	.14476	.14593	.14710	.14827	.14943	.15060
.6	.15176	.15292	.15409	.15524	.15640	.15756	.15871	.15987	.16102	.16217
.7	.16332	.16447	.16562	.16677	.16791	.16905	.17020	.17134	.17248	.17361
.8	.17475	.17589	.17702	.17816	.17929	.18042	.18155	.18267	.18380	.18493
.9	.18605	.18717	.18830	.18942	.19054	.19165	.19277	.19389	.19500	.19611
9.0	2.19722	.19834	.19944	.20055	.20166	.20276	.20387	.20497	.20607	.20717
.1	.20827	.20937	.21047	.21157	.21266	.21375	.21485	.21594	.21703	.21812
.2	.21920	.22029	.22138	.22246	.22354	.22462	.22570	.22678	.22786	.22894
.3	.23001	.23109	.23216	.23324	.23431	.23538	.23645	.23751	.23858	.23965
.4	.24071	.24177	.24284	.24390	.24496	.24601	.24707	.24813	.24918	.25024
.5	.25129	.25234	.25339	.25444	.25549	.25654	.25759	.25863	.25968	.26072
.6	.26176	.26280	.26384	.26488	.26592	.26696	.26799	.26903	.27006	.27109
.7	.27213	.27316	.27419	.27521	.27624	.27727	.27829	.27932	.28034	.28136
.8	.28238	.28340	.28442	.28544	.28646	.28747	.28849	.28950	.29051	.29152
.9	.29253	.29354	.29455	.29556	.29657	.29757	.29858	.29958	.30058	.30158

SOURCE: From *Handbook of Probability and Statistics*, 2d ed., 1968, The Chemical Rubber Co., Cleveland, Ohio, with permission from the publisher.

TABLE B.11 Common Logarithms of Numbers (Base 10)

N	0	1	2	3	4	5	6	7	8	9
10	0000	0043	0086	0128	0170	0212	0253	0294	0334	0374
11	0414	0453	0492	0531	0569	0607	0645	0682	0719	0755
12	0792	0828	0864	0899	0934	0969	1004	1038	1072	1106
13	1139	1173	1206	1239	1271	1303	1335	1367	1399	1430
14	1461	1492	1523	1553	1584	1614	1644	1673	1703	1732
15	1761	1790	1818	1847	1875	1903	1931	1959	1987	2014
16	2041	2068	2095	2122	2148	2175	2201	2227	2253	2279
17	2304	2330	2355	2380	2405	2430	2455	2480	2504	2529
18	2553	2577	2601	2625	2648	2672	2695	2718	2742	2765
19	2788	2810	2833	2856	2878	2900	2923	2945	2967	2989
20	3010	3032	3054	3075	3096	3118	3139	3160	3181	3201
21	3222	3243	3263	3284	3304	3324	3345	3365	3385	3404
22	3424	3444	3464	3483	3502	3522	3541	3560	3579	3598
23	3617	3636	3655	3674	3692	3711	3729	3747	3766	3784
24	3802	3820	3838	3856	3874	3892	3909	3927	3945	3962
25	3979	3997	4014	4031	4048	4065	4082	4099	4116	4133
26	4150	4166	4183	4200	4216	4232	4249	4265	4281	4298
27	4314	4330	4346	4362	4378	4393	4409	4425	4440	4456
28	4472	4487	4502	4518	4533	4548	4564	4579	4594	4609
29	4624	4639	4654	4669	4683	4698	4713	4728	4742	4757
30	4771	4786	4800	4814	4829	4843	4857	4871	4886	4900
31	4914	4928	4942	4955	4969	4983	4997	5011	5024	5038
32	5051	5065	5079	5092	5105	5119	5132	5145	5159	5172
33	5185	5198	5211	5224	5237	5250	5263	5276	5289	5302
34	5315	5328	5340	5353	5366	5378	5391	5403	5416	5428
35	5441	5453	5465	5478	5490	5502	5514	5527	5539	5551
36	5563	5575	5587	5599	5611	5623	5635	5647	5658	5670
37	5682	5694	5705	5717	5729	5740	5752	5763	5775	5786
38	5798	5809	5821	5832	5843	5855	5866	5877	5888	5899
39	5911	5922	5933	5944	5955	5966	5977	5988	5999	6010
40	6021	6031	6042	6053	6064	6075	6085	6096	6107	6117
41	6128	6138	6149	6160	6170	6180	6191	6201	6212	6222
42	6232	6243	6253	6263	6274	6284	6294	6304	6314	6325
43	6335	6345	6355	6365	6375	6385	6395	6405	6415	6425
44	6435	6444	6454	6464	6474	6484	6493	6503	6513	6522
45	6532	6542	6551	6561	6571	6580	6590	6599	6609	6618
46	6628	6637	6646	6656	6665	6675	6684	6693	6702	6712
47	6721	6730	6739	6749	6758	6767	6776	6785	6794	6803
48	6812	6821	6830	6839	6848	6857	6866	6875	6884	6893
49	6902	6911	6920	6928	6937	6946	6955	6964	6972	6981
50	6990	6998	7007	7016	7024	7033	7042	7050	7059	7067
51	7076	7084	7093	7101	7110	7118	7126	7135	7143	7152
52	7160	7168	7177	7185	7193	7202	7210	7218	7226	7235
53	7243	7251	7259	7267	7275	7284	7292	7300	7308	7316
54	7324	7332	7340	7348	7356	7364	7372	7380	7388	7396
N	0	1	2	3	4	5	6	7	8	9

TABLE B.11 Common Logarithms of Numbers (*Cont'd*)

N	0	1	2	3	4	5	6	7	8	9
55	7404	7412	7419	7427	7435	7443	7451	7459	7466	7474
56	7482	7490	7497	7505	7513	7520	7528	7536	7543	7551
57	7559	7566	7574	7582	7589	7597	7604	7612	7619	7627
58	7634	7642	7649	7657	7664	7672	7679	7686	7694	7701
59	7709	7716	7723	7731	7738	7745	7752	7760	7767	7774
60	7782	7789	7796	7803	7810	7818	7825	7832	7839	7846
61	7853	7860	7868	7875	7882	7889	7896	7903	7910	7917
62	7924	7931	7938	7945	7952	7959	7966	7973	7980	7987
63	7993	8000	8007	8014	8021	8028	8035	8041	8048	8055
64	8062	8069	8075	8082	8089	8096	8102	8109	8116	8122
65	8129	8136	8142	8149	8156	8162	8169	8176	8182	8189
66	8195	8202	8209	8215	8222	8228	8235	8241	8248	8254
67	8261	8267	8274	8280	8287	8293	8299	8306	8312	8319
68	8325	8331	8338	8344	8351	8357	8363	8370	8376	8382
69	8388	8395	8401	8407	8414	8420	8426	8432	8439	8445
70	8451	8457	8463	8470	8476	8482	8488	8494	8500	8506
71	8513	8519	8525	8531	8537	8543	8549	8555	8561	8567
72	8573	8579	8585	8591	8597	8603	8609	8615	8621	8627
73	8633	8639	8645	8651	8657	8663	8669	8675	8681	8686
74	8692	8698	8704	8710	8716	8722	8727	8733	8739	8745
75	8751	8756	8762	8768	8774	8779	8785	8791	8797	8802
76	8808	8814	8820	8825	8831	8837	8842	8848	8854	8859
77	8865	8871	8876	8882	8887	8893	8899	8904	8910	8915
78	8921	8927	8932	8938	8943	8949	8954	8960	8965	8971
79	8976	8982	8987	8993	8998	9004	9009	9015	9020	9025
80	9031	9036	9042	9047	9053	9058	9063	9069	9074	9079
81	9085	9090	9096	9101	9106	9112	9117	9122	9128	9133
82	9138	9143	9149	9154	9159	9165	9170	9175	9180	9186
83	9191	9196	9201	9206	9212	9217	9222	9227	9232	9238
84	9243	9248	9253	9258	9263	9269	9274	9279	9284	9289
85	9294	9299	9304	9309	9315	9320	9325	9330	9335	9340
86	9345	9350	9355	9360	9365	9370	9375	9380	9385	9390
87	9395	9400	9405	9410	9415	9420	9425	9430	9435	9440
88	9445	9450	9455	9460	9465	9469	9474	9479	9484	9489
89	9494	9499	9504	9509	9513	9518	9523	9528	9533	9538
90	9542	9547	9552	9557	9562	9566	9571	9576	9581	9586
91	9590	9595	9600	9605	9609	9614	9619	9624	9628	9633
92	9638	9643	9647	9652	9657	9661	9666	9671	9675	9680
93	9685	9689	9694	9699	9703	9708	9713	9717	9722	9727
94	9731	9736	9741	9745	9750	9754	9759	9763	9768	9773
95	9777	9782	9786	9791	9795	9800	9805	9809	9814	9818
96	9823	9827	9832	9836	9841	9845	9850	9854	9859	9863
97	9868	9872	9877	9881	9886	9890	9894	9899	9903	9908
98	9912	9917	9921	9926	9930	9934	9939	9943	9948	9952
99	9956	9961	9965	9969	9974	9978	9983	9987	9991	9996
N	0	1	2	3	4	5	6	7	8	9

TABLE B.12 Powers and Roots

n	n^2	$\sqrt{n}$	$\sqrt{10n}$	n	n^2	$\sqrt{n}$	$\sqrt{10n}$
1.00	1.0000	1.00000	3.16228	**1.50**	2.2500	1.22474	3.87298
1.01	1.0201	1.00499	3.17805	1.51	2.2801	1.22882	3.88587
1.02	1.0404	1.00995	3.19374	1.52	2.3104	1.23288	3.89872
1.03	1.0609	1.01489	3.20936	1.53	2.3409	1.23693	3.91152
1.04	1.0816	1.01980	3.22490	1.54	2.3716	1.24097	3.92428
1.05	1.1025	1.02470	3.24037	1.55	2.4025	1.24499	3.93700
1.06	1.1236	1.02956	3.25576	1.56	2.4336	1.24900	3.94968
1.07	1.1449	1.03441	3.27109	1.57	2.4649	1.25300	3.96232
1.08	1.1664	1.03923	3.28634	1.58	2.4964	1.25698	3.97492
1.09	1.1881	1.04403	3.30151	1.59	2.5281	1.26095	3.98748
1.10	1.2100	1.04881	3.31662	**1.60**	2.5600	1.26491	4.00000
1.11	1.2321	1.05357	3.33167	1.61	2.5921	1.26886	4.01248
1.12	1.2544	1.05830	3.34664	1.62	2.6244	1.27279	4.02492
1.13	1.2769	1.06301	3.36155	1.63	2.6569	1.27671	4.03733
1.14	1.2996	1.06771	3.37639	1.64	2.6896	1.28062	4.04969
1.15	1.3225	1.07238	3.39116	1.65	2.7225	1.28452	4.06202
1.16	1.3456	1.07703	3.40588	1.66	2.7556	1.28841	4.07431
1.17	1.3689	1.08167	3.42053	1.67	2.7889	1.29228	4.08656
1.18	1.3924	1.08628	3.43511	1.68	2.8224	1.29615	4.09878
1.19	1.4161	1.09087	3.44964	1.69	2.8561	1.30000	4.11096
1.20	1.4400	1.09545	3.46410	**1.70**	2.8900	1.30384	4.12311
1.21	1.4641	1.10000	3.47851	1.71	2.9241	1.30767	4.13521
1.22	1.4884	1.10454	3.49285	1.72	2.9584	1.31149	4.14729
1.23	1.5129	1.10905	3.50714	1.73	2.9929	1.31529	4.15933
1.24	1.5376	1.11355	3.52136	1.74	3.0276	1.31909	4.17133
1.25	1.5625	1.11803	3.53553	1.75	3.0625	1.32288	4.18330
1.26	1.5876	1.12250	3.54965	1.76	3.0976	1.32665	4.19524
1.27	1.6129	1.12694	3.56371	1.77	3.1329	1.33041	4.20714
1.28	1.6384	1.13137	3.57771	1.78	3.1684	1.33417	4.21900
1.29	1.6641	1.13578	3.59166	1.79	3.2041	1.33791	4.23084
1.30	1.6900	1.14018	3.60555	**1.80**	3.2400	1.34164	4.24264
1.31	1.7161	1.14455	3.61939	1.81	3.2761	1.34536	4.25441
1.32	1.7424	1.14891	3.63318	1.82	3.3124	1.34907	4.26615
1.33	1.7689	1.15326	3.64692	1.83	3.3489	1.35277	4.27785
1.34	1.7956	1.15758	3.66060	1.84	3.3856	1.35647	4.28952
1.35	1.8225	1.16190	3.67423	1.85	3.4225	1.36015	4.30116
1.36	1.8496	1.16619	3.68782	1.86	3.4596	1.36382	4.31277
1.37	1.8769	1.17047	3.70135	1.87	3.4969	1.36748	4.32435
1.38	1.9044	1.17473	3.71484	1.88	3.5344	1.37113	4.33590
1.39	1.9321	1.17898	3.72827	1.89	3.5721	1.37477	4.34741
1.40	1.9600	1.18322	3.74166	**1.90**	3.6100	1.37840	4.35890
1.41	1.9881	1.18743	3.75500	1.91	3.6481	1.38203	4.37035
1.42	2.0164	1.19164	3.76829	1.92	3.6864	1.38564	4.38178
1.43	2.0449	1.19583	3.78153	1.93	3.7249	1.38924	4.39318
1.44	2.0736	1.20000	3.79473	1.94	3.7636	1.39284	4.40454
1.45	2.1025	1.20416	3.80789	1.95	3.8025	1.39642	4.41588
1.46	2.1316	1.20830	3.82099	1.96	3.8416	1.40000	4.42719
1.47	2.1609	1.21244	3.83406	1.97	3.8809	1.40357	4.43847
1.48	2.1904	1.21655	3.84708	1.98	3.9204	1.40712	4.44972
1.49	2.2201	1.22066	3.86005	1.99	3.9601	1.41067	4.46094

TABLE B.12 Powers and Roots (Cont'd)

n	n^2	$\sqrt{n}$	$\sqrt{10n}$	n	n^2	$\sqrt{n}$	$\sqrt{10n}$
2.00	4.0000	1.41421	4.47214	2.50	6.2500	1.58114	5.00000
2.01	4.0401	1.41774	4.48330	2.51	6.3001	1.58430	5.00999
2.02	4.0804	1.42127	4.49444	2.52	6.3504	1.58745	5.01996
2.03	4.1209	1.42478	4.50555	2.53	6.4009	1.59060	5.02991
2.04	4.1616	1.42829	4.51664	2.54	6.4516	1.59374	5.03984
2.05	4.2025	1.43178	4.52769	2.55	6.5025	1.59687	5.04975
2.06	4.2436	1.43527	4.53872	2.56	6.5536	1.60000	5.05964
2.07	4.2849	1.43875	4.54973	2.57	6.6049	1.60312	5.06952
2.08	4.3264	1.44222	4.56070	2.58	6.6564	1.60624	5.07937
2.09	4.3681	1.44568	4.57165	2.59	6.7081	1.60935	5.08920
2.10	4.4100	1.44914	4.58258	2.60	6.7600	1.61245	5.09902
2.11	4.4521	1.45258	4.59347	2.61	6.8121	1.61555	5.10882
2.12	4.4944	1.45602	4.60435	2.62	6.8644	1.61864	5.11859
2.13	4.5369	1.45945	4.61519	2.63	6.9169	1.62173	5.12835
2.14	4.5796	1.46287	4.62601	2.64	6.9696	1.62481	5.13809
2.15	4.6225	1.46629	4.63681	2.65	7.0225	1.62788	5.14782
2.16	4.6656	1.46969	4.64758	2.66	7.0756	1.63095	5.15752
2.17	4.7089	1.47309	4.65833	2.67	7.1289	1.63401	5.16720
2.18	4.7524	1.47648	4.66905	2.68	7.1824	1.63707	5.17687
2.19	4.7961	1.47986	4.67974	2.69	7.2361	1.64012	5.18652
2.20	4.8400	1.48324	4.69042	2.70	7.2900	1.64317	5.19615
2.21	4.8841	1.48661	4.70106	2.71	7.3441	1.64621	5.20577
2.22	4.9284	1.48997	4.71169	2.72	7.3984	1.64924	5.21536
2.23	4.9729	1.49332	4.72229	2.73	7.4529	1.65227	5.22494
2.24	5.0176	1.49666	4.73286	2.74	7.5076	1.65529	5.23450
2.25	5.0625	1.50000	4.74342	2.75	7.5625	1.65831	5.24404
2.26	5.1076	1.50333	4.75395	2.76	7.6176	1.66132	5.25357
2.27	5.1529	1.50665	4.76445	2.77	7.6729	1.66433	5.26308
2.28	5.1984	1.50997	4.77493	2.78	7.7284	1.66733	5.27257
2.29	5.2441	1.51327	4.78539	2.79	7.7841	1.67033	5.28205
2.30	5.2900	1.51658	4.79583	2.80	7.8400	1.67332	5.29150
2.31	5.3361	1.51987	4.80625	2.81	7.8961	1.67631	5.30094
2.32	5.3824	1.52315	4.81664	2.82	7.9524	1.67929	5.31037
2.33	5.4289	1.52643	4.82701	2.83	8.0089	1.68226	5.31977
2.34	5.4756	1.52971	4.83735	2.84	8.0656	1.68523	5.32917
2.35	5.5225	1.53297	4.84768	2.85	8.1225	1.68819	5.33854
2.36	5.5696	1.53623	4.85798	2.86	8.1796	1.69115	5.34790
2.37	5.6169	1.53948	4.86826	2.87	8.2369	1.69411	5.35724
2.38	5.6644	1.54272	4.87852	2.88	8.2944	1.69706	5.36656
2.39	5.7121	1.54596	4.88876	2.89	8.3521	1.70000	5.37587
2.40	5.7600	1.54919	4.89898	2.90	8.4100	1.70294	5.38516
2.41	5.8081	1.55242	4.90918	2.91	8.4681	1.70587	5.39444
2.42	5.8564	1.55563	4.91935	2.92	8.5264	1.70880	5.40370
2.43	5.9049	1.55885	4.92950	2.93	8.5849	1.71172	5.41295
2.44	5.9536	1.56205	4.93964	2.94	8.6436	1.71464	5.42218
2.45	6.0025	1.56525	4.94975	2.95	8.7025	1.71756	5.43139
2.46	6.0516	1.56844	4.95984	2.96	8.7616	1.72047	5.44059
2.47	6.1009	1.57162	4.96991	2.97	8.8209	1.72337	5.44977
2.48	6.1504	1.57480	4.97996	2.98	8.8804	1.72627	5.45894
2.49	6.2001	1.57797	4.98999	2.99	8.9401	1.72916	5.46809

TABLE B.12 Powers and Roots (*Cont'd*)

n	n^2	$\sqrt{n}$	$\sqrt{10n}$	n	n^2	$\sqrt{n}$	$\sqrt{10n}$
3.00	9.0000	1.73205	5.47723	3.50	12.2500	1.87083	5.91608
3.01	9.0601	1.73494	5.48635	3.51	12.3201	1.87350	5.92453
3.02	9.1204	1.73781	5.49545	3.52	12.3904	1.87617	5.93296
3.03	9.1809	1.74069	5.50454	3.53	12.4609	1.87883	5.94138
3.04	9.2416	1.74356	5.51362	3.54	12.5316	1.88149	5.94979
3.05	9.3025	1.74642	5.52268	3.55	12.6025	1.88414	5.95819
3.06	9.3636	1.74929	5.53173	3.56	12.6736	1.88680	5.96657
3.07	9.4249	1.75214	5.54076	3.57	12.7449	1.88944	5.97495
3.08	9.4864	1.75499	5.54977	3.58	12.8164	1.89209	5.98331
3.09	9.5481	1.75784	5.55878	3.59	12.8881	1.89473	5.99166
3.10	9.6100	1.76068	5.56776	3.60	12.9600	1.89737	6.00000
3.11	9.6721	1.76352	5.57674	3.61	13.0321	1.90000	6.00833
3.12	9.7344	1.76635	5.58570	3.62	13.1044	1.90263	6.01664
3.13	9.7969	1.76918	5.59464	3.63	13.1769	1.90526	6.02495
3.14	9.8596	1.77200	5.60357	3.64	13.2496	1.90788	6.03324
3.15	9.9225	1.77482	5.61249	3.65	13.3225	1.91050	6.04152
3.16	9.9856	1.77764	5.62139	3.66	13.3956	1.91311	6.04979
3.17	10.0489	1.78045	5.63028	3.67	13.4689	1.91572	6.05805
3.18	10.1124	1.78326	5.63915	3.68	13.5424	1.91833	6.06630
3.19	10.1761	1.78606	5.64801	3.69	13.6161	1.92094	6.07454
3.20	10.2400	1.78885	5.65685	3.70	13.6900	1.92354	6.08276
3.21	10.3041	1.79165	5.66569	3.71	13.7641	1.92614	6.09098
3.22	10.3684	1.79444	5.67450	3.72	13.8384	1.92873	6.09918
3.23	10.4329	1.79722	5.68331	3.73	13.9129	1.93132	6.10737
3.24	10.4976	1.80000	5.69210	3.74	13.9876	1.93391	6.11555
3.25	10.5625	1.80278	5.70088	3.75	14.0625	1.93649	6.12372
3.26	10.6276	1.80555	5.70964	3.76	14.1376	1.93907	6.13188
3.27	10.6929	1.80831	5.71839	3.77	14.2129	1.94165	6.14003
3.28	10.7584	1.81108	5.72713	3.78	14.2884	1.94422	6.14817
3.29	10.8241	1.81384	5.73585	3.79	14.3641	1.94679	6.15630
3.30	10.8900	1.81659	5.74456	3.80	14.4400	1.94936	6.16441
3.31	10.9561	1.81934	5.75326	3.81	14.5161	1.95192	6.17252
3.32	11.0224	1.82209	5.76194	3.82	14.5924	1.95448	6.18061
3.33	11.0889	1.82483	5.77062	3.83	14.6689	1.95704	6.18870
3.34	11.1556	1.82757	5.77927	3.84	14.7456	1.95959	6.19677
3.35	11.2225	1.83030	5.78792	3.85	14.8225	1.96214	6.20484
3.36	11.2896	1.83303	5.79655	3.86	14.8996	1.96469	6.21289
3.37	11.3569	1.83576	5.80517	3.87	14.9769	1.96723	6.22093
3.38	11.4244	1.83848	5.81378	3.88	15.0544	1.96977	6.22896
3.39	11.4921	1.84120	5.82237	3.89	15.1321	1.97231	6.23699
3.40	11.5600	1.84391	5.83095	3.90	15.2100	1.97484	6.24500
3.41	11.6281	1.84662	5.83952	3.91	15.2881	1.97737	6.25300
3.42	11.6964	1.84932	5.84808	3.92	15.3664	1.97990	6.26099
3.43	11.7649	1.85203	5.85662	3.93	15.4449	1.98242	6.26897
3.44	11.8336	1.85472	5.86515	3.94	15.5236	1.98494	6.27694
3.45	11.9025	1.85742	5.87367	3.95	15.6025	1.98746	6.28490
3.46	11.9716	1.86011	5.88218	3.96	15.6816	1.98997	6.29285
3.47	12.0409	1.86279	5.89067	3.97	15.7609	1.99249	6.30079
3.48	12.1104	1.86548	5.89915	3.98	15.8404	1.99499	6.30872
3.49	12.1801	1.86815	5.90762	3.99	15.9201	1.99750	6.31664

TABLE B.12 Powers and Roots (*Cont'd*)

n	n^2	$\sqrt{n}$	$\sqrt{10n}$	n	n^2	$\sqrt{n}$	$\sqrt{10n}$
4.00	16.0000	2.00000	6.32456	4.50	20.2500	2.12132	6.70820
4.01	16.0801	2.00250	6.33246	4.51	20.3401	2.12368	6.71565
4.02	16.1604	2.00499	6.34035	4.52	20.4304	2.12603	6.72309
4.03	16.2409	2.00749	6.34823	4.53	20.5209	2.12838	6.73053
4.04	16.3216	2.00998	6.35610	4.54	20.6116	2.13073	6.73795
4.05	16.4025	2.01246	6.36396	4.55	20.7025	2.13307	6.74537
4.06	16.4836	2.01494	6.37181	4.56	20.7936	2.13542	6.75278
4.07	16.5649	2.01742	6.37966	4.57	20.8849	2.13776	6.76018
4.08	16.6464	2.01990	6.38749	4.58	20.9764	2.14009	6.76757
4.09	16.7281	2.02237	6.39531	4.59	21.0681	2.14243	6.77495
4.10	16.8100	2.02485	6.40312	4.60	21.1600	2.14476	6.78233
4.11	16.8921	2.02731	6.41093	4.61	21.2521	2.14709	6.78970
4.12	16.9744	2.02978	6.41872	4.62	21.3444	2.14942	6.79706
4.13	17.0569	2.03224	6.42651	4.63	21.4369	2.15174	6.80441
4.14	17.1396	2.03470	6.43428	4.64	21.5296	2.15407	6.81175
4.15	17.2225	2.03715	6.44205	4.65	21.6225	2.15639	6.81909
4.16	17.3056	2.03961	6.44981	4.66	21.7156	2.15870	6.82642
4.17	17.3889	2.04206	6.45755	4.67	21.8089	2.16102	6.83374
4.18	17.4724	2.04450	6.46529	4.68	21.9024	2.16333	6.84105
4.19	17.5561	2.04695	6.47302	4.69	21.9961	2.16564	6.84836
4.20	17.6400	2.04939	6.48074	4.70	22.0900	2.16795	6.85565
4.21	17.7241	2.05183	6.48845	4.71	22.1841	2.17025	6.86294
4.22	17.8084	2.05426	6.49615	4.72	22.2784	2.17256	6.87023
4.23	17.8929	2.05670	6.50384	4.73	22.3729	2.17486	6.87750
4.24	17.9776	2.05913	6.51153	4.74	22.4676	2.17715	6.88477
4.25	18.0625	2.06155	6.51920	4.75	22.5625	2.17945	6.89202
4.26	18.1476	2.06398	6.52687	4.76	22.6576	2.18174	6.89928
4.27	18.2329	2.06640	6.53452	4.77	22.7529	2.18403	6.90652
4.28	18.3184	2.06882	6.54217	4.78	22.8484	2.18632	6.91375
4.29	18.4041	2.07123	6.54981	4.79	22.9441	2.18861	6.92098
4.30	18.4900	2.07364	6.55744	4.80	23.0400	2.19089	6.92820
4.31	18.5761	2.07605	6.56506	4.81	23.1361	2.19317	6.93542
4.32	18.6624	2.07846	6.57267	4.82	23.2324	2.19545	6.94262
4.33	18.7489	2.08087	6.58027	4.83	23.3289	2.19773	6.94982
4.34	18.8356	2.08327	6.58787	4.84	23.4256	2.20000	6.95701
4.35	18.9225	2.08567	6.59545	4.85	23.5225	2.20227	6.96419
4.36	19.0096	2.08806	6.60303	4.86	23.6196	2.20454	6.97137
4.37	19.0969	2.09045	6.61060	4.87	23.7169	2.20681	6.97854
4.38	19.1844	2.09284	6.61816	4.88	23.8144	2.20907	6.98570
4.39	19.2721	2.09523	6.62571	4.89	23.9121	2.21133	6.99285
4.40	19.3600	2.09762	6.63325	4.90	24.0100	2.21359	7.00000
4.41	19.4481	2.10000	6.64078	4.91	24.1081	2.21585	7.00714
4.42	19.5364	2.10238	6.64831	4.92	24.2064	2.21811	7.01427
4.43	19.6249	2.10476	6.65582	4.93	24.3049	2.22036	7.02140
4.44	19.7136	2.10713	6.66333	4.94	24.4036	2.22261	7.02851
4.45	19.8025	2.10950	6.67083	4.95	24.5025	2.22486	7.03562
4.46	19.8916	2.11187	6.67832	4.96	24.6016	2.22711	7.04273
4.47	19.9809	2.11424	6.68581	4.97	24.7009	2.22935	7.04982
4.48	20.0704	2.11660	6.69328	4.98	24.8004	2.23159	7.05691
4.49	20.1601	2.11896	6.70075	4.99	24.9001	2.23383	7.06399

TABLE B.12 Powers and Roots (*Cont'd*)

n	n^2	$\sqrt{n}$	$\sqrt{10n}$	n	n^2	$\sqrt{n}$	$\sqrt{10n}$
5.00	25.0000	2.23607	7.07107	**5.50**	30.2500	2.34521	7.41620
5.01	25.1001	2.23830	7.07814	5.51	30.3601	2.34734	7.42294
5.02	25.2004	2.24054	7.08520	5.52	30.4704	2.34947	7.42967
5.03	25.3009	2.24277	7.09225	5.53	30.5809	2.35160	7.43640
5.04	25.4016	2.24499	7.09930	5.54	30.6916	2.35372	7.44312
5.05	25.5025	2.24722	7.10634	5.55	30.8025	2.35584	7.44983
5.06	25.6036	2.24944	7.11337	5.56	30.9136	2.35797	7.45654
5.07	25.7049	2.25167	7.12039	5.57	31.0249	2.36008	7.46324
5.08	25.8064	2.25389	7.12741	5.58	31.1364	2.36220	7.46994
5.09	25.9081	2.25610	7.13442	5.59	31.2481	2.36432	7.47663
5.10	26.0100	2.25832	7.14143	**5.60**	31.3600	2.36643	7.48331
5.11	26.1121	2.26053	7.14843	5.61	31.4721	2.36854	7.48999
5.12	26.2144	2.26274	7.15542	5.62	31.5844	2.37065	7.49667
5.13	26.3169	2.26495	7.16240	5.63	31.6969	2.37276	7.50333
5.14	26.4196	2.26716	7.16938	5.64	31.8096	2.37487	7.50999
5.15	26.5225	2.26936	7.17635	5.65	31.9225	2.37697	7.51665
5.16	26.6256	2.27156	7.18331	5.66	32.0356	2.37908	7.52330
5.17	26.7289	2.27376	7.19027	5.67	32.1489	2.38118	7.52994
5.18	26.8324	2.27596	7.19722	5.68	32.2624	2.38238	7.53658
5.19	26.9361	2.27816	7.20417	5.69	32.3761	2.38537	7.54321
5.20	27.0400	2.28035	7.21110	**5.70**	32.4900	2.38747	7.54983
5.21	27.1441	2.28254	7.21803	5.71	32.6041	2.38956	7.55645
5.22	27.2484	2.28473	7.22496	5.72	32.7184	2.39165	7.56307
5.23	27.3529	2.28692	7.23187	5.73	32.8329	2.39374	7.56968
5.24	27.4576	2.28910	7.23878	5.74	32.9476	2.39583	7.57628
5.25	27.5625	2.29129	7.24569	5.75	33.0625	2.39792	7.58288
5.26	27.6676	2.29347	7.25259	5.76	33.1776	2.40000	7.58947
5.27	27.7729	2.29565	7.25948	5.77	33.2929	2.40208	7.59605
5.28	27.8784	2.29783	7.26636	5.78	33.4084	2.40416	7.60263
5.29	27.9841	2.30000	7.27324	5.79	33.5241	2.40624	7.60920
5.30	28.0900	2.30217	7.28011	**5.80**	33.6400	2.40832	7.61577
5.31	28.1961	2.30434	7.28697	5.81	33.7561	2.41039	7.62234
5.32	28.3024	2.30651	7.29383	5.82	33.8724	2.41247	7.62889
5.33	28.4089	2.30868	7.30068	5.83	33.9889	2.41454	7.63544
5.34	28.5156	2.31084	7.30753	5.84	34.1056	2.41661	7.64199
5.35	28.6225	2.31301	7.31437	5.85	34.2225	2.41868	7.64853
5.36	28.7296	2.31517	7.32120	5.86	34.3396	2.42074	7.65506
5.37	28.8369	2.31733	7.32803	5.87	34.4569	2.42281	7.66159
5.38	28.9444	2.31948	7.33485	5.88	34.5744	2.42487	7.66812
5.39	29.0521	2.32164	7.34166	5.89	34.6921	2.42693	7.67463
5.40	29.1600	2.32379	7.34847	**5.90**	34.8100	2.42899	7.68115
5.41	29.2681	2.32594	7.35527	5.91	34.9281	2.43105	7.68765
5.42	29.3764	2.32809	7.36206	5.92	35.0464	2.43311	7.69415
5.43	29.4849	2.33024	7.36885	5.93	35.1649	2.43516	7.70065
5.44	29.5936	2.33238	7.37564	5.94	35.2836	2.43721	7.70714
5.45	29.7025	2.33452	7.38241	5.95	35.4025	2.43926	7.71362
5.46	29.8116	2.33666	7.38918	5.96	35.5216	2.44131	7.72010
5.47	29.9209	2.33880	7.39594	5.97	35.6409	2.44336	7.72658
5.48	30.0304	2.34094	7.40270	5.98	35.7604	2.44540	7.73305
5.49	30.1401	2.34307	7.40945	5.99	35.8801	2.44745	7.73951

TABLE B.12 Powers and Roots (*Cont'd*)

n	n^2	$\sqrt{n}$	$\sqrt{10n}$	n	n^2	$\sqrt{n}$	$\sqrt{10n}$
6.00	36.0000	2.44949	7.74597	**6.50**	42.2500	2.54951	8.06226
6.01	36.1201	2.45153	7.75242	6.51	42.3801	2.55147	8.06846
6.02	36.2404	2.45357	7.75887	6.52	42.5104	2.55343	8.07465
6.03	36.3609	2.45561	7.76531	6.53	42.6409	2.55539	8.08084
6.04	36.4816	2.45764	7.77174	6.54	42.7716	2.55734	8.08703
6.05	36.6025	2.45967	7.77817	6.55	42.9025	2.55930	8.09321
6.06	36.7236	2.46171	7.78460	6.56	43.0336	2.56125	8.09938
6.07	36.8449	2.46374	7.79102	6.57	43.1649	2.56320	8.10555
6.08	36.9664	2.46577	7.79744	6.58	43.2964	2.56515	8.11172
6.09	37.0881	2.46779	7.80385	6.59	43.4281	2.56710	8.11788
6.10	37.2100	2.46982	7.81025	**6.60**	43.5600	2.56905	8.12404
6.11	37.3321	2.47184	7.81665	6.61	43.6921	2.57099	8.13019
6.12	37.4544	2.47386	7.82304	6.62	43.8244	2.57294	8.13634
6.13	37.5769	2.47588	7.82943	6.63	43.9569	2.57488	8.14248
6.14	37.6996	2.47790	7.83582	6.64	44.0896	2.57682	8.14862
6.15	37.8225	2.47992	7.84219	6.65	44.2225	2.57876	8.15475
6.16	37.9456	2.48193	7.84857	6.66	44.3556	2.58070	8.16088
6.17	38.0689	2.48395	7.85493	6.67	44.4889	2.58263	8.16701
6.18	38.1924	2.48596	7.86130	6.68	44.6224	2.58457	8.17313
6.19	38.3161	2.48797	7.86766	6.69	44.7561	2.58650	8.17924
6.20	38.4400	2.48998	7.87401	**6.70**	44.8900	2.58844	8.18535
6.21	38.5641	2.49199	7.88036	6.71	45.0241	2.59037	8.19146
6.22	38.6884	2.49399	7.88670	6.72	45.1584	2.59230	8.19756
6.23	38.8129	2.49600	7.89303	6.73	45.2929	2.59422	8.20366
6.24	38.9376	2.49800	7.89937	6.74	45.4276	2.59615	8.20975
6.25	39.0625	2.50000	7.90569	6.75	45.5625	2.59808	8.21584
6.26	39.1876	2.50200	7.91202	6.76	45.6976	2.60000	8.22192
6.27	39.3129	2.50400	7.91833	6.77	45.8329	2.60192	8.22800
6.28	39.4384	2.50599	7.92465	6.78	45.9684	2.60384	8.23408
6.29	39.5641	2.50799	7.93095	6.79	46.1041	2.60576	8.24015
6.30	39.6900	2.50998	7.93725	**6.80**	46.2400	2.60768	8.24621
6.31	39.8161	2.51197	7.94355	6.81	46.3761	2.60960	8.25227
6.32	39.9424	2.51396	7.94984	6.82	46.5124	2.61151	8.25833
6.33	40.0689	2.51595	7.95613	6.83	46.6489	2.61343	8.26438
6.34	40.1956	2.51794	7.96241	6.84	46.7856	2.61534	8.27043
6.35	40.3225	2.51992	7.96869	6.85	46.9225	2.61725	8.27647
6.36	40.4496	2.52190	7.97496	6.86	47.0596	2.61916	8.28251
6.37	40.5769	2.52389	7.98123	6.87	47.1969	2.62107	8.28855
6.38	40.7044	2.52587	7.98749	6.88	47.3344	2.62298	8.29458
6.39	40.8321	2.52784	7.99375	6.89	47.4721	2.62488	8.30060
6.40	40.9600	2.52982	8.00000	**6.90**	47.6100	2.62679	8.30662
6.41	41.0881	2.53180	8.00625	6.91	47.7481	2.62869	8.31264
6.42	41.2164	2.53377	8.01249	6.92	47.8864	2.63059	8.31865
6.43	41.3449	2.53574	8.01873	6.93	48.0249	2.63249	8.32466
6.44	41.4736	2.53772	8.02496	6.94	48.1636	2.63439	8.33067
6.45	41.6025	2.53969	8.03119	6.95	48.3025	2.63629	8.33667
6.46	41.7316	2.54165	8.03741	6.96	48.4416	2.63818	8.34266
6.47	41.8609	2.54362	8.04363	6.97	48.5809	2.64008	8.34865
6.48	41.9904	2.54558	8.04984	6.98	48.7204	2.64197	8.35464
6.49	42.1201	2.54755	8.05605	6.99	48.8601	2.64386	8.36062

TABLE B.12 Powers and Roots (*Cont'd*)

n	n^2	$\sqrt{n}$	$\sqrt{10n}$
7.00	49.0000	2.64575	8.36660
7.01	49.1401	2.64764	8.37257
7.02	49.2804	2.64953	8.37854
7.03	49.4209	2.65141	8.38451
7.04	49.5616	2.65330	8.39047
7.05	49.7025	2.65518	8.39643
7.06	49.8436	2.65707	8.40238
7.07	49.9849	2.65895	8.40833
7.08	50.1264	2.66083	8.41427
7.09	50.2681	2.66271	8.42021
7.10	50.4100	2.66458	8.42615
7.11	50.5521	2.66646	8.43208
7.12	50.6944	2.66833	8.43801
7.13	50.8369	2.67021	8.44393
7.14	50.9796	2.67208	8.44985
7.15	51.1225	2.67395	8.45577
7.16	51.2656	2.67582	8.46168
7.17	51.4089	2.67769	8.46759
7.18	51.5524	2.67955	8.47349
7.19	51.6961	2.68142	8.47939
7.20	51.8400	2.68328	8.48528
7.21	51.9841	2.68514	8.49117
7.22	52.1284	2.68701	8.49706
7.23	52.2729	2.68887	8.50294
7.24	52.4176	2.69072	8.50882
7.25	52.5625	2.69258	8.51469
7.26	52.7076	2.69444	8.52056
7.27	52.8529	2.69629	8.52643
7.28	52.9984	2.69815	8.53229
7.29	53.1441	2.70000	8.53815
7.30	53.2900	2.70185	8.54400
7.31	53.4361	2.70370	8.54985
7.32	53.5824	2.70555	8.55570
7.33	53.7289	2.70740	8.56154
7.34	53.8756	2.70924	8.56738
7.35	54.0225	2.71109	8.57321
7.36	54.1696	2.71293	8.57904
7.37	54.3169	2.71477	8.58487
7.38	54.4644	2.71662	8.59069
7.39	54.6121	2.71846	8.59651
7.40	54.7600	2.72029	8.60233
7.41	54.9081	2.72213	8.60814
7.42	55.0564	2.72397	8.61394
7.43	55.2049	2.72580	8.61974
7.44	55.3536	2.72764	8.62554
7.45	55.5025	2.72947	8.63134
7.46	55.6516	2.73130	8.63713
7.47	55.8009	2.73313	8.64292
7.48	55.9504	2.73496	8.64870
7.49	56.1001	2.73679	8.65448

n	n^2	$\sqrt{n}$	$\sqrt{10n}$
7.50	56.2500	2.73861	8.66025
7.51	56.4001	2.74044	8.66603
7.52	56.5504	2.74226	8.67179
7.53	56.7009	2.74408	8.67756
7.54	56.8516	2.74591	8.68332
7.55	57.0025	2.74773	8.68907
7.56	57.1536	2.74955	8.69483
7.57	57.3049	2.75136	8.70057
7.58	57.4564	2.75318	8.70632
7.59	57.6081	2.75500	8.71206
7.60	57.7600	2.75681	8.71780
7.61	57.9121	2.75862	8.72353
7.62	58.0644	2.76043	8.72926
7.63	58.2169	2.76225	8.73499
7.64	58.3696	2.76405	8.74071
7.65	58.5225	2.76586	8.74643
7.66	58.6756	2.76767	8.75214
7.67	58.8289	2.76948	8.75785
7.68	58.9824	2.77128	8.76356
7.69	59.1361	2.77308	8.76926
7.70	59.2900	2.77489	8.77496
7.71	59.4441	2.77669	8.78066
7.72	59.5984	2.77849	8.78635
7.73	59.7529	2.78029	8.79204
7.74	59.9076	2.78209	8.79773
7.75	60.0625	2.78388	8.80341
7.76	60.2176	2.78568	8.80909
7.77	60.3729	2.78747	8.81476
7.78	60.5284	2.78927	8.82043
7.79	60.6841	2.79106	8.82610
7.80	60.8400	2.79285	8.83176
7.81	60.9961	2.79464	8.83742
7.82	61.1524	2.79643	8.84308
7.83	61.3089	2.79821	8.84873
7.84	61.4656	2.80000	8.85438
7.85	61.6225	2.80179	8.86002
7.86	61.7796	2.80357	8.86566
7.87	61.9369	2.80535	8.87130
7.88	62.0944	2.80713	8.87694
7.89	62.2521	2.80891	8.88257
7.90	62.4100	2.81069	8.88819
7.91	62.5681	2.81247	8.89382
7.92	62.7264	2.81425	8.89944
7.93	62.8849	2.81603	8.90505
7.94	63.0436	2.81780	8.91067
7.95	63.2025	2.81957	8.91628
7.96	63.3616	2.82135	8.92188
7.97	63.5209	2.82312	8.92749
7.98	63.6804	2.82489	8.93308
7.99	63.8401	2.82666	8.93868

TABLE B.12 Powers and Roots (*Cont'd*)

n	n²	√n	√10n
8.00	64.0000	2.82843	8.94427
8.01	64.1601	2.83019	8.94986
8.02	64.3204	2.83196	8.95545
8.03	64.4809	2.83373	8.96103
8.04	64.6416	2.83549	8.96660
8.05	64.8025	2.83725	8.97218
8.06	64.9636	2.83901	8.97775
8.07	65.1249	2.84077	8.98332
8.08	65.2864	2.84253	8.98888
8.09	65.4481	2.84429	8.99444
8.10	65.6100	2.84605	9.00000
8.11	65.7721	2.84781	9.00555
8.12	65.9344	2.84956	9.01110
8.13	66.0969	2.85132	9.01665
8.14	66.2596	2.85307	9.02219
8.15	66.4225	2.85482	9.02774
8.16	66.5856	2.85657	9.03327
8.17	66.7489	2.85832	9.03881
8.18	66.9124	2.86007	9.04434
8.19	67.0761	2.86182	9.04986
8.20	67.2400	2.86356	9.05539
8.21	67.4041	2.86531	9.06091
8.22	67.5684	2.86705	9.06642
8.23	67.7329	2.86880	9.07193
8.24	67.8976	2.87054	9.07744
8.25	68.0625	2.87228	9.08295
8.26	68.2276	2.87402	9.08845
8.27	68.3929	2.87576	9.09395
8.28	68.5584	2.87750	9.09945
8.29	68.7241	2.87924	9.10494
8.30	68.8900	2.88097	9.11043
8.31	69.0561	2.88271	9.11592
8.32	69.2224	2.88444	9.12140
8.33	69.3889	2.88617	9.12688
8.34	69.5556	2.88791	9.13236
8.35	69.7225	2.88964	9.13783
8.36	69.8896	2.89137	9.14330
8.37	70.0569	2.89310	9.14877
8.38	70.2244	2.89482	9.15423
8.39	70.3921	2.89655	9.15969
8.40	70.5600	2.89828	9.16515
8.41	70.7281	2.90000	9.17061
8.42	70.8964	2.90172	9.17606
8.43	71.0649	2.90345	9.18150
8.44	71.2336	2.90517	9.18695
8.45	71.4025	2.90689	9.19239
8.46	71.5716	2.90861	9.19783
8.47	71.7409	2.91033	9.20326
8.48	71.9104	2.91204	9.20869
8.49	72.0801	2.91376	9.21412
8.50	72.2500	2.91548	9.21954
8.51	72.4201	2.91719	9.22497
8.52	72.5904	2.91890	9.23038
8.53	72.7609	2.92062	9.23580
8.54	72.9316	2.92233	9.24121
8.55	73.1025	2.92404	9.24662
8.56	73.2736	2.92575	9.25203
8.57	73.4449	2.92746	9.25743
8.58	73.6164	2.92916	9.26283
8.59	73.7881	2.93087	9.26823
8.60	73.9600	2.93258	9.27362
8.61	74.1321	2.93428	9.27901
8.62	74.3044	2.93598	9.28440
8.63	74.4769	2.93769	9.28978
8.64	74.6496	2.93939	9.29516
8.65	74.8225	2.94109	9.30054
8.66	74.9956	2.94279	9.30591
8.67	75.1689	2.94449	9.31128
8.68	75.3424	2.94618	9.31665
8.69	75.5161	2.94788	9.32202
8.70	75.6900	2.94958	9.32738
8.71	75.8641	2.95127	9.33274
8.72	76.0384	2.95296	9.33809
8.73	76.2129	2.95466	9.34345
8.74	76.3876	2.95635	9.34880
8.75	76.5625	2.95804	9.35414
8.76	76.7376	2.95973	9.35949
8.77	76.9129	2.96142	9.36483
8.78	77.0884	2.96311	9.37017
8.79	77.2641	2.96479	9.37550
8.80	77.4400	2.96648	9.38083
8.81	77.6161	2.96816	9.38616
8.82	77.7924	2.96985	9.39149
8.83	77.9689	2.97153	9.39681
8.84	78.1456	2.97321	9.40213
8.85	78.3225	2.97489	9.40744
8.86	78.4996	2.97658	9.41276
8.87	78.6769	2.97825	9.41807
8.88	78.8544	2.97993	9.42338
8.89	79.0321	2.98161	9.42868
8.90	79.2100	2.98329	9.43398
8.91	79.3881	2.98496	9.43928
8.92	79.5664	2.98664	9.44458
8.93	79.7449	2.98831	9.44987
8.94	79.9236	2.98998	9.45516
8.95	80.1025	2.99166	9.46044
8.96	80.2816	2.99333	9.46573
8.97	80.4609	2.99500	9.47101
8.98	80.6404	2.99666	9.47629
8.99	80.8201	2.99833	9.48156

TABLE B.12 Powers and Roots (*Cont'd*)

n	n^2	$\sqrt{n}$	$\sqrt{10n}$
9.00	81.0000	3.00000	9.48683
9.01	81.1801	3.00167	9.49210
9.02	81.3604	3.00333	9.49737
9.03	81.5409	3.00500	9.50263
9.04	81.7216	3.00666	9.50789
9.05	81.9025	3.00832	9.51315
9.06	82.0836	3.00998	9.51840
9.07	82.2649	3.01164	9.52365
9.08	82.4464	3.01330	9.52890
9.09	82.6281	3.01496	9.53415
9.10	82.8100	3.01662	9.53939
9.11	82.9921	3.01828	9.54463
9.12	83.1744	3.01993	9.54987
9.13	83.3569	3.02159	9.55510
9.14	83.5396	3.02324	9.56033
9.15	83.7225	3.02490	9.56556
9.16	83.9056	3.02655	9.57079
9.17	84.0889	3.02820	9.57601
9.18	84.2724	3.02985	9.58123
9.19	84.4561	3.03150	9.58645
9.20	84.6400	3.03315	9.59166
9.21	84.8241	3.03480	9.59687
9.22	85.0084	3.03645	9.60208
9.23	85.1929	3.03809	9.60729
9.24	85.3776	3.03974	9.61249
9.25	85.5625	3.04138	9.61769
9.26	85.7476	3.04302	9.62289
9.27	85.9329	3.04467	9.62808
9.28	86.1184	3.04631	9.63328
9.29	86.3041	3.04795	9.63846
9.30	86.4900	3.04959	9.64365
9.31	86.6761	3.05123	9.64883
9.32	86.8624	3.05287	9.65401
9.33	87.0489	3.05450	9.65919
9.34	87.2356	3.05614	9.66437
9.35	87.4225	3.05778	9.66954
9.36	87.6096	3.05941	9.67471
9.37	87.7969	3.06105	9.67988
9.38	87.9844	3.06268	9.68504
9.39	88.1721	3.06431	9.69020
9.40	88.3600	3.06594	9.69536
9.41	88.5481	3.06757	9.70052
9.42	88.7364	3.06920	9.70567
9.43	88.9249	3.07083	9.71082
9.44	89.1136	3.07246	9.71597
9.45	89.3025	3.07409	9.72111
9.46	89.4916	3.07571	9.72625
9.47	89.6809	3.07734	9.73139
9.48	89.8704	3.07896	9.73653
9.49	90.0601	3.08058	9.74166

n	n^2	$\sqrt{n}$	$\sqrt{10n}$
9.50	90.2500	3.08221	9.74679
9.51	90.4401	3.08383	9.75192
9.52	90.6304	3.08545	9.75705
9.53	90.8209	3.08707	9.76217
9.54	91.0116	3.08869	9.76729
9.55	91.2025	3.09031	9.77241
9.56	91.3936	3.09192	9.77753
9.57	91.5849	3.09354	9.78264
9.58	91.7764	3.09516	9.78775
9.59	91.9681	3.09677	9.79285
9.60	92.1600	3.09839	9.79796
9.61	92.3521	3.10000	9.80306
9.62	92.5444	3.10161	9.80816
9.63	92.7369	3.10322	9.81326
9.64	92.9296	3.10483	9.81835
9.65	93.1225	3.10644	9.82344
9.66	93.3156	3.10805	9.82853
9.67	93.5089	3.10966	9.83362
9.68	93.7024	3.11127	9.83870
9.69	93.8961	3.11288	9.84378
9.70	94.0900	3.11448	9.84886
9.71	94.2841	3.11609	9.85393
9.72	94.4784	3.11769	9.85901
9.73	94.6729	3.11929	9.86408
9.74	94.8676	3.12090	9.86914
9.75	95.0625	3.12250	9.87421
9.76	95.2576	3.12410	9.87927
9.77	95.4529	3.12570	9.88433
9.78	95.6484	3.12730	9.88939
9.79	95.8441	3.12890	9.89444
9.80	96.0400	3.13050	9.89949
9.81	96.2361	3.13209	9.90454
9.82	96.4324	3.13369	9.90959
9.83	96.6289	3.13528	9.91464
9.84	96.8256	3.13688	9.91968
9.85	97.0225	3.13847	9.92472
9.86	97.2196	3.14006	9.92975
9.87	97.4169	3.14166	9.93479
9.88	97.6144	3.14325	9.93982
9.89	97.8121	3.14484	9.94485
9.90	98.0100	3.14643	9.94987
9.91	98.2081	3.14802	9.95490
9.92	98.4064	3.14960	9.95992
9.93	98.6049	3.15119	9.96494
9.94	98.8036	3.15278	9.96995
9.95	99.0025	3.15436	9.97497
9.96	99.2016	3.15595	9.97998
9.97	99.4009	3.15753	9.98499
9.98	99.6004	3.15911	9.98999
9.99	99.8001	3.16070	9.99500

SOURCE: Reprinted with permission of The Macmillan Company from Macmillan *Selected Mathematics Tables* by E. R. Hedrick. Copyright © 1936 by The Macmillan Company, renewed by Dorothy H. McWilliams, Clyde L. Hedrick and Elisabeth B. Miller.

Index

Acceptance region, 286
Admissible decision function, 254, 258
Alpha risk, 247
Alternate hypothesis, 243

Bayes decision function, 264, 325, 331
   with normal probability, 362
Bayes theorem, 38, 327
Bayesian statisticians, 279
Bernoulli distribution, 120
Best unbiased estimator, 219
Beta risk, 247
Binomial distribution, 122
Borel field, 16

Cauchy distribution, 389
Central limit theorem, 150, 156
Chi-square distribution, 376
Classical decision function, 279
Complement, 14
Composite hypothesis, 243, 351
   Bernoulli case of, 294
   continuous case of, 305
   two-sided, 313

Compound event, 12
Conditional expected loss, 256
   posterior, 335, 370
   prior, 332, 366
Conditional probability, 28
Conditional probability function, 78
Confidence coefficient, 234
Confidence interval, 234
   mean, 234
   proportion of, 237
   variance of, 385
Continuity correction factor, 159
Correlation, 90
Covariance, 84
Critical region, 286
Cumulative probability function, 54

Decision function, 246
   admissible, 254, 258
   Bayes, 264, 325, 331
   classical, 279
   minimax, 260
   most powerful, 291
   pure, 249

Decision function (*continued*)
  randomized, 249, 251
  uniformly most powerful, 298
Decision space, 246
Density function, 59
Disjoint event, 25
Disjoint sets, 471
Domain, of a function, 42

Efficient estimator, 217
Efficient set, 255
"Equally-likely approach," 19
Error probabilities, 247, 287
Estimator, 213
  best unbiased, 219
  efficient, 217
  maximum likelihood, 228
  most efficient, 218
  sufficient, 219
Event, 9
  Borel field of, 16
  compound, 12
  dependent, 34
  disjoint, 25
  independent, 33
  mutually exclusive, 12
  occurrence of, 10
  simple, 12
Expected value, 62
  of perfect information, 340, 368
  of sample information, 373
Exponential distribution, 137

F distribution, 395
Function, 42
  of random variables, 96

Gauss Markoff theorem, 433
Generalized likelihood-ratio test, 317
"Goodness-of-fit," 404

Hypothesis, 242
  alternate, 243
  composite, 243
  null, 242
  simple, 243
Hypothesis testing, 242

Image set, 42
Independent events, 33

Induction, 2
Insufficient reason, principle of, 19
Intersection, between sets, 471
Interval estimate, 212, 232

Joint probability, 25
Joint probability function, 72

Large sampling, theory of, 402
Law of large numbers, 200
Lead time, 162
Least favorable prior probability, 268
Least square method, 422
Likelihood assessment, 15
Likelihood function, 226
Likelihood-ratio test, 280, 287
Linear dependence, 83
Loss function, 256

Marginal probability function, 75
Maximum likelihood estimate, 226, 228
Median, 215
Mid-range, 215
Minimal complete class, 255
Minimax decision function, 260
Most efficient estimator, 218
Most powerful decision function, 291
Multivariate sampling, 413
Mutually exclusive events, 12

Neyman-Pearson lemma, 290, 292
Normal distribution, 143
Normalized loss function, 368
Null hypothesis, 242

Objective school, 20
Optimal sample size, 341

Parameter, of a probability function, 181
Perfect information, 340
  expected value of, 340, 368
Personalistic approach, 21
Point estimate, 212, 222
Poisson distribution, 128
Population, 2
Posterior conditional expected loss, 335, 370
Posterior probability function, 327, 329
Power function, 296, 307

Prediction, with regression equation, 438
Preposterior analysis, 348, 359, 371
Preposterior probability function, 348, 372
Principle of insufficient reason, 19
Prior conditional expected loss, 332, 366
Prior probability, 264
  least favorable, 268
Prior probability function, 326
Probability, 15
  of a compound event, 22
  conditional, 28
  joint, 25
Probability function, 50
  conditional, 78
  of a continuous random variable, 56
  cumulative, 54
  of a discrete random variable, 52
  joint, 72
  marginal, 75
  posterior, 327, 329
  preposterior, 348, 372
  prior, 326
  of a sample proportion, 181
Pure decision function, 249

Random numbers, 178
Random phenomenon, 5
Random variable, 42, 45
  Bernoulli distributed, 120
  binomially distributed, 122
  continuous, 48
  discrete, 48
  exponentially distributed, 137
  normally distributed, 143
  Poisson distributed, 128
  standard normal, 144
  uniformly distributed, 135
Randomized decision function, 249, 251
Randomized strategy, 251
Range, of a function, 42
Rao and Blackwell, theorem of, 219
Rejection region, 286
Relative frequency approach, 20

Sample, 2
Sample information, expected value of, 373

Sample mean, 188
Sample proportion, 182
Sample size, 207
  optimal, 341
Sample space, 6
Sampling distribution, 181
Sampling error, 194
Sampling techniques, 169
Scalar product of random variables, 97
Set, 467
Simple event, 12
Simple hypothesis, 243
  continuous case of a, 285
Simple random sample, 165
Simple regression, 414, 416
Small sample problem, 390
Standard deviation, 65
Standard error of mean, 192, 199
Standard error of proportion, 185, 198
Standard normal variable, 144
Statistical decision, 2
Statistical decision theory, 3
Statistical dependence, 34, 82
Statistical description, 1
Statistical independence, 33, 80
Statistical induction, 2
Statistics, 1, 2
Student-t distribution, 388
Subjective school, 20
Subset, 470
Sufficient estimator, 219
Sum of random variables, 100, 105
Sum of the squares, 421

Tchebycheff Inequality, 69
Testing hypothesis, 242
  of two variances, 399
  of a variance, 383
Type I error, 245
Type II error, 245

Unbiased estimator, 213
Unconditional expected loss, 264
Uniform distribution, 135
Uniformly most powerful decision function, 298
Union of two sets, 471

Variance, of a random variable, 64